9th International Conference on Future Environment and Energy (ICFEE 2019)

AA006959

IOP Conference Series: Earth and Environmental Science
Volume 257

Osaka, Japan
9 – 11 January 2019

Editors:

Akira Kondo
Jae K. Park
Jeong-Hun Park

ISBN: 978-1-5108-8714-5
ISSN: 1755-1307

Printed from e-media with permission by:

Curran Associates, Inc.
57 Morehouse Lane
Red Hook, NY 12571

Some format issues inherent in the e-media version may also appear in this print version.

This work is licensed under a Creative Commons Attribution 3.0 International Licence.
Licence details: http://creativecommons.org/licenses/by/3.0/.

No changes have been made to the content of these proceedings. There may be changes to pagination and minor adjustments for aesthetics.

Printed with permission by Curran Associates, Inc. (2021)

For permission requests, please contact the Institute of Physics at the address below.

Institute of Physics
Dirac House, Temple Back
Bristol BS1 6BE UK

Phone: 44 1 17 929 7481
Fax: 44 1 17 920 0979

techtracking@iop.org

Additional copies of this publication are available from:

Curran Associates, Inc.
57 Morehouse Lane
Red Hook, NY 12571 USA
Phone: 845-758-0400
Fax: 845-758-2633
Email: curran@proceedings.com
Web: www.proceedings.com

TABLE OF CONTENTS

ENVIRONMENTAL AND CHEMICAL ENGINEERING

RENEWABLE ENERGY

BUILDING ENERGY EFFICIENCY AND ENERGY MANAGEMENT

COMBUSTION SCIENCE AND POWER ENGINEERING

BATTERY TECHNOLOGY AND PHOTOVOLTAIC SYSTEM OPTIMIZATION

POWER SYSTEM MANAGEMENT

TRAFFIC ENGINEERING

IOP Conf. Series: Earth and Environmental Science **257** (2019) 011001 doi:10.1088/1755-1315/257/1/011001

PREFACE

We are honored to bring you this collection of articles from the 2019 9th International Conference on Future Environment and Energy (ICFEE 2019) which was held in Osaka, Japan from January 9 to 11, 2019. ICFEE 2019 is dedicated to issues related to Future Environment and Energy.

The primary focus of the conference is to bring together academic scientists, engineers, and industry researchers to share their experiences and research results, and discuss the practical challenges encountered and the solutions adopted. The ICFEE 2019 served as a good platform for the scientific community where 90 participants met, exchanged ideas and research results, and examined the way forward regarding future environment and energy.

The former Vice-President of the International Water Resources Association, Prof. Ngai Weng Chan did the Opening Remarks for the conference in the first day, with Prof. Kondo Akira from Osaka University, Japan doing the Opening Remarks for the conference on the second day. There were four Keynote Speeches. Prof. Orawan Siriratpiriya from Aquatic Resources Research Institute, Chulalongkorn University (ARRIC) presented the 1st Keynote Speech titled "Closed Materials Looped with Low Carbon Technologies to Sustainable Society" on 9th January. This was followed by the 2nd Keynote Speaker Prof Ngai Weng Chan from the School of Humanities, Universiti Sains Malaysia, Malaysia who spoke on "Sustainable Urban Drainage as a Viable Measure of Coping with Heat and Floods Due to Climate Change". On the 2nd day 10th January, the 3rd Keynote Speaker was Prof Kondo Akira from Osaka University, Japan who spoke on "Simulation in City Block Environment by CFD Coupled with Meteorological and Air Quality Model". This was followed by the 4th Keynote Speaker Prof. Jae K. Park from University of Wisconsin-Madison, USA who presented his ideas on the talk titled "Life Cycle Analysis of Photovoltaic Solar Panels". All the four keynote speakers focused on the latest information and most innovative developments in their respective expertise areas on future environment and energy. For the conference proper, there were nine sub-sessions and one poster session with various interesting topics as follows: Environmental Management and Sustainable Development, Atmospheric Monitoring and Water Quality Analysis, Environmental and Chemical Engineering, Building Energy Efficiency and Energy Management, Life Cycle Analysis of Photovoltaic Solar Panels, Combustion Science and Power Engineering, Urban Engineering and Transportation Engineering, Battery Technology and Photovoltaic System Optimization, Geological and Civil Engineering.

ICFEE 2019 was indeed a golden opportunity for academics, students, researchers and engineers to interact with the experts and specialists to get their advice or consultation on state-of-the-art developments in future environment and energy, technical matters, as well as research methodologies and innovative strategies in the field of environmental and energy management. On the third day of the conference, we had a

Content from this work may be used under the terms of the Creative Commons Attribution 3.0 licence. Any further distribution of this work must maintain attribution to the author(s) and the title of the work, journal citation and DOI.
Published under licence by IOP Publishing Ltd

one-day visit in Osaka city.

This proceeding presents a selection from papers submitted to the conference from participants from universities, research institutes and industry. All papers were subjected to peer-review by conference committee members and international reviewers. The papers selected were based on their quality and their relevancy to the conference. The volume tends to present to the readers the recent advances in the field of Future Environment and Energy and various related areas, such as Energy and Environment, Renewable Energy, Advanced Energy Technologies, Fuels and Combustion, Benefits of the Conference, etc.

Finally, it is appropriate that we record our thanks to our organizing committees and reviewers for their work in securing a substantial input of papers from 13 countries and in encouraging participation from those areas. We are also indebted to those who served as Keynote Speakers, Conference Chairmen and Session Chairmen. Without their support, the conference could not have been as successful as it was. We also acknowledge the authors who have contributed to this volume, without whose expert inputs, this compilation of proceedings would not have been possible. We would like to thank all the conference participants for their support to ICFEE 2019.

Prof. Ngai Weng Chan
Universiti Sains Malaysia, Malaysia
January 17, 2019

IOP Conf. Series: Earth and Environmental Science **257** (2019) 011002 doi:10.1088/1755-1315/257/1/011002

Organizing Committee

Conference General Co-Chairs

Prof. Akira Kondo, Osaka University, Japan

Prof. Orawan Siriratpiriya, Environmental Research Institute of Chulaongkorn University, Thailand

Prof. Ngai Weng Chan, Universiti Sains Malaysia, Malaysia

Program Co-Chairs

Prof. Vladan Babovic, National University of Singapore

Prof. Je-Lueng Shie, National I-Lan University, Taiwan

Assoc. Prof. Anja Pfennig, HTW University of Applied Sciences Berlin, Gemany

Guest Editors

Prof. Akira Kondo, Osaka University, Japan

Prof. Jae K. Park, University of Wisconsin-Madison, USA

Prof. Jeong-Hun Park, Chonnam National University, South Korea

Organizing Chair

Ms. Tracy Yang, Hong Kong Chemical, Biological & Environmental Engineering Society, Hong Kong

Publication Chair

Mr Issac Lee, Hong Kong Chemical, Biological & Environmental Engineering Society, Hong Kong

International Technical Committee

Prof. Ken Nagasaka, Tokyo University of Agriculture and Technology, Japan

Prof. Enrico Sciubba, University of Roma Sapienza, Italy

Prof. Enrique Romero Cadaval, University of Extremadura, Spain

Prof. Roberto San Jose, Technical University of Madrid, Spain

Prof. Hui Tong Chua, University of Western Australia

Prof. Jeong-Hun Park, Chonnam National University, South Korea

Prof. Sheng-Yuan Chu, National Cheng Kung University, Taiwan

Prof. Kazutaka Itako, Kanagawa Institute of Technology, Japan

Prof. Ali Cheknane, University Of Laghouat. Algeria

Prof. Tanakorn Wongwuttanasatian, Khon Kaen University, Thailand

Prof. Ruey-Fang Yu, National United University, Taiwan

Prof. Yanpeng Cao, Zhejiang University, China

Assoc. Prof. Mathias Cehlin, University of Gävle, Sweden

Assoc. Prof. Amnart Suksri, Khon Kaen University, Thailand

Content from this work may be used under the terms of the Creative Commons Attribution 3.0 licence. Any further distribution of this work must maintain attribution to the author(s) and the title of the work, journal citation and DOI.
Published under licence by IOP Publishing Ltd

IOP Conf. Series: Earth and Environmental Science **257** (2019) 011002 doi:10.1088/1755-1315/257/1/011002

Assoc. Prof. Md Maruf Mortula, American University of Sharjah, UAE

Assoc. Prof. Khurram Shahzad, University of the Punjab, Pakistan

Assoc. Prof. Maikel Mendez, Instituto Tecnológico de Costa Rica, Costa Rica

Assoc. Prof. Wim F. Fuls , University of Cape Town, South Africa

Assoc. Prof. Hao Zhang, Southwest University, China

Assoc. Prof. Luiz Carlos,Pereira da Silva university of campinas

Assoc. Prof. Wei Li, Zhejiang University, China

Assoc. Prof. Willyanto Anggono, Petra Christian University, Indonesia

Dr. Yao Yu, North Dakota State University, USA

Dr. Qing Zhu, Lawrence Berkeley National Lab, USA

Dr. Hen Friman, Holon Institute of Technology, Holon, Israel

Dr. Xiaolin Wang, University of Tasmania, Australia

Dr. Gerardo Reyes, Lakehead University, Canada

Dr. Viktor Kochkodan, Hamad bin Khalifa University, Qatar

Dr. Jarotwan Koiwanit, King Mongkut's Institute of Technology Ladkrabang, Thailand

Dr. Arpapong Changjan, Pathumwan Institute of Technology, Thailand

Dr. Sutinee Sinutok, Prince of Songkla University, Thailand

Dr Wong Yee-Shian, University Malaysia Perlis (UniMAP), Malaysia

Dr. Winardi Dwi Nugraha, Diponegoro University, Indonesia

Dr. Nuttakorn Intaravicha, Pathumwan Institute of Technology, Thailand

Dr. Atsushi Shiota, Kyushu Institute of Technology, Japan

Dr. Rahul S. Todmal, Vidya Pratishthan's ASC College, India

Dr. Chumnan Boonyaputthipong, Khon Kaen University, Thailand

Dr. Loo Joo Ling, Universiti Tunku Abdul Rahman (UTAR), Malaysia

Dr. Lee Hwang Sheng, Universiti Tunku Abdul Rahman (UTAR), Malaysia

Dr. Teng Shao, Northwestern Polytechnical University, China

Dr. A. S. Devakumar, University of Agricultural Sciences, India

IOP Conf. Series: Earth and Environmental Science **257** (2019) 011003 doi:10.1088/1755-1315/257/1/011003

Peer review statement

All papers published in this volume of *IOP Conference Series: Earth and Environmental Science* have been peer reviewed through processes administered by the proceedings Editors. Reviews were conducted by expert referees to the professional and scientific standards expected of a proceedings journal published by IOP Publishing.

Content from this work may be used under the terms of the Creative Commons Attribution 3.0 licence. Any further distribution of this work must maintain attribution to the author(s) and the title of the work, journal citation and DOI.
Published under licence by IOP Publishing Ltd

Assessment of aerosol direct effects on $PM_{2.5}$ and O_3 air quality in Continental Southeast Asia

Giang Tran Huong Nguyen[1,2], Hikari Shimadera[1], Tomohito Matsuo[1] and Akira Kondo[1]

1 Graduate School of Engineering, Osaka University, Suita, Osaka 565-0871, Japan
2 Department of Environment and Natural Resources, Da Lat University, Vietnam

Abstract. An online coupled modeling system composed of Weather Research and Forecasting (WRF) model and Community Multiscale Air Quality (CMAQ) model was applied to assess aerosol direct effects on air quality with the focus on fine particulate matter ($PM_{2.5}$) and ozone (O_3) in Continental Southeast Asia where has suffered significant air pollution recently due to rapid development. The results showed that, in four focused countries including Cambodia, Laos, Thailand, and Vietnam, the direct effects moderately decreased shortwave radiation, temperature, planetary boundary layer (PBL) height, and wind speed by -9.08%, -0.44°C, -10.27%, and -2.21% in dry season, and -2.37%, -0.04°C, -2.05%, and -0.57% in wet season, respectively. Consequently, $PM_{2.5}$ concentration was found to increase by +10.51% in dry season and +1.44% in wet season. O_3 concentration was decreased by -2.76% in dry season while slightly increased by +0.56% in wet season. The increasing effect of aerosols on $PM_{2.5}$ concentration was caused by the more stable atmospheric condition. The increase or decrease in O_3 concentration depended on the responses of atmospheric dynamics as well as photolysis rates of photochemical reactions to direct effects.

1. Introduction

Environmental management cannot keep pace with continuously growths of economic and population at a rapid rate, thus it results in significant air pollution in Southeast Asia. Air quality in Continental Southeast Asia varies widely over space and time. In rural areas, biomass burning is the most dominant source of air pollutants followed by vehicle emission, whereas, in urban areas, vehicle and industrial emissions are the most important sources. During the dry season, from November to April, agricultural biomass burning and forest fire frequently cause severe particulate pollution not only in the local areas but also across the whole region and beyond [1]. Adverse air pollution has undesirable effects on human health and ecosystems. The most severe health effects can be attributed to fine particulate matter ($PM_{2.5}$) and ozone (O_3) [2]. Hence reduction of O_3 and $PM_{2.5}$ concentrations has become one of the key objectives of air quality control in the region.

In order to understand sources, transport, and chemical transformation of air pollutants for further air quality control, scientists have widely used air quality models. The air quality model traditionally used is called offline model which is decoupled treatments of meteorology and chemistry. Recently, online model which allows to facilitate feedbacks between chemistry and meteorology [3] has been developed rapidly and considered as a very promising way for future atmospheric simulation. In line with this trend, United States Environmental Protection Agency (US EPA) coupled the Weather Research and Forecasting (WRF) model and the Community Multiscale Air Quality (CMAQ) model to create a two-way online coupled modeling system. The essence of this interaction is focused on the direct radiative effects of scattering and absorbing aerosol in the troposphere that in turn affect

Content from this work may be used under the terms of the Creative Commons Attribution 3.0 licence. Any further distribution of this work must maintain attribution to the author(s) and the title of the work, journal citation and DOI.
Published under licence by IOP Publishing Ltd

radiation calculations in the meteorological model, WRF [4]. This development not only helps to more realistically simulate the atmosphere but also gives a chance in investigating aerosol direct effects on the air quality. Aerosol direct effects can cause increase or decrease in air pollutant concentrations through change in meteorological condition. Assessment of aerosol impacts is thus important for supporting the development of pollution control strategies. In addition, this topic is interesting since aerosol direct effects on air quality is far from being well understood in Continental Southeast Asia.
Therefore, this paper aimed to evaluate the impacts of aerosol direct effects on meteorology and air quality in Continental Southeast Asia targeted on four countries, namely Cambodia, Laos, Thailand and Vietnam. Focused pollutants are $PM_{2.5}$, and O_3. To fulfill the objectives, WRF and CMAQ models were coupled in both one-way and two-way online for the whole historical year 2014.

2. Methodology

2.1. Modelling framework and simulation design
Modeling domain together with locations of observation sites is presented in Figure 1. The focused domain covered almost part of continental region of Southeast Asia. Four countries was targeted including Cambodia, Laos, Thailand and Vietnam. The domain had a horizontal resolution of 24 km x 24 km, and 98 grid cells in each direction.
The online coupled WRF-CMAQ modeling system was applied for air quality simulations in the study region. The datasets used for the modeling system included geological data from USGS (United States Geological Survey); RTG_SST_HR (real time global high resolution sea surface temperature), and FNL (final analysis) meteorological data which were provided by the United States National Centers for Environmental Prediction (NCEP); chemical initial and boundary condition data produced by the global chemistry transport model, MOZART (Model Ozone and Related Chemical Tracer); HTAP (Hemispheric Transport of Air Pollution), FINN (Fire Inventory from National Center for Atmospheric Research), and Aerocom (Aerosol Comparisons between Observations and Models) emission databases, and biogenic volatile organic compounds (BVOCs) emission data outputted from MEGAN (Model of Emissions of Gases and Aerosols from Nature).

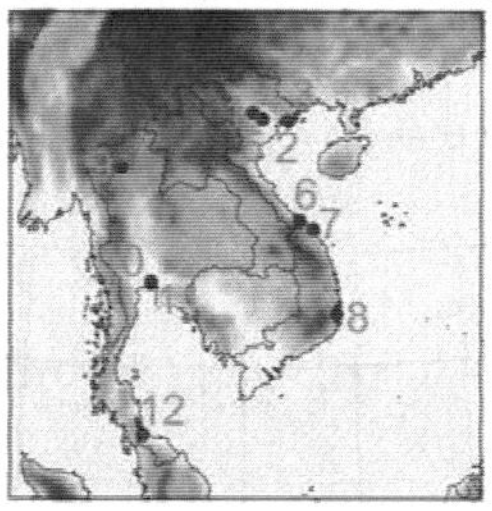

Figure 1. Modeling domain and locations of observation sites.
To capture aerosol direct effects, two simulations were designed: one-way online simulation (without feedback from CMAQ to WRF) and two-way online simulation (with feedback from CMAQ to WRF) for entire historical year 2014. Since the feedback in two-way simulation was implemented to represent aerosol direct effects, their impacts on meteorology and air quality were calculated as the differences between the results of two-way online simulation and one-way online simulation, and the contribution percentages of direct effects were calculated as the ratio of these differences to the one-way simulation results.

2.2. Model evaluation
Model performance was evaluated by statistical calculation of model results against observations. Hourly observed radiation, temperature, wind speed, pressure, and relative humidity were gathered from 7 stations located in Viet Nam. Hourly observed $PM_{2.5}$ and O_3 concentrations were gathered from 10 stations, 6 stations located in Viet Nam, and 4 stations located in Thailand (Figure. 1). Beside traditional statistical metrics using correlation coefficient (R), mean bias (B_{MB}), mean absolute gross error (E_{MAGE}), this study was also followed the evaluation method proposed by Yu et al. (2006) using

normalized mean bias factor (B_{NMBF}) and normalized mean absolute error factor (E_{NMAEF}) [5].

3. Results and discussion

3.1. Model performance against observation

In general, the coupled WRF-CMAQ model showed a good performance. For meteorological simulation, modeled temperature, radiation, humidity, and pressure were highly consistent with observations as shown by R values (Table 1). The model overestimated the observation by the factor of 1.43 (two-way online simulation) for radiation. No significant overestimation or underestimation was found for temperature, humidity and pressure. For wind speed, its correlation with observation was lower than those of the other variables and the model overestimated the observation by the factor of 1.74 (two-way online simulation). For air quality simulation, the correlation coefficients for $PM_{2.5}$ and O_3 were acceptable. However, overestimations were found in model calculation for the two pollutants. The model overestimated $PM_{2.5}$ concentration by the factor of 1.26 (two-way online simulation) and O_3 concentration by the factor of 1.57 (two-way online simulation).

Table 1. Statistical evaluation for model performance.

Parameter	n	$\overline{O}$	Run type	$\overline{M}$	R	B_{MB}	E_{MAGE}	B_{NMBF}	E_{NMAEF}
Temperature (°C)	42471	25.59	One-way	24.35	0.92	-1.24	2.07		
			Two-way	24.09	0.92	-1.50	2.20		
Radiation (W/m^2)	42471	111.15	One-way	174.61	0.76	63.46	94.66	0.57	0.85
			Two-way	158.76	0.77	47.61	82.43	0.43	0.74
Humidity (%)	42471	79.46	One-way	80.00	0.62	0.54	9.51	0.01	0.12
			Two-way	80.80	0.63	1.34	9.39	0.02	0.12
Pressure (mBar)	42471	1009.48	One-way	1006.99	0.88	-2.49	3.79	0.00	0.00
			Two-way	1006.99	0.88	-2.49	3.83	0.00	0.00
Wind speed (m/s)	2920	1.39	One-way	2.50	0.48	1.12	1.40	0.81	1.01
			Two-way	2.41	0.51	1.02	1.32	0.74	0.95
$PM_{2.5}$ ($\mu g/m^3$)	59259	29.82	One-way	35.40	0.57	5.58	20.81	0.19	0.70
			Two-way	37.56	0.58	7.74	22.17	0.26	0.74
O_3 (ppb)	50246	21.57	One-way	34.24	0.54	12.67	17.23	0.59	0.80
			Two-way	33.91	0.55	12.35	17.01	0.57	0.79

3.2. Meteorology response to aerosol direct effects

Under the aerosol direct effects, ground-level shortwave radiation, temperature, PBL height, and wind speed showed decreasing trends of mean values (Figure 2). The reduction in these meteorological variables were -5.99%, -0.24°C, -7.02%, and -1.43% on average, and in the range of -15.38% to -1.31%, -0.67°C to +0.05°C, -21.03% to +0.67%, and -6.96% to +1.69%, respectively, for the whole year in four focused countries. In dry season (from November to April), these contribution were -9.08%, -0.44°C, -10.27%, and -2.21% on average and in the range of -25.87% to -0.76%, -1.25°C to +0.05°C, -36.80% to +2.84%, and -11.89% to +2.86%, respectively. In wet season (from May to October), they were -2.37%, -0.04°C, -2.05%, and -0.57% on average and in the range of -8.36% to +2.52%, -0.25°C to +0.12°C, -9.66% to +3.64%, and -3.44% to +1.23%, respectively. These results showed that the direct effects were larger in dry season than those in wet season. The reason was that $PM_{2.5}$ concentration in dry season was higher than that in wet season. Among four countries, Northern Laos suffered the highest effects and Cambodia suffered the lowest effects from aerosols, there had no much difference in the effects between Thailand and Vietnam.

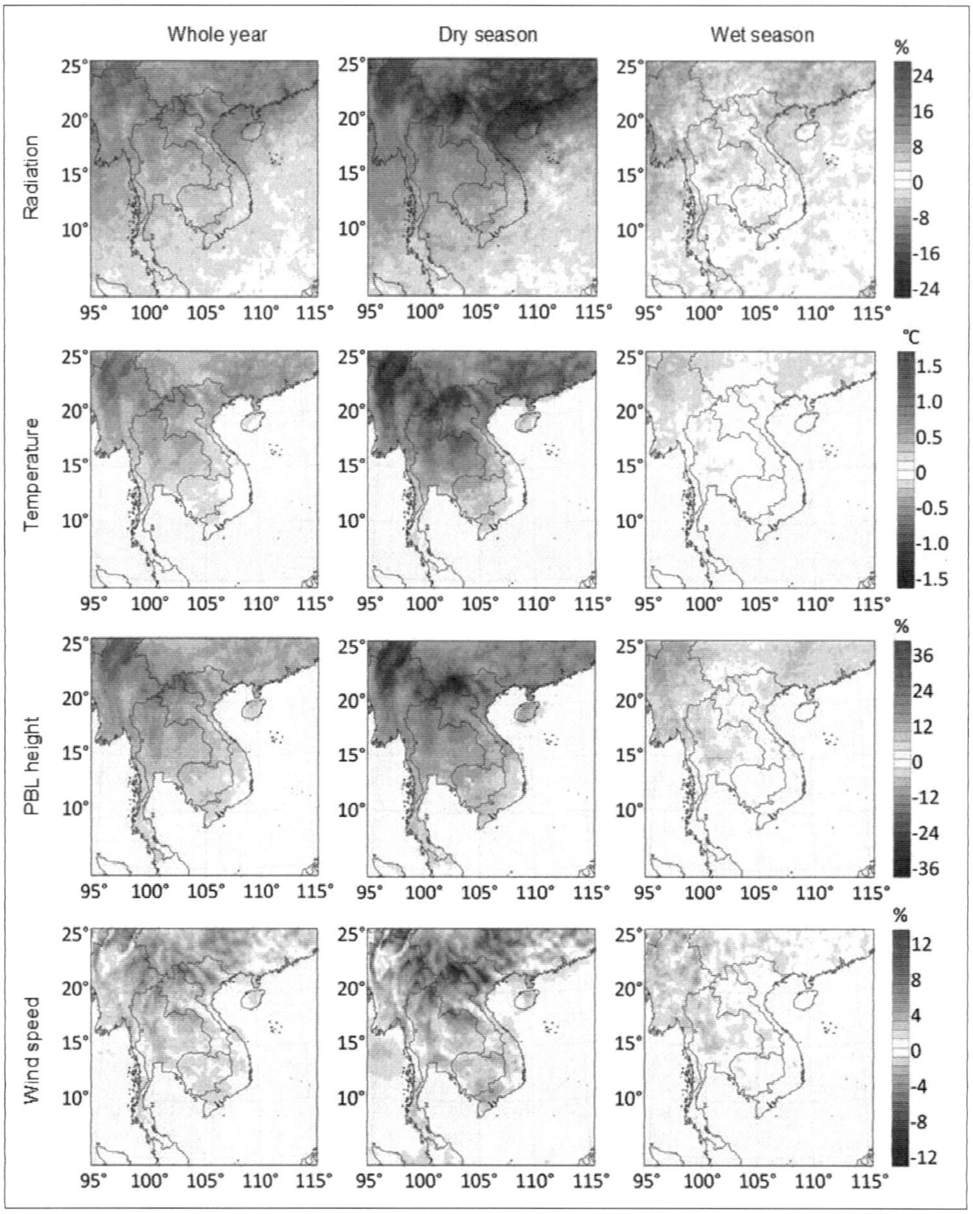

Figure 2. Spatial distributions of contributions of aerosol direct effects to ground-level meteorological variables.

3.3. $PM_{2.5}$ response to aerosol direct effects

The spatial distributions of contribution percentages of aerosol direct effects to ground-level $PM_{2.5}$ concentration are shown in Figure 3. Over four focused countries, aerosols mostly produced increasing effects on $PM_{2.5}$ concentrations with contribution percentage of +8.56% on average and in the range from -4.75% to +44.74% for the whole year. In both dry season and wet season, $PM_{2.5}$ concentrations were increased with the percentage of +10.51% on average and in the range from -5.50% to +52.08%, and +1.44% on average and in the range from -5.15% to +9.82%, respectively. Similarly to meteorology, the effects on $PM_{2.5}$ were larger in dry season than those in wet season because of lower $PM_{2.5}$ loadings in wet season. The contribution percentages were high in Northern Laos in dry season where was found to have high $PM_{2.5}$ concentration. In contract with other countries, in Cambodia where was found to have low $PM_{2.5}$ concentration, contribution percentage was negative with the value of -0.05% in wet season. The average contribution percentages in Thailand and Vietnam were quite similar. The results indicated that the effects on $PM_{2.5}$ concentration were highly associated with $PM_{2.5}$ concentration level; the effects were large at high $PM_{2.5}$ periods and locations.

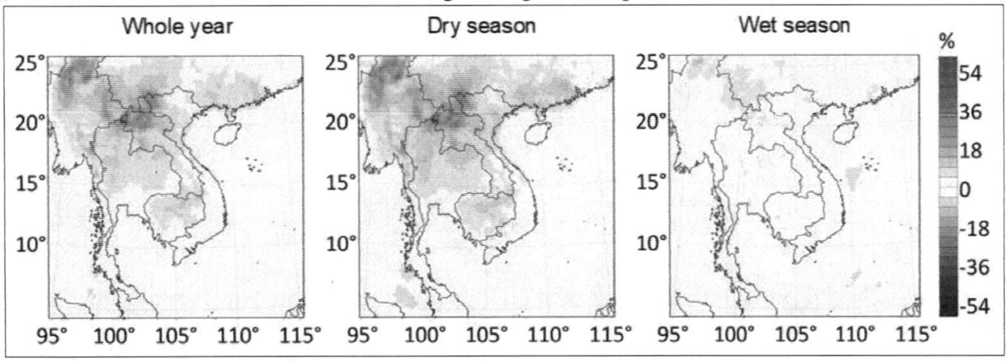

Figure 3. Spatial distributions of contribution percentages of aerosol direct effects to ground-level PM2.5 concentration.

Table 2. Correlation coefficients between the response of $PM_{2.5}$ concentration and the responses of meteorological variables.

	Variable	Radiation, %	Temperature, °C	PBL height, %	Wind speed, %
Whole year	$PM_{2.5}$, %	-0.74	-0.76	-0.81	-0.68
Dry season	$PM_{2.5}$, %	-0.72	-0.79	-0.81	-0.69
Wet season	$PM_{2.5}$, %	-0.32	-0.41	-0.37	-0.24

The increase in $PM_{2.5}$ concentration could be explained as followed. Shortwave radiation is scattered by aerosols, thus shortwave radiation arriving the ground surface is decreased, and then temperature near the ground surface is decreased. On the other hand, temperature in the atmosphere at high level is raised as aerosols absorbed shortwave radiation and heat is released. Thus, temperature inversion is enhanced that leads to increase in the atmospheric stability and decrease in PBL height. Therefore, the vertical dispersion of pollutants is reduced. Besides, the horizontal transportation is also reduced because of reducing in wind speed. As a result, $PM_{2.5}$ accumulates more and its concentration is increased. This explanation was confirmed by high negative correlations between the responses of $PM_{2.5}$ concentration and the responses of shortwave radiation, temperature, PBL height, and wind speed (Table 2). Note that, the correlation coefficients in wet season were low because in this season, $PM_{2.5}$ concentration was low, thus the direct effects were not significant, so that, the relationship could not be clearly observed.

3.4. O_3 response to aerosol direct effects

The spatial distributions of contribution percentages of aerosol direct effects to ground-level O_3 concentration are shown in Figure 4. Aerosol direct effects produced a slightly decreasing effect on O_3 concentration with the percentage of -1.51% on average and in the range from -3.86% to +1.47% for the whole year in four focused countries. In dry season, O_3 concentration was decreased with the percentage of -2.76% on average and in the range from -7.27% to +1.75% while, in wet season, O_3 concentration was slightly increased with percentage of +0.56% on average and in the range from -1.85% to +3.38%. There had not much difference in contribution percentages among four countries.

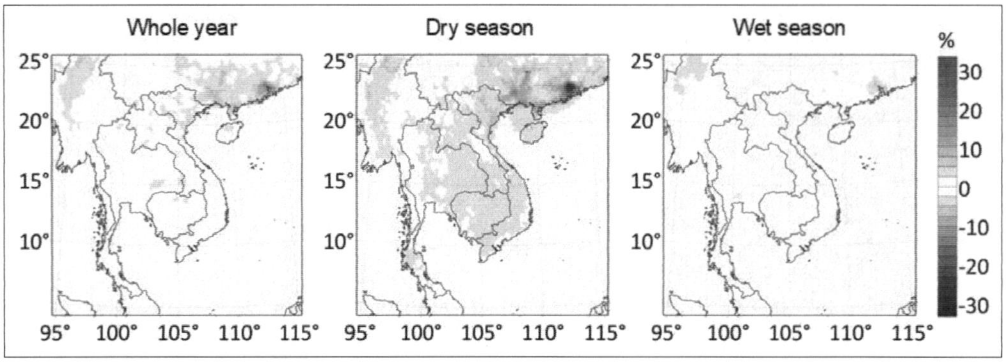

Figure 4. Spatial distributions of contribution percentages of aerosol direct effects to ground-level O3 concentration.

In order to explain the response of O_3 concentration to aerosol direct effects, both atmospheric dynamics as well as photolysis rates of photochemical reactions should be taken into account. First, change in atmospheric dynamics due to direct effects, more stabilized atmosphere and shallower PBL height as mentioned above, leads to increase in NO_x concentration. In case of VOC-limited regime, the increase in NO_x concentration enhances titration by NO as in equation (3) then causing a decrease in O_3 concentration. But in case of NO_x-limited regime, the increase in NO_x concentration facilitates O_3 formation as in equations (1) and (2) then causing an increase in O_3 concentration. Second, reducing in photolysis rate due to the decrease in surface shortwave radiation caused by direct effects reduces the generation of O_3 as in equation (1) then causing a decrease in O_3 concentration [6]. Therefore, the overall response of O_3 concentration to direct effects results from the competition of two opposing effects, a decreasing effect and an increasing effect. In this study, it was found that, in four focused countries, the mean ratio of non-methane hydrocarbon to nitrogen oxides ($NMHC/NO_x$) in wet season, 71.35 (ppbC/ppb), was higher than that in dry season, 33.85 (ppbC/ppb). In addition, in wet season, the direct effects on shortwave radiation were less than those in dry season (Figure 2) that led to less reduction in photolysis rate compared to dry season. Thus, the increasing effect was slightly larger in wet season while the decreasing effect was larger in dry season.

$$NO_2 + h\nu \rightarrow NO + O \quad (1)$$
$$O + O_2 \rightarrow O_3 \quad (2)$$
$$NO + O_3 \rightarrow NO_2 + O_2 \quad (3)$$

4. Conclusion

The coupled meteorology and atmospheric chemistry WRF-CMAQ model together with utilized data, in overall, was able to reproduce the observation acceptably. Aerosol direct effects were recognizable in the region, particularly, under high $PM_{2.5}$ pollution conditions. Dry season suffered larger impacts than wet season. Regarding meteorology, aerosol produced decreasing effects on shortwave radiation, temperature, PBL height, and wind speed. In term of air quality, the inclusion of aerosol direct effects resulted in an increase in $PM_{2.5}$ concentration and a decrease in O_3 concentration for the whole year. The effects on O_3 concentration were not significantly compared with those on $PM_{2.5}$ concentration.

In closing, this application of the online coupled WRF-CMAQ modeling system demonstrated the model performance and presented a first using this system to study aerosol direct effects in the region. The results of this study can provide valuable information for developing effective pollution control strategies in the regions.

5. References

[1] Nguyen D H, Ho Q B and Ngo X Q 2016 Modelling and prediction of air pollutant transport during the 2014 biomass burning and forest fires in peninsular Southeast Asia *Environ Monit Assess* 188: 106

[2] EEA 2012 *Air quality in Europe - 2012 report* European Environmental Agency report No 4/2012 Doi: 10.2800/55823

[3] Gerald R N, John A, Pyle J and Zhang F 2014 *Encyclopedia of Atmospheric Sciences* 2nd Edition Elsevier
[4] Wong D C, Pleim J, Mathur R, Binkowsk F, Otte T, Gilliam R, Pouliot G, Xiu A and Kang D 2012 WRF-CMAQ two-way coupled system with aerosol feedback: software development and preliminary results *Geosci. Model Dev.*, **5**, 299-312
[5] Yu S, Eder B, Dennis R, Chu S H and Schwartz S E 2006 New unbiased symmetric metrics for evaluation of air quality models *Atmos. Sci. Let.* **7**: 26–34
[6] Xing J, Wang J, Mathur R, Wang S, Sarwar G, Pleim J, Hogrefe C, Zhang Y, Jiang J, Wong D C and Hao J 2017 Impacts of aerosol direct effects on tropospheric ozone through changes in atmospheric dynamics and photolysis rates *Atmos. Chem. Phys.*, **17**, 9869–9883

IOP Conf. Series: Earth and Environmental Science **257** (2019) 012002 doi:10.1088/1755-1315/257/1/012002

Air pollution modeling over Shanghai and Guangzhou

TY Zhang[1], Z Chen[2], Y Zhu[3], RH Zong[4]

1 School of Statistics, Jiangxi University of Finance and Economics, Qingshanhu, Nanchang, China
2 School of Life Science and Technology, University of Electronic Science and Technology of China, Chen Du, Sichuan, China
3 School of Electronic Engineering, Beijing University of Posts and Telecommunications, Beijing, China
4 School of Computer Science, Sichuan University, Chengdu, Sichuan, China
Email: ztycreative@outlook.com

Abstract. As one of the most prominent pollutants that threaten human health over big cities, fine particle matter (PM2.5) has largely attracted public and researches' attention. Critical challenges are unresolved regarding how to effectively predict atmospheric PM2.5 concentrations. Here, our team aimed to capture the PM2.5 high-frequency dynamics over Shanghai and Guangzhou using advanced machine learning technique. Our results showed that PM2.5 concentration could be forecasted with historical PM2.5 record and meteorology forcings including temperature, humidity, precipitation, pressure, wind speed, and due point temperature. Our sensitivity analyses further revealed that the perdition was robust against critical model parameters across cities. The underlying sink processes could be different despite of their similar temporal features.

1. Introduction:

China has been quickly growing during the recent decades in terms of economic and social development. Accompanied with the rapid industrialization, emissions of air pollutants (e.g., fine particle matter and N_2O) have been largely increased, most of which sink remotely and generate negative impacts on human health and environmental conditions [1]. Furthermore, land surface properties (e.g., natural forest cover versus urban area) are greatly modified due to the rapid urbanization. Land cover changes modify land surface energy balance and could induce regional climate changes that increase the residence time of air pollutant [2]. Mega cities, such as Beijing, Shanghai, and Guangzhou, have been suffering from periodic sever air pollution events, in which fine particle matter (PM) pollution, especially PM2.5 is one of the most prominent ones [3]. PM2.5 is defined as particles less than 2.5 micrometers in diameter. It could strongly impact human health especially respiratory system and lead to higher mortality. For example, in toxicological studies, PM2.5 is cytotoxic to cell death which may cause respiratory inflammation [4].

Multiple factors convolute together and contribute a high ambient PM2.5 condition. They include anthropogenic factors, such as, industrial emission with the relatively long-term influence (monthly or yearly) and vehicle emission with a clear diurnal pattern (e.g., high during rush hours). In addition, environmental factors could also contribute to severe PM2.5 pollution. Meteorological driver such as high pressure could stabilize the boundary condition, and prevent PM2.5 from being removed out of atmosphere. High wind speed may have a double-edged effects. It could either decrease the concentration of the PM2.5 by quickly propagating the pollutant to adjacent regions or increase the

Content from this work may be used under the terms of the Creative Commons Attribution 3.0 licence. Any further distribution of this work must maintain attribution to the author(s) and the title of the work, journal citation and DOI.
Published under licence by IOP Publishing Ltd

PM2.5 emission by blowing up fine dust from the land surface. Humidity and precipitation negatively affect the density of particulate matters by inducing wet deposition [5,6].
To tackle the critical challenge of understanding and predicting PM2.5 over large cities, previous efforts have been focused on three major avenues:1) physical modeling; 2) simple statistical inference; 3) advanced machine learning techniques. Physical models were developed based on inner physical processes of particulate matter formation, emission, transport, chemical reaction, and removal. Statistical models could also be applied to infer PM2.5 dynamics, mainly based on regressed relationship between target pollutant (PM2.5) concentration and drivers (e.g., wind speed, precipitation, and air temperature). However, the underlying processes behind PM2.5 dynamics is extreme complex, thus restricts the accuracy of the simple statistical inference prediction. Recently, many researchers used advanced machine learning algorithms to reproduce this complex processes from a data-driven perspective. For example, existing studies used neural network techniques to either directly predict the air pollutant concentration or classify the air pollutant level [7,8]. In this work, our team apply advanced neural network architecture, called Recurrent Neural Network (RNN) to forecast the PM2.5 concentration over Shanghai and Guangzhou. Unlike other feedforward neural networks models, RNN mechanistically considers the historical information or internal memory of the system itself, which is more appropriate to time series tasks like PM2.5 concentrations.

2. Methodology

2.1. Air pollutant and relevant meteorology data

The data is continuous high-frequency measurements of atmospheric PM2.5 concentration collected over the cities of Shanghai and Guangzhou [9]. The temporal period covers from 2011 to 2015. High-frequency measurements of meteorology drivers were also used, including temperature, pressure, relative humidity, precipitation, wind speed, and dew point (temperature that induces the saturated water). At the data pre-processing stage, the missing values and extreme values (outliers) were removed, so that the data can be safely used for model training. The reason to choose Guangzhou and Shanghai to simultaneously model PM2.5 is that their annual variation as well as a seasonal cycle of PM2.5 are relatively consistent (Figure 1).

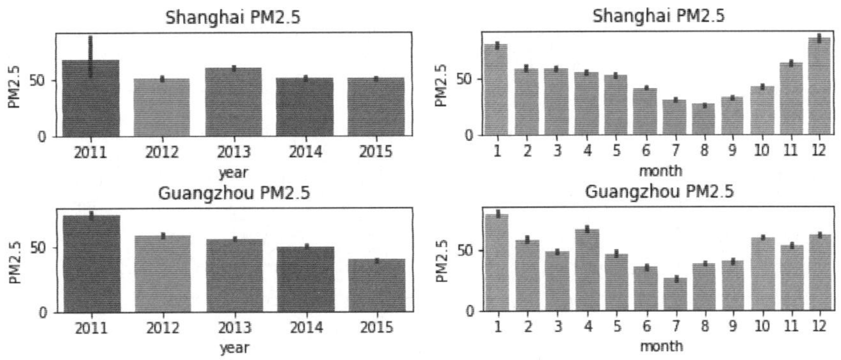

Figure 1. Time series of PM2.5 concentration over Shanghai and Gugnagzhou from 2011 to 2015. The left panels show annual variation and the right panels show seasonal dynamics

2.2. Predictive modeling

In this study recurrent neural network was employed to model atmospheric PM2.5 concentration, and particularly with long-short-term-memory (LSTM) technique first proposed by Hochreiter et al. [10]. LSTM was originally designed to help conserve historical information within a neural network, which cannot be handled by feedforward neural networks, such as convolutional neural network or fully connected neural network..Similarly, here our aim is to conserve a certain fraction of historical information about PM2.5 concentrations as well as relevant meteorology factors to help understand and predict future fate of PM2.5.

The first step in the LSTM is to decide which information will be discarded from the cell state. This decision is accomplished by forget gate layer. This gate will read previous time step output h_{t-1} and current time step input x_t to output a value between zero and one, and later on will be applied to previous time step cell state C_{t-1}. Here, one means "complete reservation" and zero means "complete abandonment".

$$f_t=\sigma(W_f \cdot [h_{t-1}, x_t]) \tag{1}$$

The next step determines which newly input information should be stored in the cell state, which is calculated in two steps. First, the sigmoid neural network layer (input gate) decides how much new information in terms of percentage will be used. Then, a tanh neural network layer creates a candidate value vector of in internal state. Finally, two components are combined to generate a new update of internal state.

$$i_t = \sigma(W_o \cdot [h_{t-1}, x_t]) \tag{2}$$

$$\tilde{C}_t = \tanh(W_c \cdot [h_{t-1}, x_t]) \tag{3}$$

The new internal state C_t is updated using previous internal state C_{t-1} and the newly generated candidate state $\tilde{C}_t$.

$$C_t = f_t * C_{t-1} + i_t * \tilde{C}_t \tag{4}$$

Finally, the model output is based on the new internal cell state. First, a sigmoid neural network layer determines which part of the cell state will be outputted. Then, running the cell state through tanh activation function (get a value between -1 and 1) and multiplying it with the output from previous sigmoid layer will get the output at current time step h_t:

$$o_t = \sigma(W \cdot [h_{t-1}, x_t]) \tag{5}$$

$$h_t = o_t * \tanh(C_t) \tag{6}$$

The optimization method is Adam (Adaptive Moment Estimation), a momemtent based stochastic gradient descent method, to train LSTM, and (Mean Squared Error) MSE to evaluate the optimization and prediction.

$$RMSE = \sqrt{\frac{1}{N}\sum(\hat{z}_t - z_t)^2} \tag{7}$$

where $\hat{z}_t$ is the prediction value, z_t is the real value

3. Results and discussion

3.1. Model performance at Shanghai and Guangzhou

Two independent Recurrent Neural Network (RNN) models at Shanghai and Guangzhou was trained, using first 80% of the observed time series. Input features are historical PM2.5 and meteorology forcings; and target variable is the present time step PM2.5 concentration. Figure 2 showed the evaluation of the two models using the rest 20% data. In general, the results demonstrated that LSTM was an effective modeling tool to capture the temporal evaluation of atmospheric PM2.5 concentration over Shanghai (Figure 2 left panel) and Guangzhou (Figure 2 right panel).

In addition, sensitivity test of the model performance against the number of LSTM layers answers the question of how deep the LSTM network is necessarily needed, in order to best capture the observed PM2.5 concentrations. Figure 2 indicated that three LSTM layers was sufficient enough to predict PM2.5 dynamics, while increase the number of LSTM layers did not significantly benefit the overall model performance and sometime even decreased the model performance due to introducing more parameters (Figure 1 left panel). Then the number of LSTM layers was fixed at three, while another sensitivity test of the model performance against the number of neuron at each LSTM layer was

IOP Conf. Series: Earth and Environmental Science **257** (2019) 012002 doi:10.1088/1755-1315/257/1/012002

conducted for how much internal LSTM complexity is required. Figure 3 showed that increasing the number of neurons from ten to fifty did not improve model performance at all.

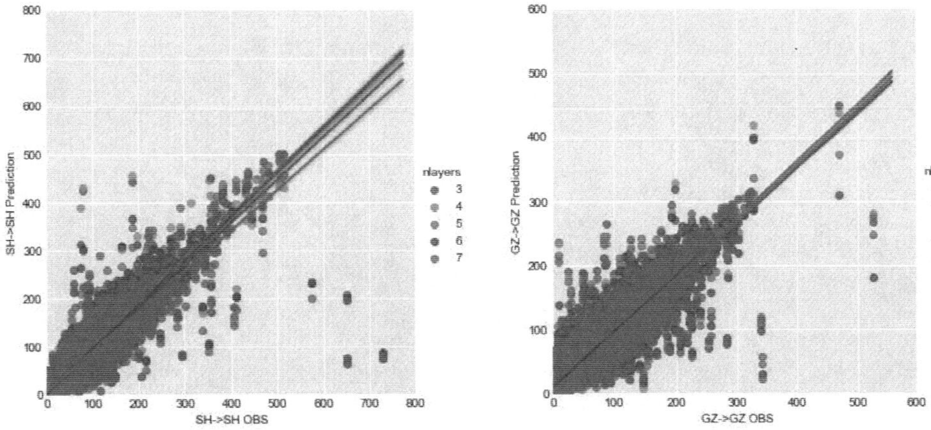

Figure 2. Model evaluation at Shanghai and Guangzhou. The overall performance has been tested against the number of layers in LSTM.

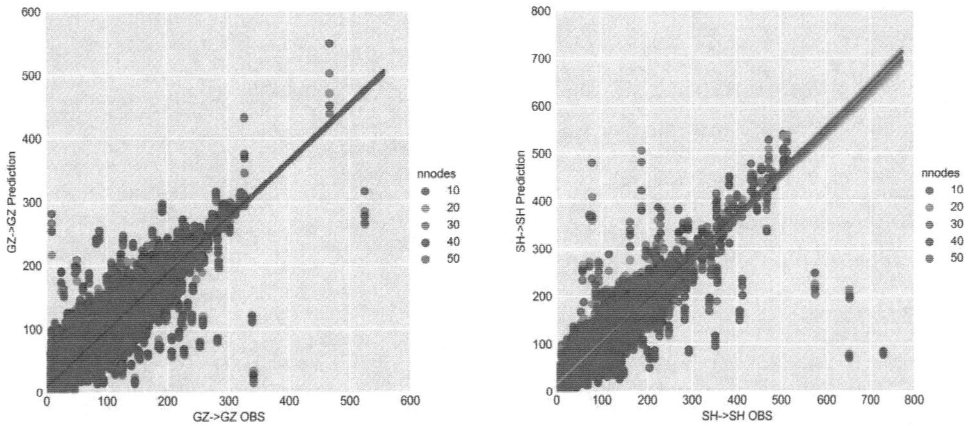

Figure 3. Model evaluation at Shanghai and Guangzhou. The overall performance has been tested against the number of nodes used in each LSTM layer.

3.2. Model performance across cities

The observed temporal patterns of PM2.5 concentration are generally similar at Shanghai and Guangzhou (Figure 1). However, the underlying mechanisms that govern the emergent patterns could be different. Fortunately, it was able to test this hypothesis by applying the RNN model trained at one city (Shanghai or Guangzhou) to another (Guangzhou or Shanghai). Figure 3 summarized the cross-evaluation results of RNN models. Firstly, applying RNN trained at Guangzhou to Shanghai generally underestimated PM2.5 concentration (Figure 4 left two panels). Further such underestimation is sensitive to the number of LSTM layers but not number of neurons per LSTM layer. More LSTM layers will prevent the model being effectively applied to other cities. Secondly, applying RNN trained at Shanghai to Guangzhou could either overestimate or underestimate PM2.5 concentrations (Figure 4 right two panels) according to the different model setup (number of LSTM, number of neurons per LSTM). For example, the RNN with five LSTM layers and 30 neurons per LSTM layer (blue dots) tended to overestimate PM2.5 at Guangzhou, while an underestimation was observed for the RNN with seven LSTM layers (purple dots, upper right panel). In summary, it indicated that although observed PM2.5 dynamics were quite similar across different cities, having a general model that is successful at different cities could be challenging because the underlying predictive relationship and PM2.5 sink and source processes could be largely different.

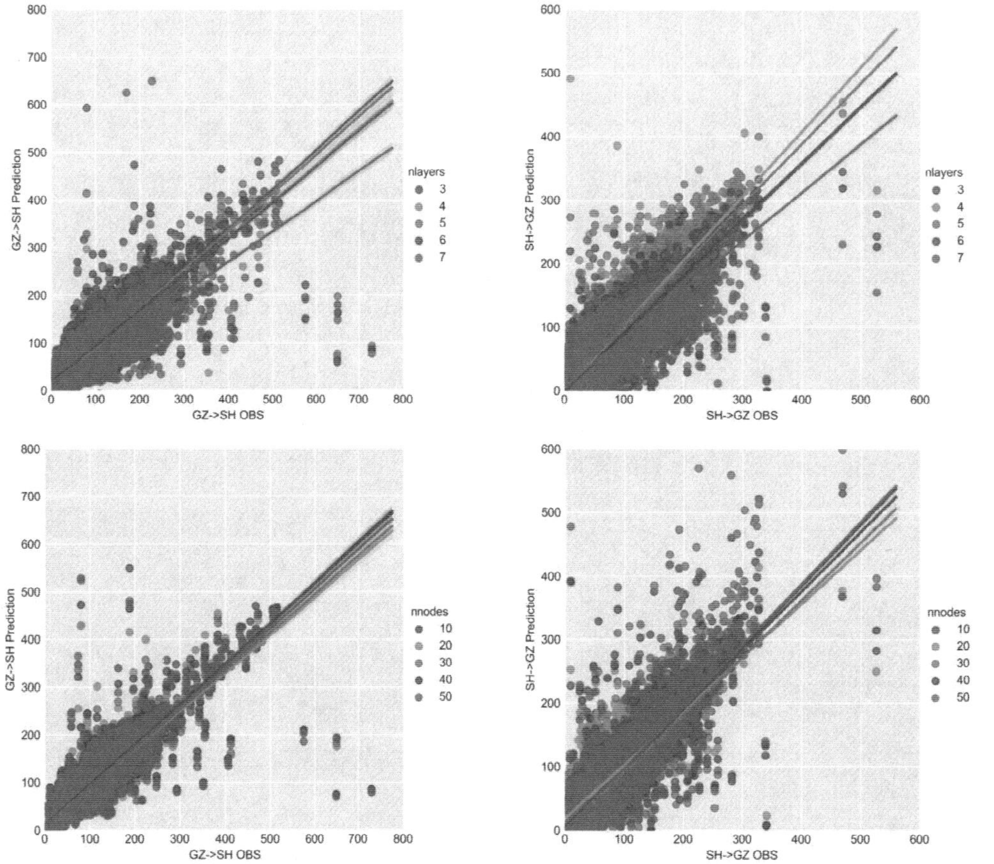

Figure 4. The comparison of the trend in cross-prediction by changes in layers and nodes

4. Conclusion and Limitation

4.1. Conclusion

Predictive modeling of important air pollutant is challenging due to the complex nature of the pollutant source/sink dynamics. In this paper, a recurrent neural network architecture called long-short-term-memory (LSTM) was employed to model the atmospheric PM2.5 concentrations over Shanghai and Guangzhou. The model successfully forecasted the PM2.5 dynamics over the two cities, furthermore, the performance was quite robust and stable with different neural network settings. More importantly, cross-city application suggested that although the observed PM2.5 temporal features are similar, their underlying source/sink processes could be different.

4.2. Limitation and future work

This study is limited by computational resources. In the RNN modeling, the number of the epochs and batch size was constant, although the training of RNN could be sensitive to both of them. Future work will be carried out for a full set of values for epochs and batch size in order to find the generally optimal combination of these hyperparameters. Secondly, this study only tested our RNN framework at two cities, while the methodology should also be applicable to other big cities. It is possible for us to expand our modeling efforts to other cities of interest, such as New York or Los Angeles.

5. Reference

[1] Wang, J, et al. (2017). Large-scale transport of PM 2.5 in the lower troposphere during winter cold surges in China. Scientific reports, 7(1), 13238.

[2] Kalnay, Eugenia et al., (2003). Impact of urbanization and land-use change on climate. Nature 423.6939: 528.

[3] Lu, F, et al. (2015). Systematic review and meta-analysis of the adverse health effects of ambient PM2. 5 and PM10 pollution in the Chinese population. Environmental research, 136, 196-204.

[4] Xing Y F, et al. (2016) The impact of PM2.5 on the human respiratory system[J]. J Thorac Dis,, 8(1):E**69**.

[5] Yang L et al., (2011). Seasonal and diurnal variations in N(2)O concentrations and fluxes from three eutrophic rivers in Southeast China[J]. JOURNAL OF GEOGRAPHICAL SCIENCES ,20(5):820-832.

[6] Gibson M et al., (198). Ground effects on pressure fluctuations in the atmospheric boundary layer. Journal of Fluid Mechanics, 86.3: 491-511.

[7] Fu, M et al. (2015). Prediction of particular matter concentrations by developed feed-forward neural network with rolling mechanism and gray model. Neural Comput. Appl, 26, 1789–97.

[8] Kleine D, et al. Modeling (2017). PM2. 5 urban pollution using machine learning and selected meteorological parameters. Journal of Electrical and Computer Engineering

[9] Liang X, et al. (2016). PM2.5 data reliability, consistency, and air quality assessment in five Chinese cities[J]. Journal of Geophysical Research Atmospheres, 121.

[10] Hochreiter S, et al., (1997). Long Short Term Memory, Neural computation,, 9(8): 1735-80.

IOP Conf. Series: Earth and Environmental Science **257** (2019) 012003 doi:10.1088/1755-1315/257/1/012003

Analysis of environmental behaviour of sulfa-drugs in the Lake Biwa and Yodo River Basin by using Multimedia Models

S Li[1], H Shimadera[1], T Matsuo[1], A Kondo[1]

1 Graduate School of Engineering, Osaka University, Suita, Osaka 565-0871, Japan

Abstract. This study focuses on some sulfa-drugs that have been heavily used in livestock husbandry and emitted into the Lake Biwa and Yodo River Basin. Emissions of sulfa-drugs were estimated using the national statistical data on veterinary drug usage and livestock excrement in Japan. One-box and distributional type multimedia fate models were utilized with the estimated emissions in order to simulate the behaviour of sulfa-drugs in four environmental media (atmosphere, soil, water and sediment) in the study period from 2005 to 2015 in the basin. Based on the calculated results, sulfa-drugs emitted to the soil media were efficiently transported to water, which resulted in the largest amount in water among the environmental media and high concentrations in and around areas with large farmland proportion.

1. Introduction

Chemicals which are the active substances of pharmaceutical and personal care products (PPCPs) have been detected in trace amounts from the environment [1-2]. Veterinary medicines, as one of the PPCPs, are used in large quantities on agriculture industry. There is an increasing concern about the ecological consequences of chronic exposure to the chemicals. Sulfa-drugs, which are typical antibiotics, are frequently found in the environment media, especially water bodies, such as river [3-4] and lake [5].When a sulfa-drug is used for human beings, the chemical substances contained therein are metabolized in the body, and then released into the sewage as a metabolite. Those left without being treated in the sewage treatment plant are released into the river. On the other hand, when a veterinary medicine is used for livestock, after metabolized in their body, the chemical substances are discharged as manure. Most of manure is reused in farmland after composting treatment and liquid fertilization treatment while some of the manure are not subject to carbonization or incineration or wastewater purification treatment and even discharge to the environment directly such as field piling or digging. This study focuses on sulfa-drugs and chooses 2 typical substances which belong to sulfonamide type synthetic antibacterial agents, and are sold in large quantities among veterinary drugs.

2. Model description

2.1. Multimedia model

In order to evaluate the environmental risk, it is necessary to estimate the concentration of chemical substances in the environment. One of the methods is numerical simulation based on a multimedia fate prediction model that expresses the behavior of chemical substances in and between plurality of environmental media. In this study, the kinetics of sulfa-drugs for livestock discharged into the environment were analyzed using one-box and distributional type multimedia models (OMM and DMM, respectively) for the Lake Biwa and Yodo River basin (LBYRB). In the multimedia model, the

Content from this work may be used under the terms of the Creative Commons Attribution 3.0 licence. Any further distribution of this work must maintain attribution to the author(s) and the title of the work, journal citation and DOI.
Published under licence by IOP Publishing Ltd

environment is divided into four main media, atmosphere, soil, water and sediment and some sub-media. The relationship and the mass transfer between all media are shown in figure 1. The substances movement caused by advection in the air and water media from the adjacent area is took into the consideration. The atmosphere media consists of gas phase and particle phase, and is divided into the upper layer and the lower layer corresponding to the mixed layer. The water media consists of liquid phase and suspended solid (SS) phase, and considers outflow from soil and equilibrium with sediment.

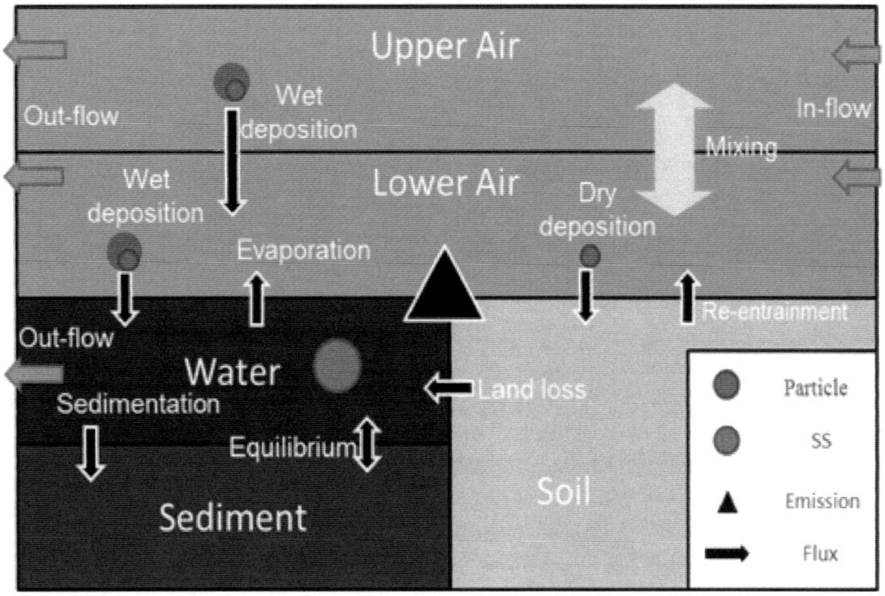

Figure 1. Structure of multimedia model.

2.2. One-box type model

The OMM assumed that the LBYRB is a huge three-dimension box which is shown in figure 2, and the concentration in each media was uniform. This imaginary box was about 111 km in the north-south direction, about 74 km in the east-west direction, and 2 km in the height direction. The inflow and outflow of air and water were considered as constant (with the average value of the whole LBYRB).

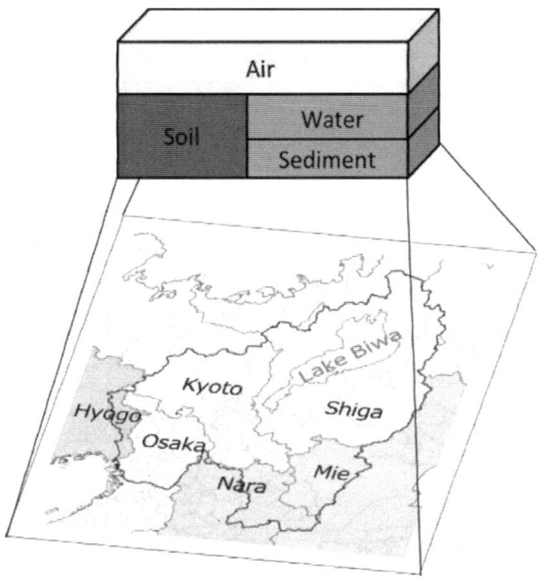

Figure 2. One-box multimedia model

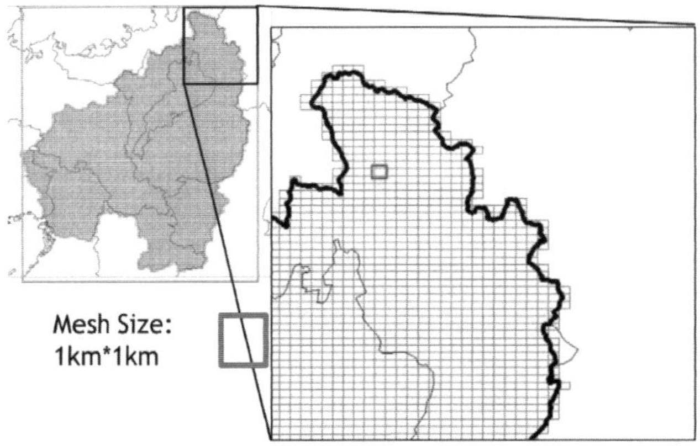

Figure 3. Distributional multimedia model

2.3. Distributional type model

In the OMM, the assumption of the uniform spatial distribution of chemicals leads to an uncertainty and an inaccuracy in simulated environmental fate. In order to interpret the spatial distribution of different chemicals in different environmental media, the DMM was developed as a mathematical tool. In the DMM, the LBYRB was divided into a grid with the size of 1 km (figure 3) and each of these grids was assumed as a three-dimensional box whose structure is the same as the one in OMM and the fluxes and mass transportation inside the compartment are affected by the 4 adjacent grids (if exist) in four directions (up down left right). The value of the inflow and outflow in each compartment which interact with the adjacent compartments were calculated dynamically using the meteorology and hydrology data.

3. Emission estimate

3.1. Target chemicals

This study focused on two kinds of sulfa-drugs, sulfamethoxazole (SMX) and sulfadimethoxine (Albon) which are widely used in Japan as an antibacterial agent for livestock production since they are synthetic drugs which show persistent degradability in nature compared to drugs made of natural extracts. According to statistical data by the Ministry of Agriculture, Forestry and Fisheries (MAFF), SMX is used in a larger amount than any other sulfur antimicrobial agents [3]

3.2. National usage

Japan began statistics on veterinary drug sales since 2005 based on the promulgation of the MAFF Ordinance No. 62 Animal Medicines Regulation Rules. In this research, the amount of 2 chemicals that were released into the environment was estimated based on the national sales amount of pure active substance sulfa-drug (an example shown in table 1) from 2005-2015 [3].

Table 1. Sales amount (active substance) and proportion by animal in 2005

Substance	Amount of pure active substance (kg)	Beef Cattle (%)	Dairy Cow (%)	Pig (%)	Broiler (%)	Layer (%)
SMX	60,223.7	0.0	0.0	79.7	16.1	4.2
Albon	18,092.8	17.0	17.8	40.9	9.8	7.2

3.3 LBYRB usage

A sulfa-drug administered to domestic animals is eventually discharged outside their body as excrement after organ metabolism. In this study, it was assumed that all sulfa-drugs administered were finally excreted into the environment, therewas no residual in livestock's organ and the usage of sulfa-drugs and excrement volume per livestock were uniform across the whole country. Based on the ratio of 6 prefectures (Osaka, Kyoto, Hyogo, Shiga, Mie, Nara) included in LBYRB and national livestock excreta (table 2) [4-7], the sulfa-drug emission in the LBYRB, E_{lcoal} were calculated by using the equation (1).

$$E_{local} = \sum_{i=0}^{n} E_{i,all} \cdot \frac{S_{i,local}}{S_{i,all}} \quad (1)$$

where i is the type of animal, E_{all} is national sulfa-drug emission, S_{all} and S_{local} are livestock excreta in the nationwide and LBYRB, respectively. The annual emissions in different prefectures are shown in figure 4.

Table 2. Excrement amount in different place (unit: 1000 ton)

	Cattle	Cow	Pig	Hens
National	23570	24420	22380	7450
Mie	123	244	278	260
Shiga	56	168	12	18
Kyoto	108	64	26	85
Osaka	27	6	15	4
Hyogo	251	468	50	235
Nara	73	29	15	22
Local	638	979	396	624

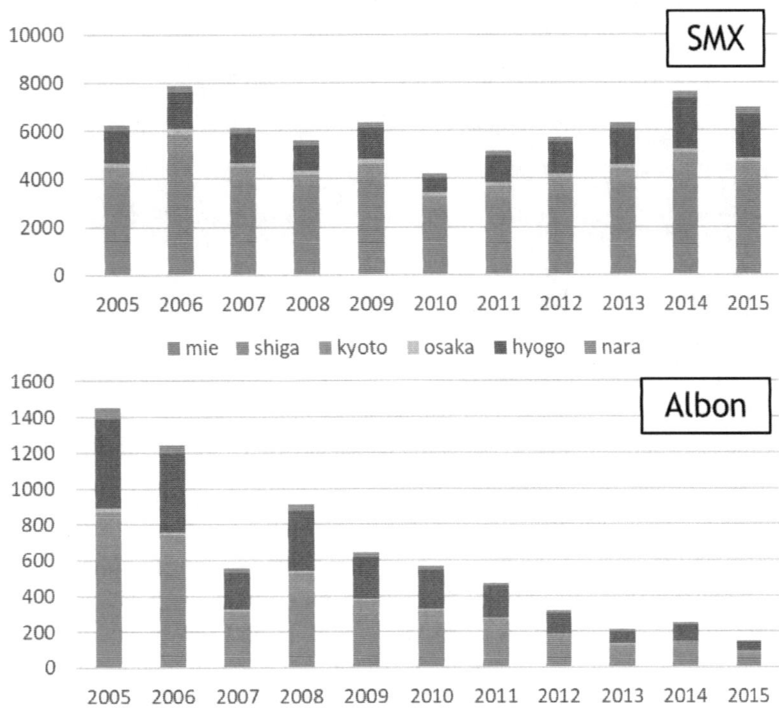

Figure 4. Estimated annual usage of sulfa-drug in 6 prefectures (unit: mol)

The spatial distribution of sulfa-drug emission in the LBYRB was estimated by using the formula (2), based on the farmland promotion according to the land use fragmented mesh data from the National Land Numerical Information System [8].

$$E_{mesh} = E_{all} \cdot P_{mesh} \tag{2}$$

The spatial distribution (figure 5) shows that most emissions of the sulfa-drugs were concentrated in the east and south of the LBYRB. In terms of the temporal trend, the Albon was decreasing while SMX was basically the same.

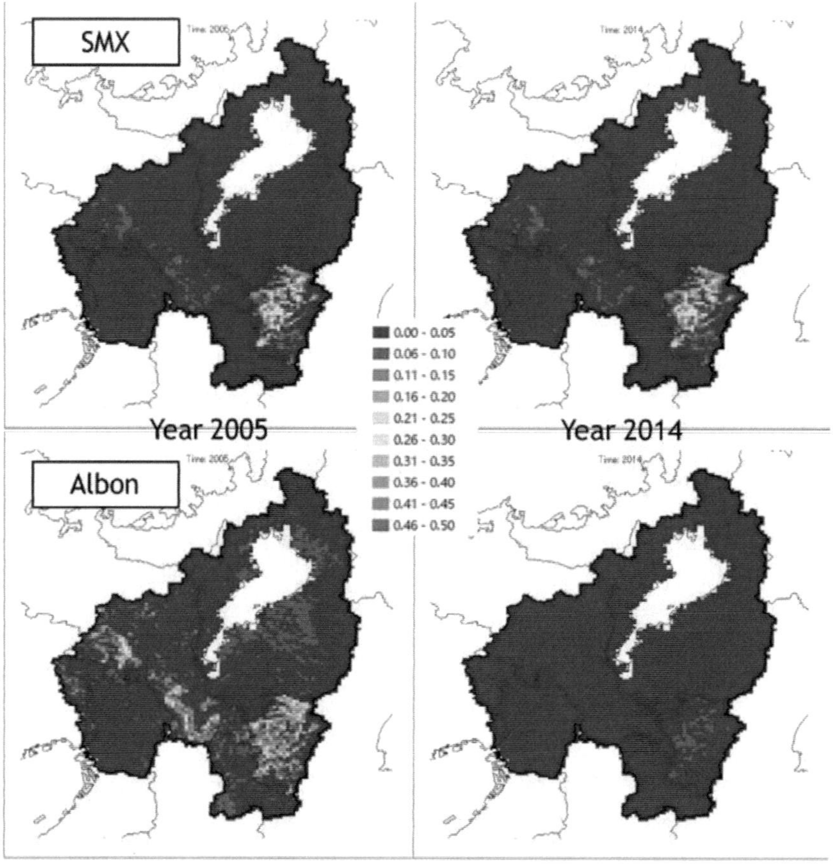

Figure 5. Distribution of estimated sulfa-drug emission in LBYRB (unit: mol)

4. Results

4.1. One-box type model

The results of the OMM were shown in figures 6, 7 and 8. The emission and environmental mole amount of SMX (figure 6) indicated that the environment residual is only about 7% of that year's emission. Most of the residual, around 81%, were dissolved in water media, about 17% was absorbed in soil and only a few was deposited in the sediment (figure 7). Meanwhile, considering the volume of the environment media, the annual average concentration of chemicals (figure 8) was around 18.5 ng/L in water, 9.1 ng/L in soil, 4.3 ng/L in sediment, which was a quite low-level pollution concentration.

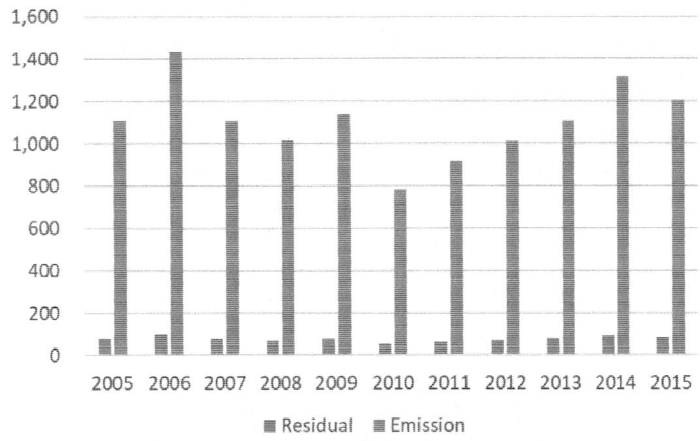

Figure 6. Amount of SMX emission and residual in environment (unit: mol)

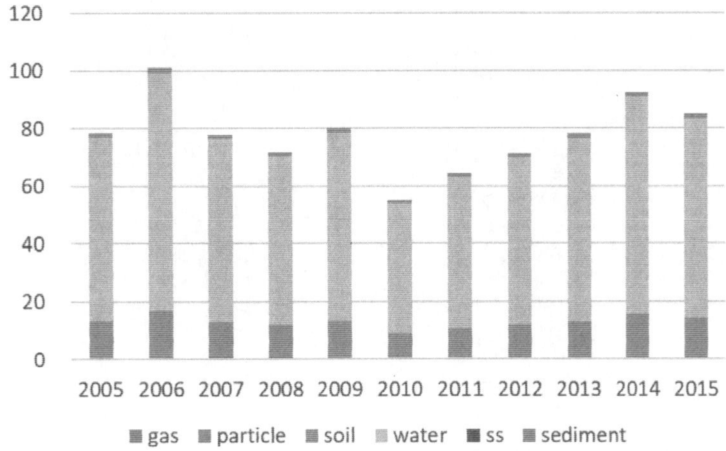

Figure 7. SMX residual in different environment media (unit: mol)

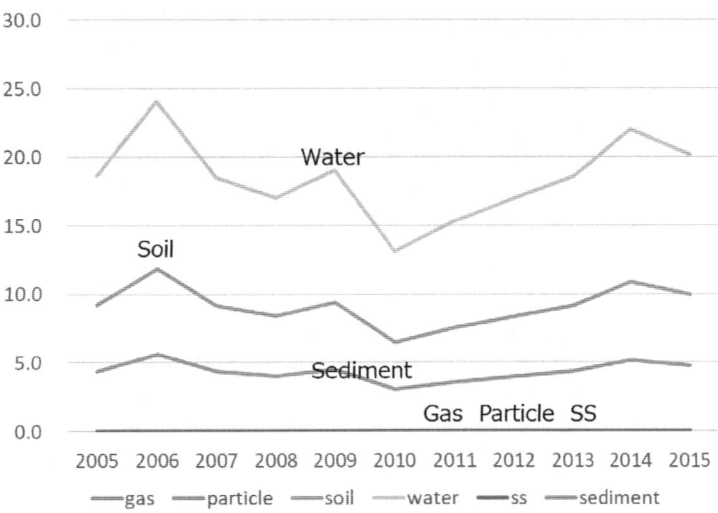

Figure 8. Concentration of SMX in different environment media (unit: ng/L)

4.2. Distributional type model

The spatial distribution of annual average concentration in soil media (figure 9) shows that the sulfa-drug was mainly concentrated in same place where the emission was concentrated as well, with the maximum SMX concentration of 600 ng/L and the maximum Albon concentration of 150 ng/L. Meanwhile, the distribution of annual average concentration in water media (figure 10) is different from that in soil. Although the emission was all assumed to be discharged into the soil, because of the fluxes between water and soil such as rainfall land loss, some chemicals move to the water media. Some mesh with large rivers had a lower level of the sulfa-drug concentration because the water flow was large, bringing a lot of clean water which dilute the chemicals. The annual concentration of Albon was decreasing while SMX was basically the same both in soil and water media that were consistent with the results of emissions estimation.

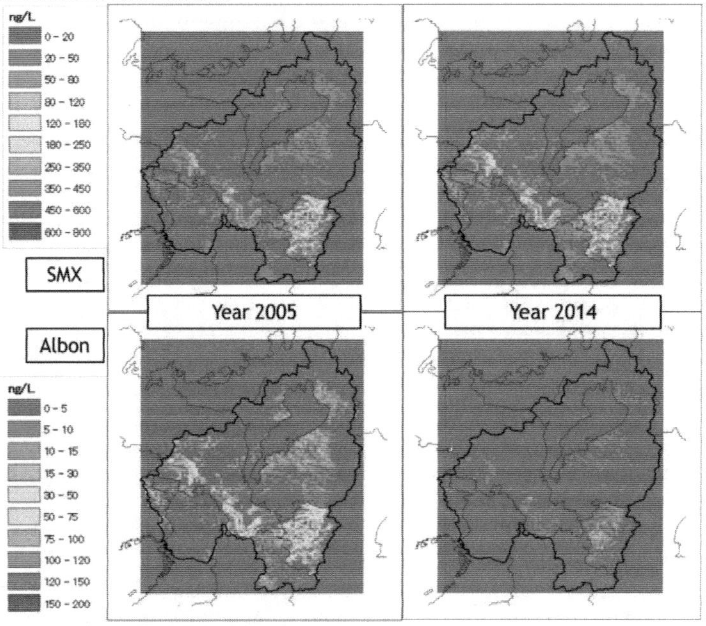

Figure 9. Distribution of sulfa-drag concentration in soil in LBYRB

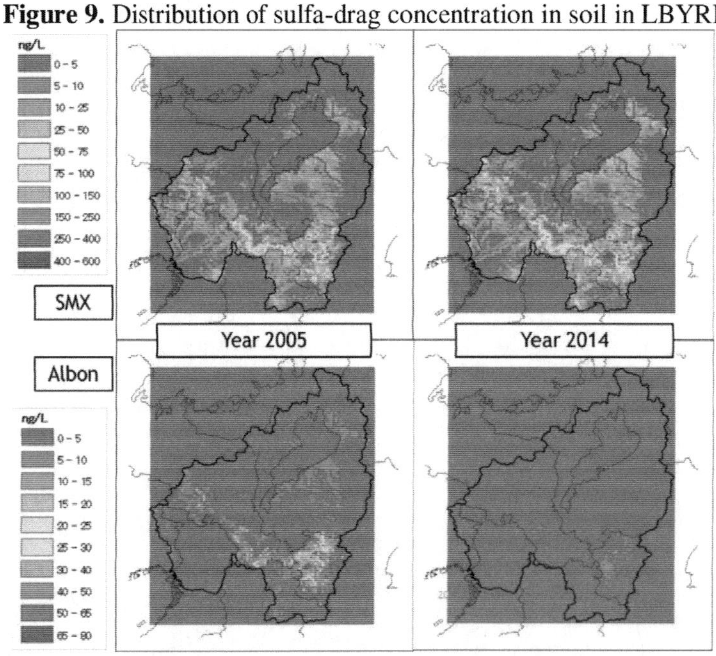

Figure 10. Distribution of sulfa-drag concentration in water in LBYRB

IOP Conf. Series: Earth and Environmental Science **257** (2019) 012003 doi:10.1088/1755-1315/257/1/012003

5. Conclusion

Two types of multimedia fate models (OMM and DMM) were successfully utilized in order to analysis the environmental behaviour of two kinds of sulfa-drugs (SMX and Albon) which are widely used in agriculture and husbandry. According to the result of the OMM simulation, around 93% of the sulfa-drugs emission in the Lake Biwa and Yodo River Basin (LBYRB) were quickly flowed out from the LBYRB to the ocean. Most of the residuals were dissolved in water, while the minority were absorbed in the soil or deposited in the sediment media. In the DMM simulation, according to the spatial distribution, most emission were occurred in the area where the farmland proportion are large. In both the simulations, the environmental concentration was correlated to the annual emission from the government statistics. Limitations of the current work that need to be solved include 1. Evaluating the model with the observation data in the area; 2. Improving the accuracy of fundamental data which were used during the simulation; 3. Reducing the error caused by assumptions and simplifications in methodology. The above limitations require further studies in order to evaluate the potential risk of long-term exposure of such a low level concentration of these sulfa-drugs and judge if that can affect human health and leads to security problem. People need to be alert to the adverse environmental impacts of animal-use drugs discharged into the environment.

References

[1] Ikumi T, Yusuke Y, Kei-ichiro K, Saori Y, Norihide N, Vimal K, Yutaka K, Kumiko K, Norihisa T and Hiroshi Y. Contribution of pharmaceuticals and personal care products (PPCPs) to whole toxicity of water samples collected in effluent-dominated urban streams, Ecotoxicology and Environmental Safety, Vol. 144, 2017, PP 338-350

[2] Qian S, Yan L, Mingyue L, Muhammad A, Min L, Hongjie W, Anyi H and Chang-Ping Y. PPCPs in Jiulong River estuary (China): Spatiotemporal distributions, fate, and their use as chemical markers of wastewater, Chemosphere, Vol. 150, 2016, PP 596-604

[3] Fatima Tamtam, Fabien Mercier, Barbara Le Bot, Joëlle Eurin, Quoc Tuc Dinh, Michel Clément and Marc Chevreuil. Occurrence and fate of antibiotics in the Seine River in various hydrological conditions, Science of The Total Environment, Vol. 393, Issue 1, 2008, PP 84-95

[4] Toshiya S. Prediction of actual presence and environmental concentration of human medicines in water environment, Ann. Rep. Tokyo Metr. Inst. Pub. Health, 63, 69-81, 2012

[5] Xiaohui Liu, Shaoyong Lu, Wei Guo, Beidou Xi and Weiliang Wang. Antibiotics in the aquatic environments: A review of lakes, China, Science of The Total Environment, Vol. 627, 2018, PP 1195-1208

[6] Ministry of Agriculture Forestry and Fisheries, Annual Report of Sales Amount and Sales Volume of Veterinary drugs, Quasi-drugs and Medical Devices, http://www.maff.go.jp/nval/iyakutou/hanbaidaka/index.html

[7] Livestock environmental measures, Ministry of Agriculture, Forestry and Fisheries, Status of occurrence and management of livestock manure.

[8] Mie Prefectural Government, Plan to promote the use of livestock excreta, Production agricultural income statistics from Ministry of Agriculture, Forestry and Fisheries, 2014.

[9] Shiga Prefectural Government, Prefectural plan to promote the use of livestock excreta, Production agricultural income statistics from Ministry of Agriculture, Forestry and Fisheries, 2016

[10] Osaka Prefectural Government, Osaka prefecture livestock waste utilization promotion plan, Production agricultural income statistics, Ministry of Agriculture, Forestry and Fisheries, 2016

[11] Ministry of Land, Infrastructure, Transport and Tourism, National Land Numerical Information, http://nlftp.mlit.go.jp/ksj-e/index.html

IOP Conf. Series: Earth and Environmental Science **257** (2019) 012004 doi:10.1088/1755-1315/257/1/012004

Numerical simulation of heavy rainfall in August 2014 over Japan under Pseudo Global Warming Conditions

Y Minamiguchi[1], H Shimadera[1], T Matsuo[1] and A Kondo[1]

1 Graduate School of Engineering, Osaka University, 2-1 Yamadaoka, Suita, Osaka 565-0871, Japan
E-mail: minamiguchi@ea.see.eng.osaka-u.ac.jp

Abstract. Climate change as a consequence of global warming may enhance severe precipitation events. This study investigated the impact of global warming on summertime heavy rainfall using the Weather Research and Forecasting model with the pseudo global warming (PGW) method. WRF simulations were conducted for August 2014 over a domain covering Japan, in which two big typhoons attacked Japan and rain fronts frequently passed over, in a baseline and two PGW (RCP4.5 and RCP8.5) conditions. The analysis showed that the future climate led to a larger amount of precipitation than the past climate during the study period in Japan. The mean increase rate of upward moisture supply from the surface of the entire modelling domain was 2.8% K^{-1} in RCP4.5 condition and 3.3% K^{-1} RCP8.5 condition. The mean increase rate of 2-m specific humidity was 6.7% K^{-1} in both RCP4.5 and RCP8.5 conditions, which is comparable to the Clausius-Clapeyron relationship. Therefore, the water supply from the lateral boundaries contributed to the increase in humidity largely and precipitation subsequently. The increases in precipitation related to typhoons were larger than rain fronts, in particular, the changes were remarkable over the paths of typhoons. These results show global warming will significantly increase the summertime heavy precipitation over Japan in the future.

1. Introduction

In recent decades, climate change is proceeding as a consequence of global warming. According to the Intergovernmental Panel on Climate Change (IPCC) Fifth Assessment Report, the climate change causes severe precipitation events with more intense and more frequent rainfall [1]. This is because the saturation pressure of water vapour increases due to the air temperature rise under the global warming, and the water vapour supplied from the sea increases due to the rise in sea surface temperature. Based on the Clausius-Clapeyron (CC) relationship, the atmosphere can hold more moisture in warmer air temperatures with a rate of about 6–7% K^{-1}, and extreme precipitation intensities can also increase by this rate because of global warming [2–5]. While daily extreme precipitation intensities typically increase with a warming atmosphere by within the rate expected from the CC relationship [6,7], changes in shorter duration precipitation extremes may well exceed the rate [6].

The Pseudo Global Warming (PGW) method, which is a dynamical downscaling method proposed by Kimura and Kitoh [8], allows the projections of regional climate change of a specific precipitation event. Several previous studies have applied the method to analyses on Baiu rain front [9] and typhoons [10–12], and reported that global warming should have a large impact on precipitation. In this study, we focused on recent heavy precipitation events observed in August 2014, in which Typhoons Nakri and Halong attacked Japan and rain fronts frequently passed over [13]. The total precipitation of the month was the largest on record for August in western Japan. During this period,

Content from this work may be used under the terms of the Creative Commons Attribution 3.0 licence. Any further distribution of this work must maintain attribution to the author(s) and the title of the work, journal citation and DOI.
Published under licence by IOP Publishing Ltd

torrential rain and storms caused serious damage in Japan such as landslides, inundations, floods, and other disasters. Typhoon Nakri occurred in the east of the Philippines on July 29, 2014, and approached the Okinawa and Amami Islands from July 31 to August 1. The typhoon caused heavy rain in Japan even after it turned into a tropical depression in the Yellow Sea on August 3. Typhoon Halong, which occurred on the east of Guam on July 29, approached the Daito Islands on August 7. It approached and passed through the Shikoku and Kinki regions in western Japan from August 9 to 10, causing the highest peak precipitation in the series of events. In the last half of the month, rain fronts stagnated near Japan and led to frequent occurrence of intense rainfall events. On August 20, landslides occurred at Hiroshima city following an intense rain of more than 100 mm h^{-1}. During these events, warm and humid air was continuously supplied from south of Japan. In order to evaluate the impact of global warming on record-high summertime precipitation over Japan, this study conducted numerical simulations of the precipitation events in August 2014 using the Weather Research and Forecasting (WRF) model [14] version 3.7 with the PGW method.

2. Methods

2.1. Model configurations

This study focused on the series of heavy rainfall events in August 2014. Figure 1 shows the WRF modelling domain on the Lambert conformal conic projection map, which has an area of 2250 km × 2250 km and covers almost all of Japan. The horizontal resolution and the number of grid cells are 5km and 450 × 450, respectively. The vertical layers consisted of 40 sigma-pressure coordinated layers from the surface to 100 hPa. Figure 1a also shows average sea surface temperature in the study period, which tended to be higher than that in August of 1981–2010 according to the Japan Meteorological Agency (JMA) [15]. As shown in Figure 1b, there are 147 meteorological observatories for validation of WRF simulation in the domain. The area of Japan was divided into eight regions consisting of Hokkaido, Tohoku, Kanto, Chubu, Kinki, Chugoku, Shikoku, and Kyushu. The numbers of observatories in these regions are 18, 17, 17, 30, 11, 14, 9, and 31, respectively.

Static geographical data, objective analysis data, and physical parameterizations used in this study were similar to those used by Shimadera et al. [16]. Topography and land use were derived from the 30-s resolution data of the United States Geological Survey and the 100-m resolution National Land Numerical Information data of the Geospatial Information Authority of Japan, respectively. Initial and boundary conditions in the atmosphere were derived from the mesoscale model grid point value (MSM GPV) data by JMA [17]. Daily sea surface temperature was derived from the high-resolution, real-time, global sea surface temperature analysis data (RTG_SST_HR) by the U.S. National Centres for Environmental Prediction (NCEP) [18]. Initial conditions of variables on the ground surface and in soil layers were derived from the final operational global analysis (FNL) data by NCEP [19]. The MSM GPV data have spatial resolutions of 0.0625° (longitude) × 0.05° (latitude) for ground level data and 0.125° (longitude) × 0.1° (latitude) for pressure level data and a temporal resolution of 3 hours. The RTG_SST_HR data have a spatial resolution of 0.0833° and a temporal resolution of 24 hours. The FNL data have a spatial resolution of 1° and a temporal resolution of 6 hours. The physical parameterizations used in this study include the Yonsei University scheme [20] for the planetary boundary layer parameterization, the WRF single-moment 6-class microphysics scheme [21], the Noah land surface model [22], the rapid radiative transfer model [23] for long wave radiation, and the Dudhia scheme [24] for the shortwave radiation.

Meteorological fields in August 2014 were produced by repeating 5-day simulations with 4-day (from 1st to 4th day) spin-up 31 times. During the spin-up period, the grid nudging for horizontal wind components was applied with a nudging coefficient of 1.0×10^{-4} s^{-1} in the entire simulation domain with the MSM GPV data.

2.2. Experimental design

In order to estimate the sensitivity of precipitation to global warming in the heavy rainfall in August 2014, WRF simulations were conducted in one baseline condition and two PGW conditions (RCP4.5 and RCP8.5). The baseline simulation should reproduce meteorological fields accurately with initial and boundary conditions prepared from MSM GPV, NCEP FNL and RTG_SST_HR datasets. The PGW conditions were prepared with the global bias-corrected climate model output data from the Community Earth System Model version 1 by the U.S. National Centre for Atmospheric Research [25]. We used datasets for a 20th century simulation (20THC) and two concomitant Representative Concentration Pathway (RCP) future scenarios (RCP45 and RCP85), which are based on the IPCC AR5.

In order to develop PGW fields, averaged climate conditions were calculated from 30-year data for two months (July and August) during 1971–2000 in 20THC for the past and during 2071–2100 in RCP4.5 and RCP8.5 for the future. Then, anomalies of global warming were calculated as the difference between the averaged future and past climate conditions, and were added to the regional climate dataset of the baseline condition. Thus, sets of the PGW conditions were constructed for various meteorological factors, for example, air temperature, humidity, sea surface temperature, sea level pressure and wind speed. Figure 2 illustrates spatial distributions of surface air temperature and sea surface temperature differences in the future conditions (RCP45 and RCP85) from the past condition (20THC). The mean increase amounts over the entire domain of surface air temperature and sea surface temperature from 20THC were 1.4 K and 1.3 K in RCP45 condition, and 3.2 K and 2.7 K in RCP85 condition, respectively. The increase in air temperature was larger than in sea surface temperature, and in regard to spatial distributions, the change in high latitude was larger than in low latitude.

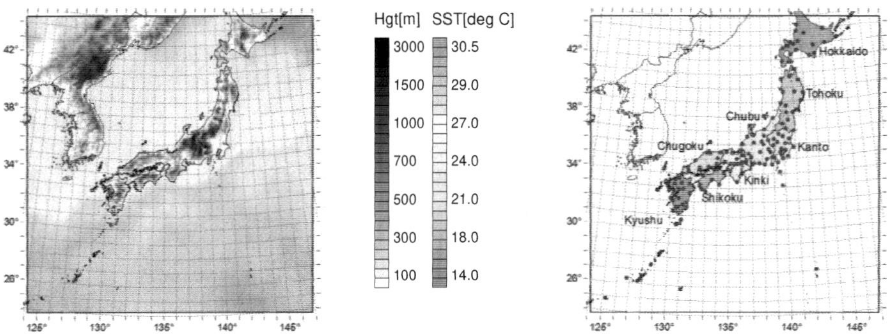

(a) (b)

Figure 1. Modelling domains showing (a) topography of land areas and average sea surface temperature of August 2014; (b) locations of meteorological observatories and division of regions for analyses of precipitation.

3. Results and discussion

3.1. Model Performance

The result of the WRF simulation in the baseline condition was compared with observation data. The index of agreement for all the meteorological observatories was 0.96 for air temperature, 0.95 for specific humidity, 0.85 for wind speed and 0.74 for precipitation. The values indicate high correlations and small biases between the observed and simulated meteorological variables. Figure 3 shows comparisons between observed and simulated monthly precipitations at all the observatories in August 2014. Although the model tended to underestimate at the observatories where the monthly values over approximately 500 mm were observed, most of the simulated values (85%) were within a factor 2 of the observed values. Figure 4 shows spatial distributions of the observed and simulated monthly precipitations in August 2014. The observed precipitation was obtained from the Radar/Raingauge-Analyzed Precipitation data by JMA [26]. The Radar/Raingauge-Analyzed Precipitation data have a spatial resolution of 1 km and a temporal resolution of 30 minutes, and cover areas within a few hundred kilometers from the coastal lines of Japan. Although the simulation slightly underestimated the largest amount of monthly precipitation in the Shikoku region caused by Typhoons Nakri and Halong, it approximately reproduced the spatial variation pattern over the land area in Japan.

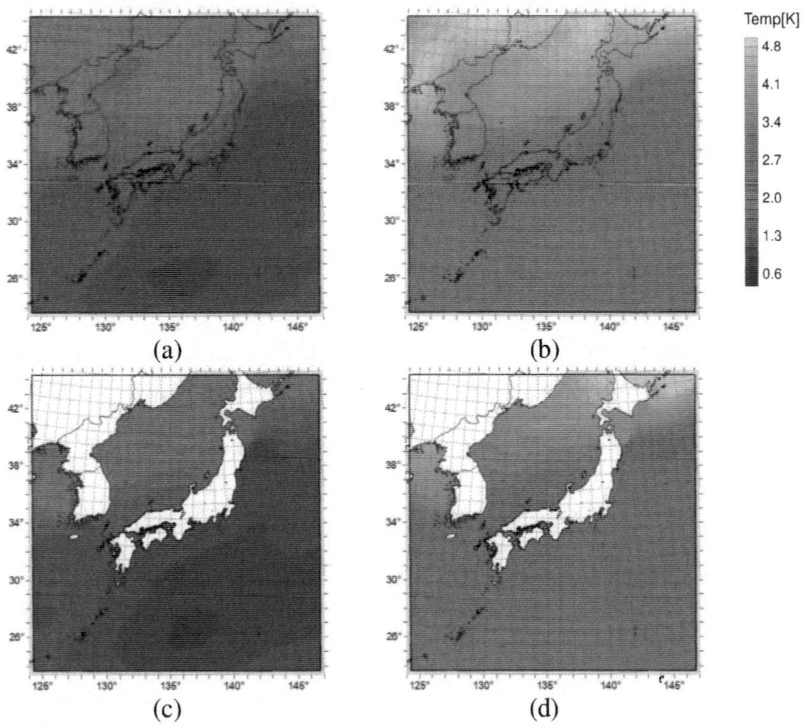

Figure 2. Spatial distributions of surface air temperature differences of the 30-year means for July and August from the past period, 1971-2000, in (a) RCP4.5 and (b) RCP8.5 (2071-2100), and of sea surface temperature differences of the 30-year means for July and August fro m the past period in (c) RCP4.5 and (d) RCP8.5.

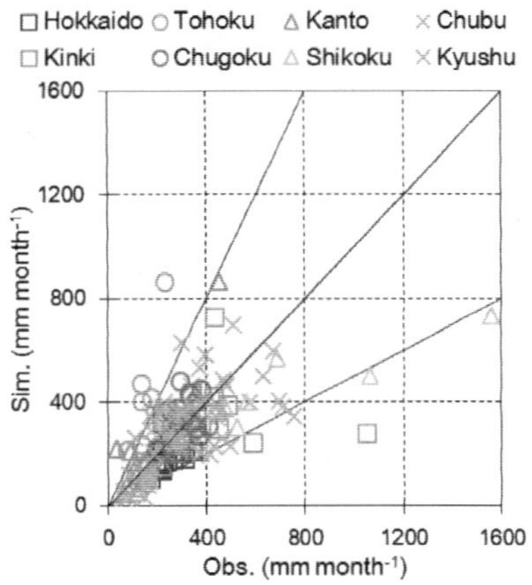

Figure 3. Scatter plots of monthly precipitation at the meteorological observatories in August 2014 for the observation and the simulation results in the baseline condition. The plots are classified as each region.

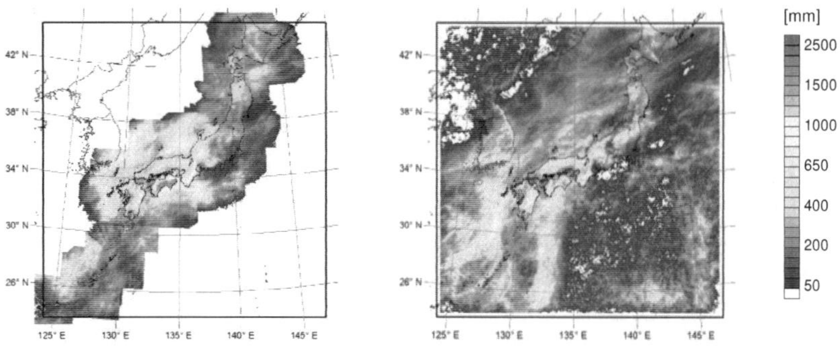

(a) (b)

Figure 4. Spatial distributions of monthly precipitation in August 2014 obtained from (a) Radar/Raingauge-Analyzed Precipitation data and (b) baseline condition. The Radar/Raingauge-Analyzed Precipitation data covers areas within a few hundred kilometers from the coastal lines of Japan.

3.2. Sensitivity under PGW conditions

The sensitivity of precipitation to global warming was analysed in terms of temporal and spatial variations. Figure 5 shows a time series comparison of daily precipitations averaged for all the meteorological observatories in the baseline and PGW conditions. Figure 6 shows monthly precipitation averaged for the meteorological observatories over the land of Japan and each of the eight regions (Figure 1b) in August 2014. Figure 7 illustrates spatial distributions of daily precipitation difference from the baseline condition in the PGW conditions for the first and last halves of the simulation period. Table 1 shows 2-m air temperature, 2-m specific humidity, monthly precipitation and upward water vapour flux averaged over the land area of Japan, the sea area, and the entire area in the modelling domain. The mean increase amounts of precipitation were 67 ± 34 and 95 ± 38 mm $month^{-1}$ under RCP4.5 and RCP8.5 conditions, respectively, with being significantly larger than the bias between observed and simulated monthly precipitation in the baseline condition ($P < 0.001$). The increase in precipitation under the PGW conditions, especially under RCP8.5 condition, were remarkable over southern to western Japan including the Kinki and Shikoku regions where there was a large amount of precipitation associated with Typhoons Nakri and Halong during the first half of August 2014. By contrast, there were smaller changes in the amount of precipitation related to rain fronts during the last half of the month. As a result, the mean increase amounts of precipitation under RCP8.5 condition were 4.0 ± 1.7 and 2.2 ± 1.4 mm d^{-1} K^{-1} over the land area of Japan in the first and last halves, respectively. The large difference between the two periods was probably due to the difference in the total precipitation (larger in the first half and smaller in the last), because the difference in the corresponding increase rates (8.6 and 7.5% K^{-1} for the first and last halves, respectively) was less clear.

The mean increase rates of upward moisture supply from the surface of the entire area of the modelling domain to 2-m air temperature increase were 2.8% K^{-1} and 3.3% K^{-1} in RCP4.5 and RCP8.5 conditions, respectively, and the rates of 2-m specific humidity change were larger with the values being 6.7% K^{-1} in both PGW conditions. Based on the CC relationship, as mentioned above, the atmosphere can hold more moisture in warmer air temperatures with a rate of about 6–7% K^{-1}, namely the 2-m humidity increase rates were comparable to the rate of the CC relationship. These results suggested that the most of the increased amount of water were transported from the lateral boundaries, particularly the south one with warm and humid air inflows. The rates of the precipitation change were still larger than the change of the water flux and of the humidity, furthermore localized or in short durations, the rates were much larger than the CC relationship. Additionally, the increasing rate of precipitation to 2-m air temperature in RCP4.5 was larger than in RCP8.5. This is because the saturation pressure of water vapour increases exponentially with air temperature, which resulted in a larger amount of precipitable water under higher temperature condition. It suppressed the increase in precipitation in this research, and it should also enhance the potential for extreme precipitation.

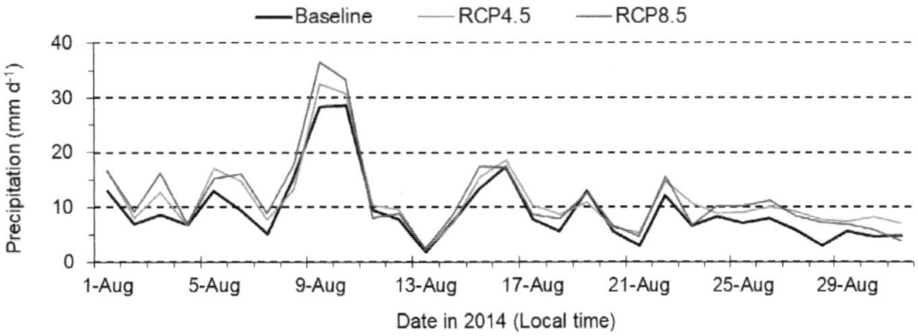

Figure 5. Temporal variations of daily precipitation averaged for all the meteorological observatories in August 2014 for the comparison of the baseline, RCP4.5, and RCP8.5 conditions.

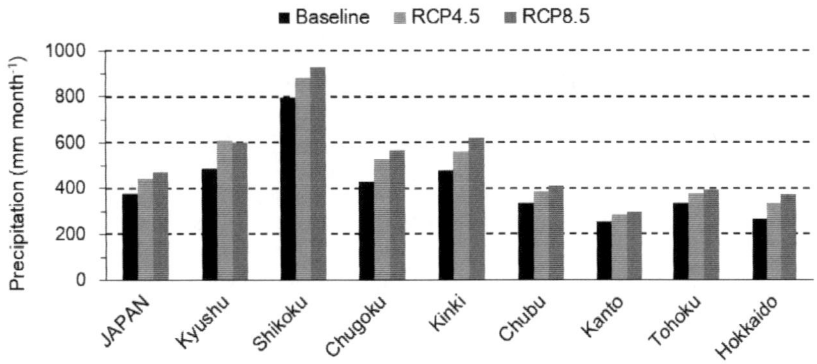

Figure 6. Monthly precipitation averaged over the land area of Japan and each region in August 2014 for the comparison of the baseline, RCP4.5, and RCP8.5 conditions.

4. Conclusions

This study was conducted to investigate the impact of global warming on simulated precipitation fields in the period. Numerical experiments under pseudo global warming conditions were conducted for examining the sensitivity of simulated heavy precipitation to global warming.

Precipitation generally increased under pseudo global warming conditions. The analysis showed that the future climate led to a larger amount of precipitation than the past climate during the study period in Japan. The mean increase rate of upward moisture supply from the surface of the entire area of the modelling domain was 2.8% K^{-1} in RCP4.5 condition and 3.3% K^{-1} in RCP8.5 condition. The mean increase rate of 2-m specific humidity was 6.7% K^{-1} over the entire domain in both RCP4.5 and RCP8.5 conditions, which is comparable to the Clausius-Clapeyron relationship. Therefore, the water supply from the boundary contributed largely to the increase in humidity, and subsequent precipitation. The increases in precipitation related to typhoons were larger than rain fronts, in particular, the changes were remarkable over the paths of typhoons. These results show global warming will significantly increase the summer-time heavy precipitation over Japan in the future.

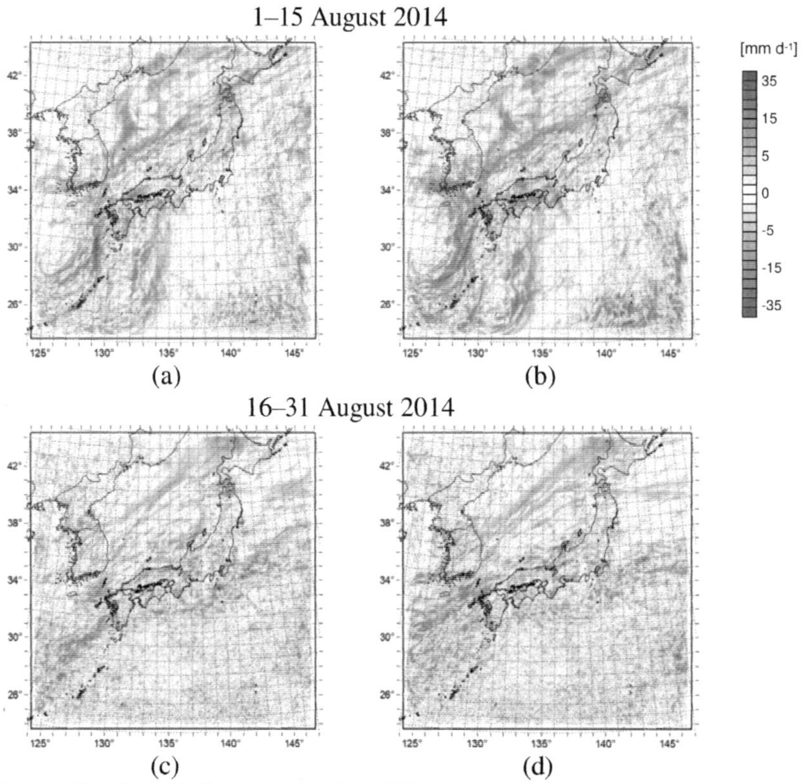

Figure 7. Mean daily precipitation difference from the baseline condition in (a) RCP4.5 and (b) RCP8.5 conditions during the first half of August 2014, and (c) RCP4.5 and (d) RCP8.5 conditions during the last half of August 2014.

Table 1. 2-m air temperature, 2-m specific humidity, monthly precipitation and upward surface water vapour flux averaged over land area of Japan, sea area, and entire area in the modelling domain in the baseline, RCP4.5, and RCP8.5 conditions.

Area	Baseline	RCP4.5	RCP8.5
2-m air temperature (°C)			
Japan	22.2	23.6	25.3
Sea	25.7	27.1	28.6
Domain	24.9	26.3	27.9
2-m specific humidity (g kg^{-1})			
Japan	15.1	16.6	18.4
Sea	18.1	19.6	21.6
Domain	17.2	18.7	20.6
Precipitation (mm month^{-1})			
Japan	376	444	471
Sea	214	251	277
Domain	217	255	279
Upward surface water vapour flux (mm month^{-1})			
Japan	90	91	96
Sea	109	113	120
Domain	107	111	118

5. References

[1] Intergovernmental Panel on Climate Change 2014 *CLIMATE CHANGE 2013: The Physical Science Basis* (Geneva: Intergovernmental Panel on Climate Change) p 7
[2] Boer G J 1993 *Clim. Dyn.* **8** 225–239
[3] Allen M R and Ingram W J 2002 *Nature* **419** 224–232
[4] Trenberth K E, Dai A, Rasmussen R M and Parsons D B 2003 *Bull. Am. Meteorol. Soc.* **84** 1205–17
[5] Pall P, Allen M R and Stone D A 2007 *Clim. Dyn.* **28** 351–63
[6] Lenderink G and van Meijgaard E 2008 *Geosci.* **1** 511–4
[7] Nayak S and Dairaku K 2016 *Hydrol. Res. Lett.* **10** 139–44
[8] Kimura F and Kitoh A 2007 *Final Rep.* (Kyoto: Research Institute for Humanity and Nature) pp 43−6
[9] Kawase H, Yoshikane T, Hara M, Kimura F, Yasunari T, Ailikun B, Ueda H and Inoue T 2009 *J. Geo. Res.* **114** D24110
[10] Ito R, Takemi T and Arakawa O 2016 *SOLA* **12** 100−5
[11] Takemi T, Ito R and Arakawa O 2016 *Hydrol. Res. Lett.* **10** 88−94
[12] Kanada S, Takemi T, Kato M, Yamasaki S, Fudeyasu H, Tsuboki K, Arakawa O and Takayabu I 2017 *J. Clim.* **30** 6017−36
[13] Japan Meteorological Agency 2014 *Prompt report on heavy precipitation in August 2014* (Tokyo: Japan Meteorological Agency) p 1 (in Japanese)
[14] Skamarock W C, Klemp J B, Dudhia J, Gill D O, Baker D M, Duda M G, Huang X –Y, Wang W and Powers J G 2008 *A description of the advanced research WRF version 3; Technical Note TN-475+STR* (Boulder: National Center for Atmospheric Research)
[15] Japan Meteorological Agency http://www.data.jma.go.jp/kaiyou/shindan/index.html (accessed on 8 August 2018)
[16] Shimadera H, Kondo A, Shrestha K L, Kitaoka K and Inoue Y 2015 *Adv. Meteorol.* **2015** 379361
[17] Japan Meteorological Business Support Center http://www.jmbsc.or.jp/jp/online/file/f-online10200.html (accessed on 8 August 2018)
[18] Gemmill W, Katz B and Li X 2007 Daily real-time, global sea surface temperature high-resolution analysis: RTG_SST_HR, NOAA/NWS/NCEP/MMAB Office Note http://polar.ncep.noaa.gov/sst/ (accessed on 8 August 2018)
[19] National Centers for Environmental Prediction/National Weather Service/NOAA/U.S. Department of Commerce 2000 *NCEP FNL Operational Model Global Tropospheric Analyses, continuing from July 1999* (Boulder: National Center for Atmospheric Research) https://doi.org/10.5065/D6M043C6 (accessed on 8 August 2018)
[20] Hong S -Y, Noh Y and Dudhia J 2006 *Mon. Weather Rev.* **134** 2318–41
[21] Hong S -Y and Lim J -O J 2006 *J. Korean Meteor. Soc.* **42** 129–51
[22] Chen F and Dudhia J 2001 *Mon. Weather Rev.* **129** 569–85
[23] Mlawer E J, Taubman S J, Brown P D and Iacono M J 1997 *J. Geophys. Res.* **102** 16663–82
[24] Dudhia J 1989 *J. Atmos. Sci.* **46** 3077–107
[25] Monaghan A J, Steinhoff D F, Bruyere C L and Yates D 2014 *NCAR CESM Global Bias-Corrected CMIP5 Output to Support WRF/MPAS Research* https://doi.org/10.5065/D6DJ5CN4 (accessed on 8 August 2018)
[26] Japan Meteorological Business Support Center http://www.jmbsc.or.jp/jp/offline/cd0100.html (accessed on 8 August 2018)

IOP Conf. Series: Earth and Environmental Science **257** (2019) 012005 doi:10.1088/1755-1315/257/1/012005

Understanding and modeling climate impacts on ecosystem dynamics with FLUXNET data and artificial intelligence

NY Zhu[1], XL Yu[2], SR Zhang[3], ZS Liu[4], YW Tong[5]

1 Graduate study of art and science, Columbia University, New York City, USA
2 School of Science, Beijing University of Posts and Telecommunications, Beijing, China
3 School of Biological Science and Medical Engineering, Southeast University, Nanjing, Jiangsu, China
4 College of Agriculture, Shandong Agricultural University, Taian, Shandong, China
5 School of economics and management, Nanchang University, Nanchang, China
E-mail: nanyan.zhu@columbia.edu

Abstract. Since preindustrial era, the radiative energy balance of the earth system has been largely perturbed by anthropogenic activities such as CO_2 emissions from fossil fuel burning. As a net effect, global temperature increasingly warms up and will further increase in the future if CO_2 concentration in the atmosphere keeps going up. Plants sequestrate a large amount of atmospheric CO_2 via photosynthesis, thus greatly mediate the global warming. In this study, we aim to model the temporal dynamics of photosynthesis for various different vegetation types and further understand controlling factors of photosynthesis machinery. Our results showed that the photosynthesis and its interactions with climate drivers, such as temperature, precipitatin, radiation, and vapor pressure deficit, has an internal system memory about 14 days. Thus, the predictive model could be best trained with historical data of the past two weeks and could best predict future temporal evolution of photosynthesis in the following two weeks. Our leave-one-out experiment also showed that temperature and solar radiation dramatically control grassland and forest photosynthesis activity.

1. Introduction

Climate change and global warming have been threatening the development and stability of both natural ecosystems and human society [1]. The dynamics of green house gases, such as CO_2, play pivotal roles in controlling the radiative forcing and the energy balance of the whole earth system [2]. Therefore, understanding the full cycle (sources and sinks) of atmospheric CO_2 is critically important in order to better control the growth of CO_2 concentrations in the future. Nowadays, photosynthesis pathway is the largest land surface CO_2 sink that sequestrate atmospheric CO_2 into vegetation biomass and store them as living biomass. The magnitude of photosynthetic C sequestration pathway is around 120-130 Pg C per year (1Pg = 10^{15} g), globally [3]. Ecosystem respiration, including autotrophic and heterotrophic, consumes most of the photosynthetic carbon sink, cycles it back into atmosphere, and leading to a much smaller net carbon sink compared with gross carbon input from photosynthesis. The balance between photosynthesis carbon input and ecosystem reparation carbon output determines the fate of atmospheric CO2 concentration and mitigate the anthropogenic CO_2 emissions, such as from fossil fuel and biomass burning [4].

Given the significant role of photosynthesis carbon uptake in determining the global carbon cycle, atmospheric CO_2 concentrations, radiative balance of earth system, and global warming, it is critically

Content from this work may be used under the terms of the Creative Commons Attribution 3.0 licence. Any further distribution of this work must maintain attribution to the author(s) and the title of the work, journal citation and DOI.
Published under licence by IOP Publishing Ltd

important to gain mechanistic understand of photosynthesis machinery and its relationship to climate factors such as temperature, precipitation, and radiation. Observational network has been established globally to measure and understand various different components of land surface carbon cycle including photosynthesis [5]. At leaf scale, photosynthesis reaction is carried out by RUBISCO enzyme, which combines CO_2 and water molecules to generate carbohydrate products. It occurs at two stages: 1) first light-dependent reactions to capture the energy of light store in ATP and NADPH; 2) second,, light-independent reactions capture and reduce carbon dioxide. This biological reaction relies on substrate concentration (CO_2 and water), activity of temperature-sensitive RUBISCO enzyme, and driven by solar energy. Thus, theoretically, one is able to build an effective photosynthesis model that takes all those important factors into account and predict the magnitude of land surface photosynthesis given relevant climate drivers.
Historically, mechanistic models have been used to study the dynamics of land surface photosynthesis and its relevance to the fate of global warming. These models are based on either Monteith law of light use efficiency [6] or Farquhar photosynthesis modeling framework [7]. Alternatively, photosynthesis rate could also be modeled with data-driven machine learning models that are trained by a large amount of observed photosynthesis data. The latter approach is gaining popularity because the growth of observational network of photosynthesis such as FLUXNET and the advancing of effective machine learning techniques. In this study, the objective is to model the temporal dynamics of plant photosynthesis with two different artificial neural network architectures. With appropriate design model experiments, we also aim to explore and understand the controlling of different climate drivers on photosynthesis.

2. Methodology

2.1. FLUXNET Data

We use Eddy Covariance data collected from four different sites from FLUXNET observational network [5]. It measures plant Gross Primary Productivity (GPP, hereafter we refer photosynthesis to be GPP) and micrometeorological measurements of temperature, precipitation, radidation, vapor pressure deficit. At present, hundreds of sites are operating on a long-term and continuous basis. In this study, we include both boreal and temperate forested sites and grass sites (Table 1). Figure 1 shows the probability density distribution of GPP and climate drivers at the four sites (Red: DE-Tha, Green: NL-Loo, Blue: AT-Neu, Orange: US-Var).

Table 1. Site information of four FLUXNET towers

Name	Latitude (N)	Longitude (E)	Elevation (m)	Land cover type	MAP (mm^{-1} yr^{-1})	MAT (C)	GPP (gC day^{-1})
DE-Tha	50.96	13.56	380	Boreal Evergreen Needleleaf tree	334±814	6.8±8.2	6±6
NL-Loo	52.16	5.74	25	Temperate Evergreen Needleleaf tree	420±988	8.8±7.9	5±4
AT-Neu	47.11	11.31	970	Grasslands	419±827	10.1±6.4	4±3
US-Var	38.41	-120.95	129	Grasslands	282±980	15.8±6.8	2±3

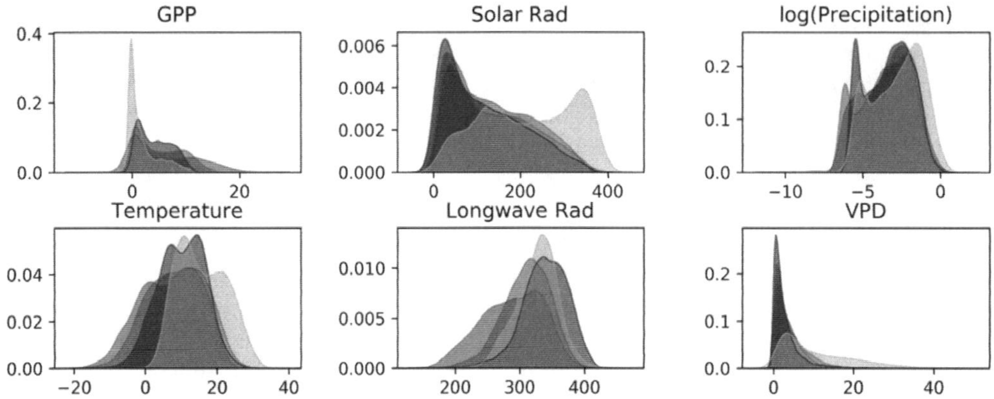

Figure 1. Probability density of observed variables used in this study at four FLUXNET towers.

2.2. Artificial Intelligence modeling frameworks

We first build an fully connected artificial neural network that has three hidden layers and ten nodes each layer to model the temporal dynamics of GPP using temperature, precipitation, radiation, and vapor pressure deficit. Eqn. 1 denotes the general mathematical design of each hidden layer neuron.

$$y = f(\sum_{i=1}^{n} x_i \cdot w_i) \qquad (1)$$

where f is rectifier activation function. W_i are neural network parameters for each input features. The input features (x_i) in represent those climate drivers; and the target output (y_i) is GPP.

Secondly, we establish a memory based Recurrent Neural Network (RNN) (Figure 2) to model GPP based on not only climate drivers but also the historical memory of the GPP and climate drivers.

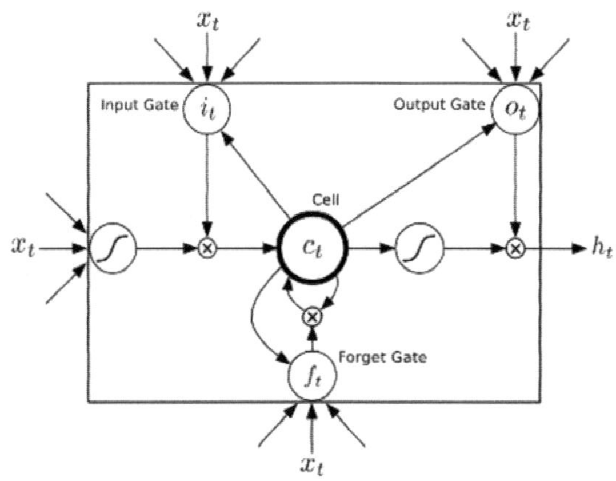

Figure 2. Modeling structure of Recurrent Neural Network

A implementation structure of RNN is denoted by Eqn. 2 - 7 (Figure 2). Given an input sequence x = (x_1,...,x_T) a standard recurrent neural network computes the hidden vector sequence h = (h_1,..., h_T) and output vector sequence y = (y_1,..., y_T) by iterating the following equations from t = 1 to T:

$$h_t = H(W_{ih}x_t + W_{hh}h_{t-1} + b_h) \qquad (2)$$

$$y_t = W_{ho}h_t + b_o \qquad (3)$$

$$i_t = \sigma(W_{xi}x_t + W_{hi}h_t - 1 + W_{ci}c_{t-1} + b_i) \qquad (4)$$

$$c_t = f_t c_{t-1} + i_t \tanh(W_{xc} x_t + W_{hc} h_{t-1} + b_c) \tag{5}$$

$$o_t = \sigma(W_{xo} x_t + W_{ho} h_t - 1 + W_{co} c_t + b_o) \tag{6}$$

$$h_t = o_t \tanh(c_t) \tag{7}$$

where σ is the logistic sigmoid function, and i, f, o, and c are respectively the input gate, forget gate, output gate, and cell activation vectors. W and x are model parameters and input at each time step.

3. Results and discussion

We first used fully connected Artificial Neural Network (ANN) to model the dynamics of GPP across the fours different sites (trained independently). We found that the model performance at a forest site was better than at a grassland site in general (Figure 3). Furthermore, our leave-one-out experiment showed that removing temperature from input features dramatically degraded model performance the Flux AT-Neu grass site, which suggested a significant temperature control at this site; while removing shortwave radiation from input features at another forest site (FLUX DE-Tha) led to a much worse model performance suggested a significant energy control. ANN model was generally successful at three sites but failed at FLUX US-Var grassland site (Figure 3 right panel, red bars mean negative correlation).

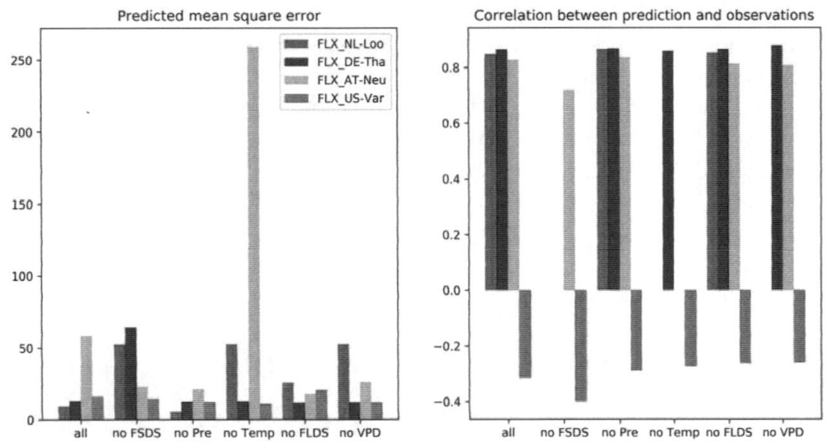

Figure 3. ANN model performance of predicted men square error (left) and prediction-observation correlation (right) at four FLUXNET towers with different combination of climate drivers.

Secondly, we employed Recurrent Neural Network (RNN) to model the dynamics of GPP and conducted a leave-one-out experiment for climate drivers (similar to ANN modeling). Compared with ANN, our RNN model predicted a consistently low mean square error and high model-data correlation (Figure 3 and 4). At forested sites (green and blue bars), temperature, vapor pressure deficit and shortwave radiation played a more important role in controlling GPP; while at grassland sites (orange and rad bars) temperature significantly affect the performance of RNN .

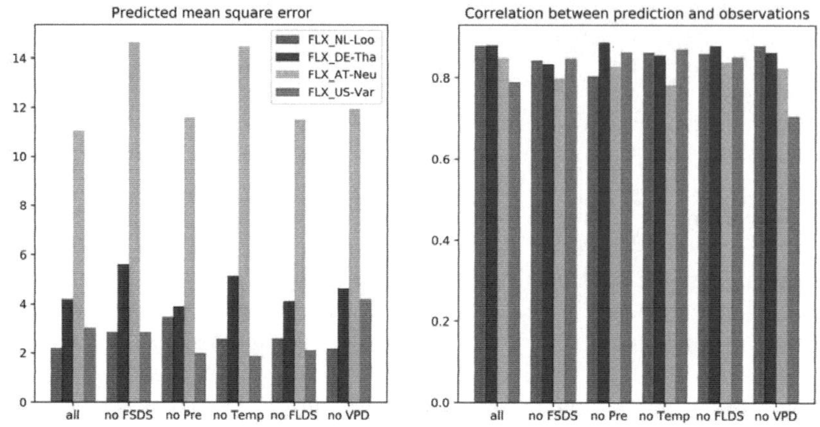

Figure 4. RNN model performance of predicted men square error (left) and prediction-observation correlation (right) at four FLUXNET towers with different combination of climate drivers.

Given that RNN model performed much better than ANN model, we then conducted experiment with RNN model to explore the possible internal memory of the photosynthesis process itself. Our hypothesis is that time series is often self-explained to some extent, which corresponding to the dependency of future state on historical memory of the system. An optimal length of memory is helpful to understand the internal linkage of photosynthesis process across time. Our experiment implied that photosynthesis has a internal memory length of two weeks (Figure 5). Feeding historical memory longer than two weeks into RNN resulted a worse performance compared with 14 days memory length.

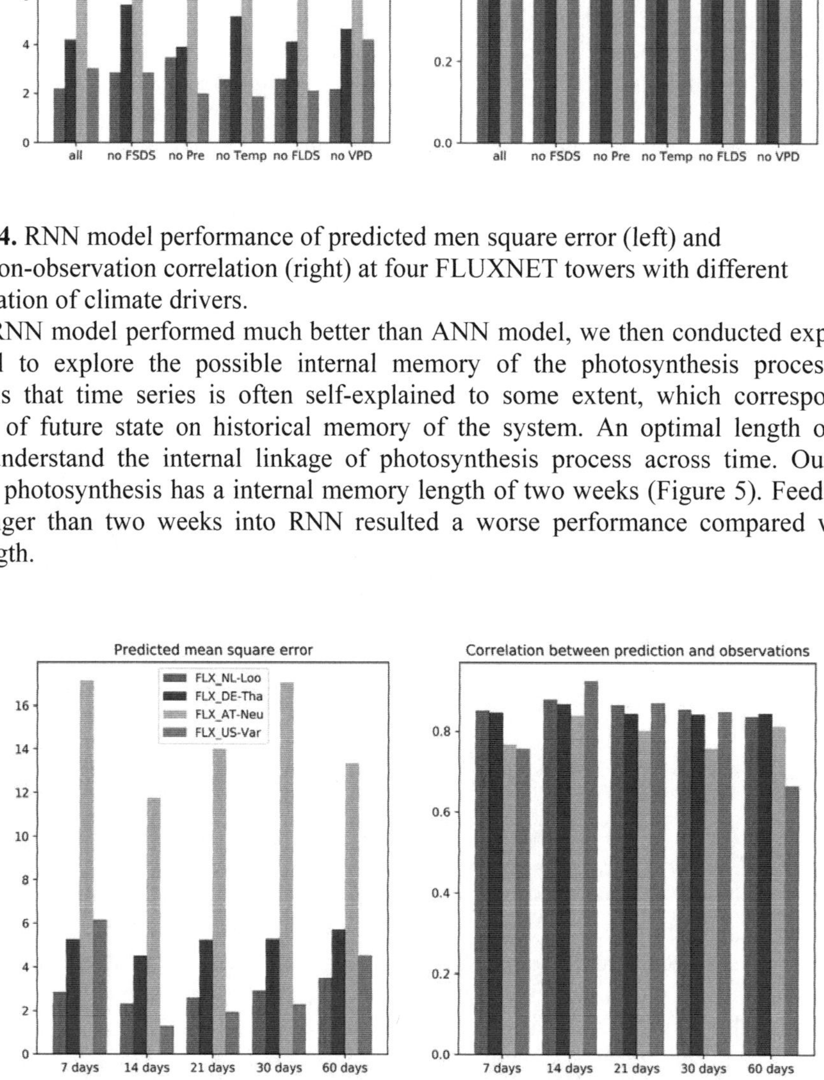

Figure 5. RNN model performance of predicted men square error (left) and prediction-observation correlation (right) at four FLUXNET towers with different lengths of historical memories (from 7 days to 60 days) that RNN considers.

Finally, predictive length of photosynthesis process is critically important to understand the reliability of RNN model in the future time step. We further conducted a modeling experiment that predicted different lengths of future time step from 1 day to up to 30 days. Figure 6 revealed that our RNN modeling framework was able to predict future GPP in two weeks at most. Predictions that were longer than two weeks had large bias and small model-date correlation. Our results on the optimal historical system memory was quite consistent with the model longest temporal predictability, which were both about two weeks.

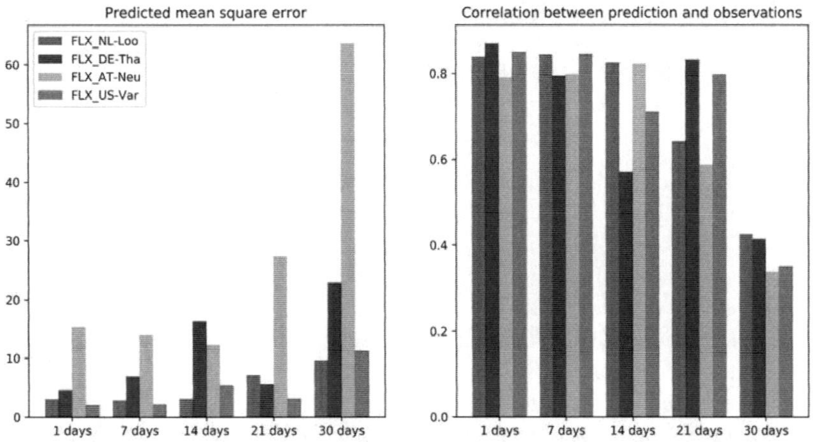

Figure 6. RNN model performance of predicted men square error (left) and prediction-observation correlation (right) at four FLUXNET towers with different lengths predictive lengths (from 1 days to 30 days).

4. Conclusion

Green house gas emissions such CO_2 could dramatically warm up climate system via positive radiative forcing effect. Fortunately, terrestrial ecosystems are able to mitigate the anthropogenic CO2 emissions via photosynthesis. In this study, we aim to model the dynamics of plant photosynthesis activity with advanced machine learning frameworks. Our results showed that system memory is critically important for improving model predictability especially at grassland site. Our memory-based neural network model was able to successfully capture the temporal dynamics of plant photosynthesis at all four sites of interest. Our modeling experiment also demonstrated a clear internal system memory of the photosynthesis machinery.

5. Reference:

[1] Stips, A., et al. (2016). On the causal structure between CO 2 and global temperature. Scientific reports, 6, 21691.

[2] Falkowski, P. et al. (2000). The global carbon cycle: a test of our knowledge of earth as a system. science, 290(5490), 291-296.

[3] Beer, C et al., (2010). Terrestrial gross carbon dioxide uptake: global distribution and covariation with climate. Science, 1184984.

[4] Friedlingstein, P et al., (2010). Update on CO2 emissions, Nat. Geosci., 3, 811–812

[5] Baldocchi, D et al., (2001). FLUXNET: A new tool to study the temporal and spatial variability of ecosystem-scale carbon dioxide, water vapor, and energy flux densities. Bulletin of the American Meteorological Society, 82(11), 2415-2434.

[6] Medlyn, B. E. (1998). Physiological basis of the light use efficiency model. Tree physiology, 18(3), 167-176.

[7] Farquhar, G. V. et al., (1980). A biochemical model of photosynthetic CO2 assimilation in leaves of C3 species. Planta, 149(1), 78-90.

IOP Conf. Series: Earth and Environmental Science **257** (2019) 012006 doi:10.1088/1755-1315/257/1/012006

Prototype of horizontal bio-trickling air scrubber (HBAS) for ammonia removal in swine farm

Jirarat Pinthong[1], Vorapot Kanokkantapong[1,2*], Teerawit Poopa[3] and Suphanut Jiebna[4]

1 Department of Environmental Science, Faculty of Science, Chulalongkorn University, Bangkok, Thailand
2 Center of Excellence on Hazardous Substance Management (HSM), Chulalongkorn University, Bangkok, Thailand
3 Department of Environmental Health, Faculty of Public and Environmental Health, Huachiew Chalermprakiet University, Samutprakarn, Thailand
4 Amcovet Co. Ltd., Bangkok, Thailand
* E-mail: vorapot.ka@chula.ac.th

Abstract. Modelling of Bio-Trickling Air Scrubber (HBAS) system is one of the methods to treat biological odor and gas emission in air treatment of manufacturing facility management. A cheaper way of efficient gas elimination is using sprinkling water in the direction of air flow in the media microbial adhesion. Samples of ammonia gas was continuously collected throughout swine production for 20 weeks. The experimental systems consisted of six different conditions by varying the amount of scrubbed water, the space between water spray and media, and the amount of media materials. Microorganism species in media, which affected to ammonia removal, were also classified. The results indicated that the highest efficiency of ammonia removal was 51.9% at the experimental condition as follows: 1) water spray rate of turn on: turn off - 60 : 30 seconds, 2) distance between water spray and media of 1 meter, and 3) a half of media microbial adhesion area. The treated ammonia concentration was lower than the standard value of the chemical in working atmosphere as determined by the American Conference of Governmental Industrial Hygienist (ACGIH TLV), giving ammonia not more than 25 ppm. Additionally, media bio-scrubber could classify five bacteria species (*Brachybacterium* sp., *Micromonospora* sp., *Methylobacterium* sp., *Rhodococcus rubber* and *Cellulosimicrobium* sp.), which ammonia was used as a natural resource. The main mechanism in gas removal was proposed by the liquid adsorption, whereas the microorganism adhering to the media did not affect in statistical signification. Improvement of removal efficiency could be achieved by an increase of gas retention time in the system and the space between water spray and media.

1. Introduction

Over the past several years, the demand of pork has been increasing significantly, which has led to the higher need for pig production. At present, the large scale of pig farming has a density of 400-1,000 pigs/house and provides an impact on environmental pollution, especially air pollution due to high odors and gases. There are many sources of pollution such as feed, manure and wastewater. Directly exhausted from the house to the atmosphere without treatment results in odor's high concentration of gases, which can have a negative effect on both health and people's quality of life in the community near the farm. Chemical composition of pig house gas is a group of volatile organic compounds (VOCs) such as ketone, carboxylic acid, alcohol, aromatic hydrocarbon, hydrogen sulfide, methane

Content from this work may be used under the terms of the Creative Commons Attribution 3.0 licence. Any further distribution of this work must maintain attribution to the author(s) and the title of the work, journal citation and DOI.
Published under licence by IOP Publishing Ltd

and ammonia etc. [1]. Ammonia is the important air pollutant because it is a strong gas causing respiratory irritation, coughing, searing, and lethargy [2]. Global release of ammonia was found that 39% of them came from agricultural sources while 15% belong to pig production [3]. In addition, more than 50% of gas production from pig farm is ammonia [4]. The study also found that the amount of ammonia released from swine production was 79.2 g per day and concentration ranged from 649 to 3,751 mg per hour in 500 kg of live pigs [5]. The concentration measured from the house is from 5 to 18 ppm [6] and the smell of ammonia is a nuisance to the farmer and the surrounding community.

At present, the improved farm management is a closed system including the house's ventilation system that can reduce the odor problem caused by ammonia; however, but it does not solve the problem at all. The technology to tackle the odor and gases concerns from the pig house is to treat the gases released from the house. One of the most widely used technologies is the air scrubber method, which the liquid is sprayed through the air in order to drain and absorb the gas. The system of bio-trickling air scrubber is the new application of microorganisms concept that functions as gas treatment at low concentration using media to adsorb and spray water to create moisture in the media. Principles of bio-trickling system for the treatment of gases from swine house composes of ventilation system, liquid distributor, washing water trap and packing media etc. Ogawa et al. (2011) studied the microbiological-based air cleaning system in a two-step process: 1) the process of splashing water to filter the ammonia and 2) nitrification process which changes the ammonia to nitrile by microorganism [7]. The result of ammonia removal from pig houses was 55-95%. Melse et al. (2012) studied the efficiency of ammonia removal using packed-bed bio-trickling filter from pig house [8]. The results showed that the ammonia removal efficiency was 85% with an average rate 24 g NH_3 m^{-3}/h. Furthermore, mass balance between ammonia concentration in the air vented with the ammonia dissolved in water showed that nitrogen was eliminated in ammonium and nitrate form for 86% and released in nitrous oxide form for 5%. Kafle et al. (2015) studied the use of wood bark as an intermediary of biofilters to remove the ammonia and hydrogen sulfide from piglet nursery [9]. Based on the depth of media and moisture content of biofilters, the amount of water required was between 3.8-556 L/m^3/d, depending on the age of pigs. This study was able to reduce the odor from ammonia and hydrogen sulfide by 73.5-76.9% and 95.2-97.9%, respectively. The optimization was found that the media should have more than 254 mm depth and 35-50% of moisture content. Heyden et al. (2015) studied the design of equipment used for air scrubbers and biofilters in animal houses, including air flow, liquid rate and intermediate layers [10]. The study revealed that installation of air scrubbers is preferable to the back of ventilation and the direction of gas flow through the middle layer with co-current and counter-current. The flow of cross-current is shorter contact time, which does not affect the efficiency of the treatment. It can be seen that the bio-trickling system may be applied as a multi-stage system to optimize odor and gas treatment [11-15].

This work is intended to develop an ammonia removal model from a pig farm with Horizontal Bio-trickling Air Scrubber (HBAS) by integrating physical treatment with air scrubber and bio-trickling, which is a new way to handle odor from pig farming in Thailand. The purposes are to decrease the smell of pig farms, grow plants with high nitrogen water, minimize the odor problems in the community and promote the environmental image of the entrepreneurs.

2. Materials and methods

2.1. Research location

Pig farm of American Market Co., Ltd. located at amphoe Phrayaeng, Chachoengsao province, Thailand was selected in this study. There are 35,000 fattening pigs with a total of 6 experimental conditions from 20 farms.

2.2. Experimental method

A study on the efficiency of ammonia treatment from pig farm using the Horizontal Bio-Tricking Air Scrubber (HBAS) was designed in a pig farm with 900-1,000 pigs per house. Daily ammonia concentration data was collected continuously throughout the pig age from entering house to capturing pigs (20 weeks of age). Water absorption coefficients for ammonia absorption and area of cross section of plastic media were studied. pH of ammonia (electrometric method), BOD (dillution water method) and COD (titration method) in ponds from the scrubber system were carried out. Wastewater from the pig manure was analyzed by colorimetric analysis and the colonies of isolated ammonia-decomposing bacteria from dust in the house since the first week, week 10 and week 20. The work was divided into 6 experimental groups to measure the concentration of ammonia at various points (before entering the house, out of the fan and out of the house after media) in the day. Optimum condition of scrubbing for ammonia removal was conducted by adjusting the open and closing time of water pump and the distance between nozzle and media. Type and total of deodorized microorganisms on the medium were also determined.

2.3. Designing prototype system of Horizontal Bio-trickling Air Scrubber (HBAS)
In the pig house, temperature was controlled by exchanging heat with water using cooling pad. The air was transferred from the front of the house to rear by ventilation fans with ventilation rate relating to temperature and humidity control in the house depending on the age of pig. The house was approximately 10 meters wide and 100 meters long. Ammonia gas was vented through the HBAS system at the back of the house with 8-meter length. Plastic materials used was honeycomb pads with 0.4 meters thick and 3.6 square meters of installed space as shown in Figures 1 and 2.

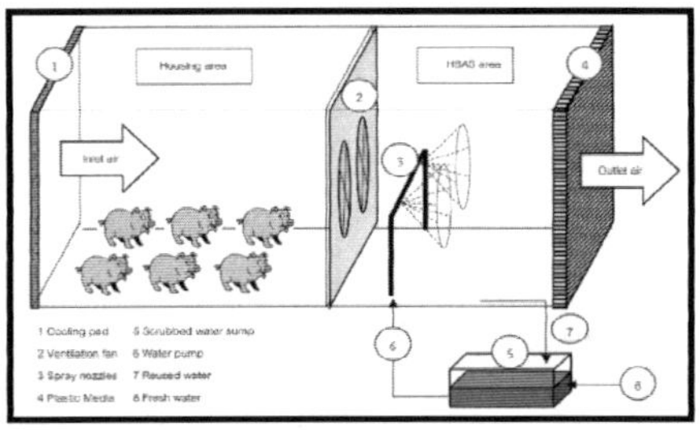

Figure 1. Design of HBAS system with 1 meter distance spray after fan.

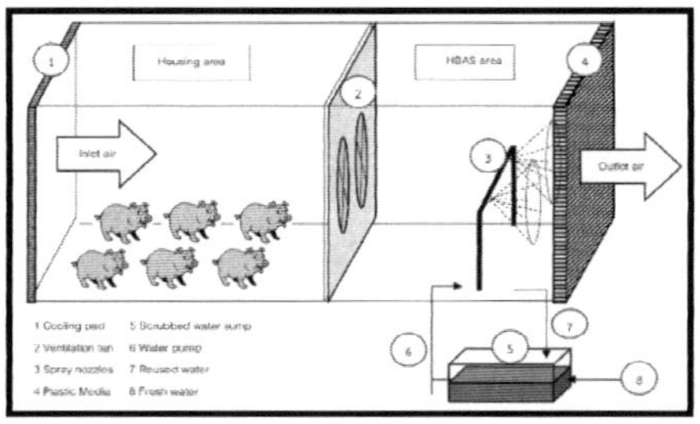

Figure 2. Design of HBAS system with 1 meter distance spray before media.

2.4. Sampling

2.4.1 Air sampling. Air samples of ammonia gas were collected from inlet air at the end of the house, before the ventilator, and air outlet from HBAS system for twice a day (9.00 am and 4.00 pm).

2.4.2 Media sampling. Dust samples were collected for analysis in three areas: left, middle, and right of the media. The samples were collected at three locations which each point depends on the size of sample areas as shown in Figures 3 and 4. Swab test was used as a technique to detect bacteria on surface. Phosphate buffer pH 7.4 was applied to sterile laboratory swab. Dust mark on the media was determined at various points throughout the area and then returned to the phosphate buffer pH 7.4.

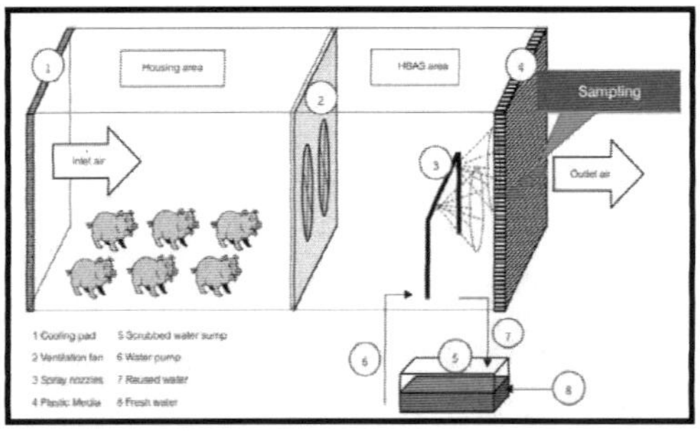

Figure 3. Point of dust collector on the media.

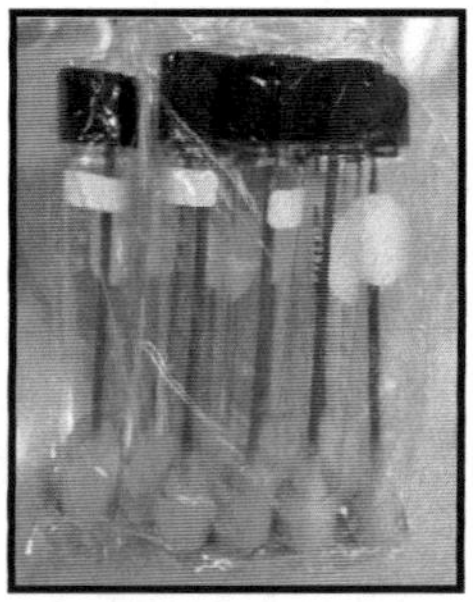

Figure 4. Experimental kit for swab sampling.

2.5. Microbiological study

2.5.1. Bacterial separation from dust samples After collecting, separation of bacteria from swab kit by shaking with vortex mixer provided a suspension solution to count the bacteria that used ammonia as a resource.

2.5.2. Bacteria counting Classifications of taxonomic classes of selected bacteria were studied for morphology such as colour and colonies of bacteria. Moreover, study of species levels and selected strains were analyzed for the identification of a variety of bacterial species by sequencing of 16s-rDNA.

2.6. Tools

Aeroqual portable gas collection model ver. 200 was used to collect samples of ammonia. Analysis of water parameters from scrubber and pig manure were carried out using standard method of water and wastewater analysis.

2.7. Statistic study and translation

Loading rate and removal efficiency (%) of ammonia were calculated using statistic of Least Significant Difference (LCD) at the confidence level of 90%.

3. Results and discussion

The efficiency of ammonia treatment was carried out from pig farms with HBAS system at various control conditions. The study was conducted throughout the life cycle of pigs from 5 to 20 weeks with 112 days samplings. The sampling was divided into 2 areas i.e., the side of swine latrines with wastewater from house washing and workers walkaway. Two samples were collected at 9 am and 4 pm. The details of 6 experimental samples were shown in Table 1.

Table 1. A series of experiments for this research.

Experiments	**Control conditions of the experiment**		
	Water spray range	**Area installation of media filter**	**Water supply rate**
1	1 meter after ventilation fans	full area	Open 60 seconds Close 30 seconds
2	1 meter after ventilation fans	full area	Open 30 seconds Close 30 seconds
3	1 meter after ventilation fans	half area	Open 60 seconds Close 30 seconds
4	1 meter after ventilation fans	half area	Open 30 seconds Close 30 seconds
5	1 meter before media filter	full area	Open 60 seconds Close 30 seconds
6	1 meter before media filter	full area	Open 30 seconds Close 30 seconds

3.1. Efficiency of ammonia treatment

From collecting the data, it was found that the measuring point played a significant role in ammonia concentration. The samples in swine latrines side generated higher ammonia concentrations than the walkways in all day. As a result, the concentration of excreta from pigs, especially in the back of the house, was higher than the front of the house. It has been observed that many traces of several samples from walkway cannot be identified because the concentration was lower than detection limit. At the front of HBAS area, inlet air from walkway was mixed with the air from the latrines, therefore: the outlet air concentrations were higher than those found in walkway inside the house. Table 2 shows the efficiency of gaseous removal and the treated concentration of ammonia. It was found that the maximum ammonia treatment efficiency at sampling point of latrines was found in experimental set 3 (51.9% in the afternoon). The treated ammonia gas in all experiments showed the lower concentration than 2 ppm, which was not harmful to their community according to ACGIH TLV standard.

Table 2. The efficiency of ammonia treatment and the treated concentration

Set	Sampling points	Efficiency (%)	Treated concentration (ppm)
1	Latrines-morning	48.1±24.6	1.95±1.59
	Latrines-afternoon	45.3±24.8	2.02±1.48
	Corridors-morning	34.9±33.5	1.26±1.08
	Corridors-afternoon	20.6±28.6	1.43±1.30
2	Latrines-morning	48.4±25.9	0.83±0.57
	Latrines-afternoon	49.7±23.8	0.96±0.59
	Corridors-morning	30.5±37.1	0.58±0.43
	Corridors-afternoon	14.8±29.1	0.70±0.41
3	Latrines-morning	46.9±23.4	0.85±0.54
	Latrines-afternoon	51.9±21.3	1.39±1.01
	Corridors-morning	26.8±32.3	0.62±0.40
	Corridors-afternoon	23.4±29.8	1.00±0.77
4	Latrines-morning	44.0±29.4	0.67±0.69
	Latrines-afternoon	42.5±28.9	0.80±0.72
	Corridors-morning	18.7±28.7	0.63±0.85
	Corridors-afternoon	16.1±28.6	0.67±0.57
5	Latrines-morning	33.0±24.6	0.82±0.76
	Latrines-afternoon	33.6±26.6	1.01±0.81
	Corridors-morning	8.4±16.7	0.67±0.70
	Corridors-afternoon	15.3±27.0	0.66±0.56
6	Latrines-morning	38.2±29.8	0.67±0.55
	Latrines-afternoon	37.6±32.3	0.61±0.33
	Corridors-morning	13.5±21.4	0.68±0.59
	Corridors-afternoon	20.1±26.3	0.50±0.26

The results also revealed that workers' activities played a vital role in ammonia concentration. The ammonia tended to accumulate from the afternoon to evening due to pig house cleaning every morning resulting in washed away of pig manure. The overall concentration of gas was likely to increase with the age of pigs. However, the highest concentration of treated ammonia with HBAS was 6.60 ppm, which was still lower than the ACGIH TLV (2011) standard (not exceeding 25 ppm).

3.2. Water analysis from treatment system

The analysis of water from the treatment system was also considered. Two water samples were collected from scrubber system and pig manure once a week during whole 20 weeks. The samples were analyzed for pH, conductivity, COD, BOD, and ammonia. Unfortunately, there was pandemic situation in swine farm in week 12 to 15 of the experiment. Hence, it was not able to collect the sampling water for analysis every week.

3.2.1. Water samples from scrubber system. The surprising results showed that pH values were not changed during the experiment with ranged from 6 to 7 although the theory predicates that dissolved ammonia resulted in increasing pH of water. Due to the design of experiment, fresh water was constantly replenished to the system; therefore, pH of water from scrubber system did not increase. Ogawa et. al., (2011) revealed that controlling pH less than 7.5 resulted in the efficiency of ammonia removal of 55-95%. There was no correlation between the age of the pigs and the concentration of ammonia in the scrubbed water as shown in Figure 5 because of replenished water. However, BOD values were very low in experimental sets 1 to 6 which ranging from 1.00-17.0, 1.00-28.0, 1.00-17.0, 2.00-45, 1.00-27.0 and 1.00-30.0 mg/L, respectively. According to the data, it can be estimated that water from the scrubber system was less polluting and can be reused in the new system.

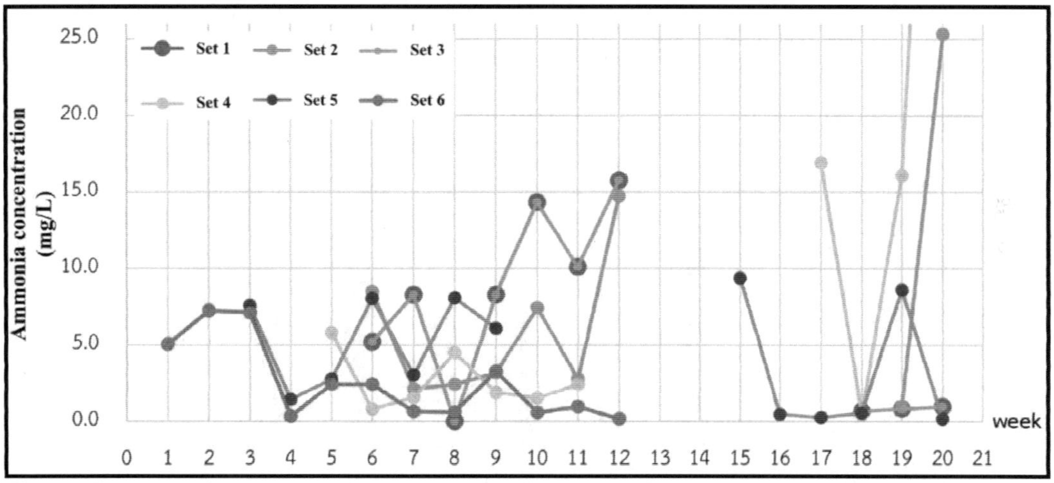

Figure 5. Ammonia concentration in the sampling water from scrubber system of experimental sets 1 to 6

3.2.2. Water samples from pig manure. pH values of pig manure were 6-8. The results showed that ammonia, COD, BOD and conductivity in the sampling water from pig manure were higher than scrubber system. The reason behind this is the older the pigs became, the more excreta they had. The ammonia concentrations of wastewater from pig manure in experimental sets 1 to 6 were 321-3,471, 588-1,862, 103-2,232, 387-3,361, 117-1,678 and 51.8-1,993 mg/L, respectively. This ammonia concentration had enough carbon source to use as a biogas system of the farm.

3.3. Microorganism in the medium

3.3.1. Total bacteria. The total bacteria in all samples by spread plate method on nutrient agar were determined at 3 points. It was found that the total number of bacteria in the house was equal to $2.10x10^9±2.03x10^9$, $3.80x10^9±3.50x10^9$ and $9.22x10^8±9.81x10^8$ CFU/mL, respectively, as shown in Figure 6. These numbers indicated the possible trend of bacteria accordance with Zhao et. al., (2011) [16] and Seedorf et. al., (1998) [17], which reported that general number of bacteria counted in swine house was between $1.00x10^{11}$ and $2.00x10^{16}$ CFU/mL.

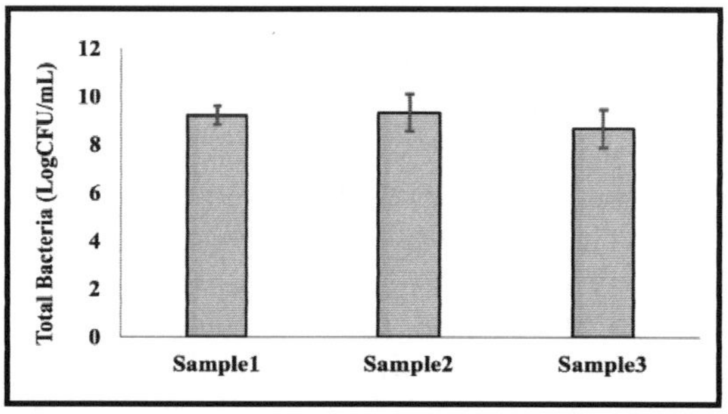

Figure 6. Total isolated bacteria from the dust in pig houses.

3.3.2. Classification of taxonomic species of bacteria using ammonia as a source The number of bacteria using ammonia as a source of food in the sample by spread plate on Burk's N-free medium at week-0 was equal to $1.06x10^5±6.61x10^4$ CFU/mL. After 10 weeks, samples were collected from each house to count the number of bacteria in experimental sets 1 to 6, which was equal to $3.20x10^7±3.16x10^7$, $3.51x10^4±1.57x10^4$, $5.22x10^4±3.56x10^4$, $3.41x10^5±5.30x10^5$, $1.56x10^6±9.10x10^5$, and $1.38x10^6±2.34x10^6$ CFU/mL, respectively. After 20 weeks, the number of bacteria was $1.55x10^4±1.08x10^4$, $4.33x10^4±5.79x10^4$, $2.58x10^5±$ $2.76x10^5$, $3.23x10^5±5.29x10^5$, $5.90x10^6±9.61x10^5$, and $5.19x10^5$ $±6.19x10^5$ CFU/mL, respectively, as shown in Figure 7.

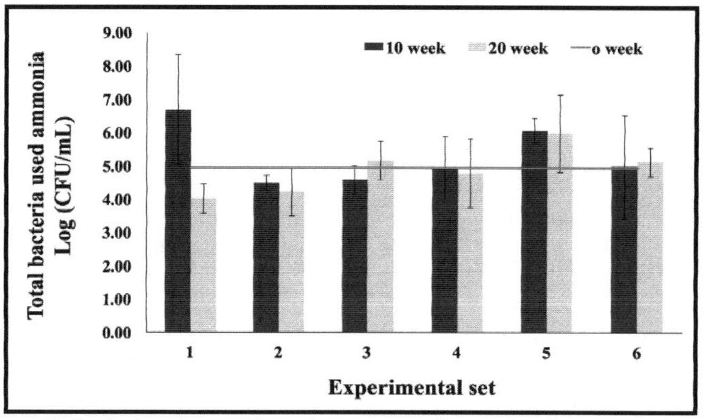

Figure 7. Isolation bacteria using ammonia as food sources in each experimental set over a period of 10 and 20 weeks.

For separation of bacteria using ammonia as a source by spread plate method on nutrient agar, the results showed that bacteria over 10 and 20 weeks presented different colour and bacterial colonies. At the morphological study, 10 isolates of bacteria were selected for screening to analyze the 16s rDNA nucleotide sequences in the level of species. The results were similar to the bacterial nucleotides in the database that revealed 5 specified species consisted of *Brachybacterium* sp., *Micromonospora* sp., *Methylobacterium* sp., *Rhodococcus rubber*, and *Cellulosimicrobium* sp. It was also discovered that 5 species in swine house, especially of *Brachybacterium* sp., which was investigated in poultry manure, pig manure and organic fertilizer similar to the research of Tian et. al., (2016) [18].

IOP Conf. Series: Earth and Environmental Science **257** (2019) 012006 doi:10.1088/1755-1315/257/1/012006

4. Conclusion

The study of odor and gas removal from swine houses with Horizontal Bio-trickling Air Scrubber (HBAS) system presented the average efficiency of ammonia treatment for all of experiments at 43.9%. The highest ammonia treatment efficiency was 51.9% with the following factors: opening and closing rate of water spray (60 seconds and 30 seconds, respectively), distance of head nozzle 1 meter behind the fan and a half media filter area. From the study, the main mechanism affecting the efficiency of gas removal was the absorption mechanism of gas into liquid. In addition, factors influencing the efficiency of the treatment were water spray rate and distance of the water dispenser to the media filter. Increasing water spray duration or rate was likely to result in higher efficiency. The distance of spray nozzle and media filter affected the growth and adhesion of microorganism. Although microbial degradation treatment was not the main mechanism, it also affected the removal efficiency of ammonia. The HBAS system features to promote speedy efficient on atmospheric odor treatment, and then favors to develop for animal industrial facility management in the future.

5. References

[1] Blazy V, Guardia A, Benoist JC, Daumoin M, Guiziou F, Lemasle M, Wolbert D and Barrington S 2015 *Chemical Engineering Journal* **276** 398

[2] Erisman JW and Schaap M 2004 *Environmental Pollution* **129** 159

[3] Galloway JN, Dentener FJ, Capone DG, Boyer EW, Howarth RW, Seitzinger Sp, Asner GP, Cleveland CC, Green PA, Holland EA, Karl DM, Michaels AF, Porter JH, Townsend AR and Vorosmarty CJ 2004 *Biogeochemistry* **70** 153

[4] Olivier JGJ, Bouwman AF, Van der Hoek KW and Berdowski JJM 1990 *Environmental Pollution* **102** 135

[5] Misselbrook TH, Van der Weerden TJ, Pain BF, Jarvis SC, Chambers BJ, Smith KA, Philips VR and Demmers TGM 2000 *Atmospheric Environment* **34** 871

[6] Groot Koerkamo PWG, Merz JHM, Uenk GH, Phillips VR, Holden MR, Sneath RW, Short JL, White RP, Hartung J, Seedorf J, Schroder M, Linkert KH, Pedersen S, Takai H, Johnsen JO and Wathes CM 1998 *journal of Agricultural Engineering Research* **70** 79

[7] Ogawa H, Dahl PJ, Suzuki T and Takai H 2011 *Biosystems Engineering* **109** 108

[8] Melse RW, Estelles F, Calvet S and Ogink NWM 2012 *Environmental Engineering Science* 2012 **29(6)** 1

[9] Kafle GK, Chen L and Neibling H 2015 *Journal of Environmental Management* **147** 164

[10] Heyden MLM, Oehmichen J, Nichting S and Volberda HW 2015 *Global Strategy Journal* **5** 154

[11] Melse RW and Mol G 2004 *Water Science and Technology* **50(4)** 275

[12] Rashmi MB, Kavyashree BA, Ramya N, Sanjay UAM, Chandan KB and Shilpa BS 2015 *International Advanced Research Journal in Science, Engineering and Technology* **2(7)** 110

[13] Cloirec PL, Malhautier L, Gracian C, Roux JC and Fanlo JL 2003 *Chemosphere* **50(1)** 145

[14] Xie Q, Kim H, Kim YJ and Chung JS 2002 *Journal of the Air and Waste Management Association* **52(12)** 1389

[15] Pan JR, Chung YC, Huang C and Tseng CP 2000 *Chemosphere* **41(3)** 329

[16] Zhao Y, Aarnink AJA, De Jong MCM, Going NWM and Groot Koerkamp PWG 2011 *Transactions of the ASABE* **54(1)** 285

[17] Seedorf J, Hartung J, Schro M, Linkert KH, Phillips VR and Holden MR 1998 *Journal of Agricultural Engineering Research* **70** 97

[18] Tien T, Yang Q, Zhangm H and Gao Y 2016 *Journal of Environmental Research and Public Health* **13** 662

Acknowledgements

This works was funded by Research and Researchers for Industries, The Thailand Research Fund (RDG59I0002).

IOP Conf. Series: Earth and Environmental Science **257** (2019) 012007 doi:10.1088/1755-1315/257/1/012007

Research progress in association between indoor air environment and elderly cardiovascular diseases

Yang Lv[1], Zhimeng Wang[1], and Bin Chen[1]

1 School of civil engineering, Dalian University of Technology, Dalian, China
E-mail: lvyang@dlut.edu.cn

Abstract. The relationship between indoor air environment and elderly with cardiovascular diseases (CVD) was reviewed through literature research in this paper. In heat and humidity environment, studies have shown that changes in heat environment such as high temperature, cold and step-changes were highly associated with cardiovascular diseases in the elderly. The influence of air humidity on cardiovascular diseases was uncertain. In indoor air quality, PM10, PM2.5, CO, SO2, NO2 and ozone had correlation with cardiovascular diseases in the elderly, with the higher incidence and mortality resulting from higher concentrations of pollutants. It is significant to establish the relationship between indoor air environment and elderly cardiovascular diseases in different climatic zones. The multi-parameter joint action of indoor air environment and comprehensive characterization and evaluation method will be the future research orientation. In addition, the combined field of medicine and engineering may provide more research ideas for environmental and health issues in the future.

1. Introduction

In China, the policy "Healthy China 2030" points out that health is the necessary requirement for promoting the overall development of people and the basic conditions for economic and social development [1]. As an important index component of the "Healthy China", the building bears the corresponding responsibility. Nowadays, the proportion of the elderly over 65 years old in China accounts for more than 10% [2] of the total population.

As the problem of population aging in China is becoming more and more serious [3]–[5], China has become one of the fastest aging countries in the world. Modern people spend more than 80% time indoors [6], especially longer for elderly people. It is of great significance to study the correlation between the health problems of the elderly and the indoor environment.

According to survey of human diseases via the World Health Organization (WHO) (Figure 1), cardiovascular diseases (CVD) had become the main cause of death in elderly people. Studies showed that nearly half of the elderly die from [7] cardiovascular diseases. In China, the deaths caused by cardiovascular diseases was about 2.6 million each year and the mortality rate of cardiovascular diseases ranked first. The mortality rate of cardiovascular diseases in rural and urban areas both had reached over 40% in China [8]. It was common to cardiovascular diseases in the elderly, which was a group of diseases of the circulatory system including heart, blood vessel and so on. Cardiovascular diseases generally resulted from hyperlipidemia, atherosclerosis, hypertension, blood viscosity, and other reasons. The main symptoms of cardiovascular diseases were hypertension, coronary heart disease, myocardial infarction, heart failure, stroke and so on [9]–[11].

Content from this work may be used under the terms of the Creative Commons Attribution 3.0 licence. Any further distribution of this work must maintain attribution to the author(s) and the title of the work, journal citation and DOI.
Published under licence by IOP Publishing Ltd

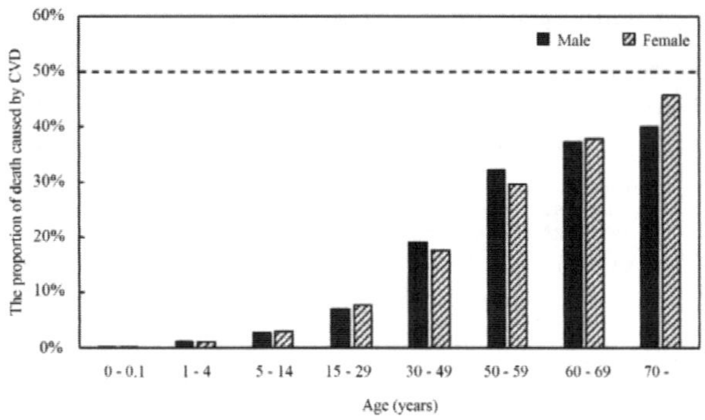

Figure 1. The proportion of deaths caused by cardiovascular diseases of all ages in 2015 in the world.

In the world, due to the different speed of economic development among countries, and the difference of living standard, lifestyle, eating habits and cultural background, the incidence and mortality of cardiovascular diseases were also different. In the 1960s, JW Brant [12] started to study the relationship between air pollution and cardiovascular diseases. Recently, more and more scholars paid their attention on the association between the atmospheric environment and the incidence or mortality of cardiovascular diseases. The quantified indexes of cardiovascular diseases often used incidence and mortality and the quantified indexes of indoor air environment often used temperature and humidity, air pollutant concentration and other environmental parameters. The researches of these two kinds of parameters can establish the association between the environment and cardiovascular diseases.

In the 1980s, Yoshino Hiroshi [13] began to study the impact of the indoor environment on the elderly stroke, and the research on the relationship between the indoor air environment and the health of the elderly has also drawn the attention of other scholars. Studies had shown that the indoor air environmental factors affecting the elderly cardiovascular diseases mainly include indoor thermal and humid environment [14]–[16] and air quality [17]–[20]. The thermal and humid environment included temperature, humidity, and air quality mainly included PM_{10}, $PM_{2.5}$ and CO, SO_2, NO_2, ozone and other pollutants, we will be based on the above factors to analyze the elderly cardiovascular diseases.

2. Effect of thermal and humid environment on CVD

Research on literature, temperature and relative humidity were the important factors affecting the incidence of cardiovascular diseases. Under the environment of high temperature, low temperature or temperature step-changes, cardiovascular diseases of elderly people were in a state of vulnerability. Relative humidity on the impact of cardiovascular diseases showed uncertain results.

2.1. Effect of temperature on CVD

The research on the association between high temperature and cardiovascular diseases was extensive, but mainly concentrated in the field of epidemiology. The research methods mainly included time series method, case-crossover study and meta-analysis. In 2005, Patz J.A. et al. [21] analyzed data from around the world and concluded that after the temperature was higher than the limit, the mortality rate of cardiovascular diseases rose with increasing temperature and there was a steep change at a higher temperature. In 2012, the meta-analysis study of Mengersen K. et al. [22] showed that every 1 °C increase in temperature during high temperatures, the risk of death resulting from cardiovascular diseases among the elderly increased by 2% to 5%. Chinese scholars also had similar conclusions on the impact of summer hot weather on the death of cardiovascular diseases. Zhang Xia [23] and Zheng Shan [24] pointed out that heat waves led to an increase in the incidence, mortality and hospitalization of cardiovascular diseases, especially for people over 65 years old. Cheng et al. [25] and Zeng et al. [26] found that the number of deaths from cardiovascular diseases increased significantly when the maximum daily temperature was 36 ° C. Each rise of 1 °C increased the risk of death caused by cardiovascular among the elderly by 4.3%.

Similar to the atmospheric environment, Anderson et al. [15] believed that there was the most direct correlation between indoor thermal environment and health outcomes and posed health risk to susceptible people like the elderly, when the indoor temperature exceeds a certain threshold. Hwang et al. [27] found that elderly people were in an uncomfortable state when their indoor temperature was higher than 27.1 °C in summer. Wu Jing [28] pointed out that the test subjects maintained a stable blood pressure with the indoor temperature ranged 18 °C to 20 °C. When the temperature rose, the thermal activity of the central hub of test subjects led to vasodilation of blood vessels. Liu et al. [29] conducted an investigation and analysis of the thermal comfort of the elderly in Chongqing and found that when the indoor temperature was higher than 27.78 °C, they were in an uncomfortable state. Therefore, when the indoor air appeared high temperature for elderly people whose arterial wall elasticity had declined, vasodilation of blood vessels easily lead to rupture of the vessel wall to trigger thrombosis, causing cardiovascular diseases [30]. The cold environment also has an impact on the incidence and mortality of cardiovascular diseases in the elderly. The studies of Yoshino Hiroshi et al. [14], [31], [32] had shown that in the cold area of winter, promoting the indoor temperature could reduce stroke mortality, the mean indoor temperature in case group (elderly people died of stroke) was 1.3 °C lower than that in control group (healthy elderly people with other environmental factors similar to the case group), and the elderly people in the room with a higher temperature appeared a relatively healthy state. In 2015, Yoshino Hiroshi [33] continued to investigate the study and found that the conclusion was basically consistent with the results of research 30 years ago. The study also found that the temperature of the period without heating was 5 - 20 °C lower than the period with heating in the same room. After stopping heating, elderly people's blood pressure was higher than the blood pressure during heating, which was likely to cause cardiovascular diseases like hypertension. Saeki et al. [34], [35] pointed out that cold can trigger raising blood pressure, leading to the development of cardiovascular diseases, and heating can improve the situation. For the elderly, nocturnal blood pressure was positively correlated with bed temperature and weakly correlated with outdoor temperature.

Table 1. Climatic Regions in China

Partition	Mean temperature in the coldest month [36]	Mean temperature designed in center heating [37]
Severe cold area	-10 °C	18 °C
Cold area	-10 - 0 °C	18 °C
Hot-summer and cold-winter zone	0 - 10 °C	N/A
Hot-summer and warm-winter area	10 °C	N/A
Temperate zone	0 - 13 °C	N/A

Due to the vast territory of China, the area in different climates varies greatly in the indoor and outdoor environment in winter. As shown in Table 1, in northern China (severe cold and cold areas), central heating was used in winter, causing the large temperature difference between indoor and outdoor. According to Xiong's study [38], a large step-change temperature, such as exposure from 20 °C to 0 °C or 20 °C to -10 °C, can easily cause physical discomfort, especially the skin temperature decreases with the decrease of ambient temperature. Then, the skin temperature will cause vasoconstriction of blood vessels [28], so that cardiovascular diseases can easily occur. In southern China (hot-summer and cold-winter zone, hot-summer and warm-winter area and temperate zone), there is a high correlation between indoor and outdoor temperatures due to lack of central heating in winter. The survey on the living environment of elderly people in Shanghai of Yao Xinling [39] found that the utilization rate of winter air conditioners in the room was only 10.2%, which made the occurrence of cold weather have a greater impact on indoor temperature. Therefore, the risk of cardiovascular diseases would rise. In winter in China, the impact of the outdoor thermal environment on elderly cardiovascular diseases was dependent on the relationship between indoor and outdoor thermal environment in which elderly people live. At the same time, large step-change temperature and cold environment can easily lead to cardiovascular disease for the elderly.

2.2. Effect of relative humidity on CVD

In the study of the effect of air relative humidity on cardiovascular diseases, Joel Schwartz et al. used time series to study 12 cities in the United States [16], [40], which showed no obvious correlation between relative humidity and cardiovascular diseases. Pan et al. [41] also used the time series to study the number of cardiovascular diseases in Beijing, which also found no significant association between the relative humidity and the incidence of cardiovascular diseases. Cui et al. [42] concluded that the relative humidity and incidence of cardiovascular disease in Beijing were positively correlated via case-crossover study. Wang [43] and Dong et al. [44] found that in Shandong and Guangzhou relative humidity and incidence of cardiovascular disease were both negatively correlated. In addition, Guo et al. [45] found that the relative humidity below 60% prone to occurring cardiovascular diseases, while Meng [46] pointed out that the optimal value of relative humidity ranging from 55% to 60%, exceeding which will lead to increase in the incidence of cardiovascular diseases in Lanzhou.

All in all, we can see different conclusions about the relationship between relative humidity and cardiovascular diseases in different regions or regions. The relative humidity had an uncertain association with the impact of cardiovascular diseases.

3. Effect of air quality on CVD

3.1. Effects of particulate matter on CVD

The particulate matter of air pollutants on cardiovascular diseases was most widely studied. Numerous studies had shown that increased risk of cardiovascular diseases in the elderly was associated with increase of atmospheric particulate matter [18], [20], [47]. Not only outdoor rise of particulate matter, indoor rise of particulate matter also increased the risk of cardiovascular diseases in the elderly. Weinhold et al. [48] in 2011 found that indoor chronic particulate exposure was positively correlated with hypertension, and $PM_{2.5}$ in indoor air increased blood pressure. In 2012, Allen Ryan et al. [49] found that $PM_{2.5}$ in indoor air aggravated atherosclerosis. In 2014, Acevedo et al. [50] found that the effect of second hand smoking (SHS) on cardiovascular diseases resulted from $PM_{2.5}$. Soppa et al. [51] in 2017 found that indoor particulate intake will increase blood pressure via randomized cross-control study.

Indoor pollution caused by indoor combustion, smoking, cooking and other activities had led to the indoor staff showing signs of cardiovascular diseases such as increased blood pressure and increased atherosclerosis [48], [50], [51]. Wang [52] measured the $PM_{2.5}$ in the dwellings of elderly with cardiovascular diseases in Beijing and found that 62% exceeded the standard, and indoor pollution should not be ignored. The main sources of indoor particulate matter were outdoor atmospheric particulates and various types of indoor combustion, cooking, people, pets and other activities. Outdoor pollutants entered inside buildings mainly through ventilation and infiltration [53]. Zhao et al. [54] in Beijing found that outdoor $PM_{2.5}$ contribution rate of indoor $PM_{2.5}$ reached 54% - 96%. Outdoor air pollution, combined with indoor combustion, cooking and other sources of particulate matter, would make the indoor pollution of particulate matter on elderly cardiovascular diseases more serious than the atmospheric environment.

Currently, in the study of parameter representation of cardiovascular diseases, epidemiology was often characterized by incidence and mortality of diseases, which was very difficult to reflect the human health when studying mechanism of cardiovascular diseases of health damage effect on indoor air environment. In the medical diagnosis, the diagnosis of cardiovascular need to be judged according to a variety of blood test indicators, so scholars had used blood test parameters in the study of particulate matter and cardiovascular diseases. For instance, Chuang et al. [55] established the relationship of indoor $PM_{2.5}$ and cardiovascular disease through the survey blood test parameters C-reactive protein. In addition, blood test parameters such as total cholesterol, triglycerides, high-density lipoprotein cholesterol, and low-density lipoprotein cholesterol [56] also reflect the severity of cardiovascular diseases (Table 2). It will be possible to combine the medical science and engineering via the parameters above to more deeply study association of indoor air and cardiovascular diseases.

Table 2. Blood Test Parameters of CVD [56]

Parameters	Normal range	Abnormal symptoms
Total cholesterol	Adult: 110 - 230 mg/dL Children: 120 - 200 mg/dL	*Increase*: hyperlipidemia, atherosclerosis, diabetes, hypertension
Triglyceride (TG)	Adult: < 150 mg/dL Children: < 100 mg/dL	*Increase*: hyperlipidemia
High density lipoprotein cholesterol (hdl-c)	Male: 1.16 - 1.42 mmol/L Female: 1.29 - 1.55 mmol/L	*Decrease*: cerebrovascular disease, coronary heart disease, hypertriglyceridemia
Low density lipoprotein cholesterol (ldl-c)	Youth mean: 2.7 mmol/L. Middle-aged and elderly people mean: 3.37 mmol/L. More than 4.14 mmol/L was significantly.	*Increase*: coronary atherosclerotic heart disease (coronary heart disease), hyperlipoproteinemia, diabetes.

3.2. Effects of other air pollutants on CVD

The air pollutants of NO_2, and SO_2 were harmful to human health. More and more studies have shown that SO_2 and NO_2 can be used as inducers or risk factors of cardiovascular diseases. Joel Schwartz et al. [18], [20] reported that there was the correlation between NO_2 and hospital admission rate of cardiovascular diseases in the elderly. Each increase of NO_2 by 5.1 ppb would increase heart failure by 6.9%. Meanwhile, the study pointed out that SO_2 was only associated with leukocyte, and NO_2 was associated with platelets and fibrinogen. These reasons might be the pathogenesis of cardiovascular diseases. Not only combustion of indoor kitchen cooking but also the body's metabolism would both produce SO_2 and NO_2 [57]. Yoshino [58] believed that it was unhealthy for people indoors to use oil-fired stoves for heating because of the combustion. Mitter et al. [59] also believed that there was a positive correlation between the risk of indoor fossil fuel combustion and the mortality rate of cardiovascular diseases. Pan et al. [60] analyzed the air pollution in Beijing and the death of residents with cardiovascular diseases, then he found that the relationship between air pollutants and the death of cardiovascular diseases was statistically significant. Each increase of SO_2 and NO_2 by 10 g/m^3 would increase the risk of cardiovascular disease death by 0.4% and 1.3%, respectively.

CO was also a major causative agent of cardiovascular diseases [57]. Tsai et al. [61] found that CO has a 2-day lag effect on cardiovascular diseases. In the study of CO on the impact of cardiovascular diseases, Joel Schwartz et al. [40] found that CO had a consistently damage effect on ischemic heart disease, myocardial infarction and heart failure. Zhang et al. [62] conducted a study on the association of cardiovascular disease and CO in elderly people in Taiyuan. The results showed that the increase of average death rate of cardiovascular diseases by 100 g/m^3 resulted to the OR value by 1.006.

In addition, studies recently had shown that ozone on cardiovascular diseases also had an impact. In 2015, Gu [63] studied the combined effects of PM_{10}, NO_2 and ozone to enhance the mortality of cardiovascular diseases through the study of cross-control model. In 2016, Bedada et al. [64] concluded that ozone increased the risk of myocardial infarction via the generalized additive model. In 2017, Zhang et al. [65] concluded that ozone increased the risk of cardiovascular diseases and expounds the mechanism of action by conducting tests on human body physiological indices.

All in all, PM_{10}, $PM_{2.5}$ and NO_2, SO_2, CO and ozone were all reflected that the higher concentration, the greater risk of cardiovascular diseases occurred. In addition, there were many sources of indoor air pollutants, and a variety of pollutants would have a combined effect, which also would make the model of association of indoor air quality and elderly cardiovascular diseases more complicated and challenging.

4. Conclusions

Studies have shown that in the indoor thermal and humid environment, high temperature, cold and large temperature step-changes on the elderly cardiovascular diseases were related, specifically over the high temperature threshold, reflecting the higher the temperature, the more risk of cardiovascular diseases. When in the cold and large temperature step-changes environment, the incidence of cardiovascular diseases would rise. Relative humidity on the impact of cardiovascular diseases was still uncertain. In the indoor air quality, there is a certain correlation between PM_{10}, $PM_{2.5}$, and CO, SO_2, NO_2 and ozone and elderly cardiovascular diseases, reflecting the higher concentration, the higher incidence of cardiovascular diseases.

We believed that in the study of indoor thermal and humidity environment and elderly cardiovascular disease, we must fully consider the seasonal and regional impact on the study results. The extreme weather in summer and winter and the difference between north and south in our country both affected the research results of the correlation between indoor thermal and humid environment and the elderly with cardiovascular diseases. Therefore, it was of great significance to establish the different correlation between indoor environment and elderly cardiovascular diseases in different climates. In the indoor air quality, the combined effect of various pollutants on elderly cardiovascular diseases research were more meaningful. The comprehensive indoor thermal environment and air quality characterization and evaluation methods would become the future research trends.

In this research field, we should actively promote the combination of medicine science and engineering. Changes in environmental parameters provided a reference for medical diagnosis, and pathogenic mechanism of research provided more support for the indoor environment and health problems. In addition, genetic, diet and other factors in the field of research of association of indoor air environment and elderly cardiovascular diseases should also be emphasized.

5. References

[1] "Healthy *China 2030" Outline*, Central Committee of the Communist Party of China and State Council of the People's Republic of China, 2016.

[2] National Bureau of Statistics of China, *China Statistical Yearbook 2017*. Beijing, China: China Statistics Press, 2017.

[3] M. Chen and G. Hao, "Research on Regional Difference Decomposition and Influence Factors of Population Aging in China," *China Population, Resources and Environment*, vol. 24, pp. 136-141, 2014. (In Chinese with English abstract)

[4] H. Wang, S. Zhao, E. Zeng, C. Ma, J. Wang and L. Duan, "The Social Development and the Coping Strategies for China's Aging Population," *Chinese Journal of Social Medicine*, vol. 31, pp. 75-77, April 2014. (In Chinese with English abstract)

[5] L. He and W. Shi, "Quantitative analysis of population aging on mortality disparities for major non-communicable diseases in China," *Chin J Dis Control Prev*, vol. 20, pp. 121-124, 2016. (In Chinese with English abstract)

[6] N. E. Klepeis, W. C. Nelson, W. R. Ott, J. P. Robinson, A. M. Tsang, P. Switzer, et al., "The National Human Activity Pattern Survey (NHAPS): a resource for assessing exposure to environmental pollutants," *Journal of Exposure Analysis & Environmental Epidemiology*, vol. 11, pp. 231, 2001.

[7] World Health Oragnization (WHO), "World health statistics 2016: Monitoring health for the SDGs Sustainable Development Goals," *Geneva Switzerland WHO*, vol. 41, pp. 293-328, 2016.

[8] S. Hu, *Report on Cardiovascular Diseases in China 2016*. Beijing, China: Encyclopedia of China Publishing House, 2016. (In Chinese)

[9] M. Rao, *Guidelines for Prevention and Treatment of Cerebrovascular Diseases in China*. Beijing, China: People's Medical Publishing House, 2007. (In Chinese)

[10] China High Blood Pressure Prevention Guidelines Revision Committee, "Guidelines for the Prevention and Treatment of Hypertension in China 2010," *Chinese Journal of Cardiovascular Disease*, vol. 39, pp. 701-708, 2011. (In Chinese)

[11] Chinese Society of Cardiology and Editorial Board of Chinese Journal of Cardiovascular Diseases. "Guidelines for Chinese Cardiovascular Disease Prevention," *Chinese Journal of Cardiovascular Disease*, vol. 39, pp. 263-279, 2011. (In Chinese)

[12] J. W. Brant and S. R. Hill, "Human Cardiovascular Diseases and Atmospheric air pollution in Los Angeles, California," *Air & Water Pollution*, vol. 8, pp. 259, 1964.

[13] F. Hasegawa, H. Yoshino, S. I. Akabayashi, "Investigation on Indoor Thermal Environment of Detached Wooden Houses in City Areas of Tohoku District in Winter," *Transactions of the Architectural Institute of Japan*, vol. 326, pp. 91-102, 1983.

[14] F. Hasegawa, "A survey study on the room temperature in winter in various houses in the Tohoku district," *Report of the Architectural Institute of Japan*, pp. 18-26, vol. 371, 1987. (In Japanese)

[15] M. Anderson, C. Carmichael, V. Murray, A. Dengel and M. Swainson, "Defining indoor heat thresholds for health in the UK," *Perspectives in Public Health*, vol. 133, pp. 158-164, 2013.

[16] J. Schwartz and J. A. Patz, "Hospital Admissions for Heart Disease: The Effects of Temperature and Humidity," *Epidemiology*, vol. 15, pp. 755, 2004.

[17] J. Cho, Y. J. Choi, M. Suh, J. Sohn, H. Kim, S. Cho, et al., "Air pollution as a risk factor for depressive episode in patients with cardiovascular disease, diabetes mellitus, or asthma," *Journal of Affective Disorders*, vol. 157, pp. 45, 2014.

[18] J. Schwartz, "Air pollution and blood markers of cardiovascular risk," *Environmental Health Perspectives*, vol. 109, pp. 405, 2001.

[19] J. Shi, D. Yuan and Z. Zhao, "Residential indoor PM2.5 sources, concentration and influencing factors in China," *J Environ Health*, col. 32, pp. 825-829, Sep. 2015. (In Chinese with English abstract)

[20] A. G. Barnett, G. M. Williams, J. Schwartz, T. L. Best, A. H. Neller, A. L. Petroeschevsky, et al., "The Effects of Air Pollution on Hospitalizations for Cardiovascular Diseasein Elderly People in Australian and New Zealand Cities, " *Environmental Health Perspectives*, vol. 114, pp. 1018-1023, 2006.

[21] J. A. Patz, D. Campbelllendrum, T. Holloway and J. A. Foley, "Impact of regional climate change on human health," *Nature*, vol. 438, pp. 310, 2005.

[22] W. Yu, K. Mengersen, X. Wang, X. Ye, Y. Guo, X. Pan, et al, "Daily average temperature and mortality among the elderly: a meta-analysis and systematic review of epidemiological evidence," *International Journal of Biometeorology*, vol. 56, pp. 569-581, 2012.

[23] X. Zhang and Q. Liu, "Research Progress of the Effect of High Temperature and Heat Wave on Cardiovascular and Cerebrovascular Diseases," *Chin J Public Health*, vol. 30, pp. 242-243, Feb. 2014. (In Chinese)

[24] S. Zheng, M. Wang, K. Shang, S. He, L. Yin, T. Li, et al., "A case-crossover analysis of heat wave and hospital emergency department visits for cardiovascular diseases in 3 hospitals in Beijing," *Journal of hygiene Research*, vol. 45, pp. 246-251, 2016. (In Chinese with English abstract)

[25] Y. Cheng, Y. Jin, Y. Li, J. Gong, N. Yang, M. Wang, et al., "Impacts of High Temperature on Death Caused by Cerebrocardiovascular Diseases in Wuhan," *J Environ Health*, vol. 26, pp. 224-225, March 2009. (In Chinese with English abstract)

[26] W. Zeng, W. Ma, Y. Zhang, T. Liu, Y. Luo, J. Xiao, et al., "Moiflcation Effect of Latitude on Relationship between High Temperature and Mortality Risk among Elderly: a Mete-Analysis," *J Environ Health*, vol. 29, pp. 639-642, July 2012.

[27] R. L. Hwang and C. P. Chen, "Field study on behaviors and adaptation of elderly people and their thermal comfort requirements in residential environments," *Indoor Air*, vol. 20, pp. 235-245, 2010.

[28] J. Wu, "Research papers on the relationship between indoor air velocity and human comfort and physiological stress," M.S. thesis, Chinese, Chongqing Univ., Chongqing, China, 2005.

[29] H. Liu, Y. Wu, H. Zhang and X. Du, "Field study on elderly people's adaptive thermal comfort evaluation in naturally ventilated residential buildings in summer," *HV&AC*, vol. 45, pp. 50-58, 2015. (In Chinese with English abstract)

[30] J. Chen, "A new trend of prevention and treatment of atherosclerosis," *Chinese Circulation Journal*, vol. 16, pp. 163, 2001. (In Chinese)

[31] H. Yoshino, M. Momiyama, T. Sato and K. Sasaki, "Study of Cerebrovascular Disease and the Indoor Thermal Environment in Two Selected Towns in Miyagi Prefecture, Japan," Ethnic Hgiene, vol. 55, pp. 294-305, 1989. (In Japanese with English abstract)

[32] H. Yoshino, M. Momiyama, T. Sato and K. Sasaki, "Relationship between cerebrovascular disease and indoor thermal environment in two selected towns in Miyagi Prefecture, Japan," *Journal of Thermal Biology*, vol. 18, pp. 481-86, 1993.

[33] K. Hasegawa, H. Yoshino and T. Goto, "Study on association between indoor environment factors and cerebral vascular accident - Influence of indoor temperature on the blood pressure fluctuation of elderly persons," presented at the JSSI & JSSE Joint Conference, Matsumoto, Japan, Sep. 13-16, 2015. (In Japanese)

[34] K. Saeki, K. Obayashi, J. Iwamoto, N. Tone, N. Okamoto, K. Tomioka, et al., "Stronger association of indoor temperature than outdoor temperature with blood pressure in colder months," *Journal of Hypertension*, vol. 32, pp. 1582, 2014.

[35] K. Saeki, K. Obayashi and N. Kurumatani, "Short-term effects of instruction in home heating on indoor temperature and blood pressure in elderly people: a randomized controlled trial," *Journal of Hypertension*, vol. 33, pp. 2338-2343, 2015.

[36] *Code for thermal design of civil buildings*, Chinese Standard GB50176-2016.

[37] Y. Lu, *Practical Handbook of heating and air conditioning design*, 3rd ed. Beijing, China: China Building Industry Press, 2008.

[38] J. Xiong, Z. Lian and H. Zhang, "Effects of exposure to winter temperature step-changes on human subjective perceptions," *Building & Environment*, vol. 107, pp. 226-234, 2016.

[39] X. Yao, "Investigation and analysis of indoor thermal environment in welfare centers for the aged in Shanghai," *HV&AC*, vol. 41, pp. 66-70, 2011. (In Chinese with English abstract)

[40] A. L. F. Braga, A. Zanobetti and J. Schwartz, "The effect of weather on respiratory and cardiovascular deaths in 12 U.S. cities," *Environmental Health Perspectives*, vol. 110, pp. 859, 2002.

[41] L. Liu, X. Pan, Y. Zheng, H. E. Wichmann and U. Franck, "Relationship between Temperature, Relative Humidity and Wind Speed an,d Cardio-cerebral Diseases Hospital Emergency Room Visit Number in a Certain District of Beijing: a Time-Series Analysis," *J Environ Health*, vol.25, pp. 578-82, July, 2008. (In Chinese with English abstract)

[42] M. Cui, L. Yu, J. Zhang D. Zhang and X. Wang, "Association between meteorological factors and the incidence of acute cardiovascular disease," *Chin J Emerg Med*, vol. 23, pp. 465-469, April 2014. (In Chinese with English abstract)

[43] S. Wang, D. Liu and X. Gao, "Study on the relation between blood pressure and weather factor in patients with essential hypertension," Chin J Clinicians (Electronic Edition), vol. 5, pp. 1570-1574, March 2011. (In Chinese with English abstract)

[44] H. Dong, X. Li, H. Liu, G. Lin, Yan Li and K. Li, "Time series study on the association of daily meteorological factors with mortality of cardiovascular disease in Guangzhou from 2012 to 2013," Journal of Tropical Diseases and Parasitology, vol.14, pp. 155-159, 2016. (In Chinese with English abstract)

[45] D. Guo, X. Wang and R. Wang, "Relationship between the occurrence of acute myocardial infarction and meteorological factors," *Chinese Journal of Integrative Medicine on Cardio/Cerebrovascular Disease*, vol. 9, pp. 1423-1424, 2011. (In Chinese)

[46] J. Meng, "Meteorological factors in three cities of Gansu Province and residents of cardiovascular system disease day admission number of time series," M.S. thesis, Chinese, Lanzhou Univ., Gansu, China, 2013.

[47] J. Wu, L. Zhou, B. Xu, Q. Wang, Z. Ding, X. Chen, et al., "Air particular matter pollution and cardio-cerebrovascular diseases death: a case-crossover study," *J Environ Health*, vol. 31, pp. 569-571, 2014. (In Chinese with English abstract)

[48] B. Weinhold., "Indoor PM pollution and elevated blood pressure: cardiovascular impact of indoor biomass burning," *Environmental Health Perspectives*, vol. 119, pp. A442, 2011.

[49] R. W. Allen, S. D. Adar, E. Avol, M. Cohen, C. L. Curl, T. Larson, et al., "Modeling the Residential Infiltration of Outdoor PM2.5 in the Multi-Ethnic Study of Atherosclerosis and Air Pollution (MESA Air)," *Environmental Health Perspectives*, vol. 120, pp. 824-830, 2012.

[50] V. Acevedo-Bolton, W. R. Ott, K. C. Cheng, R. T. Jiang, N. E. Klepeis and L. M. Hildemann, "Controlled experiments measuring personal exposure to PM2.5 in close proximity to cigarette smoking," *Indoor Air*, vol. 24, pp. 199, 2014.

[51] V. J. Soppa, R. P. F. Schins, F. Hennig, M. J. Nieuwenhuijsen, B. Hellack, U. Quass, et al., "Arterial blood pressure responses to short-term exposure to fine and ultrafine particles from indoor sources - A randomized sham-controlled exposure study of healthy volunteers," *Environmental Research*, vol. 158, pp.225-232, 2017.

[52] Y. Wang, W. Huang, T. Wang, Y. Chen, Y. Su and L. Zhang, "Characterization and assessment of exposure to PM2.5 and CO for a cardiovascular elderly panel in summer in Beijing," *China Environmental Science*, vol. 29, pp. 1005-1008, 2009. (In Chinese with English abstract)

[53] S. Chai, "Air conditioning and non-air-conditioned room with the concentration of particulate matter changes," M.S. thesis, Chinese, Donghua Univ., Shanghai, China, 2006.

[54] W. Ji and B. Zhao, "Contribution of outdoor-originating particles, indoor-emitted particles and indoor secondary organic aerosol (SOA) to residential indoor PM2.5 concentration: A model-based estimation," *Building & Environment*, vol. 90, pp. 196-205, 2015.

[55] H. C. Chang, K. F. Ho, L. Y. Lin, T. Y. Chang, G. B. Hong, C. M. Ma, et al., "Long-term indoor air conditioner filtration and cardiovascular health: A randomized crossover intervention study," *Environment International*, vol. 106, pp. 91-96, 2017.

[56] Y. Jiang and L. Yu, *Pathology and pathophysiology*, Beijing, China: Peking Union Medical College Press, 2012.

[57] Y. Zhu, *Building Environmental Science*, Beijing, China: China Building Industry Press, 2010.

[58] K. Hasegawa, "Investigation and study on the relationship between the stroke and the living environment and the relationship between Yamagata prefectures and Japan," *Japanese Journal of Public Health*, vol. 32, pp. 181-193, 1985. (In Japanese)

[59] X. Gao, Z. Bai, Y. You, J. Miao and B. Liu, "Investigation of Human Exposure Levels of Volatile Organic Compounds and Effects on Health," *J Environ Health*, vol. 23, pp. 300-303, 2006. (In Chinese with English abstract)

[60] M. Yang and X. Pan, "Time-series Analysis of Air Pollution and Cardiovascular Mortality in Beijing, China," *J Environ Health*, vol. 25, pp. 294-297, 2008. (In Chinese with English abstract)

[61] D. H. Tsai, J. L. Wang, K. J. Chuang and C. C. Chan, "Traffic-related air pollution and cardiovascular mortality in central Taiwan," *Science of the Total Environment*, vol. 408, pp. 1818-1823, 2010.

[62] X. Zhang, Z. Zhang, B. Su, B. Feng and H. Li, "Acute Effects of Ambient Air CO Pollution on Cerebrocardiovascular Diseases Mortality in People Aged 65 Years and over in Taiyuan City," *J Environ Health*, vol. 25, pp. 672-675, 2008. (In Chinese with English abstract)

[63] Y. Gu, "Study on the Interaction between PM10, NO2 and O3 Pollution on the Mortality Risk in Guangzhou," M.S. thesis, Chinese, Jinan Univ., Guangdong, China, 2015.

[64] G. B. Bedada, A. Raza, B. Forsberg, T. Lind, P. Ljungman, G. Pershagen, et al., "Short-term Exposure to Ozone and Mortality in Subjects With and Without Previous Cardiovascular Disease," *Epidemiology*, vol. 27, pp. 663, 2016.

[65] D. B. Day, J. Xiang, J. Mo, F. Li, M. Chung, J. Gong, et al., "Association of Ozone Exposure With Cardiorespiratory Pathophysiologic Mechanisms in Healthy Adults" *Jama Internal Medicine*, vol. 177, pp. 1344-1353, 2017.

Acknowledgments
This paper received February 25, 2018. This work was supported in part by the China Environmental Protection Project under Grant 201509063, and the National Nature Science Foundation of China under Grant 51308088, 51578103 and 91743102.

Experimental study of membrane-based electrostatic precipitators with high filtration accuracy

Jing Chen[1], Fengjun Xie[2], Yijing Kang[2], Shunquan Chen[1], Zhaolin Feng[2]

1 Guangdong Key Laboratory of Membrane Science and Membrane Separation, Guangzhou Institute of Advanced Technology, Chinese Academy of Science, Guangzhou
2 GuangZhou Guangyi Air Treatment Engineering Co. Ltd, Guangzhou
E-mail: Jing.chen@giat.ac.cn

Abstract. Electrostatic precipitators (ESPs) with the membrane-based collecting electrode play an important role on the welding flue gas cleaning process. The research has been conducted on both bench and pilot scales with novel electrostatic precipitation using PTFE microporous membranes. Two types of membranes collector were used for testing — the charged and uncharged PTFE microporous membranes substrate. Tests were performed to measure the collection efficiency of the membranes. The result indicated that uncharged membrane is effective as charged membrane plates as a collection surface in a bench-scale ESP for the capture of PM2.5, but the charged membrane has a higher efficiency when particle size is lower than 0.3μm, which is more suitable for collecting fine particulates. The preliminary data of the pilot-scale experimental also indicated that charged membranes exhibited excellent particulate capture efficiencies for welding flue gas, and the collection efficiency was reached 99.1%, which can achieve an ultra-low concentration emission of flue gas less than 1 mg/m^3.

1. Introduction

Fine particulates, particles with aerodynamic diameters less than 2.5μm, also known as PM2.5, are considered a significant problem. It has been estimated that 2.1 million people premature deaths in the worldwide scale every year may be attributed to respiratory distresses associated with the effects of fine particulate matter. The World Health Organization (WHO) estimated in 2005 that fine particulate air pollution (PM2.5), causes about 3% of mortality from cardiopulmonary disease, about 5% of mortality from cancer of the trachea, bronchus, and lung, and about 1% of mortality from acute respiratory infections in children under 5 years, worldwide. Therefore, controlling the discharge of fine particulates, especially PM2.5, is an important measure to improve air quality and protect public health and the environment.

Electrostatic precipitators (ESPs) are widely used to reduce the emissions of smoke, fumes and dust [1] because of the low pressure drop, low energy consumption, high collection efficiency and long service life. However, many current ESPs cannot be adequate to address the challenges of fine particulate collection [2]. Electrostatic precipitators exhibit reduced collection efficiencies for particles in the diameter range of 0.1-2.0 μm due to inherent charging mechanism limitations on particles in this size range [3]. Researchers at Ohio University developed and patented a novel membrane ESP that successfully addresses many of the problems associated with ESPs [4-6]. Membranes made from light, strong, and corrosion-resistant advanced materials were used to replace steel collecting plates. Such replacement facilitates tension-based rapping, which shears the adhered particles layer from the collector surface more effectively than hammer-based rapping.

Content from this work may be used under the terms of the Creative Commons Attribution 3.0 licence. Any further distribution of this work must maintain attribution to the author(s) and the title of the work, journal citation and DOI.
Published under licence by IOP Publishing Ltd

Extensive research has been conducted at our works in the last three years to develop a new electrostatic precipitation in which heavy metal plates are replaced by light and thin membranes made from advanced, mainly fibre-based and other similar materials. Membrane applications could be beneficial to both dry and wet ESPs, and possess great potential for reducing the cost and increasing particle capture efficiency. Although applicable to numerous industries, these changes may allow the implementation of advanced technologies to reduce fine particulate, as well as acid aerosol emissions from coal-burning power plants, and thus help to meet the proposed WHO PM2.5 standard. It is expected that the knowledge and experience from this study could be used in the near future to develop hybrid and retrofitted ESPs. The research has been conducted on both bench and pilot scales with novel electrostatic precipitation using PTFE microporous membranes. In the following article, some recent results related to particle collection and removal efficiencies, selection of membrane materials and other relevant issues, as well as comparisons with existing ESP technologies are presented.

2. Experimental facility

2.1. Bench-scale Experimental

The bench-scale experimental ESP, schematically shown in Figure 1 and imaged in Figure 2, was used to examine the performance of different electrode structures and different collecting membrane in a controlled environment simulating conditions experienced in typical welding fume-dust. The experimental facility consisted of four different parts: an ash fluidizer, an inlet section, a test section, and an outlet. The ash fluidizer introduced dispersed fly ash into the test section of the precipitator by blowing compressed air underneath a fine screen supporting the fly ash.

For the experiments, two types of electrodes structures were used. One of the electrodes structure is a wire-plate electrode module and the other is a line-cylinder electrode module as shown in Figure 3 and Figure 4. Several charging electrodes were located in the main test section for each module. The charging electrodes were made of 1/4-in. copper tubes with protruding nails to enhance corona current. The grounding terminals for the pre-charging electrodes were located at the precipitator wall across from the pre-charging electrodes.

Two types of membranes were used for testing - the charged and uncharged PTFE microporous membranes collecting plate. Test for particles collection efficiency was conducted. A controlled feeding system was used to introduce a known particle concentration into the test section. Particulates removed in the precipitator test section and the initial loading mass were used to determine the actual collection efficiency, which was then compared the collection efficiency of those two different membranes observed in the experiments.

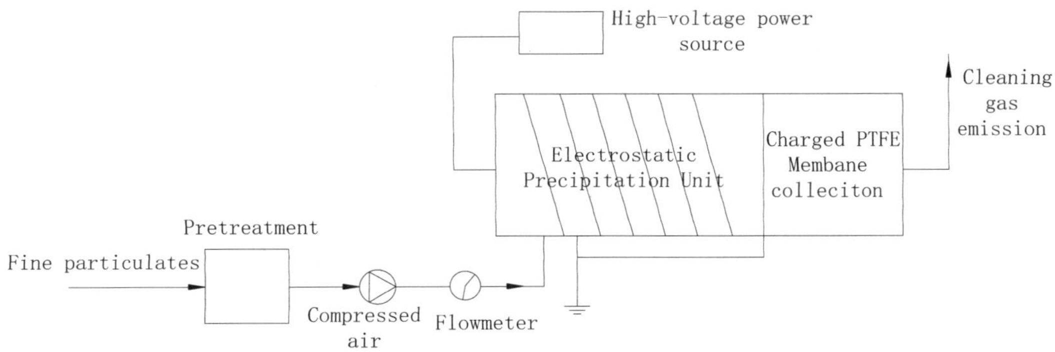

Figure 1. Diagram of the electrostatic precipitator test section

Figure 2. Image of the bench-scale experimental ESP test section

Figure 3. Wire-plate electrode structure

Figure 4. Line-cylinder electrode structure

2.2. Pilot-scale Experimental

A pilot test section was built at a welding workshop of Guangzhou zhengjian packaging materials technology co. LTD (Guangdong Province, China) to demonstrate the membrane-based precipitator technology developed at our work. A 0.60×0.6×1.5m precipitator box with two 1.2×0.59m charged PTFE membranes, was constructed as schematically shown in Figure 5 and imaged in Figure 6.

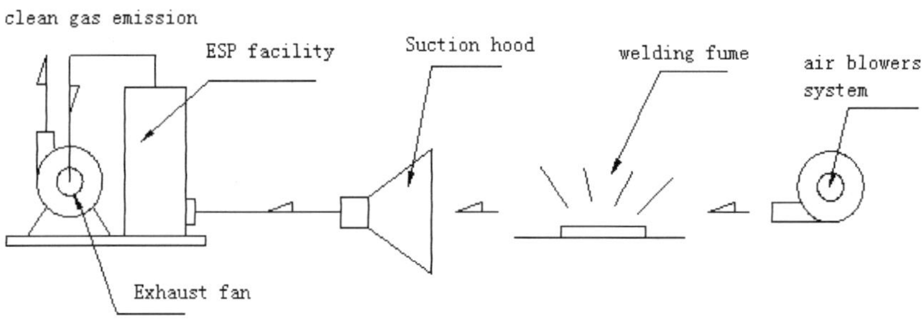

Figure 5. Pilot membrane-based precipitator configuration

Figure 6. View of the pilot-scale precipitator test facility at a welding workshop

2.3. Analysis and measurement

Particle size distribution at the inlet and outlet to the precipitator test section was determined from isokinetic sampling of the particulate suspension and analysing the sample using the following equipment: TSI DustTrak (8532), Laser airborne particle counter(Y09-6), Malvern Mastersizer 3000, Atomic absorption spectrometer (HITACHI Z-5000), and Scanning electron microscope (SEM, Zeiss LEO1530VP).
The collection efficiency of PM2.5 is calculated as the following equation:

$$\eta = \frac{M_0 - M_1}{M_0}$$

Where M_0 is the ash mass concentration at the 0 kV (mg/m^3) and M_l is the escaping ash mass concentration at the 25 kV (mg/m^3)

3. Results

3.1. Bench scale collection experiments

A series of collection efficiency and particle analysis tests were performed in the previously described precipitator unit using both uncharged and charged membrane collection plate. Membranes behaviour was examined under various conditions at gas flow velocities ranging between 1-3m/min, voltage of 25kV and current flow of about 0.4-1.8mA. The collection efficiency results using uncharged and charged membranes are shown in Table 1. These data indicate that charged membrane is a better collector than uncharged membrane, especially when particle size is lower than 0.3μm, and the collection efficiency of the uncharged membrane is 39.46%, and the charged membrane is 82.24%. And for both uncharged and charged membrane, the collection efficiency for PM2.5 is 92.43% and 93.78%, respectively. This is proved that membrane collection plates can promote condensation and more efficient capture of fine particulates. Charged membrane can achieve corona power levels that are several times higher than uncharged membrane, which greatly enhances charging and collection of submicron-sized particulates.

Table 1. Comparison of purification efficiency of electrostatic precipitator with charged and uncharged membrane collection plates

Test parameter	Collection efficiency			
Particle size	0.3μm	0.5μm	1.0μm	PM2.5
Uncharged membrane collection plate	39.46%	50.64%	86.41%	92.43%
Charged membrane collection plate	82.24%	86.41%	95.70%	93.78%
Collection efficiency increased by charged membrane	42.78%	35.77%	9.29%	1.35%

The data in Table 2 shows the relationship between the collection efficiency, wind resistance, and filtration velocity and operation time. For a long time running experimental, the collection efficiency of the charged membrane based ESP is relatively stable, but the wind resistance increase and filtration velocity decrease with the operation time. The results of the experiment show that when the membrane is running continuously for 6-7 hours, the wind resistance and filtration velocity reach their ultimate value. At this time, the collection efficiency is also reduced. This is suggested that the membrane need to clean or replace.

3.2. Pilot-tests results

Basic testing was performed to examine the collection efficiency of the charged membrane substrate in the pilot-scale electrostatic precipitator. Here, a China Industry Standard (HJ 836-2017 Stationary source emission—Determination of mass concentration of particulate matter at low concentration—Manual gravimetric method) was used to determine the particle collection efficiency. Long-time concentration detection of welding fume and Mn oxides were conducted at the inlet and outlet of the ESP facility, the data are shown in Table 3. The results showed that approximately 99.1% of the particles were removed from the welding fume gas stream. And for Mn and its compounds (Calculate by MnO_2), the removal efficiency is about 98.6%. The results show a highly unexpected and encouraging collection efficiency, which can be reasonably explained by particle agglomeration. The data show that the charged membrane based ESP is high efficient way to treat and purify heavy metals in flue gas, which can achieve an ultra-low concentration emission of flue gas less than 1 mg/m^3.Based on our experimental, the charged membrane based ESPs are especially suitable for collection of fine particulates.

Table 2. The relationship between the collection efficiency, wind resistance, and filtration velocity with operation time

The collection efficiency of charged membrane based ESP change with time					Wind resistance and filtration velocity change with time	
Time (h)	Particle size				Wind resistance (Pa)	Filtration velocity (m/min)
	0.3μm	0.5μm	1.0μm	PM2.5		
0	80.38%	80.30%	98.66%	96.68%	200	3.35
1	80.87%	89.13%	99.36%	96.60%	220	3.17
2	81.24%	90.08%	99.64%	94.26%	240	3.25
3	75.29%	85.68%	99.32%	96.50%	240	3.00
4	77.64%	87.01%	99.60%	94.62%	350	2.23
5	77.14%	80.88%	99.17%	94.69%	400	2.03
6	59.79%	62.75%	90.70%	94.82%	660	1.12
7	53.94%	51.53%	89.26%	94.12%	700	0.55
Average efficiency	73.29%	78.42%	96.96%	95.29%	/	/

Table 3. Collection efficiency of the pilot-scale ESP with charged membrane collector

Sample Locations	number of samples	Rapping test results（mg/m^3）	
		welding fume (The total dust)	Mn and its compounds (Calculate by MnO_2)
Tubing entry points	F1-1，A3-1	63.75	1.43
Tubing outlet points	F1-2，A3-2	0.58	0.02
collection efficiency		99.1%	98.6%

Particle samples were collected by air sample membrane at both inlet and outlet point of the flue pipe, the collecting time was 30 seconds. The collected samples and their surface morphology and characteristic were observed by scanning electron microscopy (SEM), as shown in Figure 7. Figure 7(a) shows the samples collected at the entrance of the flue pipe, from the SEM picture, we can see that welding fume particles are uniform distributed on the surface of the sample membrane. The fine particles deposited on the membrane surface are mostly spherical, and the single particle size was mostly below 2.5μm (PM2.5). After deposition, the particles are agglomerated together, and have a certain adsorption effect on other particles due to their larger specific surface area. Figure 7(b) shows the samples collected at the outlet of the flue pipe, and the sample surface is almost clean, which indicated that the membrane-based ESP have a high collection efficiency for fine particulates. This result is consistent with the previous description.

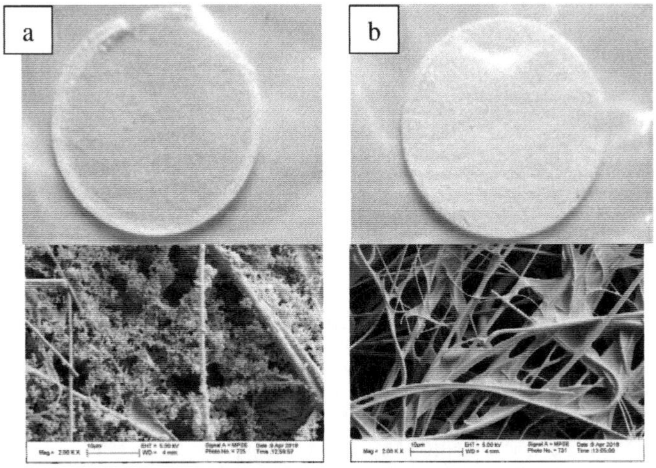

Figure 7. Sample membrane and its surface under the scanning electron microscopy (SEM). (a) Samples collected at the entrance for 30 seconds; (b) Samples collected at the outlet for 30 seconds

4. Conclusion

The research conducted has indicated some key possibility for the use of light and smooth PTFE porous membrane as replacements collector for ESPs facility. PTFE porous membrane materials possess superior properties, such as low price, mechanical strength and resistance to temperature, combustion and corrosion. Membrane based ESP facility also showed excellent collection efficiency and separation properties that are extremely important in ESP applications, and in turn could resolve many of the PM2.5 regulation requirements for China ambient air quality standards (GB3095-2012). A comparison of bench-scale experimental was taken for both uncharged and charged membrane collection plates. The result indicated that uncharged membrane is effective as charged membrane plates as a collection surface in a bench-scale ESP for the capture of PM2.5, but the charged membrane has a higher efficiency when particle size is lower than 0.3μm, which is more suitable for collecting fine particulates. The preliminary data of the pilot-scale experimental also indicated that charged membranes exhibited excellent particulate capture efficiencies for welding flue gas, and the collection efficiency was reached approximately 99.1%. As a collection plate, membrane cleaning and membrane regeneration were not quantified during this work and should be the subject of future work. In addition, alternative rapping techniques, specifically ones that do not induce side forces (shaking), must be explored in future work to further minimize re-entrainment.

5. References

[1] Jaworek A, Krupa A, Czech T 2004 Modern electrostatic devices and methods for exhaust gas cleaning: A brief review *J. Electrost.* **65** 133-155.
[2] Altman R, Offen G, Buckley W, Ray I 2001 Wet electrostatic precipitation: demonstrating promise for fine particulate control-part I. *Power Engineering* **105** 37-39.
[3] Flagan R and Seinfeld J 1998 Fundamentals of Air Pollution Engineering. *Prentice Hall, NJ: Prentice Hall* p 21.
[4] David J, Bayless D, Liming Shi 2005 Membrane- based wet electrostatic precipitator. *Journal of the air and waste management association.* **55** 784-91.
[5] Bayless D, Alam M, Radcliff R 2004 Membrane-based wet electrostatic precipitation. *Fuel Process. Technol.* **85** 781-98.
[6] Bayless D, Pasic H, Alam M 2001 Use of membrane collectors in electrostatic precipitators. *Journal of the air and waste management association.* **51** 1401-07.

Acknowledgments
This work was supported by Guangzhou Science and Technology Plan Project (NO.201604010070).

IOP Conf. Series: Earth and Environmental Science **257** (2019) 012009 doi:10.1088/1755-1315/257/1/012009

Hydrometeorological monitoring for hydropower reservoirs in remote areas

H Basri[1], LM Sidek[2], L H C Chua[3], A Z Abdul Razad[4] and M S Kamarulzaman[5]

1 PhD Candidate, Civil Engineering Department, College of Engineering, and Sustainable Technology and Environment Group, Institute of Energy Infrastructure, Universiti Tenaga Nasional, Malaysia
2 Professor, Civil Engineering Department, College of Engineering, and Sustainable Technology and Environment Group, Institute of Energy Infrastructure, Universiti Tenaga Nasional, Malaysia
3 Associate Professor, School of Engineering, Faculty of Science Engineering & Built Environment, Deakin University, 75 Pigdons Road, Waurn Ponds, 3220 VIC, Australia
4 Researcher, TNB Research Sdn Bhd, Malaysia
5 Senior SCADA & Instrumentation Engineer, Tenaga Nasional Berhad, Malaysia
E-mail: BHidayah@uniten.edu.my

Abstract. Operational hydrometric network is an essential element in hydropower reservoir operation. The data collected, to be used for reservoir management and decision making require high reliability with automatic data monitoring, processing and retrieval. However, the unique characteristic of every catchment and operational requirements raises the need to balance hydrometeorological monitoring objectives and site desirability. This paper describes the overall component of the Temengor Inflow Forecasting System developed as part of a decision support module for optimised hydropower dam operation for the Temengor Hydropower Dam system in Malaysia. The equipment used and challenges faced in deployment in a densely forested catchment in the Sungei Perak basin are discussed. The siting of the hydrometeorological network was made based on World Meterological Organisation guidelines and state of the art communication technology was utilized in overcoming accessibility to our monitoring sites. The real time data on power generation, river and reservoir level, rainfall that were collected are presented in this paper.

1. Introduction

The efficient and effective management of hydropower reservoirs is vital for hydroelectric power plant operation. The monitoring of rainfall, inflows and lake level provides benefits by optimizing hydropower generation with the goal of maximizing energy revenue, while taking into account dam safety risks. One of the critical factors that contribute to successful hydropower reservoir management is through reliable monitoring infrastructure, equipment and technology. Therefore, the continuous monitoring of inflows and lake level is an essential tool for hydropower dam operators by providing real-time data for decision making in power generation and planning [1]. Real time recording instruments provide the ideal means of recording the responses of rivers, lakes and reservoirs to short-term changes in weather [2]. Inflow forecasting simulates the run-off process in a catchment by taking all necessary elements collected from hydrometeorological data into account, providing insight into the future resource availability for hydropower generation and advance warning of water-related

Content from this work may be used under the terms of the Creative Commons Attribution 3.0 licence. Any further distribution of this work must maintain attribution to the author(s) and the title of the work, journal citation and DOI.
Published under licence by IOP Publishing Ltd

IOP Conf. Series: Earth and Environmental Science **257** (2019) 012009 doi:10.1088/1755-1315/257/1/012009

disasters such as flood. The collected information is beneficial in deriving optimal operation rule curves for long term planning and efficient water allocation for short-term hydropower plant scheduling. Nowadays, inflow monitoring techniques have become more advanced over recent years in response to increased demands for more accurate, timely and reliable information. The information, which is highly valuable and of national interest requires timely, detailed reservoir level warnings with maximum accuracy. Thus, the hydrometeorological networks in the catchment require high reliability with automatic data monitoring, processing and retrieval [3]. Monitoring of inflows through hydrological instrumentation and appropriate hydrologic modeling provide sufficient lead time to dam managers to efficiently operate hydropower plants and mitigate against floods, while maintaining appropriate lake level during monsoon and drought season. Additionally, financial advantages could be acquired through well-planned hydropower plant operations, bringing greater efficiency in power production, as well as the implementation of action plans through early warning for downstream flood risk areas effectively and ultimately reducing the occurrence of dam failure. However, based on catchment characteristic and operational requirements, consideration on some challenges and constraints are needed for efficient hydropower reservoir operation. Among the criteria considered includes adequate hydrometeorological network, forecast time required, specific operating issues, displays or warning requirements for the user interface and also communications and information technology system available.

Hydrometric network design for catchment monitoring aims to provide operators with timely, easily accessible water quantity data relevant to hydropower reservoir operation [4]. Although the installation of monitoring equipment as part of hydropower reservoir operation has been successfully implemented in many countries, however, challenges are usually site specific. For example, a review of operational and research applications demonstrates that satellite observations can improve accuracy and spatial detail in hydrological model estimation [5] in remote areas. It has been shown that rainfall, discharge and water level data collected from the hydrological network in Nam Ngum River basin can be effectively used as a management tool for hydropower reservoirs. The data were also checked for accuracy and reliability for hydropower expansion planning in the same catchment. However, limited data availability and accuracy may cause difficulty in the planning of subsequent hydropower project [6]. Taiwan Power Company use observed reservoir inflows and hydropower generation data to study an optimisation approach to find the representative operation rules which can replace operators' experience of reservoir operation and investigate climate change impacts on reservoir inflows and subsequent hydroelectric power generation [7]. Long-term, continuous, time-series collection of environmental data is fundamental to understand the environmental dynamics of aquatic systems due to climate change. Integrated system for environmental monitoring (SIMA) has successfully provided a long-term database used for limnological study for hydropower reservoir in Brazil [8]. Advances in information technology such as fibre optic technology for river monitoring provide additional sources of data. The monitoring system developed for Yangtze River Valley in China is part of a decision support system component for flood planning and management [9]. On the other hand, the Internet of Things (IoT) technology that combines wireless sensor network technology, communication technology, sensor technology and network technology were explored for hydrology monitoring system into a simple and low-cost sensor. The system consists of wireless sensor network, GPRS network and monitoring centre [10].

This paper discusses on the hydrometeorological monitoring equipment installation and requirement of optimal network challenges encountered at the selected site with high-density forest catchment in Temengor, Malaysia. The monitoring data obtained from real-time hydrometeorological data and application for hydropower reservoir operation and forecasting is discussed. The overall objective of this study were to develop a system capable of monitoring rainfall, inflows and lake level at Temengor for hydropower reservoir operated by Tenaga Nasional Berhad (TNB).

2. Temengor catchment

Sungai Perak (Sg Perak) is the second largest river system in Peninsular Malaysia. It has an approximate length of 427 km and a catchment area of 15,180 km^2, which is about 71% of Perak State.

Along the upper Sg Perak basin, a series of lakes have been created with the construction of 4 hydropower dams, namely the Temengor, Bersia, Kenering and Chenderoh Dams, collectively known as Sungai Perak Hydroelectric Scheme, shown in Figure 1. These dams are used for hydroelectric power generation, operated by Stesen-stesen Janaelektrik Sungai Perak (SSJ Sg Perak). This study focuses on the hydrometeorological network in Temengor catchment designed as part on Temengor Inflow Forecasting System (TIFS). Sg Perak Power Stations cascading hydro power plant was built in 1970 with a total generating capacity of 1,248 Megawatt from 22 unit generators, built either for peak or base load. The hydroelectric power for Temengor is generated by 4 generators with total design capacity of 348 MW. The electricity generated if fed to the Grid System through a switchyard and tower line and distributed to customers in Peninsular Malaysia [11]. The Temengor catchment is about 3,500 km^2 with catchment elevation ranging from 206 m to 2,156 m. The catchment is mostly forested with about 150 km^2 occupied by Lake Temengor and located within then Belum Forest Reserve. The Royal Belum State Park (RBSP) was gazetted as a protected area under the Perak State Parks Corporation Enactment in 2001.

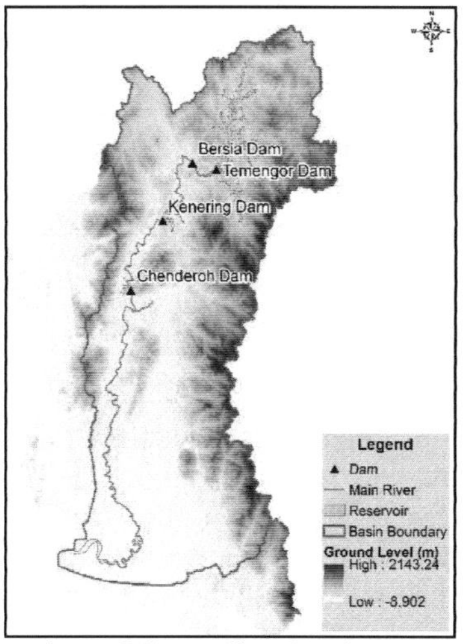

Figure 1. General Location of Cascading Hydropower Scheme in Sg Perak

3. Temengor monitoring system

The requirements of a hydrometeorological monitoring system are site specific depending on local objectives. The Temengor system was developed to support reservoir and power generation requirements taking into consideration local catchment and climate conditions. The rain gauge network required for the site should reflect the nature of the climate drivers involved in the flood-producing rainfalls and ease of operation and maintenance. The requirements of operators need to be considered in terms of warning times required, trigger levels, and the type of information needed. Operators will consider the output from the forecasting system as an input for decisions in operation of hydropower generation and respond to floods.

The high-level system requirements include inflows and lake level, which is mainly needed for advanced warning of floods into Temengor Reservoir and used in hydropower plant operation and optimisation. The operation of hydropower generation at Sg Perak Power Stations is based on the requirement by National Load Dispatch Centre (NLDC), however, additional lead time of lake level information offers advantages in operation planning while taking into consideration dam safety aspects. Therefore, a rainfall forecast would be required to give sufficient lead-time that can be available from the local meteorological service. Another important requirement for monitoring equipment is installing and upgrading the new water level and rainfall stations in Temengor catchment. Site visits were undertaken by the project team to understand the requirements for the hydrometeorology network, the data available and the proposed locations and logistic arrangement of the selected sites.

IOP Conf. Series: Earth and Environmental Science **257** (2019) 012009 doi:10.1088/1755-1315/257/1/012009

The design of the hydrometeorological network is based on a combination of rainfall, river level and reservoir level monitoring points reporting in real time, or near real time, to a central operation and control system. An adequate number of monitoring equipment is essential to report the development of an impending flood, providing sufficient time for forecasting models to produce outputs before warning dissemination and adopting operational decisions of hydropower generation [12]. Having an inadequate number and improper measurement locations can result in the poor establishment of a database for forecast model development, which may result in the less accurate predictions. Nevertheless, the challenge is to find the right balance between hydrometeorological monitoring objectives and site access. The most reliable and affordable solutions depend on proper site selection [13], and a mismatch between local conditions and appropriate technology results in poor quality data and high maintenance requirements for both field and administrative operations. It is also vital to check the data transmission method. The best system must be at minimum operating cost with high quality data transmission. The specific tasks undertaken in this stage are summarised below:

3.1. Hydrometric network design and installation
Hydrometric network for Temengor catchment composed of a group of rainfall and water level stations that are designed and operated to collect the necessary data. Observation data collected in the network will be used as in the development of predictive models for reservoir water levels.

3.2. Configuring modelling server configuration to the real-time and forecast data
The Temengor Inflow Forecast System server arrangement consists of two servers; one for telemetry (AQWEB server), and the other for forecasting. Both are installed in the Bersia Group Control Centre (BGCC). The forecasting server takes data from data storage in the telemetry server, and stores it in the hydrometric database which is then accessed by the forecasting model. This task involved configuration of the TNB forecast server with the telemetric data network, telemetric server, forecast data and TNB Supervisory Control and Data Acquisition (SCADA) inputs.

3.3. Forecast model development
This includes development of the forecast model, which consists of inflow and reservoir storage models, and generation of reservoir water level forecast ensembles.

3.4. System interface development
The task for this component involved development of web based display of the real-time data and results from the water level forecast model.
The final output from this activity is system architecture of inflow forecast system as shown in Figure 2.

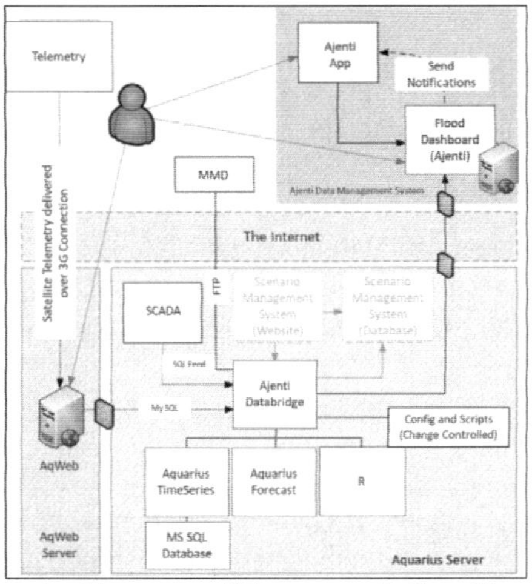

Figure 2. System Architecture of Temengor Inflow Forecasting System

The Aquarius server is located in BGCC located at Bersia Dam. This server is responsible for the collation of all the relevant data and the execution of the inflow forecasting models. This server hosts a number of key components:

3.5. Ajenti DataBridge:
this is the software that manages the data collection and forecasting services, and is pivotal in ensuring a robust and reliable inflow forecasting system. The software is designed to synchronise data between different systems using a plugin architecture that allows data to be read from databases and seamlessly synchronised with systems like Aquarius TimeSeries. The software also includes a highly configurable scheduler that can be setup to manage the execution of mission critical tasks. It allows for customised processes to be executed to manage simple to extremely complex hydropower reservoir operation scenarios. Forecast rainfall data from Malaysia Meteorological Department (MMD) also will be transmitted to Ajenti DataBridge.

3.6. Aquarius Forecast:
this software runs the core inflow forecast models. It is built on the latest .NET platform and features a versatile Application Programming Interface (API) that allows for easy and efficient interaction by third party systems like the Ajenti DataBridge. This makes Aquarius Forecast easier to manage automatically than other modelling system.

3.7. Aquarius TimeSeries:
this is a hydrographic software used around the world for managing water data. All data collected by the Ajenti DataBridge is archive into Aquarius TimeSeries. The software features data editing and quality control tools and also features a ratings editor to manage the conversion of water levels to flows. Aquarius TimeSeries retains information in the database.

3.8. R:
this is a statistical software package that allows for easy manipulation of data. This is used by the system to conduct pre-processing of data before the models are executed, and post-processing of the modelling results.

3.9. SCADA:
the SCADA system provides real-time data to the inflow forecasting system via a feed through a SQL database. This allows the models to see the latest dam level as well as various outflow values. This part of the system can be changed to use an Open Platform Communication (OPC) server in the future if required.

3.10. AqWeb:
this is the Telemetry System responsible for collecting the data from all the remote telemetry sites. The data that is captured is stored in a MySQL database on the AqWeb server. The Ajenti DataBridge uses a MySQL plugin to poll for new data. Users can access the web interface of this system to see the latest telemetry data. This access is via a 3G connection that is also used for the data feed from the satellite based telemetry stations.
In addition to the server located at BGCC, the data is synchronised into Ajenti Data Management System (ADMS). The ADMS provides both an offsite backup of all the associated data and provides a secure view of all data that is accessible via the internet or through a smart phone application.

4. Results and discussion

4.1 Operational Hydrometeorological Network in Temengor Catchment
The complete system was setup under a commercial contract. Hydrometeorological equipment installed in the study area included rain gauge, hydrostatic level measurement system for river level

IOP Conf. Series: Earth and Environmental Science **257** (2019) 012009 doi:10.1088/1755-1315/257/1/012009

monitoring and radar water level sensor for reservoir level monitoring. The monitoring equipment and its specifications are shown in Table 1.

Table 1. Hydrometeorological Monitoring Equipment for Temengor Catchment [14–18]

Equipment	Description	Specifications
 Hydrological Services Tipping Bucket Raingauge (TB3)	Rainfall measurement equipment for remote and unattended locations	• World standard 200 mm catch. • Resolution: one tip at 0.2 mm • Long term stable calibration • Accuracy : +/- 2% • Dual output signal data collection and transmission.
 Water Level Sensor (Bubble)	A stand-alone hydrostatic level measurement system that requires only a 12V DC power supply and bubble tube (capillary)	• Measurement range : 10 m • Resolution: 16 bit (digital) • Accuracy : +/- 0.05% • Fully self-contained precision piston air compressor
 Remote Terminal Unit (RTU)	Highly reliable and cost-effective remote terminal unit and data logger for data collection, communication, logging, alarming, control and analysis applications	• Built-in pressure, level and water quality transmitters, I/Os, cellular, WiFi, Ethernet, RS485 and GPS module. • Communication selections: Cellular, Wi-Fi, Ethernet
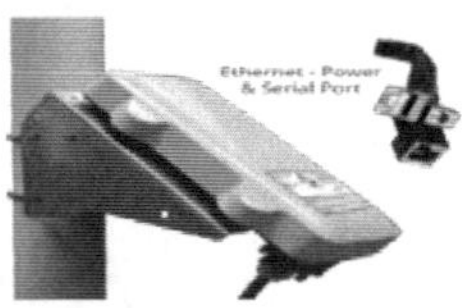 Satellite Data Transmitter (INMARSAT)	• High speed data transmission services. • Communication is achieved by interfacing RTU to a satellite transmitter, which in turn transmits the data from the field site to a receiving station.	• Specially design for remote unmanned SCADA application • Qualified for special INMARSAT m2m pricing plans. • Build in GPS receiver. • Remote configuring and debugging.
 Waterlog Radar Series – SDI-12 Water Level Sensor	• Accurately measure lake levels through non-contact transmission.	• Measuring distance: 3 m to 40 m • Frequency range 26 Ghz • Continuous operation, no warmup or "lock on"

The following criteria were adopted in designing and installing monitoring equipment in the study area.

4.1.1. All sites with telemetry should have a daily automated power reset to override communications lock-ups. This can be done with a relay or within the logger program

4.1.2. Due to the difficult access of these sites, remote interrogation for troubleshooting and making changes to the program as well as diagnostic analysis should *be enabled via telemetry*

4.1.3. It is recommended that if a dataset fails to transmit from site the system should automatically retry and send all files that have not been sent previously, not just the last file

4.1.4. Data transmission fail rate, satellite signal strength, site voltage, and electric fence voltage should be logged and transmitted to the database to provide an understanding of what the fault may be before a site visit is undertaken

4.1.5. For water level sites an automatic gas line purge is recommended on a daily basis to clear the orifice of sediment and debris

4.1.6. All issues after site commissioning should be addressed as soon as possible into the warranty period rather than leaving until the end and possibly *handing over legacy problems*

4.1.7. All instrumentation should be accessed safely and easily without the need for portable ladders and other height safety equipment

There are currently 5 telemetered rainfall stations and 4 non-telemetered rainfall stations in Temengor. The TNB Hydrology division conducts manual download from data logged at non-telemetered sites once every 3 months. The hydrometeorological stations in Temengor are shown in Figure 3. Currently, the telemetered hydrographic network is mainly confined to the north part of the catchment and does not have coverage to the south.

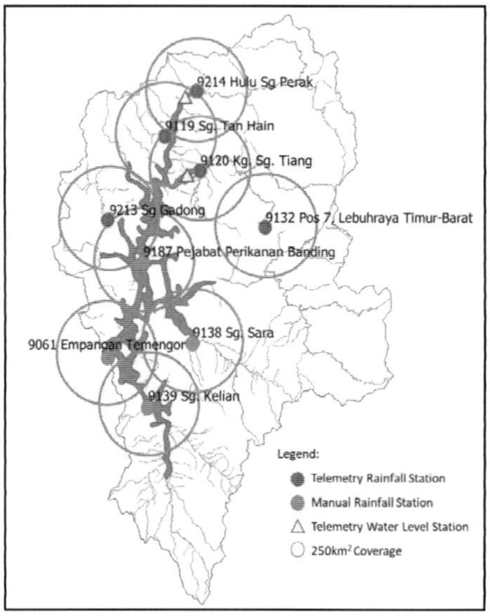

Figure 3. Temengor catchment telemetered hydrographic network

Difficulties arise when the monitoring locations are in high density forested areas. Site accessibility is crucial to ensure continuous maintenance activity. In this project, the site selection also needs to take into account surrounding flora and fauna, and approval from local authority is required to install the hydrographic stations as the site is within a reserved area.

The existing rainfall network was first analysed since sufficient rainfall data of sufficient coverage is needed to ensure success of the monitoring and forecast system. The analysis was made based on recommended minimum density of rain gauge stations for various physiographic types [19]. In this project, three existing rainfall stations were upgraded to a telemetry system (Station 9119, 9120 and 9132) and two new telemetry rainfall stations (Station 9213 and 9214) were installed to provide additional rainfall network coverage. The upgraded and new telemetered rainfall stations (Station 9119 and 9213) mostly cover rainfall occurring over the northern part of the catchment, storms occurring towards the east and south of these stations are as yet unrecorded. The locations of proposed upgrading and new installations of rain gauges were decided by an analysis of the existing data. The historical rainfall records show that the gauge 9132 located to the east part of the catchment generally records the highest rainfall, as illustrated in Figure 4. It is therefore important to provide more coverage to the east of the catchment. Nevertheless, good coverage of rainfall stations is also required to the south of the catchment, as this will ensure any storms occurring in the south are not missed and, importantly, help provide a better estimation of the spatial distribution of rainfall occurring in the catchment.

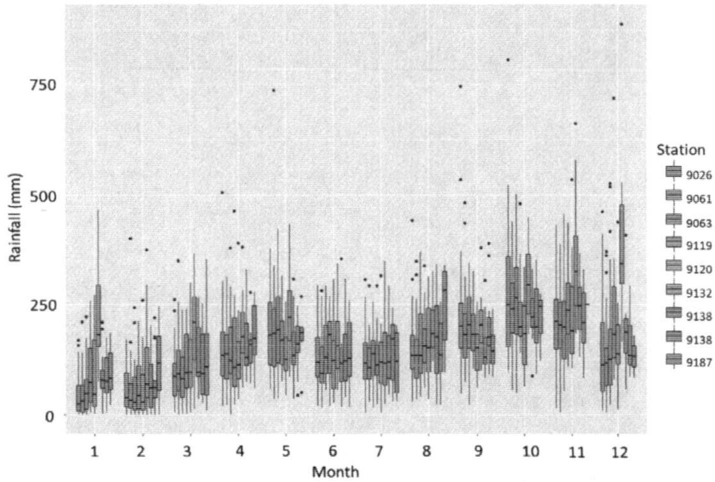

Figure 4. Distribution of monthly rainfall totals over the 12 months

The existing lake level sensor provides sub-hourly monitoring of the Temengor reservoir levels and the turbine discharge. However, the lake level data recorded by the sensor were noisy and needed to be upgraded. Thus, through this project, a radar water level sensor unit was installed on a small mounting bracket on the intake structure directly above the water surface as shown in Figure 5.

Figure 5. Lake Level Monitoring Location

Data transmission methods include terrestrial, radio and satellite-based solutions. The selection of the most suitable transmission platform depends on site suitability, system coverage availability, initial cost and operation & maintenance cost. Satellite data transmission technology has been successfully used for remote sites. The technology is able to address the difficulties in the management of hydrometric network in the Amazon [20] where telephone lines are not always present and the quality of radio transmission may be compromised. Additional challenge to this work scope is communication facilities for the transmission of data. There are no available radio and GSM network at these locations. Therefore, a communication system using satellite technology, Inmarsat, was used for data transmission from monitoring stations.

Electrical fencing was installed to protect the instrumentation at the site from being disturbed by animals and local tribesmen. All rainfall and water level stations were installed with electrical fencing for each site. An electrical fence system comprises an energizer and an insulated fence. The energizer transmits short pulses of electrical energy onto the fence line, in the form of short electrical pulses for a very short duration. The electric fence is not only a physical barrier, but is also a strong psychological barrier. Figure 6 shows a picture of a typical station.

IOP Conf. Series: Earth and Environmental Science **257** (2019) 012009 doi:10.1088/1755-1315/257/1/012009

Figure 6. Completed Electrical Fencing for existing Sg. Tan Hain Rainfall Station

4.2 Presentation of Monitoring Data

As this project aims for the operation of Temengor hydropower reservoir, the hydrometeorological monitoring data and forecast simulations are displayed through a website. The website is configured with private access by TNB engineers. The Temengor Inflow Forecast System Display (FSD), contains information on real-time measured variables and simulated outputs. Data from the hydrometeorological network are processed in a dedicated Ajenti Data Bridge server located at BGCC. The displayed variables include real-time rainfall, river level measurements, power generation and turbine release from SCADA system. Figure 7 shows a typical example of hourly rainfall data from 5 telemetry stations. The graphs shows hourly rainfall for a period of one-month, and the highest rainfall observed at Station 9120 and Station 9213, located at the upper catchment. The real-time raw rainfall inputs are read as cumulative rainfall totals at 15 minutes interval. The data will then be aggregated data to hourly time steps by automated configuration script and written in the form of 5 text files.

Figure 8 shows the power generation data, river level and reservoir level. This data is used by TNB in managing and planning hydropower plant operation. As shown in the graph, the river level corresponds with rainfall depth. With increasing inflows due to high rainfall, the graphs show increase in power generation, as flows are released through the turbines, as an effort to maintain safe reservoir level. The inflows, reservoir level and hydropower generation trend can be used as reference to optimize hydropower generation while maintaining the reservoir safety. Currently, Grid System Operator (GSO) and Single Buyer (SB) provides electricity demand information to the hydropower operators. Therefore, the availability the long term data is valuable to operation team for planning of operation rules. Furthermore, the advance reservoir level information of 5 days ahead through inflow forecast output will assist hydropower station operators and other related department in TNB to prepare short-term plan of optimized hydropower generation. These information are captured by SCADA system at BGCC that monitors, controls and set alarms at the generating facilities operating system from Sg Perak Power Stations [11]. Thus, it will complement the overall SCADA system objectives to ensure safe plant operation, lower generation cost, improve plants efficiency, and increase plant profitability. Data visualisation enables real-time monitoring of current condition in the catchment and any alert such as non-functioning stations, and the rainfall has exceeded the threshold value. For reservoir operations and decision making, model simulations results are also presented in graphical output displayed in the TIFS dashboard.

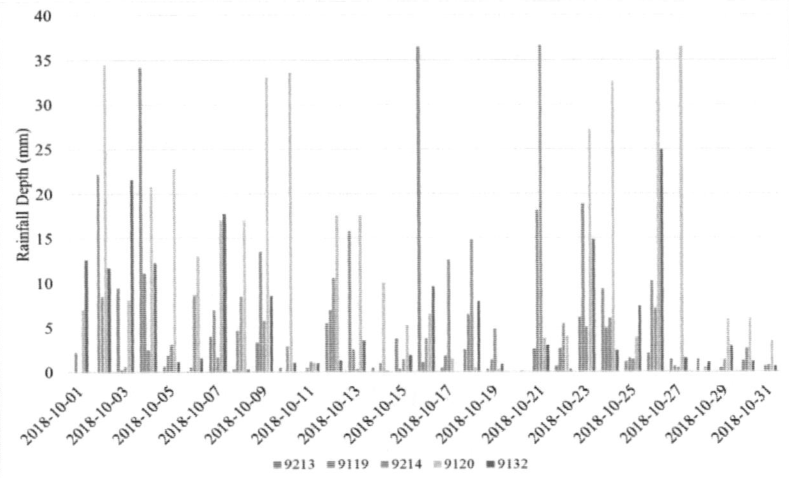

Figure 7. Rainfall Data from Five Rainfall Stations

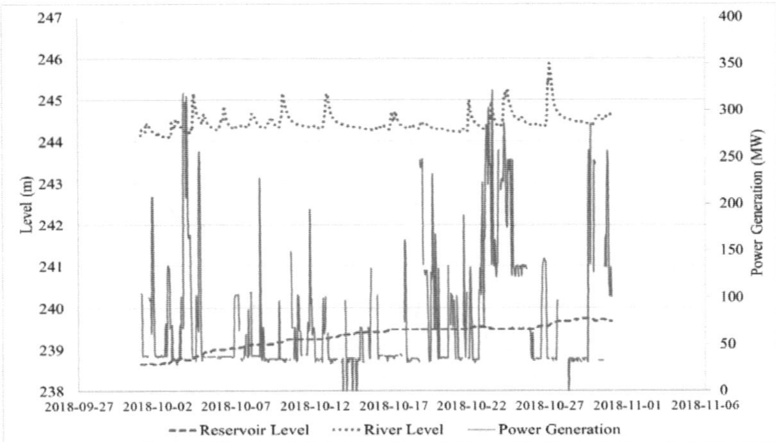

Figure 8. Typical plot of Power Output, Reservoir Level and River Level obtained from SCADA and telemetry stations

5. Conclusions

Installation of hydrometeorological monitoring equipment is an important element in implementing the hydropower reservoir monitoring framework. In any inflow forecasting system, reliable monitoring equipment is essential to provide advance information to dam owner to better planning and management of reservoir operations. Utilisation of advanced technology is beneficial in overcoming site constraint while providing the best information at site for modelling and monitoring purpose. Most importantly, it is critical to conduct periodical maintenance including equipment calibration of any equipment to ensure data reliability and longer equipment lifetime. This study demonstrates that a functional specification of equipment and system are essential for any operation, maintenance and management purposes. Additionally, continuous and strategic research is needed to verify our understanding and assumption of the underlying hydrological process occurring in the catchment. Wherever possible, research efforts are needed to support and communicate the findings for improved integrated hydropower reservoir operation and providing high reliability and availability of electricity supply throughout the country.

6. References

[1] Katole S and Bhute Y A 2017 Real Time Water Quality Monitoring System : A Review *International Journal on Recent and Innovation Trends in Computing and Communication* Vol 5 (India: Auricle Technologies Pvt. Ltd.) pp 21–5

[2] Rouen M et al 2005 *Reservoir Monitoring Freshwater Forum* Vol 23 (United Kingdom: Freshwater Biological Association) pp 20–37
[3] World Meteorological Organization 2009 *Manual on Flood Forecasting and Warning.* (Switzerland: World Meteorological Organization)
[4] Mishra A K and Coulibaly P Developments in Hydrometric Network Design : A Review 2009 *Reviews of Geophysics* Vol 47 (New York: Wiley) pp 1–24
[5] Van Dijk A I J M and Renzullo L J. 2011 Water resource monitoring systems and the role of satellite observations *Hydrology and Earth System Science* Vol 15 (Germany : Corpenicus Publications) pp 39–55
[6] Japan International Cooperation Agency 2010 Preparatory Survey on Nam Ngum 1 Hydropower Station Expansion in Lao PDR. *Final Repor*t Chapter 5 (Laos: JICA) p 5-1
[7] Yu P S et al 2014 Climate Change Impacts On Reservoir Inflows And Subsequent Hydroelectric Power Generation For Cascaded Hydropower Plants. *Hydrology Science Journal* Vol. 59-6 (London: Taylor & Francis Group) pp 1196-1212
[8] Cruz R C et al A System For Environmental Monitoring Of Hydroelectric Reservoirs In Brazil. A*n Interdisciplinary Journal of Applied Science* Vol. 8-1 (Brazil: Ambi-Agua) pp 6–17
[9] Levy J K et al Advances In Decision Support Systems For Flood Disaster Management: Challenges And Opportunities 2005 *Int. Journal of Water Resources Develepoment* Vol 21-4 (London: Taylor & Francis Group) pp 593–612
[10] Han K et al 2012 Hydrological Monitoring System Design and Implementation Based on IOT. P*hysics Procedia* Vol 33 (Dutch: Elsevier) pp 449 - 454
[11] Syazuan M K 2016 Sungai Perak Energy Management System: Improvement Of Efficiency And Reliability Of Sungai Perak Hydro Power Plant Through Modernization Of SCADA System *ARPN Journal of Engineering Applied Science* (USA: Asian Research Publishing Network) p 4036
[12] World Meteorological Organization 2013 Flood Forecasting and Early Warning. *Integrated Flood Management Tools Series* (Switzerland: World Meteorological Organization)
[13] Hamilton S 2012 *5 Essential Elements of a Hydrological Monitoring Programme* (Switzerland: World Meteorological Organization) Available from: https://public.wmo.int/en/bulletin/5-essential-elements-hydrological-monitoring-programme
[14] Hydrological Services 2014 *Instructional Manual Tipping Bucket Raingauge Model TB3* (Australia: Hydrological Services Pty Ltd)
[15] Environmental System & Services 2008 *pumppro 6150 User Manual* (Australia: Environmental System & Services)
[16] Aquas n.d. *PRO Series - RTU Controller & Data Logger User Manual* (Taiwan: Aquas Inc.)
[17] Addvalue Communications 2013 *SABRE™ Ranger M2M User's Guide.* (Singapore: Addvalue Communications Pte Ltd)
[18] Design Analysis Associate n.d *Model H-3611/3612/3613: Specification Sheet* (USA: Design Analysis Associate Inc.)
[19] World Meteorological Organization 2008 *Guide to Hydrological Practices.* 6th Edition Vol. I. (Switzerland: World Meteorological Organization)
[20] Callede J et al 2009 Use of satellite communication of hydrometric data for the whole of Brazil. *Hydrological Science Journal* Vol 41(London: Taylor & Francis Group) pp 903 - 913

Acknowledgements
The research is supported by Tenaga Nasional Berhad in collaboration with SSJ Sg Perak, TNB Research Sdn Bhd. and Entura Hydro Tasmania. The authors also would like to acknowledge the support from Malaysian Ministry of Higher Education in the form of Fundamental Research Grant Scheme 20170106FRGS.

Increased electron density and dissolved oxygen level in water through magnetic effect

Hwang Sheng Lee[1], Augustine Chung Wei Yap[1], Cho Cheng Ng[1], Nuruol Syuhadaa Mohd[2] and Joo Ling Loo[3]

1 Department of Mechanical and Material Engineering, Lee Kong Chian Faculty of Engineering & Science, Universiti Tunku Abdul Rahman (UTAR), Jalan Sungai Long, Bandar Sungai Long, Cheras, 43000 Kajang, Selangor, Malaysia
2 Department of Civil Engineering, Faculty of Engineering, University of Malaya, Lembah Pantai, 50603 Kuala Lumpur, Malaysia
3 Department of Mechatronics and BioMedical Engineering, Lee Kong Chian Faculty of Engineering & Science, Universiti Tunku Abdul Rahman (UTAR), Jalan Sungai Long, Bandar Sungai Long, Cheras, 43000 Kajang, Selangor, Malaysia
E-mail: leehs@utar.edu.my

Abstract. Water is essential in nature for the survival of all life with significant environmental and economic contribution to the world. The gradual increases in water demand and consumption have led to the discharge of inadequately treated wastewater, thereby, deteriorating water quality and triggering worldwide water quality problem. Water quality is determined by different parameters. In particular, electron density and dissolved oxygen (DO) level in water are important in monitoring and controlling the water property and quality, since both of them are associated with various chemical, biological, and physical processes in water treatment. This project developed a method to increase electron density and dissolved oxygen level in water using principle of reversed electric motor. The essential requirements to achieve this principle are magnetic field, mechanical motion by force, and electron generation. The treated water's pH, conductivity, and dissolved oxygen level showed increment of 0.40, 0.05 µS/cm, and 2.7 % saturation (0.22 mg/L), respectively whereas its oxidation-reduction potential (ORP) showed a decrement of 94 mV, indicating the increment of electron density and dissolved oxygen level in the magnetic treated water. The enhancements of these parameters are highly desirable in the applications of medical treatment, wastewater treatment, agriculture, and aquaculture industries.

1. Introduction

Water is one of Earth's most abundant natural resources as it comprises 70% of Earth's surface. It is the basic necessity for survival of all life with significant environmental and economic contribution. For examples, water is highly used in daily domestic consumption, agriculture, industries and energy production [1]. Furthermore, current global demand for water (4600 km^3 per year) would increase by 20-30% to approximately 5500-6000 km^3 per year by 2050 due to exponential global population rise [1]. As a result of the increasing demand and consumption of water, water quality begins to deteriorate and this creates severe water quality problems on a global scale. Therefore, water quality surveillance has been an important consideration in securing the clean water resource.

Water quality is determined by many parameters, depending on specific applications. Electron density and dissolved oxygen level in water are important in the surveillance and monitoring of water quality

Content from this work may be used under the terms of the Creative Commons Attribution 3.0 licence. Any further distribution of this work must maintain attribution to the author(s) and the title of the work, journal citation and DOI.
Published under licence by IOP Publishing Ltd

since they are associated with various chemical, biological, and physical processes in water treatment. Electron density in water plays fundamental role in many electron-driven processes and radical chemistry which induce significant effects in treatment of water and wastewater, radiolysis of water, remediation of wastes through photocatalytic process, denaturation of DNA, diagnosis and treatment of diseases [2-3]. In addition, electron density can also serve as an indicator of antioxidant property in water. Water containing electron with negative ORP is claimed able to combat oxidative stress due to free radicals at the cellular level by supplying more electron to the damaged cells [4-5]. However, the reaction mechanisms of radicals and ions created by electron in the water treatment system remain a challenge among research communities.

On the other hand, dissolved oxygen plays a vital role in water quality index. It can indicate the degree of water pollution by estimating the amount of organic matters and microorganisms in aquatic system through the indicators of biochemical oxygen demand (BOD) and chemical oxygen demand (COD) [6]. Many wastewater treatments involve microbial metabolic processes which require high dissolved oxygen level for enhancement of the biochemical reactions. Therefore, improved dissolved oxygen level in water is deemed important not only for wastewater treatment [6] that will enhance its treatment efficiency, operation cost and process stability, but also for agriculture [7-8] and aquaculture [9] in improving the breeding of farm animals, as well as stimulating the growth of plants and aquatic life.

Recent studies have reported the generation of electron in water through photoexcitation of plasmonic nanoelectrodes [10] and exposure to light or ionizing radiation [11-12]. In terms of improving the dissolved oxygen level in water, several conventional methods have been applied in industries to produce or supply oxygen in water through electrolysis, air diffusion, and mechanical aerators. Previous studies also reported that magnetic field was able to alter and enhance water parameters such as pH, conductivity, and dissolved oxygen [7-9, 13-14]. In this project, we developed a natural method to increase electron density and dissolved oxygen level in water using principle of reserved electric motor. The essential requirements to achieve this principle are magnetic field, mechanical motion by force, and electron generation. The principle of reversed electric motor states that when magnetic flux is disrupted by a mechanical force, current shall be induced. Hence, electron shall also be created and captured in the water stream. This can further lead to the increment of water's pH and the dissociation of hydroxyl ions for producing oxygen in water stream.

2. Methodology

2.1. Preparation of magnetic treated water

Water sample was obtained by distillation process from distiller system (Favourit W4L Water Stills). The magnetic treatment system comprised a series of magnets with magnetic field strength of 1000-1500 Gauss. Water stream was flowed through the magnets using a peristaltic pump (Masterflex L/S Economy Pump System with Easy-Load 7518-00 Pump Head) at flow rate of 1.0 ml/s.

2.2. Measurement of water parameters for magnetic treated water

The water parameters such as pH, oxidation-reduction potential (ORP), conductivity, and dissolved oxygen (DO) were measured before and after flowing through magnetic field. pH was measured using Ohaus Starter 3100 pH Bench meter equipped with STPURE electrode. ORP and conductivity were measured using Hanna Instruments HI 2550 Multiparameter pH/ORP and EC/TDS/Salinity Benchtop Meter equipped with ORP and EC electrodes, respectively. Lastly, dissolved oxygen was measured using Eutech DO 2700 meter equipped with EC620SSP BOD amperometric probe. In order to validate the effect of magnetic treatment on the flowing water, a control test was conducted by flowing the water under the same condition without magnetic effect.

3. Results and discussion

Table 1 shows the results of pH, ORP, electrical conductivity and dissolved oxygen for water after flowing through magnetic effect and without magnetic effect (control). The percentage changes were calculated based on equation (1) to indicate the increasing or decreasing trend of the measured parameters.

$$\text{Percentage change} = \frac{V_a - V_b}{V_b} \times 100\% \quad (1)$$

where V_a = value after magnetic treatment; V_b = value before magnetic treatment.

Table 1. Results of pH, ORP, conductivity and dissolved oxygen for water flowing through magnetic effect and without magnetic effect.

	Control	After magnetic treatment	Δ value (control)	Δ value (magnetic treated)	Percentage change (control)	Percentage change (magnetic treated)
pH	5.21	5.54	0.07	0.40	1.36	7.78
ORP (mV)	440	367	-21	-94	-4.56	-20.39
Conductivity (μS/cm)	1.56	1.60	0.01	0.05	0.65	3.23
DO (mg/L)	6.88	6.90	0.20	0.22	2.99	3.29
DO (% Saturation)	83.3	83.5	2.5	2.7	3.09	3.34

The comparisons of pH, ORP, conductivity and dissolved oxygen level for the control water and water flowing through magnetic effect are depicted in bar chart (Figure 1).

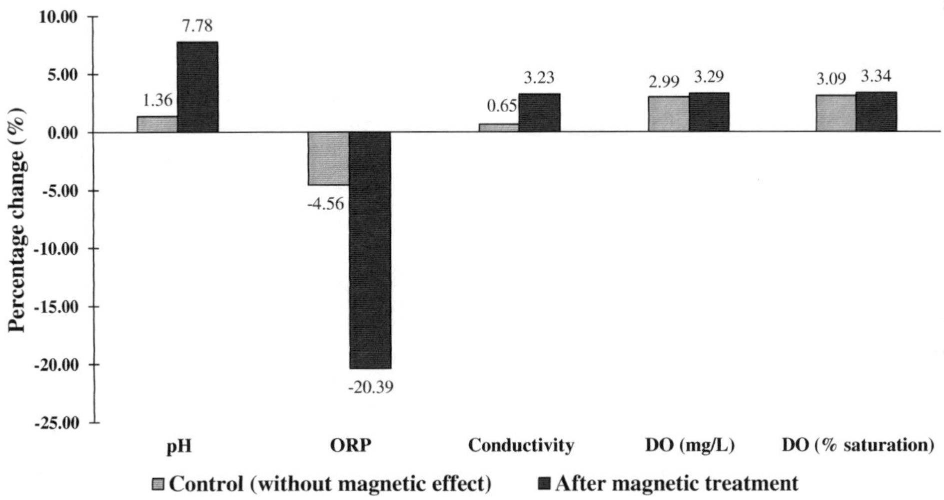

Figure 1. Comparisons of pH, ORP, conductivity and dissolved oxygen level for control water and water flowing through magnetic effect.

The results show that the pH, conductivity, and dissolved oxygen level of water are increased while ORP is decreased via magnetic effect. Most significantly, the ORP of water decreases by 20.39% (-94 mV) after subjected to magnetic effect, suggesting that the electron density in water stream is increased. ORP measures the water's potential to either donate or receive electron, indicating the availability of free electron and the oxidizing or reducing tendency of water [15]. The antioxidant property of water can then be estimated by measuring ORP [16]. When ORP of water decreases or becomes more negative, this implies that the water is a reductant and it contains excess electron. Further validation through conductivity measurement shows that the conductivity of water also increases by 3.23% (Figure 1). Since distilled water is a poor conductor that presumably does not have any impurities besides H^+ and OH^- ions, the rise in water conductivity could be highly due to the presence of electron in water. The phenomenon of free electron generation was also demonstrated by

Mohri et al. (2001) [17]. The resistivity of highly purified water, which is the reciprocal of electrical conductivity decreased due to increase in free electron after treated with a milli-gauss low frequency pulsed magnetic field [17-18].
The generation of electron in water can be explained using principle of reversed electric motor [13]. When the magnetic flux was disrupted by a mechanical motion of water, electron can be generated due to an induced current. In equilibrium condition, water molecules tend to dissociate based on equation (2), forming hydrogen cation (H^+) and hydroxyl anion (OH^-).

$$H_2O \leftrightharpoons H^+ + OH^- \quad (2)$$

Once electron is generated, the electron would have an affinity towards the H^+ cation forming metastable or weak bonding with H^+ cation due to the electron negatively-charged nature. This results in the increment of OH^- concentration and causes the pH of water increases. As a result, the equilibrium condition as shown in equation (2) is disturbed and based on Le Chatelier's principle, excess OH^- anions would further dissociate into H_2O, O_2 and electron as shown in equation (3), leading to the production of oxygen in the water stream. This was observed with the increment of dissolved oxygen level in water stream after the magnetic effect.

$$4OH^- \leftrightharpoons 2H_2O + O_2 + 4e^- \quad (3)$$

The increment of pH, electron density and dissolved oxygen level in water is highly dependent on the type of water sources and the contaminants or ionic concentration in the water. The results can postulate that higher the increment of pH in water, more electron will be captured and this can lead to more oxygen to be produced in the water.

4. Conclusions

The principle of reversed electric motor has been applied to increase electron density and dissolved oxygen level in water. Results showed that the ORP of water decreases, whereas the conductivity, dissolved oxygen level, and pH of water are increased after flowing through magnetic effect. The increased dissolved oxygen level and electron density of water is interrelated with the increased pH of water. These findings can pave the way for improving water quality and property, especially in enhancing dissolved oxygen level and electron density of water for applications in wastewater treatment, agriculture and aquaculture industries as well as medical treatment.

5. References

[1] Engin K, Michael T, Richard C and Stefan U 2018 *The United Nations World Water Development Report 2018 Nature-Based Solutions for Water* (Italy: United Nations World Water Assessment Programme) p 1–12
[2] Camaioni DM, Chipman DM, Johnson MA, Jonah CD, Kimmel GA, Miller JH, Rescigno TN and Xantheas SS 2003 *Understanding the Role of Water on Electron-Initiated Processes and Radical Chemistry* (Washington: Chemical Sciences Division, Office of Basic Energy Sciences, U.S. Department of Energy) p 1–69
[3] Alizadeh E and Sanche L 2012 *Chem. Rev.* **112** p 5578–5602
[4] Ignacio RMC, Joo KB, Lee KJ 2012 *J. Food and Drug Analysis*, **20**, p 394–397
[5] Ohta S, 2011 *Curr. Pharm. Des.* **17**(22), p 2241–2252
[6] Tu R, Jin W, Xi T, Yang Q, Han SF and Abomohra A E 2015 *J. Wat. Res* **86** p 132–138
[7] Niu WQ, Fan WT, Persaud N and Zhou XB 2013 *Pedosphere* **23** p 790–798
[8] Al-Hilali AH 2018 *Int. J. Poult. Sci.***17** p 78–84
[9] Hassan SM and Abdul Rahman R 2016 *J. Agric. Biol. Sci.***11** p 416–423
[10] Zilio P, Dipalo M, Tantussi F, C Messina G and De Angelis F 2017 *Light: Science & Applications* **6** p e17002
[11] P. Gaiduk A, Anh Pham T, Govoni M, Paesani F and Galli G 2018 *Nature Communications* **9** (247) p 1–6
[12] R. Siefermann K, Liu Y, Lugovoy E, Link O, Faubel M, Buck U, Winter B and Abel B 2010 *Nature Chemistry* **2** p 274–279

[13] Lee HS, Khoo KT, Low KO, Koon YY and Ong ZX 2017 *7th IWA-ASPIRE Conference 2017 & Water Malaysia Exhibition 2017* (Kuala Lumpur)
[14] Ibrahim IH 2006 *J. Sol.* **29** p 363–369
[15] Carlsson T and Muurinen A 2008 *Practical and Theoretical Basis for Redox-measurements in Compacted Bentonite A Literature Survey* (Finland: POSIVA OY) p 1–74
[16] Goncharuk VV, Bagrii VA, Mel'nik LA, Chebotareva RD and Bashtan SY 2010 *J. Water Chem. Techno.* **32** p 1–9
[17] Mohri K, Fukushima M and Matsumoto M 2001 *Trans. Mag. Soc. Jpn.* **1** p 22–26
[18] Esmaeilnezhad E, Choi HJ, Schaffie M, Gholizadeh M and Ranjbar M 2017 *J. Clean. Prod.* **161** p 908–921

Acknowledgement
We acknowledge Department of Mechanical and Material Engineering, Universiti Tunku Abdul Rahman (UTAR) in providing research facilities, and Ministry of Higher Education Malaysia (MOHE)'s Fundamental Research Grant Scheme (FRGS) with project number of FRGS/1/2017/TK10/UTAR/02/5 for financial support.

IOP Conf. Series: Earth and Environmental Science **257** (2019) 012011 doi:10.1088/1755-1315/257/1/012011

Study on urban ecological sustainability based on emergy theory-taking Zhenjiang city as an example

Junxue Zhang[1,2]

[1]School of Architecture, Southeast University, Nanjing 210096, PR China
[2]Key Laboratory of Urban and Architectural Heritage Conservation, Ministry of Education, PR China
E-mail: zhangjunxue2017@foxmail.com

Abstract. With the rapid development of the economy, the deterioration of the ecological environment problem has followed and the city's ecological assessment has become increasingly important. This article uses a quantitative ecological analysis method-emergy method to assess the ecological status of the city. The research object is Zhenjiang City. Through the calculation of a series of ecological sustainability parameters about Zhenjiang City, the city's ecological system is quantitatively evaluated. The results show that: 1. there is little change in renewable resources from 2012 to 2016 and a rapid increase according to the value of non-renewable resources; 2. the environmental load rate (ELR) has rapidly increased and the urban ecological sustainability has not improved; 3. Through Zhenjiang's GDP and GDP per capita grew rapidly from 2012 to 2016, it did not achieve satisfied results about industrial restructuring. In summary, the development in recent years has not achieved the expected effect and sustainable ecological development of Yangzhou City. The research significance of this article can provide a reference for the Zhenjiang municipal government to formulate better policies. Secondly, it has positive significance for the formulation of other urban policies.

1. Introduction

Along with the rapid development of China's economy, the city has experienced an explosive growth mode. The excessive use of non-renewable resources (such as petroleum and coal resources) for the city, not only cause a shortage of energy in China, but also generate numerous urban problems, even national pollution problems.

In the face of increasingly severe urban environmental issues, domestic and foreign experts have focused on the study of urban ecological sustainability, ensuring that people enjoy the economic development results while enjoying the city's good ecological environment.

This paper selects Zhenjiang City as the research object in the Yangtze River Delta region. The reasons for the selection of target city include: based on the economic sector of Jiangsu Province in terms of economic scale, Zhenjiang city is a typical medium level city.

Zhenjiang City has set a strategy for sustainable development, but there is still a tendency that sacrifices the ecological environment to develop the economy.

For this reason, the municipal government has to make specific policy adjustments to ensure that the proportion of agriculture remains unchanged, while minimizing industrial the proportion and enhancing the proportion the service industry.

The paper selects the emergy method to make the ecological assessment of Zhenjiang City. The emergy method is a world-recognized ecological assessment method [1-7], which can quantitatively

Content from this work may be used under the terms of the Creative Commons Attribution 3.0 licence. Any further distribution of this work must maintain attribution to the author(s) and the title of the work, journal citation and DOI.
Published under licence by IOP Publishing Ltd

evaluate the ecological changes of cities and provide verification basis for the implementation of government departments' policies.

2. Emergy Method and System Diagram

2.1. Emergy method introduction

The emergy is a typical "embodied energy". With the sun as the initial energy source, each object has a corresponding energy value and the unit is solar joules (sej). The emergy method originally came from the ecological field. It was a new ecological assessment method established by H.T.Odum of American ecologist in the 1980s and 1990s. It is widely used in agriculture, urban, industrial and ecological fields [8]. After 30 years of development, a complete theory has been formed and it has become a reliable ecological assessment method.

Professors of the University of Pennsylvania have integrated the emergy assessment methods into the urban field, providing a new way to explore green assessments. At present, the emergy research is developing rapidly in the urban field, involving diversification of research objects, refinement of method indicators, and localization of unit emergy value.

The basic steps of emergy analysis include basic data acquisition and data collection; emergy system diagram; emergy analysis table; emergy comprehensive index calculation; system development comprehensive evaluation and strategy analysis.

The basic calculation formula is as follows:

Solar emergy (sej) = energy (J) × unit emergy value (sej / J);

Solar emergy (sej) = substance (g) × unit emergy value (sej / g);

Solar emergy (sej) = substance ($) × unit emergy value (sej / $);

2.2. Zhenjiang city ecosystem map

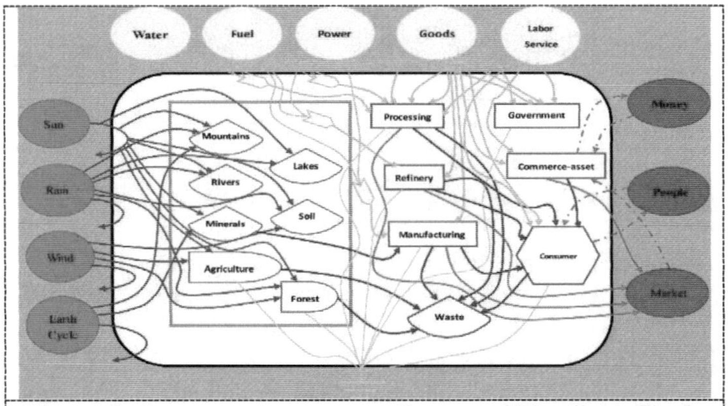

Figure 1. the emergy assessment diagram of Zhenjiang City in China.

The precondition for the evaluation of the emergy assessment is to establish the evaluation boundary conditions. Figure 1 is a schematic diagram of the emergy evaluated boundary of Zhenjiang City in China.

The left side of Figure 1 shows the types of renewable energy, including: the sun, rain, wind and geothermal energy, which are the basic natural energy sources of the city. The upper column is a non-renewable source, including: water, fuel, energy, products and services. At the right is the output, including currency, people, and market. In addition to these three items, there is also the flow of recycling system. All basic natural and industrial types of Zhenjiang city model are inside the boundary.

The diagram establishes the boundary conditions for the system assessment and provides a guarantee for the sustainability assessment of the city [9].

2.3. Emergy sustainability assessment indicator

There are 10 emergy assessment indicators, as shown in Table 1.

The value of Geobiosphere Emergy Baseline (GEB) is 12E+24 sej/year that comes from research by other scholars [10].

Table 1. Emergy assessment indicators.

index	meaning	unit
R	Renewable Energy	Sej
N	Non-renewable energy	Sej
Y	Total energy	Sej
F	Input and output emergy	Sej
Ep	Per capita emergy	Sej/person
Em	Emergy value	Sej/money
Ec	Emergy density	Sej/m2
ELR	Energy load ratio: (N + F)/R	~
EYR	Emergy yield ratio: Y/F	~
ESI	Emergy sustainability indicator: EYR/ELR	~

3. Results and Discussion

3.1. Emergy assessment of urban ecology

Table 2 shows the statistical data of the main energy and products of Zhenjiang City and the unit emergy value of various energy types. It can be divided into three categories: renewable data, non-renewable resources and others. The specific contents are shown in Table 2.

Table 2. The emergy calculation of main energy and products in Zhenjiang City

Type [10]	unit	2012	2013	2014	2015	2016	UEVs[11-13]
Renewable							
Sun radiation	J	4.15E+23	4.57E+23	4.24E+23	4.88E+23	4.03E+23	1.00E+00sej/J
Wind energy	J	7.88E+15	7.67E+15	7.56E+15	7.29E+15	7.65E+15	1.50E+03sej/J
Rain energy	J	6.36E+15	6.55E+15	6.33E+15	6.84E+15	6.48E+15	1.80E+04sej/J
Tidal energy	J	3.27E+15	3.02E+15	3.51E+15	3.16E+15	3.56E+15	1.00E+04sej/J
Non-renewable							
Coal	J	8.84E+08	5.49E+09	7.52E+09	9.49E+09	2.04E+10	1.00E+04sej/J
Oil	J	1.71E+19	3.64E+13	6.46E+13	8.68E+13	2.86E+14	6.60E+04sej/J
Electricity	J	2.46E+17	6.58E+17	8.48E+17	1.02E+18	1.47E+18	1.88E+08sej/J
Cement	kg	5.41E+09	6.17E+09	7.56E+09	8.92E+09	1.59E+10	2.59E+12sej/kg
Steel	kg	1.64E+09	1.75E+09	2.68E+09	3.35E+09	3.73E+09	6.97E+12sej/kg
Plastic	kg	1.14E+08	1.23E+08	1.46E+08	1.59E+08	1.81E+08	5.76E+12sej/kg
Paper	Kg	8.69E+07	2.71E+08	4.98E+08	7.23E+08	2.57E+08	2.38E+12sej/kg
Food	Kg	8.54E+08	8.32E+09	8.79E+09	9.26E+09	1.19E+10	8.30E+04sej/kg
Aquatic products	Kg	2.98E+07	3.69E+07	5.56E+07	7.69E+07	9.85E+07	2.00E+06sej/kg
Others							
Input money	$	5.34E+11	6.94E+11	9.83E+11	1.91E+12	2.15E+12	8.67E+12sej/ $
Output money	$	5.23E+11	7.41E+11	9.79E+11	2.91E+12	4.43E+12	1.45E+12sej/ $
Waste water	Kg	1.64E+10	4.55E+10	7.64E+10	9.22E+10	1.66E+11	8.60E+05sej/kg
Waste	Kg	8.69E+07	2.54E+08	4.05E+08	6.92E+08	7.89E+08	1.80E+06sej/kg

Based on the data of Table 2, through the basic calculated formula of emergy, the evaluation indicators of Table 3 are obtained, and the sustainability evaluation of Zhenjiang City's ecological development can be carried out through 10 emergy assessment indicators, so as to provide a strategic support of urban development.

Table 3. Sustainability index calculation results of Zhenjiang City

Index	2012	2013	2014	2015	2016	Unit
R	1.77E+25	1.63E+25	1.73E+25	1.57E+25	1.69E+25	sej
N	1.19E+26	1.42E+26	1.76E+26	2.13E+26	2.76E+26	sej
Y	1.33E+26	1.61E+26	1.89E+26	2.23E+26	2.77E+26	sej
F	1.43E+25	1.69E+25	1.84E+25	2.06E+25	2.46E+25	sej
Ep	1.08E+19	2.09E+19	4.64E+19	7.35E+19	8.45E+19	sej/person
Em	1.53E+14	2.01E+14	4.36E+14	6.49E+14	7.29E+14	sej/capita
Ec	1.59E+16	3.57E+16	4.06E+16	5.22E+16	7.21E+16	sej/m^2
ELR	7.9	10.4	10.9	19.1	16.5	none
EYR	0.93	0.95	1.02	1.13	1.63	none
ESI	0.12	0.08	0.09	0.06	0.09	none

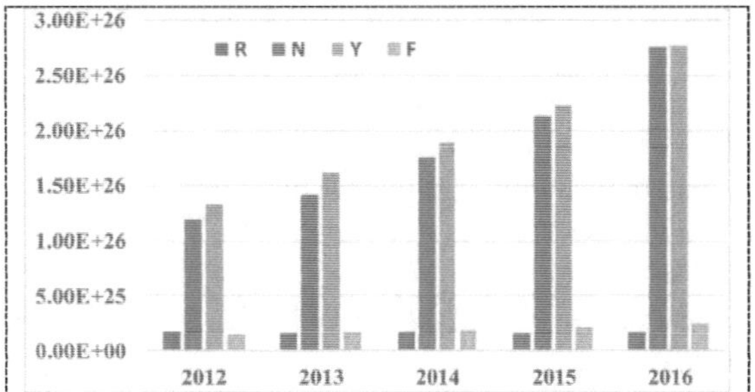

Figure 2. Emergy value changes of basic data in Zhenjiang City (2012-2016).

As the basis of the emergy calculation, the basic data needs to be sorted and calculated in advance, including: renewable resource R, non-renewable resource N, total emergy quality Y, and input and output emergy value F. The specific values are shown in Table 3. The basic data of figure 2 shows the trend in Zhenjiang from 2012 to 2016. It can be seen from Figure 2 that the renewable resources have not changed much, which is caused by the limitation of natural conditions; the amount of non-renewable resources increases with time, indicating that the city is developing rapidly and urban energy and materials are increasing rapidly. With the growth of non-renewable resources, the total emergy also increased greatly. The import and export situation showed an increasing trend from 2012 to 2016, but the change was not obvious.

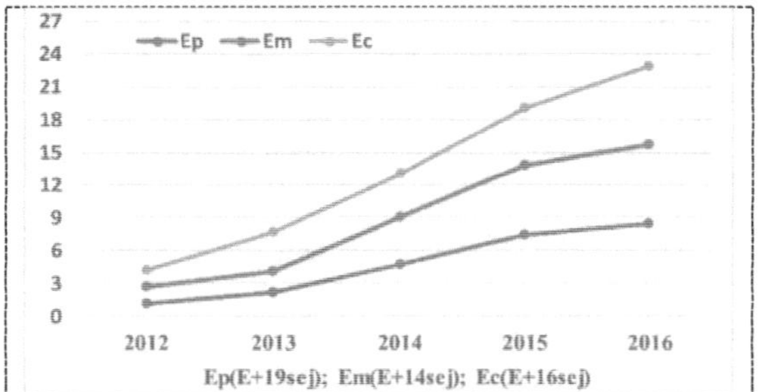

Figure 3. Average emergy of Zhenjiang City from 2012 to 2016

Figure 3 shows the mean emergy for the five years from 2012 to 16 years, including the per capita emergy, the emergy money value and emergy density. The three data are all increasing from 2012 to 2016. As been shown in Figure 3, the change of emergy density is the largest. This is that because the area of Zhenjiang city is fixed and the increase in emergy money value is directly corresponded to the emergy density. For per capita emergy and emergy money value, although the population is growing and GDP is increasing, the rate of emergy increment grows more, causing the same rapid growth in the emergy money value and per capita energy.

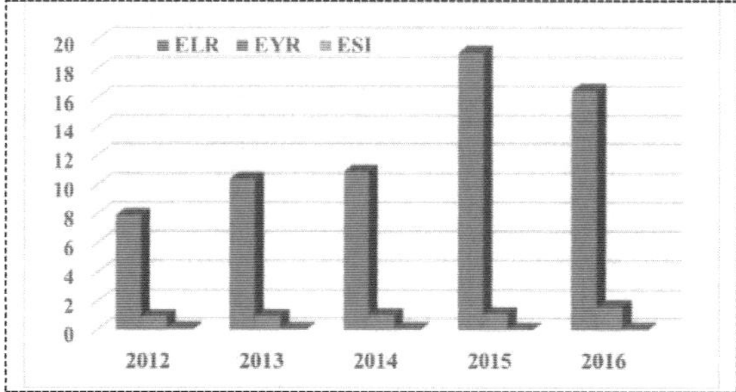

Figure 4. Evaluation of Ecological Sustainability Indicators in Zhenjiang City.

As the final indicators of the emergy evaluation, the emergy load ratio, the emergy yield ratio and the ecological sustainability indicator constitute the basic elements of the emergy method assessment. Figure 4 shows the ecological sustainability indicators of Zhenjiang City from 2012 to 2016. The most obvious market value load is increasing. The most obvious is the emergy load increment. It is the maximum value of 19.1 in 2015 and slightly decreased in 2016 and the emergy yield ratio also shows an increasing trend, but the change is not obvious. The indicators of the ecological sustainability are very low in figure 4. The indicators in Table 3 are 0.12, 0.08, 0.09, 0.06 and 0.09, respectively. The phenomenon indicates that ecological strategy of Zhenjiang city needs to be adjusted in order to improve ecological sustainability.

3.2 Urban analysis based on Gross Domestic Product (GDP)

The emergy evaluation has a direct relationship with the industrial result of the city. The proportion of the industrial structure will affect the evaluated result using the emergy method. Therefore, this paper uses the GDP proportion analysis in different industries to verify the analysis results of the above emergy assessment.

Table 4 the GDP from 2012 to 2016 (hundred million yuan)

Year[9]	GDP	Primary industry	Secondary industry	Tertiary Industry	(ten thousand yuan)
2012	2649.19	116.56	1430.56	1102.06	83636
2013	2982.33	131.22	1580.64	1270.47	92626
2014	3252.38	123.59	1661.97	1466.82	102651
2015	3502.48	140.48	1715.98	1647.26	110351
2016	3833.84	134.16	1893.51	1801.48	128803

Table 4 shows the GDP from 2012 to 2016, including the GDP share and GDP per capita of the three industries. The specific values are shown in the detailed statistics in Table 4.

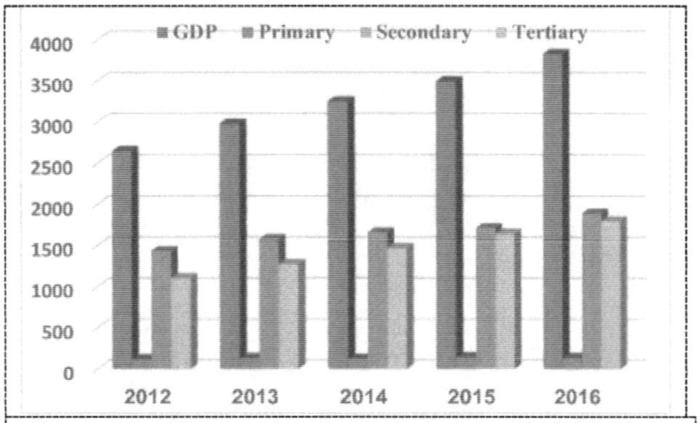

Figure 5. Changes in the proportion of GDP about three major industries from 2012 to 2016

The total GDP of Zhenjiang City shows an increasing trend and the added value is obvious. The change illustrates that the non-renewable energy in Zhenjiang City increases rapidly in figure 5, which is consistent with the trend of emergy in figure 2.

Among the three major industries, the proportion of the primary industry (agriculture) has not changed much. This is caused by several reasons, including the hardware conditions of the city and the correct guidance of the government. The second industry (industry) also shows an increased trend. However, the increased trend is smaller than that of the tertiary industry (service). This is a clear result of the government's industrial adjustment. The result is consistent with the analysis of the results in figure 3.

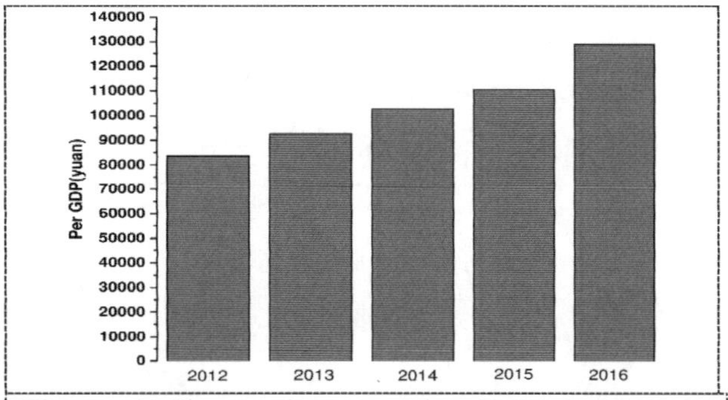

Figure 6. Changes in per capita GDP of Zhenjiang City from 2012 to 2016

Figure 6 shows the trend of per capita GDP, which the increased trend is obvious. It demonstrates that non-renewable of Zhenjiang City resources are invested heavily, driving the rapid growth of substances and the economy, which is consistent with the trend of per capita emergy in Figure 3.

3.3. Sensitivity analysis

As a vital aspect of emergy method, sensitivity analysis should be used to assess the uncertainty, in order to determine the biggest influencing factors. Non-renewable sources are the most energy consumption for the city and the results of sensitivity have been shown in Figure 7, Figure 8, Figure 9, Figure 10, Figure 11.

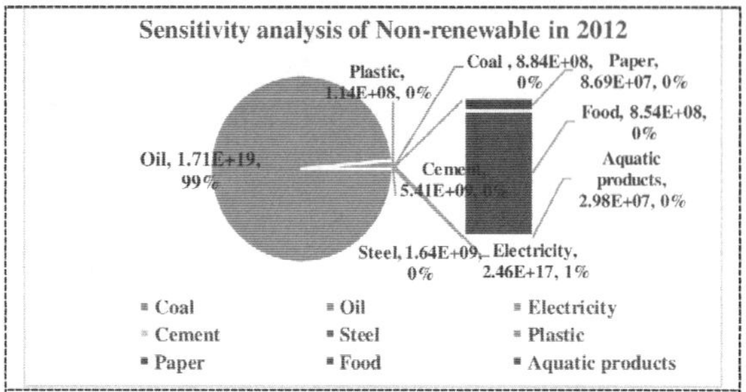

Figure 7. Sensitivity analysis of non-renewable in 2012

Based on emergy method, urban ecological analysis of this paper includes 9 non-renewable sources, involved with coal, oil, electricity, cement, steel, plastic, paper, food and aquatic products. In 2012, the main power is oil for Zhen Jiang city and the fact can be found in Figure 7. Others have very low emergy.

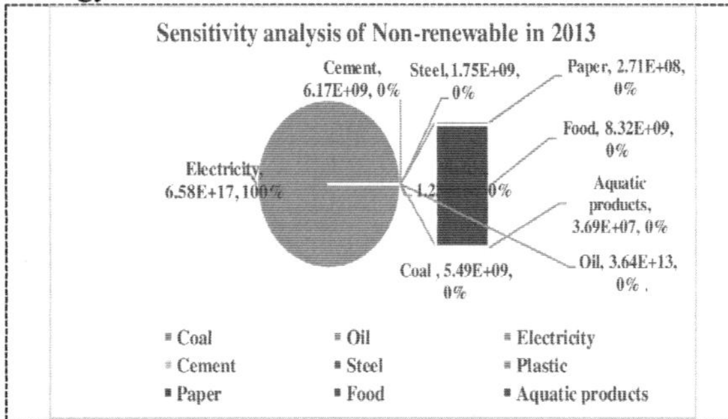

Figure 8. Sensitivity analysis of non-renewable in 2013

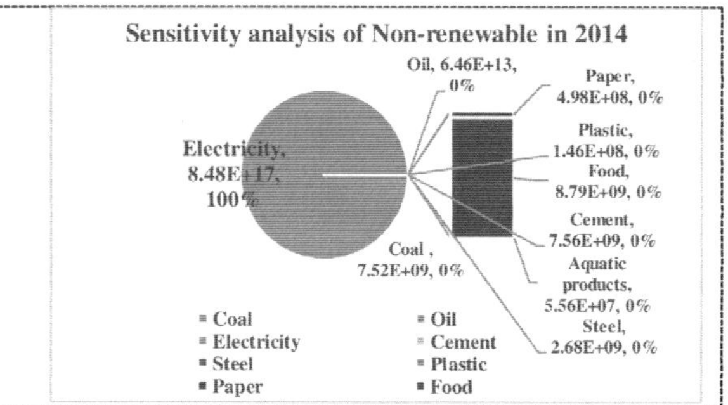

Figure 9. Sensitivity analysis of non-renewable in 2014

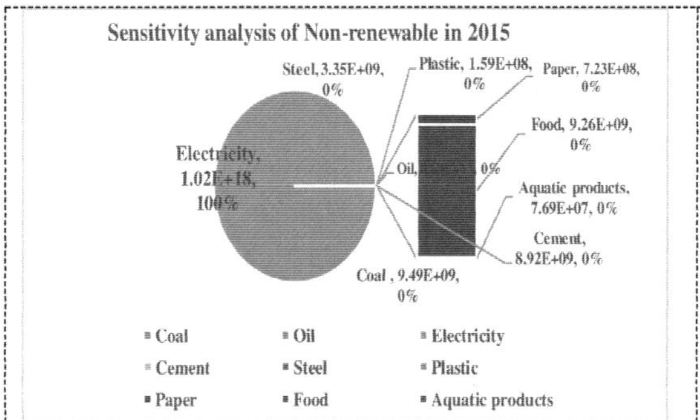

Figure 10. Sensitivity analysis of non-renewable in 2015

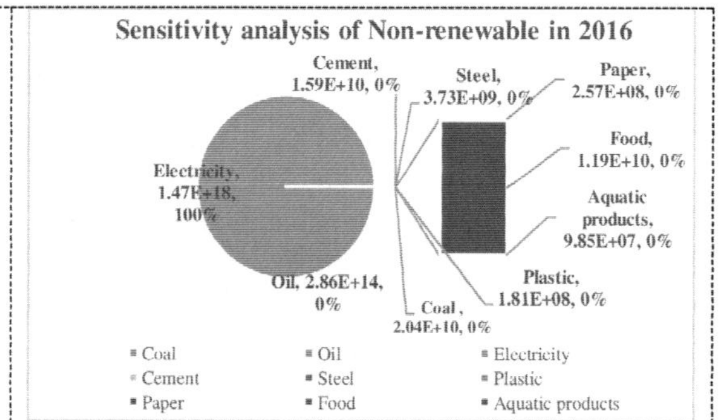

Figure 11. Sensitivity analysis of non-renewable in 2016

In addition to 2012 year, the electricity is the largest source of emergy and the results can be seen in Figure 8, Figure 9, Figure 10 and Figure 11. Electricity is the main impact factor of sensitivity analysis of non-renewable for Zhen Jiang city. In order to keep the accuracy of emergy analysis, electricity data should be collected carefully and analyse seriously.

4. Conclusion

Through the ecological sustainability analysis and the analysis of GDP verification of Zhenjiang City from 2012 to 2016, some results can be got, as following:

1. From 2012 to 2016, there had been little change in renewable resources and non-renewable resources and import and export emergy grow rapidly.

2. With the development of Zhenjiang City, the environmental load ratio (ELR) increased rapidly and the urban ecological sustainability did not been improved.
3. From 2012 to 2016, Zhenjiang's GDP and per capita GDP increased rapidly, but the industrial structure adjustment target did not obtain satisfactory results.

In short, the development from 2012 to 2016 did not reach the expected effect and ecological sustainable development in Zhenjiang City.

The research significance of this paper can provide reference for Zhenjiang Municipal Government to formulate better policies. Secondly, it has positive significance for other urban policy formulation.

Conflicts of Interest

The author declares that there is no conflict of interest regarding the publication of this article.

5. Acknowledgements

The work described in this paper was supported by the Postgraduate Research & Practice Innovation Program of Jiangsu Province (KYCX18_0104).

6. Reference

[1] Ling Shao, G.Q.Chen. Renewability assessment of a production system: Based on embodied energy as emergy. Renewable and Sustainable Energy Reviews, 57(2016)380-392.
[2] Yi-Bo Zhao, etc. An emergy-GIS method of selecting areas for sponge-like urban Reconstruction. Journal of Hydrology 564 (2018) 640 – 650.
[3] S. Mellino, M. Ripa, A. Zucaro, S. Ulgiati, An emergy–GIS approach to the evaluation of renewable resource flows: a case study of Campania Region, Italy, Ecol. Model. 271 (2014) 103-112.
[4] Jiang, Y., Zevenbergen, C., Ma, Y., 2018. Urban pluvial flooding and stormwater management: a contemporary review of China's challenges and "sponge cities" strategy. Environ. Sci. Policy 80, 132–143.
[5] Chang Liu, etc. Comparative Analysis for the Urban Metabolic Differences of Two Types of Cities in the Resource-Dependent Region Based on Emergy Theory. Sustainability 2016, 8, 635.
[6] Junying Zhang, etc. Quantifying the emergy flow of an urban complex and the ecological services of a satellite town: a case study of Zengcheng, China. Journal of Cleaner Production 163 (2017) S267-S276.
[7] Kampeng Lei, Lu Liu, Dan Hu, Inchio Lou, 2016. Mass, energy, and emergy analysis of the metabolism of Macao. Journal of Cleaner Production, 114 (2016) 160-170.
[8] Odum. H. T. Environmental Accounting, 1996. Emergy and Environmental Decision Making [M] NewYork: John Wiley, 1996. 32—34.
[9] Brown, M.T., and Ulgiati, S., "Energy quality, emergy, and transformity: H.T.Odum's contributions to quantifying and understanding systems," Ecological Modelling, 178(1), 2004; pp. 201-213.
[10] Brown, M.T., Campbell, D.E., De Vilbiss, C., Ulgiati, S., 2016. The geobiosphere emergy baseline: a 8synthesis. Ecol. Modell. 339, 92-95.
[11] http://tjj.zhenjiang.gov.cn/
[12] Brown, M.T., Herendeen, R.A., 1996. Embodied energy analysis and emergy synthesis: a comparative view. Ecol. Econ. 19, 219-235.
[13] Brown, M.T., Ulgiati, S., 1997. Emergy-based indices and ratios to evaluate sustainability: monitoring economies and technology toward environmentally sound innovation. Ecol. Eng. 9, 51-69.

Quantitative analysis on the sustainable development of four municipalities in China

Q R Shen[1], H Tian[1], X Y Han[1], H Zhang[2,*] and W Sun[1,*]

1 Key Lab of Membrane Science and Technology, College of Chemical Engineering, Beijing University of Chemical Technology, Beijing 100029, China
2 School of Chemistry and Chemical Engineering, Southwest University, Chongqing, 400715, China
E-mail: sunwei@mail.buct.edu.cn

Abstract. Due to the geographical location, development history and many other social-economic factors, the economic development of each city displays significant difference from each other. It is hard to characterize the economic development of a city by only looking at one or two measurements, which is only one aspect of a city. A snap shot of a city with all of its data information could provide its holographic image. With its evolution along time, the image can be even more impressive, but is still hard to compare with its peer cities given its multivariate nature. From system point of view, no matter how complicate the appearance of a system is, it will be always associated to its intrinsic characteristics. In this work, data from four municipalities in China are analysed quantitatively to discuss different development patterns by Pearson Correlation Coefficient and Hierarchical Clustering Analysis methods, correlation of factors and urban development patterns are clearly obtained and explained.

1. Introduction

Sustainability usually refers to the balance among economic development, social development and environment protection within a geographic region, especially the interaction between economic development and environment protection attract more attention along with the rapid urbanization in China. At the beginning, economic progress dominated while the environment cost was neglected. As the negative effect of environment became significant, such as poor air quality, people start to pay attention on the development planning with the consideration of environment constraints. During the urbanization, industry development is almost equivalent to the economic development, which gets even truer when industry is generalized to include agriculture as the first category of industry. Gross Domestic Product, GDP, is usually the measure of economic development. As the outcome of a city development, GDP is usually related to the energy cost, the composition of industry, the workforce, sometimes, the total population, while it is also associated to environment cost, most importantly, air pollution. Due to different development history and geographic condition, it is hard to evaluate the development of a city by a unique index, but the quantitative analysis still can reveal certain characteristics for a city when more data are available from various sources.

Data analysis methods are well-developed in commercial and engineering applications, among which Pearson correlation (PC) is a straightforward method to quantifying the linear relation of pair variables [1]. Zhou et al proposed an enhanced particle filter based on PCC to compare the difference of observation path and true path of particles and further provided validation that the degeneracy and sample disadvantage can be solved [2]. Xia et al evaluated the performance of thermal power unit by successfully applying PCC to assess analog signal [3]. Liu et al used PCC to identify the main factors

Content from this work may be used under the terms of the Creative Commons Attribution 3.0 licence. Any further distribution of this work must maintain attribution to the author(s) and the title of the work, journal citation and DOI.
Published under licence by IOP Publishing Ltd

that influence willingness to buy low-carbon products [4]. However, the correlation can only be investigated in each pair variables by PCC, while all variables are inevitably correlated together respect to a city development. In order to extract the correlation among multiple variables, clustering analysis is another straightforward choice. According to availability of object attributes, there are two clustering methods in practice, supervised clustering and unsupervised clustering, typical represented by k-means and hierarchical clustering analysis, HCA. Once the attribute of data is known, the data can be properly labeled, so called labeled data, and vice versa. Given the data from different cities, no priori information can be used to supervise the clustering, HCA is more suitable to analyse the data without labeling. Beaver et al analysed weather and atmospheric data by HCA, and successfully obtained the correlation between weather pattern and local air quality [5]. Ma et al classified nine districts of Beijing into three groups by HCA based on economy index such as industrial and agricultural income [6].

In this work, PCC and HCA analysis are applied to multiple data sets regarding economic development and environment conditions from four municipalities in China. The paper is organized as follows: the data is described in Section 2; the methods are shown in Section 3; the results are presented and analysed in the Section 4; the conclusion is given in Section 5.

2. Data

2.1. Data source

The data are obtained from the web of National Bureau of Statistics of China (http://data.stats.gov.cn/).

2.2. Details of data

The data considered in this work in 2012-2016 are listed in Table 1, including GDP, Urban population density, pollutant emissions, etc., from Beijing, Shanghai, Tianjin and Chongqing.

Table 1. Data information.

Variables	Description	Unit
A	GDP	USD
B	Fossil energy consumption	Ton
C	Electric power consumption	Kw·h
D	Urban population density	per square kilometer
E	Sulfur dioxide emissions	Ton
F	Nitrogen oxide emissions	Ton
G	Soot emissions	Ton
H	Number of vehicles	/

The definition of eight variables in the Table 1 was described as follow:

A- Gross Domestic Product (GDP) refers to the sum of all the final goods and services produced in a period of time. GDP estimates are commonly used to determine the economic performance of a country or region.

B- Fossil energy consumption refers to the amount of energy consumed, including coal, Coke, crude oil, gasoline, kerosene, fuel oil, natural gas.

C- Electric power consumption refers to the total annual electricity consumption in the region.

D- Urban population density refers to the density of population in urban areas which is calculated as the sum of urban population and urban temporary population divided by urban area.

E- Sulfur dioxide emissions refer to the sum of industrial SO_2 emissions and domestic SO_2 emissions over a period of time.

F- Nitrogen oxide emissions refer to the total NOx mass discharged into the atmosphere by the enterprise during fuel combustion and production processes over a period of time.

G- Soot emissions refer to the total mass of smoke, dust and industrial dust discharged into the atmosphere by the enterprise during fuel combustion and production process over a period of time.

The soot or industrial dust emission can be obtained by multiplying the exhaust air volume of the dust removal system and the dust concentration at the exit of the dust removal equipment.
H- Number of vehicles refers to the vehicles registered with the public security traffic administration department over a period of time and licensed as vehicles in certain area.

3. Methodology

Pearson correlation coefficient (PCC) and Hierarchical clustering analysis (HCA) are well applied in practice, and employed in this work. They are briefly introduced in this section.

3.1. Pearson correlation coefficient (PCC)

In statistics, the Pearson correlation coefficient is a measurement of the linear correlation between two variables *X* and *Y*, which was proposed by Karl Pearson [7].
Given two sample sets, *X* and *Y* with the same set sizes, Pearson correlation coefficient is defined as:

$$\rho_{X,Y} = \frac{\mathrm{cov}(X,Y)}{\sigma_X \sigma_Y} \quad (1)$$

where ρ refers to the Pearson correlation coefficient, cov is the covariance of *X* and *Y*, σ_X is the standard deviation of *X*, and σ_Y is the standard deviation of Y.
For one dataset $X=\{x_1, \ldots, x_n\}$ containing n values and another dataset $Y=\{y_1, \ldots, y_n\}$containing n values, r is calculated as:

$$r = \frac{\sum_{i=1}^{n}(x_i - \bar{x})(y_i - \bar{y})}{\sqrt{\sum_{i=1}^{n}(x_i - \bar{x})^2}\sqrt{\sum_{i=1}^{n}(y_i - \bar{y})^2}} \quad (2)$$

where *n* is the sample size, x_i and y_i are the individual sample points indexed with *i*, and $\bar{x}, \bar{y}$ are the sample mean. The value of *r* is between −1 and +1, where −1 represents total negative linear correlation, 0 means no linear correlation, and 1 is corresponding to total positive linear correlation.
By PCC, the correlation degree between pair variables can be calculated and the corresponding economic-environment interaction can be investigated.

3.2. Hierarchical clustering analysis (HCA)

In the method of hierarchical clustering analysis, clusters are represented through a dendrogram with node layers, where each node represents a cluster. There are two main approaches of hierarchical clustering, agglomerative and divisive, which are shown in Figure 1. For the agglomerative hierarchical clustering, each sample is initially considered as one cluster, and subsequently pairs of cluster are merged, while in divisive hierarchical clustering it starts with one cluster that includes all samples, and then recursive splits are performed. In this work, the agglomerative cluster approach is used. The most common type of hierarchical clustering algorithm proceeds as follows:
(1) Treat every single node as one cluster and calculate each pair of nodes by using Euclidean distance, which is shown as follows:

$$\mathrm{dist}(X,Y) = \sqrt{\sum_{i=1}^{n}(x_i - y_i)^2} \quad (3)$$

where x_i is the variable in cluster *X*, y_i is the same variable in cluster *Y*, and *n* is the total number of variables.
(2) Find the smallest value (closest Euclidean distance) from above and agglomerate them as a new cluster.
(3) Calculate Euclidean distance of the pairs of this new cluster from (2) and other clusters.
(4) Repeat (2) and (3) until all clusters are agglomerated as one.

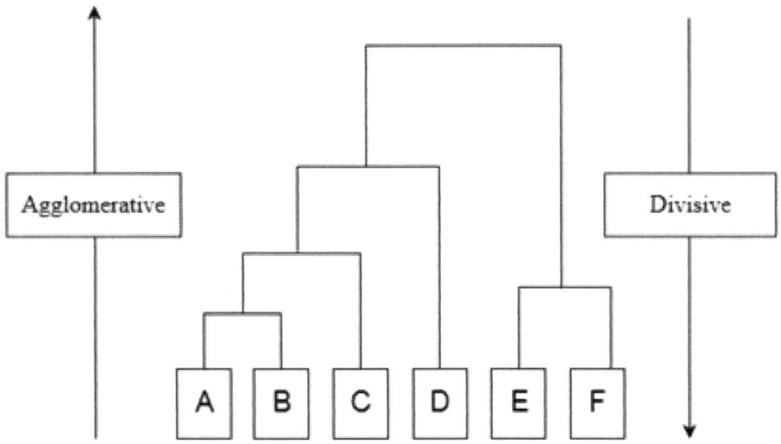

Figure 1. Hierarchical cluster analysis.

For hierarchical clustering analysis in Figure 1, there are six samples from A to F. A and B are merged to form a cluster while E and F to form another one. Then the cluster of A and B are combined with C to form a cluster and D is added to form a new cluster. Finally, this cluster united with the cluster of E and F to be one cluster.

The HCA algorithm evaluates the similarity of two clusters by calculating the Euclidean distance between them. The smaller the distance, the higher the similarity. A hierarchical clustering tree is formed by combining the two most similar clusters and repeat the process. In a clustering tree, the original data points of different categories are the lowest layer of the tree and similar clusters are combined to form upper layers. Based on HCA result, the similarity between two clusters is identified, which can be applied to the analysis of city development data to compare the pattern of each city.

4. Results

4.1. Pearson correlation coefficient

Table 2. The correlation coefficients among different variables of Beijing.

r	a	b	c	d	e	f	g	H
a	1	-0.9866	0.9894	-0.6490	-0.9485	-0.9807	-0.9820	0.9681
b		1	-0.9701	0.6191	0.9236	0.9567	0.9789	-0.9524
c			1	-0.7263	-0.9724	-0.9890	-0.9877	0.9601
d				1	0.8558	0.7806	0.7631	-0.5143
e					1	0.9912	0.9818	-0.8688
f						1	0.9914	-0.9163
g							1	-0.9208
h								1

As shown in Table 2, the Pearson correlation test of Beijing data shows negative correlations in 15 groups of variables, among which, the correlation coefficients of 12 groups of variables, such as GDP and fossil energy consumption, GDP and soot emissions, electric power consumption and sulfur dioxide emissions are close to 1. On the other side, there are 13 groups of variables with positive correlation, among which, 10 groups of variables such as GDP and electric power consumption, fossil energy consumption and sulfur dioxide emissions, urban population density and number of vehicles are strongly correlated.

It is easy to understand that GDP and number of vehicles, Electric power consumption and Urban population density are positively correlated, but some result of Pearson correlation test in Table2 needs more investigation for an acceptable explanation. In Beijing, negative correlation exists between GDP and fossil energy consumption, Urban population density, pollutant emission. The possible reason could be that the increase of GDP is not contributed by the industry with fossil fuel as raw material. According to the statistics data, the economic identities of Beijing are mainly government service, finance, education, IT and other tertiary industries, in which electricity is the major form of energy consumption. In recent years, Beijing has strictly controlled population and the high-tech technology is developed faster. Driven by the high-tech technology and finance, GDP growth is increasing, leading Urban population density and GDP negatively correlated.

Table 3. The correlation coefficients among different variables of Shanghai.

r	a	b	c	d	e	f	g	h
a	1	-0.6904	0.8519	0.6480	-0.9631	-0.9771	0.0358	0.9985
b		1	-0.2193	-0.5119	0.6038	0.6356	-0.6525	-0.6834
c			1	0.4825	-0.8914	-0.8817	-0.4540	0.8485
d				1	-0.4948	-0.5229	0.4253	0.6456
e					1	0.9981	0.1570	-0.9515
f						1	0.1091	-0.9678
g							1	0.0420
h								1

As shown in Table 3, the Pearson correlation test of Shanghai data shows negative correlations in 14 groups of variables, among which, the correlation coefficients of 10 groups of variables, such as GDP and fossil energy consumption, urban population density and sulfur dioxide emissions, electric power consumption and nitrogen oxide emissions are close to 1. On the other side, there are 14 groups of variables with positive correlation, among which, 10 groups of variables such as GDP and electric power consumption, fossil energy consumption and sulfur dioxide emissions, nitrogen oxide emissions and soot emissions are strongly correlated.

Similarly, it's straightforward that GDP and electric power consumption, urban population density and the number of vehicles are positively correlated, but it is noticeable that there is a negative correlation between GDP and pollutants such as sulfur dioxide emissions and nitrogen oxide emissions. It could be interpreted that an inverted U-shape relation exists between economic development and environment pollutants. When economy develops, the environment firstly worsens and then becomes good, which is also called Environmental Kuznet Curve [8]. As Shanghai is a highly developed region, regulations and techniques regarding pollution control are also upgraded along with its economic development, therefore, it is not too difficult to explain why sulfur dioxide and nitrogen oxide emissions have an opposite trend with GDP changes.

Table 4. The correlation coefficients among different variables of Tianjin.

r	a	b	c	d	e	f	g	h
a	1	-0.8908	0.9338	0.9614	-0.8258	-0.9327	0.0651	0.8446
b		1	-0.6727	-0.9119	0.9190	0.9621	0.1981	-0.5124
c			1	0.8676	-0.6021	-0.7546	0.2815	0.9800
d				1	-0.7585	-0.8851	0.1833	0.7646
e					1	0.9718	0.3814	-0.4495
f						1	0.2185	-0.6161
g							1	0.3887
h								1

Pearson correlation test of Tianjin data is shown in Table 4, in which one can find negative correlations in 12 groups of variables, among which, the correlation coefficients of 5 groups of variables, such as GDP and fossil energy consumption, GDP and Nitrogen oxide emissions, electric power consumption and number of Vehicles are close to 1. On the other side, there were 16 groups of variables with positive correlation, among which, 8 groups of variables such as GDP and urban population density, fossil energy consumption and polluted gas emissions, urban population density and number of vehicles are strongly correlated.

Table 5. The correlation coefficients among different variables of Chongqing.

r	a	b	c	d	e	f	g	h
a	1	-0.9419	0.8782	0.9902	-0.8862	-0.9282	-0.5490	0.9991
b		1	-0.6727	-0.9737	0.9168	0.9389	0.6674	-0.9465
c			1	0.8046	-0.6179	-0.6853	-0.1843	0.8749
d				1	-0.9324	-0.9640	-0.6458	0.9888
e					1	0.9918	0.8651	-0.8739
f						1	0.8204	-0.9182
g							1	-0.5286
h								1

As shown in Table 5, the Pearson correlation test of Chongqing data shows a negative correlation in 16 groups of variables, among which, the correlation coefficients of 9 groups of variables, such as GDP and fossil energy consumption, urban population density and nitrogen oxide emissions, sulfur dioxide emissions and number of vehicles are close to 1. On the other side, there were 12 groups of variables with positive correlation, among which, 11 groups of variables such as GDP and urban population density, fossil energy consumption and nitrogen oxide emissions, urban population density and number of vehicles are strongly correlated.

Table 4 and 5 show that it's clear Tianjin and Chongqing are quite similar in correlation results, especially in the groups of electric power consumption and sulfur dioxide emissions, nitrogen oxide emissions and soot emissions. Correlation coefficients of the three pairs of variables obtained in Tianjin and Chongqing are larger than those in Beijing. This can be perfectly explained with the different industrial structures. In an industrial structure with high percent of tertiary industry like Beijing, electric power accounts the most of the energy consumption so that the need for heavy industry is comparatively low, as a result, electric power is highly negatively correlated with pollutants emissions, while in a relatively low percent of secondary and tertiary industry like Tianjin and Chongqing, the need for heavy industry is much more which results in a smaller absolute value in correlation with electric power consumption and pollutants emissions, i.e., less correlated.

4.2. Hierarchical clustering analysis

The hierarchical clustering result about the development of the four municipalities in 2012-2016 is shown in Figure 2. The information of each sample of the clustering result is shown in Table 6. It can be seen in Figure 2 that the development of Chongqing in 2012-2016 is different from other cities, the development of Shanghai in the recent years has been kept at a steady state, while that of Beijing and Tianjin in changes since 2016. Combined with the industrial ratio variation diagram (Figure 3-Figure 6), the secondary industry of Chongqing accounts for an important part, while the tertiary industry takes a leading role in Beijing and Shanghai. In 2016, tertiary industry of Beijing contributed more than 80 percent for the first time, which may be the reason that it is different from others, while the tertiary industry of Tianjin in 2016 grows greatly compared to that in 2012-2015.The development of Shanghai was steady from 2012 to 2016, with the secondary industry gradually decreasing and the tertiary industry gradually rising. The result above shows that the development of the secondary industry is still the pillar industry, but experience a slow decrease, especially in Beijing and Shanghai. The increase of tertiary industry is more independent of the fossil energy, on the other side, requires significant amount of electricity, which is free of local emission.

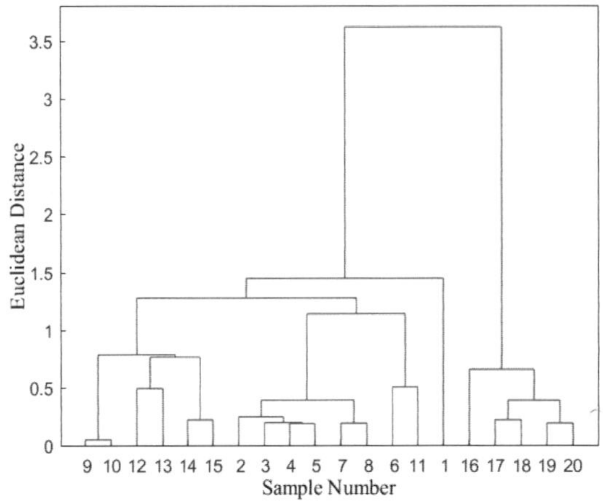

Figure 2. Hierarchical clustering analysis results.

Table 6. Sample description.

Sample	Description	Sample	Description
1	Beijing 2016	11	Tianjin 2016
2	Beijing 2015	12	Tianjin 2015
3	Beijing 2014	13	Tianjin 2014
4	Beijing 2013	14	Tianjin 2013
5	Beijing 2012	15	Tianjin 2012
6	Shanghai 2016	16	Chongqing 2016
7	Shanghai 2015	17	Chongqing 2015
8	Shanghai 2014	18	Chongqing 2014
9	Shanghai 2013	19	Chongqing 2013
10	Shanghai 2012	20	Chongqing 2012

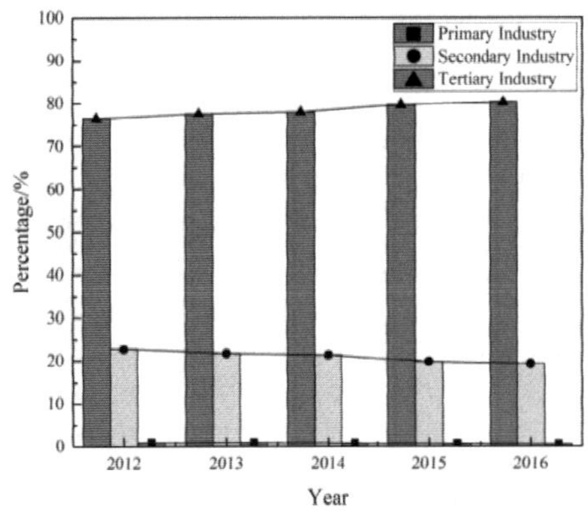

Figure 3. Industrial structure of Beijing.

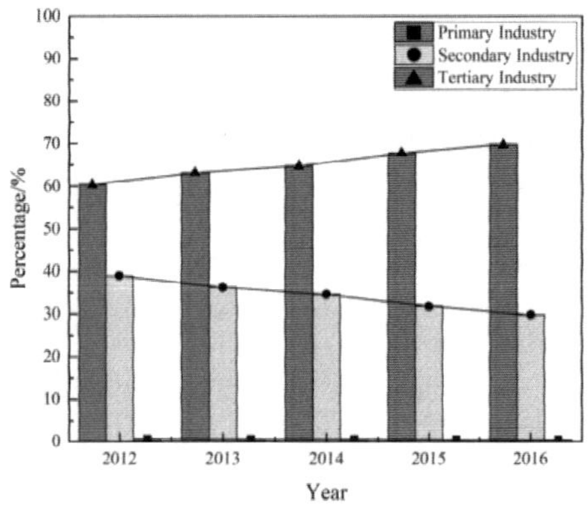

Figure 4. Industrial structure of Shanghai.

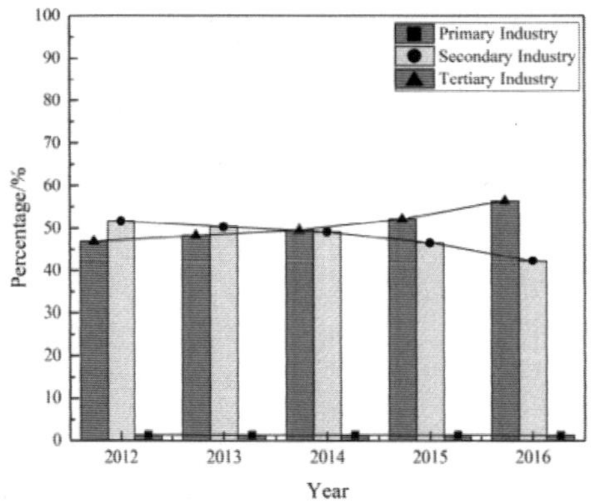

Figure 5. Industrial structure of Tianjin.

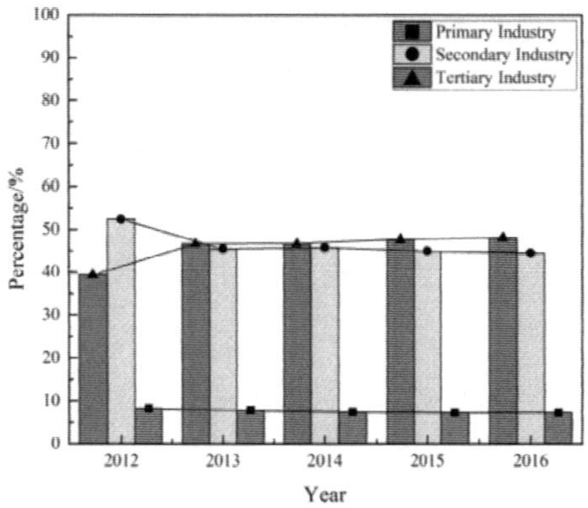

Figure 6. Industrial structure of Chongqing.

Primary and secondary industries are major part in industry structure but is confined by environment capacity because they rely heavily on resource supply and inevitably discharge to environment, while tertiary industry are mainly services, education, IT, etc. in which research and development are strongly motivated. Without any doubt, the tertiary industry provides technical support to the sustainable development of urban economy with no expense in environment. The tertiary industry is beneficial to coordinate industry development, improve the efficiency of resource allocation, and guarantee stability and economic development, in terms of society economy and environment. Due to the high pressure of metropolis such as overpopulation and overconsumption and limited natural resources, the best way for future development is to increase the proportion of tertiary industry, which is highly agree with the idea of sustainability.

5. Conclusion

In this work, the PCC and HCA are used to analyse the data of four municipalities from 2012 to 2016, and correlation within a city and urban development pattern were obtained respectively. These four cities (Beijing, Shanghai, Tianjin and Chongqing) show quite different development patterns in recent years, high-tech development of Beijing is more comprehensive. Shanghai has been kept a steady development with an increasing tertiary industry, while the secondary industry still play an important role. In Tianjin and Chongqing, the secondary industry still dominates, but their tertiary industry started to increase slightly. With the limitation of fossil fuel supply and the restriction of air emission, tertiary industry is becoming a better choice for the sustainable development in these municipal cities.

6. References

[1] Joseph L R and W A Nicewander 1988 Thirteen Ways to Look at the Correlation Coefficient *The American Statistician* **42(1)** 59-66
[2] Zhou H M, Deng Z H, Xia Y Q and Fu M Y 2016 A new sampling method in particle filter based on Pearson correlation coefficient *Neurocomputing* **216** 208-215
[3] Zhi X, Song Y X, Ma J, Zhou L J and Dong Z J 2017 Research on the Pearson correlation coefficient evaluation method of analog signal in the process of unit peak load regulation *In Electronic Measurement & Instruments (ICEMI) 13th IEEE International Conference on* pp 522-527
[4] Li Q W, Long R Y and Chen H 2017 Empirical study of the willingness of consumers to purchase low-carbon products by considering carbon labels: A case study *Journal of Cleaner Production* 161
[5] Beaver S and Palazoğlu A 2006 A cluster aggregation scheme for ozone episode selection in the San Francisco, CA Bay Area *Atmospheric Environment* **40(4)** 713-725
[6] Ma G, Li H X and Luo K 2005 Application of clustering in regional economy *International Conference on Electronic Commerce* pp 48-51

IOP Conf. Series: Earth and Environmental Science **257** (2019) 012012 doi:10.1088/1755-1315/257/1/012012

[7] Pearson K 2006 Note on Regression and Inheritance in the Case of Two Parents *Proceedings of the Royal Society of London* **58** 240-242

[8] Tajul Ariffin Masron and Yogeeswari Subramaniam 2018 The environmental Kuznets curve in the presence of corruption in developing countries *Environmental Science and Pollution Research* **25(13)** 12491-12506

IOP Conf. Series: Earth and Environmental Science **257** (2019) 012013 doi:10.1088/1755-1315/257/1/012013

Sustainable urban drainage as a viable measure of coping with heat and floods due to climate change

N W Chan[1] , M L Tan[1], A A Ghani[2] and N A Zakaria[2]

1 School of Humanities, Universiti Sains Malaysia, 11800 Penang, Malaysia
2 River Engineering & Urban Drainage Research Centre, Universiti Sains Malaysia, Engineering Campus, 14300 Nibong Tebal, Penang, Malaysia
E-mail: nwchan1@gmail.com

Abstract. Climate change is pervasive, bringing about heat and floods. To address these hazards, better coping, adaptation and resilience are needed. Cities all over the world suffer from heat and flood disasters that account for significant losses. This paper examines the incidence of heat and flood occurrence in urban areas, and examines how sustainable urban drainage systems (SUDS) is used to combat them. The methodology is based on historical event analysis, literature review and secondary data. Results indicate that heat and flood hazards in cities can be effectively reduced by using SUD systems which incorporate vegetated surfaces that absorbs and retains rain water, purifies stormwater and transfers heat. Results show that SUD system reduces flood peaks and heat, thereby reducing the incidence of flash floods and heat islands. SUD system is recommended as a viable method of heat and flood control in cities as cities using SUDS were found to cope very well with heat and floods. Overall, technical measures of flood management such as SUDS are found to be effective, affordable, aesthetically pleasing and socially acceptable. A holistic strategy combing technical SUDSs and non-technical human aspects of coping and resilience is the key towards effective management of floods in cities.

1. Introduction

Globally and nationally, climate change is considered the most pervasive environmental change currently affecting human society, especially in urban areas. In Malaysia, climate change has intensified two already deadly hazards, viz. urban heat islands (UHIs) and urban flooding. [1] The 21st century is witnessing increasing frequencies and magnitudes of these hazards as global climate change intensifies vis-à-vis rapid landuse change and urbanisation. More significantly, rapid urbanization has witnessed the unprecedented growth of cities. [2] Cities are now the preferred places of habitation, but are currently routinely affected by the impacts of heat and floods, but as climate change escalates, more severe heat and flood-related events are expected to occur. [3] [4] In order to reduce loss of lives and property damage, cities need to brace themselves by better coping and adaptation. More importantly, countries and cities need to increase their resilience to face climate-related disasters from the perspectives of water, land, energy, food and environmental sustainability. [5] As cities grow and encroach upon their hinterland, urban heat islands and flash floods are expected to intensify. For example, the larger the city, the larger is its heat island effect and the more devastating its floods. Almost every year, cities all over the world suffer from heat-related and flood-related disasters that account for a significant number of casualties, disease epidemics, property and crop damage and other intangible losses. [6]

Content from this work may be used under the terms of the Creative Commons Attribution 3.0 licence. Any further distribution of this work must maintain attribution to the author(s) and the title of the work, journal citation and DOI.
Published under licence by IOP Publishing Ltd

The term Sustainable Urban Drainage (SUD) is used to describe a type of urban drainage that provides ecological, environmental and social benefits. SUD generally takes care of the rain water or stormwater water quantity and quality (runoff quality), and the amenity value of surface water in the urban environment. This is in contrast to other types of urban drainage such as open type monsoon drains and open canals which can cause flash floods, water pollution, soil erosion, and mudflows which are unsustainable. In Malaysia, SUD can be implemented to solve environmental problems related to flooding, water pollution and excessive heat. Historically, flooding is one of the central issues affecting people and economic growth in Malaysia. [7] [4]. In Malaysia, deforestation followed by rapid land use change from forests into agriculture and then later into urban landscapes and urban sprawl over the last half a century has resulted in increased stormwater flow into receiving waters. This has increased flood magnitudes and frequencies, as conventional urban open drainage systems such as concrete monsoon drains fail to address flash floods and also cause degradation of stormwater quality. [8] Large and open monsoon drains are also exposed to garbage disposal as they are grossly polluted leading to receiving rivers being polluted as well. Consequently, SUD is needed to address flash floods via detention of rainwater at source as well as effective drainage of cities into receiving water bodies further downstream. Chan et al. (2016) [9] documented that sustainable urban drainage system (SUDS) is an innovative and aesthetic approach in the management of stormwater for urban land developments that replicate the natural drainage regime and characteristics. Via SUDS the runoff from rainfall is collected and stored temporarily to avoid flash flooding. This type of drainage also allows natural treatments to the rainwater at source prior to infiltration and the treated rainwater then flows into receiving waters via a controlled release. SUDS blends into the landscape surrounding the development and carry out water treatment with the aim to prevent pollution, flood control, groundwater recharge and environmental enhancement. Runoff is treated using enhanced natural mechanism such as filtration, infiltration, biological uptake, and settlement. It is important to understand how these techniques work together to provide the aims of sustainability in the most practical, cost effective and beneficial way. [10]

Furthermore, a SUD system that encompasses green areas, vegetation and green roofs is also effective in its cooling effect, thereby controlling heat and UHIs. Malaysia, as a tropical climate country, is already hot and humid with daytime temperatures between 30-37 degrees Celcius. Illyani Ibrahim and Azizan Abu Samah (2011) [11] showed that urbanization creates a landscape modification that leads to urban heat island in the city of Kuala Lumpur, and qualitative and quantitative analysis indicated that the land cover conditions influence the urban temperature leading to the formation of urban heat island. Norlida Mohd Dom (2016) [12] also showed that urban heat islands are contributing to climate change in Malaysia, and vice versa. Elsayed (2012) [13] examined the UHI of the City of Kuala Lumpur, and found that the intensity of the UHI of the city of Kuala Lumpur varies from 3.9 to 5.5 ° C. Comparing these values to that shown in a study by Sham (1987) [14] in 1985, the intensity was seen to have increased from 4 °C to 5.5 °C. In Penang, Nur Aili Hanim Hanafiah and Chan (2013) [15] found that George Town city was about 2 to 3 degrees Celcius warmer that its surrounding green areas. The Department of Irrigation and Drainage, Malaysia (DID) had launched the Urban Stormwater Manual of Malaysia (MSMA) in 2001 and the second edition of MSMA in August 2012. This manual promotes the application of Sustainable Urban Drainage System (SUDS) facilities like bioretention, swale, wetland and etc. Green roof is one of the SUDS facilities. However, it is not included in the latest MSMA. There is no legislation or rule provided for the green roof construction. Green roof when used together with other SUDS or BMPs facilities will intercept rainwater that can reach to the ground and consequently control the volume of rainwater on the ground. Besides, rain water quality also can be improved. According to findings by Ayub et. al. (2010),[16] a SUD system involving green swales, ponds, wetlands and green roofs provide numerous benefits such as attenuating stormwater runoff which lowers risks of urban floods and improving the urban water balance, acting as noise buffer and reduce air pollution. Moreover, green roof also creates new habitat for wildlife and biodiversity enhancement and reduction of heat in the microclimate of the area. Implementation of green roofs in SUDS in stormwater management would help not only reduce heat but also prevent floods and help increase water availability.

SUDS is consistent with objectives of stormwater management approach, controlling rain water at source and managing stormwater quantity and quality via controlling the runoff. The application of SUDS is aimed at solving three major problems commonly encountered in Malaysia namely flash floods, river pollution and water scarcity during dry periods. SUDS is an innovative way of managing stormwater in a sustainable manner without compromising the overall project cost compared to conventional open drainage system. SUDS reduces infrastructure costs, as well as maintenance costs. It's also beneficial to the environment due to its minimal impact on local ecology, and it preserves ecosystem values and functions, as well as provides benefits to society. SUDS is a viable way forward in managing water resources in the face of climate change. SUDS emphasizes a holistic approach to stormwater management, meeting the multi-objectives of runoff quantity, quality and public amenity aspects, rather than just quantity aspect of conventional approach (Figure 1). SUDS uses the concept of the stormwater management train, starting with prevention or good house-keeping measures for individual premises, and progresses through to local source control, larger downstream site and regional controls. Water could flow straight into a site control but as a general principle it is better to deal with runoff locally, returning the water to natural drainage system as near to the source as possible. Only if the water cannot be managed on site should it be conveyed elsewhere. This concept of treating and storing water locally is very beneficial for water conservation and has a huge potential on water reuse. Just as rainwater harvesting technique, SUDS can be successfully used to provide water source for non-potable use. The increased green area and enhance infiltration facilities are also important to soil moisture replenishment and ground water recharge at local level. [17] [8]

2. Methodology

The methodology is based on the following: (1) A literature review on Sustainable Urban Drainage; (2) A "Demonstration Project" on SUD in the Universiti Sains Malaysia (USM) Engineering campus whereby construction started in 2000 and the campus was completed in 2002. Data was collected on water quantity and quality after the construction of the campus was completed in 2002. Data from previous studies were documented and compared; and (3) A literature review on Urban Heat Islands and evaluation of SUD system on heat reduction, and the development of a physical model of green roof and rainfall simulator in the Universiti Sains Malaysia's Engineering campus. Results were then collected from the model to provide important data to be referred in the guidelines for the related or relevant agencies in order to implement the green roof in their developing areas.

3. Results and discussion

According to Zakaria et al. (2016), [8] there was a need for a paradigm shift in the way stormwater is managed in Malaysia in the early 2000s. The Malaysian government subsequently launched the *Urban Stormwater Management Manual for Malaysia* (MSMA) in 2000.[18] This manual was later revised and improved into a new-look manual titled the *2nd Edition of Urban Stormwater Management Manual for Malaysia* in 2011 [19] that incorporated the latest approaches in stormwater management. The introduction of MSMA marked a siginificant change in the approach of water engineers towards urban drainage and flood management, strategically transforming the stormwater management landscape in the country. The new approach gave rise to the increasing demands for green technologies to be incorporated into urban drainage and flood management that also addresses climate change issues. Engineers are now forced to consider urban drainage in a more holistic manner as they face multiple challenges to provide effective and sustainable drainage systems that can tackle multiple problems. The implementation of MSMA necessitated the need to blend new technologies or innovations into the design of drainage facilities which can provide multiple benefits rather than single benefit.

The USM Engineering Campus (Figure 1) was constructed in 2002 based on DID's MSMA approach in using the SUD system. This green and sustainable urban stormwater management system is better known as Bio-Ecological Drainage Systems (BIOECODS). It was designed by the USM's River Engineering and Urban Drainage Research Centre (REDAC), a research centre. This was the first large-scale construction using SUD system as the campus was a demonstration project to show-case the MSMA concept. BIOECODS attempt is aimed at addressing three major problems commonly encountered in Malaysia namely flash floods, river pollution and water scarcity. [8] The Design Concept of BIOECODS is based on the concept of SUDS, that fully complies with the MSMA requirements. The USM BIOECODS model comprises several important components that combine to form an effective stormwater treatment train that controls runoff quantity, preserves runoff quality, ensures safety, beautifies the landscape and reduces heat. Significantly, BIOECODS is designed to provide time for natural processes of sedimentation, filtration and biodegradation to occur, all of which reduce the pollutant loads in stormwater runoff. In addition, BIOECODS blends easily into its surrounding, adding considerably to the local amenity and/or local biodiversity. [17]

Figure 1. Typical Example of Constructed Swale in USM Engineering Campus.

In the experimental project of the USM Engineering Campus, the design of the SUD system is aimed at reducing the drainage footprint of the BIOECODS, to provide water treatment function, and to achieve aesthetic appeal. A dual layer conveyance system was introduced with a design where the surface of the swale is similar to a conventional open drain, but with water storage and treatment function that the conventional open drainage system does not possess. Typically, the surface layer resembles a grassed channel or a swale. The typical swale design has a gentle side slope, low gradient and shallow depth. Beneath the surface, the underground layer consists of a geosynthetic module enclosed in geotextile. The module is connected to the surface layer via a layer of river sand or infiltration media. Figure 1 (above) presents a cross-sectional view of the said ecological swale. During heavy rain events, runoffs initially build up on the swale surface and infiltrates into the underground module. This infiltration process provides both quantity and quality treatments to the runoff. First, the infiltration detains some of the rain water and delays flow. Then the infiltrated water is stored in subsurface module. Only after the pool of water generates enough energy will it flow downstream within the module. In the module, water is loss to adjacent ground through exfiltration of water from the side or bottom of the module. This water will percolate through the ground and are either retained as soil moisture or contribute to groundwater recharge. On the surface, swale attenuates flow by providing larger surface friction than concrete channel. [17]
In terms of water quality, the infiltration also filters the stormwater. Hence, the stormwater undergoes water quality treatment that involves three important processes. First, as water flows into the swale from impervious surfaces, grass on swale surface acts as filter media to trap out particulate pollutants. The aerobic condition of the soil promotes hydrocarbon breakdown. The second treatment involved is the infiltration of water through sand layer and into the module. Infiltration filters out particulates and some smaller solid nutrients that are attached to the runoff. The geosynthetic module is manufactured in such a way that the internal structure of the module helps to break up water flow, creating turbulence and therefore increase dissolves oxygen. Finally, both the surface and subsurface flows will combine again by both discharging into the ponds and wetlands system before discharging back into receiving waters of the Kerian River. There are also other components in the SUD system in the USM

Engineering Campus drainage design. These include "Dry Ponds", "Detention Ponds", and "Wetlands". During rain events, excess stormwater is also stored in the dry ponds which are constructed with a storage function. The dry pond is essentially a detention pond, which has been integrated with the ecological swale to temporarily store stormwater runoff. A module storage tank is placed beneath the detention pond. The outflow path of the storage module is connected to the ecological swale at the lowest point, in order to drain the dry pond system in less than 24 hours. Dry ponds diffuse flow conveyed by swales and the reduced stormwater velocities enable more effective sedimentation, filtration and infiltration water treatments. The grassed surfaces of dry ponds are able to infiltrate a substantial portion of the annual surface runoff volume due to the increased soil permeability that is created by the deep and fibrous root systems of the landscape vegetation (Figure 2). [17] Another component of the SUD system in the experimental project is the "Detention pond", which is the first community facility of the BIOECODS. The pond is primarily designed for attenuating runoff from developed areas through regulated outlet structures. The facility is typically designed to limit discharge to the pre-development stage, while storing water temporarily. On top of that, the detention pond also facilitates water quality treatment and ecological functions. With aquatic and wetland plants planted along the water fringes, the pond provides some water quality treatment and habitat for urban wildlife. Extended exposure to sunlight in a pond will also help to breakdown certain pollutants. Figure 3 shows the detention pond in USM Engineering Campus.

Figure 2. Typical Example of Constructed Dry Pond (Source: [17]).

Figure 3. Typical Example of Constructed Detention Pond (Source: [17]).

The USM BIOECODS has a wetlands component. These are considered as constructed wetlands. In terms of water quality improvements, the constructed wetland is designed as a community treatment facility. [20] Data collected showed that about 90% of the total volume of annual stormwater runoff will flow through an area supporting a healthy population of wetland plants. Water pollutants and other contaminants are shown to be removed either by direct absorption into plant tissues (soluble nutrients) or by physical entrapment and subsequent settlement on the bed of the wetlands. Apart from water quality, the wetland is also designed as a habitat area for biodiversity conservation within a development, supporting species such as small mammals, birds, fish, reptiles and plants. The end product is expected to improve the aesthetic value for surrounding areas at the most downstream end of the drainage system. [17][8]

The performance of BIOECODS based on research data collected in the USM Engineering Campus experimental site is remarkable. Since its establishment, BIOECODS has been closely monitored for its hydrological and hydraulic performances. The first BIOECODS system in USM engineering campus has been monitored with sophisticated water quantity and quality equipment to record its performance during rainfall events continuously for almost a decade. The most significant benefit of BIOECODS is its ability to reduce flow peak and flow volume. The retardation in ecological swale and detention in dry pond, wet pond and wetlands have enable BIOECODS to successfully create a stormwater system that mimics natural condition, hence reducing flood risks. Figure 4 shows an example of flow attenuation in a stretch of ecological swale while Figure 5 shows flow attenuation with detention pond during a rainfall event. Table 1 and 2 provide further examples of flow attenuation in swale and pond respectively.

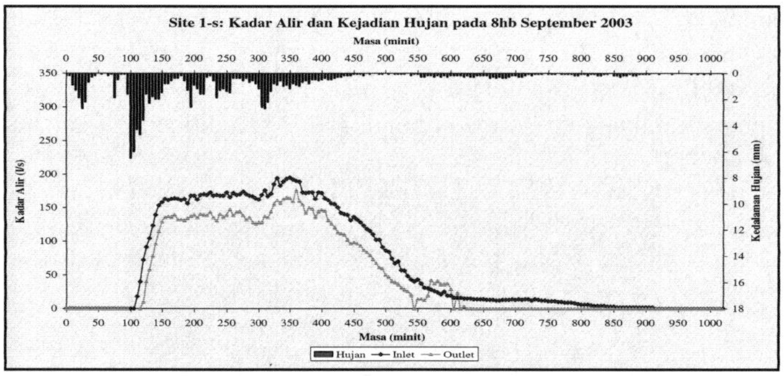

Figure 4. Example of Flow Attenuation of Ecological Swale (Source: [17]).

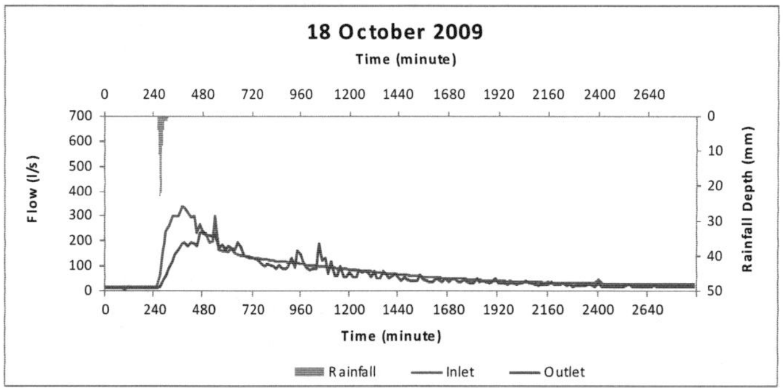

Figure 5. Example of Flow Attenuation by Detention Pond (Source: [17]).

Table 1. Example of Ecological Swale Performances for Frequent Rainfall Events

Precipitation Event	Average Rainfall Intensity (mm/hr)	Estimated (ARI)	Peak Flow (l/s)		Total Runoff Volume (m^3)		Percentage of Reduction (%)	
			Inlet	outlet	Inlet	outlet	Peak Flow	Runoff Volume
24/6/2003	11	3 months	128	91	418.5	246.6	28.9	41.1
30/8/2003	14.5	3 months	59	26	388.8	123.6	55.9	66.6
8/9/2003	13.8	5 years	195	176	4043.1	3043.2	10.0	24.1

(Source: [17]).

Table 2. Example of Detention Pond Performances for Frequent Rainfall Events

Rain Event	Total Time (min)	Rainfall Depth (mm)	Rainfall Intensity (mm/hr)	Peak Flow (m^3/s)		Volume (m^3)		Percent of Reduction (%)	
				Inlet	outlet	Inlet	Outlet	Peak Flow	Volume
14/4/2007	40	23.5	35.3	0.041	0.032	2,214	1,777	21.95	19.74
16/4/2007	70	23.2	19.9	0.034	0.026	1,545	1,200	23.53	22.33
18/10/2009	65	84.6	78.1	0.335	0.235	23,919	18,36	29.85	23.22
11/11/2009	155	171.5	66.4	0.689	0.289	38,859	18,87	50.65	51.44

(Source: [17]).

More detailed observations regarding the USM's SUDS are recorded in Ainan et al. (2004). [21] The performance of ecological swale was also verified using computer model by Abdullah et al. (2004) [22] using XP-SWMM. In both cases, the authors found that the system performed admirably to attenuate flow from the catchment. Recently, a water balance analysis by Ayub et al. (2010) [16] confirmed that the BIOECODS system actually increases groundwater recharge through infiltration. During drier days, percolated surface water eventually 'resurface' to supply much needed base flow to sustain plants and ecology within the constructed wetland. The data collected also showed substantial Water Quality Improvements. Apart from controlling the quantity of rain water flow (runoff), rain water and stormwater quality were also under control via the BIOECODS system. Regular and constant monitoring through in-situ test as well as laboratory testing for the common water quality parameters showed that the water quality is of a good class after flowing through the SUD system. The final discharge from the system is found to conform to Class IIB of National Water Quality Standard published by Department of Environment. Table 3 shows the results from a quick sample test for a 3-month ARI event on 19 April 2006. [17]

Results from this study showed that the use of this SUD system significantly reduces the pollutant loads especially particulate pollutants, i.e. sediment. For the entire system, TSS was non-detectable, indicating that even if sediments were washed into the system, they are trapped very early on by the ecological swale networks. With the use of detention ponds and wetlands, most biological activities are concentrated in this area, the biological load is rather higher, but still in Class IIB limit. However, the discharge from wetlands is significantly of better quality, indicating the success of biological treatment occurring within the ponds and wetlands. Similar testing in research by Mohd Sidek et al. (2004) [23] and Ayub et al. (2005) [24] confirmed the improvement in water quality as a result of BIOECODS.

Table 3. Water Quality Condition for Various Sampling Location of BIOECODS on 19 April 2006 (Source: Ab Ghani et al., 2004).

BIOECODS Components	Location	pH	DO (mg/l)	Temp (°C)	Turbidity (NTU)	BOD5 (mg/l)	TSS (mg/l)	TP (mg/l)	COD (mg/l)	NH3-N (mg/l)
Ecological Swale	Upstream	8.3	6	28	14	2	ND	0.1	34	0.1
	Downstream	7	6	25	23	2	ND	0.2	13	0.1
Wet Pond	Inlet	6.9	6	24	48	2	ND	0.2	7	0.1
	Outlet	6.5	7	26	74	2	ND	0.3	6	0.2
Detention Pond	Inlet	6.9	8	24	80	1	ND	0.5	22	0.4
	Outlet	6.8	8	25	10	1	ND	0.3	20	0.1
Wetland	Inlet	7	6	29	18	1	ND	0.1	13	0.1
	High Marsh	7.1	8	24	26	1	ND	0.2	16	0.1
	Micropool	7.1	7	23	15	1	ND	0.3	26	0.1
	Outlet	7.3	6	28	13	1	ND	0.3	19	0.1
Class IIB, NWQS		6 – 9	5 - 7	-	50	3	50	-	25	0.3

**note: ND denotes non-detectable

Ayub et al. (2010) [16] documented the monitoring of water balance within the constructed wetland of BIOECODS system for 2007 and 2009. It was found that the constructed wetland received considerable amount of direct precipitation as well as runoff inflow. Outflow records from the wetland however, showed significant reduction in volume, as compared to the combined inflow. Hence, water conservation and water reuse are significant benefits in BIOECODS. As evapotranspiration rate are almost constant (tropical climate) year round, the balance of water volume are translated into active storage and soil moisture or groundwater recharge. Since early 2010, a series of trials were conducted to investigate the potential water reuse in the BIOECODS system. In the first stage, the potential for non-potable use was explored. The trial involved withdrawal of water from BIOECODS recreational pond to support laboratory use in REDAC's Physical Modelling Laboratory. A simple water usage budget calculation revealed that the recreational pond was capable of supplying sufficient water demand for the physical laboratory. The laboratory requires replacement and replenishment of freshwater every week to support mainly sediment transport and scour modelling.

In terms of heat reduction, the SUD system in the USM Engineering Campus showed great potentials. A literature review on Urban Heat Islands was conducted. The review showed that cities are "hot spots" generating urban heat islands (UHIs). Green lungs such as trees and vegetation reflect solar radiation and possess lower surface and air temperatures by providing shade and transfer of heat upwards via evapotranspiration. Every 1 gram of water evapotranspirated takes away roughly 600 calories of heat. Hence, shaded surfaces, for example, experience temperatures between 1–5°C cooler than the peak temperatures of unshaded materials. The process of evapotranspiration in combination with shading and reflection, can help reduce peak summer temperatures by 1–5°C. Energy absorbed from the sun is transferred from the urban surface to the atmosphere through evapotranspiration, thereby linking the urban energy balance to the hydrological cycle. Significantly, the process of evapotranspiration is influenced by the availability of moisture which is found in vegetation cover, precipitation, irrigation, humidity, water surfaces and wind flow. It is estimated that annual global evapotranspiration consumes about 22% of the total available insolation at the top of the Earth's atmosphere. [25] When vegetation is replaced by concrete, evapotranspiration is greatly reduced and this alters the partitioning of the urban energy balance, as heat that would have otherwise been transferred upwards, stays in the concrete jungle and then contributes to the formation of the UHI. Urban planners now plan green structures to increase evapotranspiration (e.g. by planting trees, increasing vegetation cover and waterbodies) to a city, thereby replacing 'sensible heating' (Q_h) with 'latent heating' (Q_e), which in turn reduces the 'Bowen ratio' (of sensible to latent heat flux) leading to evaporative cooling ($B=Q_h/Q_e<1$). [26]

Results of the USM BIOECODS study show that water bodies (ponds, wetlands, streams) and green areas have a significant cooling effect on the micro-climate, especially at higher ambient air temperatures. Such green areas experience an average reduction of 1°C during temperatures higher than 20°C. Cooling occurred significantly during the daytime and ranged from 0.5 to 2.0°C. The cooling effect was found to be greater when temperatures are higher. It was also found that the cooling effect did not extend beyond 50 metres from the water body/green area. Cooling was significantly affected by landuse type, with road temperatures 43-50 °C compared to temperatures of surrounding trees between 28-30 °C. The average mid-day outdoor temperatures beside buildings in the USM Engineering campus was between 33-40 °C but in the BIOECODS areas with green vegetation and trees, temperatures were much lower between 28-30 °C. Road surfaces inside the campus recorded the highest temperatures of between 43-50 °C but the BIOECODS experimental area with greenery, open fields, lakes, ponds, rivers/streams, and grass, i.e. areas with reflective and evaporative surfaces, experienced effective cooling with temperatures between 28-30 °C. Hence, this study showed that SUDS with water bodies and green vegetated areas have cooling effects on their surroundings reducing high temperatures which can have a negative effect on health and wellbeing. Hence, a city that incorporates SUDS will benefit with cooling effects greatly enhanced. This can be done by careful consideration of urban design incorporating Sustainable Stormwater Management Systems.

Elsewhere, many cities adopting SUDS have benefited from the cooling effect. In Seoul, the Seoul Metropolitan Government undertook a major urban renewal project in restoring the Cheonggyecheon River via dismantling the highways which have covered the river. Despite initial criticism and skeptism, the project is now recognised as highly successful, strengthening the resilience of the eco-system and providing a useful green space for the citizens. Results of this project showed that a river as well as wind corridor was created, thereby reducing the heat island effect, improving air quality, and restoration of the natural environment. After the dismantling of the highways, the reduced traffic volume in the Cheonggyecheon area significantly reduced the concentration of fine dust (PM-10), NO_2, volatile organic compound (VOC) and other air pollutants. The heat island effect in the downtown area also declined. The temperature of the Cheonggyecheon area before the restoration was 2.2°C higher than the average of Seoul, but it declined to 1.3°C after the restoration, dropping by 8 to 18 %. The temperature of green point within the stream was 0.9°C lower than the neighbouring area. As the air-blocking elevated highway disappeared, a wind corridor was established and the creation of a stream affected the neighbouring environment. The ecosystem was also restored as wildlife fish species, birds, insects and plants increased. [27]

In Bangkok, the Chulalongkorn Centenary Park [28] was designed to hold excess rain water in large underground tanks to prevent flooding, and control heating. The park is intended not only as a welcome green space in the middle of one of the most congested cities in the world with 10 million people, but also as a place that could also retain large amounts of water, reducing monsoon flooding around Chulalongkorn University. Parks and "green roofs" planted with vegetation soak up rain during the annual monsoon and help dense urban centres like Bangkok adapt to climate change. Green architects are thinking about everything they build in the context of mitigating climate-change impact. Buildings and parks must be multi-functional and not be just about one or two aspects such as aesthetics or economics, but also about other functions such as flood control, heat management and recreational. This is Bangkok's first park in many years, a kind of "metro forest" project in a Bangkok suburb that has converted 0.8 hectares of abandoned land into a local forest with native trees, to make a start on reversing urban sprawl. The city's 11-acre Chulalongkorn Centenary Park is designed at an inclination of a three-degree angle, so that rain and floodwater flow to its lowest point, into a retention pond. At the park's highest end, a museum is topped by a green roof covered with native plants, which filter rainwater before it is stored in large tanks underground. Rainwater also flows through the park's lawn and wetlands where native vegetation filters the water, while its walkways are made of porous concrete. The park can hold up to 1 million gallons of water that can be discharged later or used in the dry season. The park also highlights the important role that green roofs and permeability of driveways and yards can play in reducing surface (water) runoff, with added benefits in reducing urban heat island effects.

Globally, the hundreds of megacities combined have become awesome in their sizes, and contribute significantly to greenhouse gas emissions and global warming. Large megacities have a ravenous appetite for energy, consuming ⅔ of the world's energy and creating over 70% of global CO_2 emissions. The implementation of SUD system in cities with green lungs and parks can have significant heat reduction effect. For example, the vegetation and greenery produced by SUD systems reduces concrete jungle, replacing dark absorbant surfaces of concrete jungle with reflective surfaces, and the increase in vegetation (especially trees) in SUD system increases the rates of evapotranspiration which transfer the heat from absorbed sunlight upwards into the atmosphere. Furthermore, the development of a physical model of green roof and rainfall simulator in the Universiti Sains Malaysia's Engineering campus depict results of heat reduction. This reduction in heat and UHIs from the model provides important data to be referred in the guidelines for the related or relevant agencies in order to implement the green roof in their developing areas. Significantly, the green areas under SUD systems are typically shown to be between 2 to 5 degrees Celcius lower than non-green areas of the city.

4. Conclusion

Sustainable Urban Drainage has been shown to be effective in performing a wide range of functions. SUD is proven to be a viable method of heat and flood control in cities. Cities using SUDS were found to cope very well with heat and floods. Overall, technical measures of flood management such as SUDS are found to be effective, affordable, aesthetically pleasing and socially acceptable. A holistic strategy combing technical SUDSs and non-technical human aspects of coping and resilience is the key towards effective management of floods in cities. The Urban Stormwater Management Manual for Malaysia, or MSMA, has been found to be an effective tool in controlling not only water quantity but also water quality. The Bio-ecological Drainage System or BIOECODS introduced in 2001 is adapted from the MSMA. Adopting the concepts of integration, control-at-source and sustainability, BIOECODS showed the way forward in USM's Engineering Campus's SUD design. The BIOECODS has innovation in design, with ecological swale, a dual layer conveyance system that minimize drainage footprint but provide additional water quantity and quality treatment. Other components such as dry ponds, wet ponds and wetlands are further evidences of the integration of stormwater facilities into surrounding landscape, adding significant values to otherwise passive open spaces. The results also showed that water bodies (ponds, wetlands, streams) and green areas have a significant cooling effect on the micro-climate, especially at higher ambient air temperatures. Cooling occurred significantly during the daytime and ranged from 0.5 to 2.0°C. It was also found that the cooling effect did not extend beyond 50 metres from the water body/green area. Cooling was significantly affected by landuse type, with road temperatures 43-50 °C compared to temperatures of surrounding trees between 28-30 °C. The average mid-day outdoor temperatures beside buildings in the USM Engineering campus was between 33-40 °C but in the BIOECODS areas with green vegetation and trees, temperatures were much lower between 28-30 °C. Road surfaces inside the campus recorded the highest temperatures of between 43-50 °C but the BIOECODS experimental area with greenery, open fields, lakes, ponds, rivers/streams, and grass, i.e. areas with reflective and evaporative surfaces, experienced effective cooling with temperatures between 28-30 °C. Hence, this study showed that SUDS with water bodies and green vegetated areas have cooling effects on their surroundings reducing high temperatures which can have a negative effect on health and wellbeing. Hence, a city that incorporates SUDS will benefit with cooling effects greatly enhanced. This can be done by careful consideration of urban design incorporating Sustainable Stormwater Management Systems. BIOECODS is currently being put on trial for water reuse by supplying non-potable water from its recreational ponds for landscape watering and laboratory use. There is a huge potential of this trial to be expanded to include other non-potable in the campus. The system is a living proof for feasibility and multi-benefits of MSMA implementation. BIOECODS also testified that the current stormwater management concept is ready to face the challenges and meet the demand for the nation, and most importantly, to secure a new source of water for the future generations.

5. References

[1] Intergovernmental Panel on Climate Change (2018) *Global warming of 1.5°C*. Switzerland: Intergovernmental Panel on Climate Change. ISBN 978-92-9169-151-7.

[2] United Nations, Department of Economic and Social Affairs, Population Division (2018) *World Urbanization Prospects: The 2018 Revision*, New York: United Nations. (Online Edition: Available from https://esa.un.org/unpd/wup/Publications Accessed 23 January 2019).

[3] Chan, N.W. (2015) Chapter II: 20 Urbanization and the Generation of Urban Heat islands in Selected Cities in Malaysia. In *Urban Sustainable Development Opportunities – Challenges.* Ho Chi Minh City: University of Social Sciences & Humanities, Vietnam National University, 218-232 (ISBN 978-604-73-3272-4).

[4] Chan, N.W. (2018) Building a Collaborative Flood Disaster Management Model for Malaysia. Paper presented at the ISEAS-Yusof Ishak Institute Singapore Workshop, 12 November 2018.

[5] Tan, M.L., Ab Latif Ibrahim, Zulkifli Yusop, Chua, V.P. and Chan, N.W. (2017) Climate change impacts under CMIP5 RCP scenarios on water resources of the Kelantan River Basin, Malaysia. *Atmospheric Research* Volume 189, 1 June 2017, Pages 1–10 (Online) (© 2017 Elsevier B.V, IF 3.377).

[6] Kerle, N. and Müller, A. (2013) Megacities and Natural Hazards. In: Bobrowsky P.T. (eds) Encyclopedia of Natural Hazards. Encyclopedia of Earth Sciences Series. Springer, Dordrecht.

[7] Chan, N.W. (1995) A Contextual Analysis of Flood Hazard Management in Peninsular Malaysia, Unpublished Ph.D. thesis, Middlesex University (UK).

[8] Zakaria, N.A., Aminuddin Ab Ghani, Ngai Weng Chan and Chun Kiat Chang (2016) Chapter 39: Sustainable Urban Drainage and Cities. In Chan, N.W., Hidefumi Imura, Akihiro Nakamura and Masazumi Ao (Editors) *Sustainable Urban Development Textbook*. Penang: Water Watch Penang & Global Cooperation Institute for Sustainable Cities, 259-270.

[9] Chan, N.W., Masazumi Ao, Nor Azazi Zakaria and Aminuddin Ab Ghani and Zullyadini A Rahaman (2016) Chapter 38: Rivers and Cities. In Chan, N.W., Hidefumi Imura, Akihiro Nakamura and Masazumi Ao (Editors) *Sustainable Urban Development Textbook*. Penang: Water Watch Penang & Global Cooperation Institute for Sustainable Cities, 248-258.

[10] Construction Industry Research and Information Association or CIRIA (2007). The SUDS Manual. CIRIA, London, UK.

[11] Illyani Ibrahim and Azizan Abu Samah (2011) Preliminary Study of Urban Heat Island: Measurement of Ambient Temperature and Relative Humidity in Relation to Landcover in Kuala Lumpur. June 2011, IEEE Geoscience and Remote Sensing Letters, DOI: 10.1109/GeoInformatics.2011.5981068.

[12] Norlida Mohd Dom (2016) Urban heat islands and the climate change in Malaysia. Paper presented in the Workshop for Capacity Building on Climate Change Impact Assessments and Adaptation Planning in the Asia-Pacific Region: Needs and Challenges for Designing and Implementing Climate Actions ; 27-28 January 2016, Manila, The Philippines.

[13] Elsayed I.S.M. (2012) A Study on the Urban Heat Island of the City of Kuala Lumpur, Malaysia. *JKAU: Met., Env. & Arid Land Agric. Sci*., Vol. 23, No. 2, pp: 121-134 (2012 A.D./1433 A.H.) DOI: 10.4197/Met. 23-2-8.

[14] Sham, S. (1987) Urbanization and the Atmospheric Environment in the Low Tropics: Experiences from the Klang Valley Region, Malaysia. Bangi: Penerbit Universit Kebangsaan Malaysia.

[15] Nur Aili Hanim Hanafiah and Chan, N.W. (2013) Kajian Keprihatinan Awam Terhadap Pulau Haba Bandar dan Amalan Mereka Terhadap Pengurusan Tekanan Haba di Pulau Pinang. In Jamaluddin Md. Jahi, Muhammad Rizal Razman, Kadir Arifin, Zuliskandar Ramli, Abdullah Sulaiman dan Emrizal (Editors) *Prosiding Seminar Serantau Ke-2 Pengurusan Persekitaran di Alam Melayu, Pekanbaru, Provinsi Riau, 6-7 Mei 2013*. Bangi: Institut Alam dan Tamadun Melayu (ATMa), Universitas Islam Riau dan Persatuan Pengurusan Persekitaran Malaysia, 204-208 (ISBN 978-983-2457-63-3).

[16] Ayub, K.R., Zakaria, N.A., Abdullah, R. & Ramli, R. (2010). Water Balance: Case Study of A Constructed Wetland As Part of the Bio-Ecological Drainage System (BIOECODS). *Water Science & Technology*, Vol. 62, No. 3, pp. 1931-1936.

[17] Ab. Ghani, A., Zakaria, N.A., Abdullah, R., Yusof, M. F., Mohd Sidek, L., A.H. Kassim & A. Ainan. (2004). BIO-Ecological Drainage System (BIOECODS): Concept, Design and Construction. The 6th International Conference on Hydroscience and Engineering (ICHE-2004), May 30th -June 3rd, Brisbane, Australia.

[18] Department of Irrigation and Drainage or DID, (2000). *Urban Stormwater Management Manual for Malaysia (MSMA)*. DID, Kuala Lumpur.

[19] http://redac.eng.usm.my/EAD/EAD512/MSMA.pdf (Accessed 24 Jan 2019)

[20] Mansor, M., Lim, P. and Shutes, R. (2002).*Constructed wetlands : Design, management and education*. Penang: Penerbit Universiti Sains Malaysia.

[21] Ainan, A., Zakaria, N.A., Ab. Ghani, A., Abdullah, R., Mohd Sidek, L., Yusof, M.F. & Wong, L.P. (2004). Peak Flow Attenuation Using Ecological Swale and Dry Pond. The 6th International Conference on Hydroscience and Engineering (ICHE-2004), May 30th -June 3rd, Brisbane, Australia.

[22] Abdullah, R., Zakaria, N.A., Ab. Ghani, A., Mohd Sidek, L., Ainan, A. & Wong, L.P. (2004). BIOECODS Modelling Using SWMM. The 6th International Conference on Hydroscience and Engineering (ICHE-2004), May 30th -June 3rd, Brisbane, Australia.

[23] Mohd Sidek, L., Ainan, A., Zakaria, N.A., Ab. Ghani, A., Abdullah, R. & Ayub, K.R. (2004). Stormwater Purification Capability of BIOECODS. The 6th International Conference on Hydroscience and Engineering (ICHE-2004), May 30th -June 3rd, Brisbane, Australia.

[24] Ayub, K. R., Mohd Sidek, L., Ainan, A., Zakaria, N. A., Ab. Ghani, A. & Abdullah, R. (2005). Storm Water Treatment using Bio-Ecological Drainage System. *International Journal River Basin Management*, Special Issue Rivers'04, IAHR & INBO, Vol. 3, No. 3, pp. 215-221.

[25] Qiu, G.Y., Li, H.Y., Zhang, Q.T., Chen,W., Liang, X.J. and Li, X.Z. (2013) Effects of evapotranspiration on mitigation of urban temperature by vegetation and urban agriculture. *J. Integr. Agric.*,12 (2013), 1307-1315.

[26] Gunawardena, K R, Wells, M.J. and Kershaw, T. (2017) Utilising green and bluespace to mitigate urban heat island intensity. *Science of the Total Environment Vol. 584-585*, 15 April 2017, 1040-1055.

[27] https://policytransfer.metropolis.org/case-studies/seoul-urban-renewal-cheonggyecheon-stream-restoration (Accessed 24 Jan 2019).

[28] https://www.chula.ac.th/en/cu-services/creative-space/cu-centenary-park/ (Accessed 24 Jan 2019)

Acknowledgements

The author graciously acknowledges the data and secondary publications related to SUD system, particularly those produced by the River Engineering and Urban Drainage Research Centre (REDAC) in USM. The Grant under the USM Account Number 304/PHUMANITI/650906/W109 is also acknowledged.

IOP Conf. Series: Earth and Environmental Science **257** (2019) 012014 doi:10.1088/1755-1315/257/1/012014

Comparison of clustering methods for identification of outdoor measurements in pollution monitoring

Xu Yang[1], Lingxi Zhu[1], Sio Lam[2], Laurie Cuthbert[2], Yapeng Wang[2]

1 Macao Polytechnic Institute, School of Public Administration, Macao SAR, China
2 Macao Polytechnic Institute, Information Systems Research Centre, Macao SAR, China
E-mail: xuyang@ipm.edu.mo

Abstract. This paper considers the problem of post-processing air pollution data to clearly identify outdoor clusters, by removing indoor data and "noise" caused by air from indoors mingling with air from outdoors. In this paper, several different clustering algorithms are compared using data from measurements in Macao. It is shown that *X-means* generally outperforms the others for this purpose and can successfully separate data modified by noise. Such a technique simplifies the collection of large data sets since the person taking the measurements does not have to make any advance decisions about what is pure outdoor, or pure indoor, data. However, it is also shown in this work that setting up suitable procedures can be quite complex.

1. Introduction

Monitoring air pollution is now routine in many cities with governments striving to reduce the well-known risks to their citizens from outdoor pollution. However, a lesser considered effect is the air quality in the home or the office, with pollution arising from gases such as radon and carbon monoxide, mould and pollen, tobacco smoke, and materials used in the building such as asbestos and formaldehyde. In a dense area like Macao, these pollutants can mix with outdoor readings with air flow through windows and doors.

Monitoring of outdoor air pollution monitoring systems has traditionally used static stations located at points where either the authorities expect there to be a problem, or where they cover particular areas of interest. Such stations tend to be fairly sparsely distributed – as an example the Guangdong-Hong Kong-Macao Pearl River Delta Regional Air-Quality Monitoring Network has only 23 stations, of which 4 are in Hong Kong and only 1 in Macao (Taipa Grande) [1].

Another approach is mobile sensing, either using specially equipped vehicles or by operatives walking round taking miniature sensors with them. Compared to static monitoring stations, such mobile stations make it easier to cover a larger area in the city and have lower cost. They also have the advantage that (i) they allow a large amount of data to be collected and (ii) the data is collected at locations where people actually are, so allowing a better picture of how people's health may be affected. These systems can use location services (like the location services in mobile phones) to record location and time information along with the pollution data, so that users can collect pollution data from both indoor and outdoor environments. However, in a large-scale deployment, it is difficult to identify whether a location is indoors or outdoors, or to expect every user to mark whether their current location is indoors or outdoors, so some means of separation has to be used.

However, the situation is more complicated than that because data can be affected by "noise". Here we define noise to be the effect of mixing data from indoor and outdoor locations – for example, sitting

Content from this work may be used under the terms of the Creative Commons Attribution 3.0 licence. Any further distribution of this work must maintain attribution to the author(s) and the title of the work, journal citation and DOI.
Published under licence by IOP Publishing Ltd

outside a restaurant may actually be outdoors, but the readings will be affected by pollutants from indoors coming through open windows and doors, and of course, *vice versa*. This sort of noise is generated by the data source itself, but there can be other sorts of noise, such as the warm-up times for sensors – good measurement practice should ensure that sensors are properly stabilised before taking readings, but it is possible that readings are taken early. [2]

In [3] and [4] it is claimed that the effect of noise in a dataset can be small if measures have been taken to minimise the noise. In this paper though, the noise can be very significant since we do not know in advance the locations to avoid because of contamination between indoor and outdoor data. It is not feasible to mark the location of every record as containing noise data or not – indeed, it would be impossible to tell. Hence there needs to be a method for identifying which points are (for example) outdoors and which represent outdoors plus noise. This is the subject of this paper.

In [5] the authors identified four techniques to enhance data analysis in the presence of high noise levels: (i) clustering-based, (ii) distance-based, (iii) an approach-based on whether an object can be considered an outlier and (iv) their proposed hyperclique technique. Here we are considering a clustering-based approach and note that the algorithm itself can detect outliers. In [6] and [7], the authors used the principle that, once the data has been clustered, noise objects are the farthest from the cluster centre. The authors in [8] took another approach and treated small clusters that are far from other major clusters, as sets of outliers. Obviously, the choice of clustering algorithm is important and can make a difference as to which objects are treated as outliers and the difficulty of deciding that may well depend on the type of dataset being considered and the characteristics of the noise.

In this paper we consider a situation where a noise cluster may be quite close to a cluster of data that is needed – for example, an indoor location and outdoors immediately adjacent to it, but with an open window in between as illustrated in Figure 1.

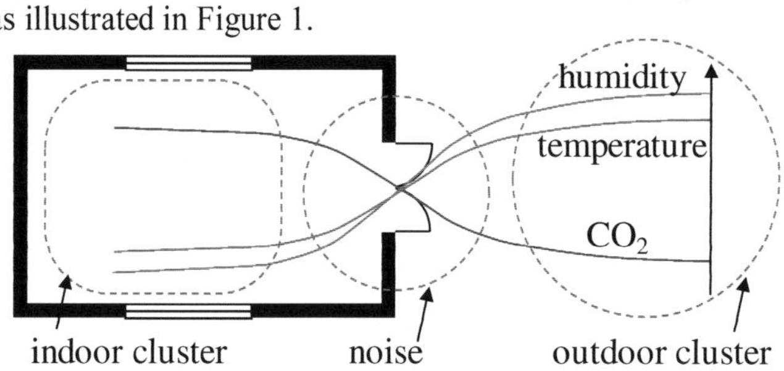

Figure 1. Illustration of the problem

In this simple case it is assumed that it is summer in Macao and the outside temperature is much hotter than indoors, where air conditioning is used; the room is filled with people, so the CO_2 level is higher than outside. Humidity in Macao is generally high so the use of air conditioning inside would lead to lower values indoors.

At other times of the year, the temperature levels will be different, and the CO_2 level will depend on the use of the room. Hence, an algorithm that is capable of separating data into noise clusters (where they exist) as well as the main indoors and outdoors, would be very useful.

Automatically determining the number of clusters has been one of the most difficult problems in data clustering. Most methods for automatically determining the number of clusters cast it into the problem of model selection. Usually, clustering algorithms are run with different numbers of cluster types K; the best value of K is then chosen based on a predefined criterion

In the set we have chosen, the X-Means and the Cascading K-Means do not need to have a value of K specified, just the range in which K is expected to lie. The algorithms themselves find the best value of K as well as the clusters. This allows a comparison of automatic choice of K to be compared with the more traditional approach mentioned above.

In this paper, the performance of the algorithms is compared, and it is shown that X-Means is clearly the best choice, although there are caveats on how the tests are conducted.

2. Clustering algorithms

Because these algorithms are generally well-known, they are not described in any detail in this paper, the comments here being restricted to a brief mention of the main features. A very comprehensive description of clustering techniques is given in [9] and that book covers all the methods below with the exception of Cascading K-Means.

2.1. Algorithms requiring k to be specified in advance

- **K-Means** is a well-known and popular iteration-based clustering method with the term stemming from [10]. The performance of the algorithm performance depends on the initial settings and there is, therefore, no guarantee of it converging to the global optimum. Because of this, there is a common practice of running the algorithm repeatedly with different starting conditions. Additionally, this algorithm is very sensitive to outliers, which have great impact on the updates of the cluster centres. However, it does have the advantage of being fast.
- **Agglomerative Hierarchical Clustering** [11] creates a hierarchical structure that reflects the order in which groups are merged or divided. With this approach, the choice of linkage algorithm is important as they do have different characteristics.
- **Bisecting K-Means** [11] combines K-Means and Hierarchical Clustering. From an initial position with all objects in a single cluster, it splits this cluster into two datasets. One is then selected for the second data partition and the algorithm repeats. This algorithm is superior to the K-Means algorithm in the selection of initial centre points. However, the initial two cluster centres are randomly selected, and this can lead to large deviations if the noise points or outliers happen to be chosen as the initial cluster centres; it is also computationally quite intensive.
- **K-Medoids** [12] is thought to be more resilient to noise and outliers than K-Means because it minimizes a sum of general pairwise dissimilarities instead of a sum of squared Euclidean distances, but this approach has a higher computational complexity.

2.2.Algorithms that determine the value of K

- **X-Means** [13] searches the space of cluster locations and number of clusters efficiently to optimize the Bayesian Information Criterion [14] or the Akaike Information Criterion [15]. In this algorithm, the number of clusters is computed by the algorithm itself, the user only having to specify the limits of the range of values that K can take.
- **Cascading K-Means** [16]. In this approach, the user specifies the range K can take, and the algorithm creates partitions forming a "cascade" from the smallest value of K to the largest. The best partition is selected using the Calinski-Harabasz [17] criterion.

3. Experiments Setting

3.1.Data normalization

Data Normalization standardizes the raw data by converting them into specific range using a linear transformation in order to generate good quality clusters and improve the accuracy of clustering algorithms. There is no universally defined rule for normalizing the datasets. [18]

In this paper, the attribute Min-Max normalization scheme [19] is applied for each dataset of each experiment. The experimental results in that paper show that Min-Max normalization leads to fewer errors than the Zero Mean and Decimal Scaling methods. In this work, we used the attribute Min-Max normalization method on the air pollution data, the results showing that this normalization improved the clustering accuracy.

The attribute Min-Max normalization method performs a linear transform on the original data so that if we know the maximum and minimum value of a given attribute v_i, it is easy to transform the attribute into a normalised value x_i within the range [0,1] by $x_i = \frac{v_i - \min(v_i)}{\max(v_i) - \min(v_i)}$.

3.2. Data Clustering Platforms

In this work, four data clustering algorithms (K-Means algorithm, agglomerative hierarchical clustering, X-Means algorithm and Cascading K means algorithm) are performed using the well-known data clustering platform *Weka* [20]. The other two algorithms (Bisecting K-Means algorithm and K-Medoids algorithm) were implemented on another popular data clustering platform, ELKI 0.7.1[21].

3.3. Evaluation Methods

A good description of evaluation methods is given in [22]. Each air pollution data set in the experiments reported here is labelled with one of five class labels: (i) indoors, (ii) indoors with noise, (iii) at door, (iv) outdoors with noise, or (v) outdoors. From [23], evaluating these clustering results requires an external criterion and we choose two well-known external indices: $Precision$ and $Recall$ [24], [25]. $Precision$ and $Recall$ describe the accuracy and completeness, respectively, of the predicted clusters. These two measures can be calculated using the following:

$$Precision(i,j) = \frac{n_{ij}}{n_j} \qquad Recall(i,j) = \frac{n_{ij}}{n_i}$$

where n_{ij} is the number of objects of class i that are in cluster j, n_j is the number of objects in cluster j, and n_i is the number of objects in cluster i.

$Precision$ is the fraction of data put in the correct cluster (so a measure of what happens to the data), and $Recall$ is the fraction of data in the cluster that belongs to it (so a measure of the cluster). Here we are interested in the outdoors cluster only and want to ensure that only outdoor data gets included – so we are looking for a $Recall$ value of 100%. However, a $Precision$ less than 100% means that some of the outdoors data will be lost to other clusters: with a large data set this is not a problem as long as the $Precision$ is not too low.

4. Experiments

These experiments were carried out using sensors that recorded values of (i) PM1.0, (ii) PM2.5, (iii) PM10, (iv)HCHO, (v) CO_2, (vi) Temperature and (vii) Humidity. Data points were recorded every 10s and each data set comprised at least 120 readings (i.e. 20 mins). The same sensor was used in each location, which meant that there was a time shift between the readings at each location, but experiments showed that the variation between sensors was likely to be greater than that over such time intervals and this is backed up by historical data.

Readings were taken in June 2018 around midday; in this season the temperature in Macao is usually hot (around 28°-34°) and there is widespread use of air conditioning indoors. CO_2 depends very much on the number of people in an indoors locations and humidity can be very variable, especially if there is cooking indoors. Our results showed that PM values did not vary much between indoors and outdoors, and HCHO readings were zero. In different seasons the variation between indoors and outdoors would be different.

Figure 2 and Table 1 show example results, averaging the data set at each location. In Figure 2 the data is normalised using the scheme explained in Section 0, whereas Table 1 shows the actual average values at each location. The CO_2 and temperature values follow the general outline of Figure 1 but notice the humidity dips inside the room, but nearer the door – this appears to be because of the air flow with the location of the air conditioning unit in the room, but it serves as an example as to the variations that might be expected in practical situations.

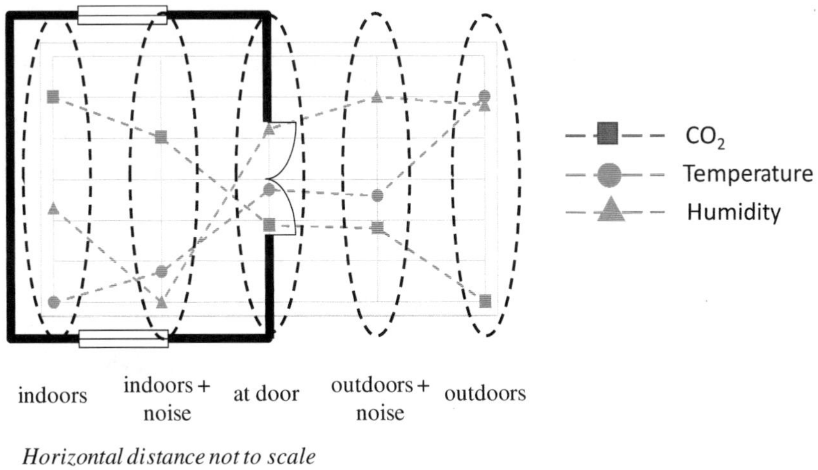

Figure 2. Example results showing difference between indoors and outdoors (normalised data values)

Table 1. Actual values at each location

	CO_2	Temperature	Humidity
	(ppm)	(°C)	(%)
Indoors	506	25.2	43.8
indoor+noise	487	26.2	35.5
noise	446	29.0	50.8
outdoor+noise	445	28.8	53.6
outdoor	410	32.2	52.9

There are 5 possible clusters in this scenario, although the "at door" and "outdoors + noise" have similar values. Initially the different algorithms were applied to data sets containing all the attributes, i.e. including PM1, PM2.5, PM10 and HCHO, with the results summarised in Table 2. Each data set overall is of the order of 800 readings).

Knowing that data samples were taken from five different locations, the value of K was set to 5 for the fixed-K methods, and the range 2-5 was offered to those two methods that determine K.

All methods had difficulty separating the data for the locations "at door" and "outdoors with noise", with K-Means Bisecting and K-Mediods mixing the data across the two clusters and Hierarchical and X-Means essentially ignoring the "at door" cluster; X-Means also determines that there are only 4 clusters. K-Means performs poorly with fairly low values for most performance measures and Cascading K-Means cannot cope at all, determining there are only 2 clusters.

Looking at the data in Figure 2, it is not really surprising that the methods have difficulty separating data between the "at door" and "outdoors + noise" since there is a lot of similarity. Hence another set of experiments was undertaken with the same data sets, but this time with the fixed value K methods using a value $K = 4$, the value determined by X-Means. These results are also shown in Table 2. K-Means Bisecting and K-Mediods have similar results to those previously for the "outdoors" cluster and K-Means is similarly as bad as the earlier results. However, the real surprise is Hierarchical that does not put any data in the outdoors cluster.

The conclusion from these tests is that applying these methods to this data set could lead to unpredictable results, although X-Means does determine the number of clusters correctly (given how similar two of them are) and gets good results for *Precision* and *Recall* of the "outdoors" cluster (the main focus) and the two indoors sets.

Table 2. Results of clustering across all data sets

	Indoors		Indoors+noise		At door		Outdoors+noise		Outdoors		*K*	
	Precision	*Recall*	*Precision*	*Recall*	*Precision*	*Recall*	*Precision*	*Recall*	*Precision*	*Recall*	*Specified*	*Found*
K-Means Bisecting	92%	100%	100%	100%	57%	63%	67%	54%	99%	**100%**	5	
	The data at the door and the noise data outside the door are mixed into two clusters											
K-Medoids	93%	100%	100%	100%	61%	75%	76%	53%	**100%**	**100%**	5	
	The data at the door and the noise data outside the door are mixed into two clusters											
K-Means	0%	0%	51%	100%	59%	64%	63%	56%	100%	57%	5	
	Indoor and noise indoor are put into the same cluster											
Hierarchical	100%	99%	100%	100%	0%	0%	51%	100%	**100%**	**100%**	5	
	Data at the door and the noise data outside the door are put into the same cluster											
X-Means	91%	100%	100%	100%	0%	0%	53%	99%	99%	**100%**	2-5	4
	Data at the door and the noise data outside the door are put into the same cluster											
Cascading K-Means	0%	0%	49%	100%	0%	0%	35%	100%	0%	0%	2-5	2
	Fails very badly											
Setting K=4												
K-Means Bisecting	92%	100%	100%	100%	0%	0%	53%	99%	99%	**100%**	4	
K-Medoids	93%	100%	100%	100%	0%	0%	52%	96%	97%	99%	4	
Hierarchical	100%	99%	100%	100%	0%	0%	34%	100%	0%	0%	4	
K-Means	0%	0%	50%	100%	0%	0%	52%	99%	100%	57%	4	
(shaded)	low value of *Precision*											

It is clear from the above discussion that the clustering might be considered a bit "hit and miss" and that while some methods give good results, a small change in configuration may cause that method to stop working. In our scenario, 5 of the 8 types of data did not change as much between location as CO_2, Temperature and Humidity. These sets of similar data reduce the effective distance between clusters, so another set of similar experiments was carried out, but this time using only those three parameters. The results of these are shown in Table 3.

Table 3. Results of clustering across temperature, CO_2 and humidity

	Indoors		Indoors+noise		At door		Outdoors+noise		Outdoors		*K*	
	Precision	*Recall*	*Precision*	*Recall*	*Precision*	*Recall*	*Precision*	*Recall*	*Precision*	*Recall*	*Specified*	*Found*
K-Means Bisecting	100%	100%	100%	100%	74%	83%	81%	71%	100%	**100%**	5	
K-Medoids	100%	100%	100%	100%	77%	97%	96%	71%	100%	**100%**	5	
K-Means	100%	100%	100%	100%	0%	0%	51%	100%	100%	55%	5	
Hierarchical	100%	97%	100%	100%	0%	0%	51%	100%	100%	**100%**	5	
X-Means	100%	100%	100%	100%	0%	0%	51%	100%	100%	**100%**	2-5	4
Cascading K-Means	100%	100%	100%	100%	0%	0%	51%	100%	100%	**100%**	2-5	4
Setting K=4												
K-Means Bisecting	100%	100%	100%	100%	0%	0%	51%	100%	100%	**100%**	4	
K-Medoids	100%	100%	100%	100%	0%	0%	51%	100%	100%	**100%**	4	
K-Means	0%	0%	51%	100%	0%	0%	51%	100%	100%	55%	4	
Hierarchical	100%	100%	100%	100%	0%	0%	51%	100%	100%	**100%**	4	
(shaded)	low value of *Precision*											

Clearly these results are much better without the dilution caused by the sets of parameters that were very similar. K-Means still produces poor *Recall* on the "outdoors" set and misses the "indoors" cluster completely when *K* is set to 4. All the methods fail to separate the "at door" and "outdoors + noise" clusters, but as before this is to be expected since the parameters are very similar.

Overall, X-Means, not only determines a suitable value of K, it also produces *Precision* and *Recall* values that as good as any other method and performs well in the clustering, for all the clusters (assuming the lumping together "at door" and "outdoors + noise" is accepted), not just "outdoors". In this case, Cascading K-Means also finds an appropriate value of K.

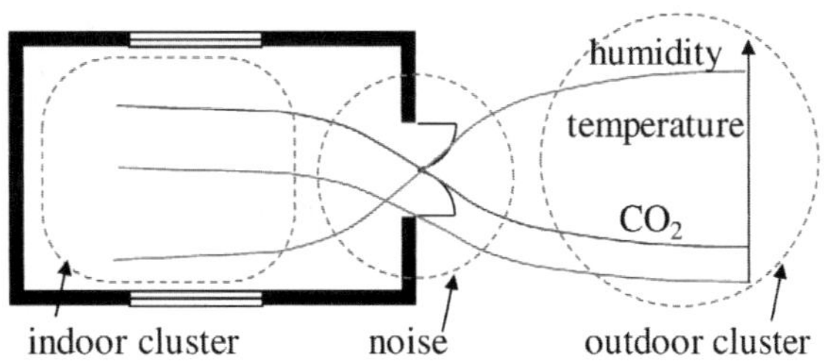

Figure 3. Likely distributions in Winter

In Winter, the temperature will be reversed with higher temperatures inside the building, although the humidity will still be lower indoors (shown in Figure 3). As mentioned earlier CO_2 levels generally depend on the occupancy so will vary enormously. Overall though, there is expected to be sufficient variation for clustering to identify successfully the "outdoors" and "indoors" cluster at least. The issue is how well this approach would work in Spring and Autumn when the temperatures indoors and out may be very similar, as may be humidity if air conditioning is not used. This is an area of further work. From Figure 4 it is clear that the temperature outdoors varies a lot with the season, but not so much the average humidity, although it was particularly low when the results for this paper were taken.

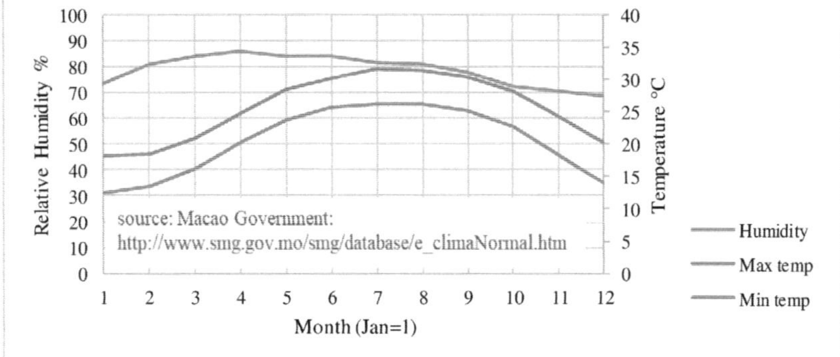

Figure 4. Average values of temperature (highest and lowest) and humidity in Macao for 1981-2010

5. Conclusions

This work has shown that it is possible to separate out noise clusters from the outdoor results required, something that is necessary in a densely populated area like Macao. However, care does need to be taken in identifying the parameters that do vary and can be used for the clustering, avoiding dilution of the data by removing parameters that are similar across all clusters. Of all the methods investigated, X-Means was clearly the best as it determined correctly the number of clusters, assuming that it is reasonable to merge "at door" with "outdoors+noise", and performed as well as others in terms of $Precision$ and $Recall$. Putting that value as K in the fixed-K methods produced no noticeable improvement.

Performing clustering over the three parameters that vary most is only used to identify where that's et of values belongs: the values of PM samples and other pollutants can than be tied to that cluster for analysis of the pollution map.

Further work will investigate the performance in different seasons in Macao and locations showing different physical characteristics.

6. References

[1] Meteorological and Geophysical Bureau M S 2018 Guangdong-Hong Kong-Macao Pearl River Delta Regional Air Quality Monitoring Network - January to March 2018 - Statistical Summary of the First Quarter Monitoring Results

[2] Fisk W J, David F and Douglas S P 2008 Accuracy of CO2 Sensors *Lawrence Berkeley Natl. Lab.* **9**

[3] Orr K 1998 Data quality and systems theory *Commun. ACM* **41** 66–71

[4] Redman T C 1998 The impact of poor data quality on the typical enterprise *Commun. ACM* **41** 79–82

[5] Xiong H, Pandey G, Steinbach M and Kumar V 2006 Enhancing data analysis with noise removal *IEEE Trans. Knowl. Data Eng.* **18** 304–19

[6] Larsen B and Aone C 1999 Fast and effective text mining using linear-time document clustering *Proceedings of the fifth ACM SIGKDD international conference on Knowledge discovery and data mining - KDD '99* KDD '99 (New York, NY, USA: ACM) pp 16–22

[7] Zhang T, Ramakrishnan R and Livny M 1996 BIRCH *Proceedings of the 1996 ACM SIGMOD international conference on Management of data - SIGMOD '96* SIGMOD '96 (New York, NY, USA: ACM) pp 103–14

[8] Portnoy L, Eskin E and Stolfo S 2001 Intrusion Detection with Unlabeled Data Using Clustering *Proceedings of ACM CSS Workshop on Data Mining Applied to Security* pp 5–8

[9] Aggarwal C C and Reddy C K 2013 *Data clustering: algorithms and applications* (CRC press)

[10] MacQueen J B 1967 Some methods for classification and analysis of multivariate observations *Proceedings of the 5th Berkeley Symposium on Mathematics Statistics and Probability* (Berkeley, Calif.: University of California Press) pp 281–97

[11] Steinbach M 2000 A Comparison of Document Clustering Techniques *KDD Work.* **400** 1–2

[12] Mirkin B G 2005 Clustering for data mining: a data recovery approach (Chapman and Hall/CRC)

[13] Pelleg D, Pelleg D, Moore A W and Moore A W 2000 X-means: Extending K-means with efficient estimation of the number of clusters *Proc. Seventeenth Int. Conf. Mach. Learn. table contents* **1** 727–734

[14] Schwarz G 1978 Estimating the Dimension of a Model *Ann. Stat.* **6** 461–4

[15] Akaike H 1974 A new look at the statistical model identification *IEEE Trans. Autom. Control* **19** 716–23

[16] Milligan G W and Cooper M C 1985 An Examination of Procedures for Determining the Number of Clusters in a Dataset *Psychometrika* **50** 159–79

[17] Calinski T and Harabasz J 1974 A dendrite method for cluster analysis *Commun. Stat. - Theory Methods* **3** 1–27

[18] Visalakshi N K and Thangavel K 2009 Impact of normalization in distributed k-means clustering *Int. J. Soft Comput.* **4** 168–72

[19] Saranya C and Manikandan G 2013 A study on normalization techniques for privacy preserving data mining *Int. J. Eng. Technol.* **5** 2701–4

[20] Witten I H, Frank E and Hall M A 2011 Data Mining: Practical Machine Learning Tools and Techniques (Google eBook) (Morgan Kaufmann)

[21] Schubert E, Koos A, Emrich T, Züfle A and Klaus Arthur Schmid A Z 2015 [ELKI 0.7] A Framework for Clustering Uncertain Data *Proc. VLDB Endow.* **8** 1976–89

[22] Jain A K 2010 Data Clustering: 50 Years Beyond K-Means1 *Pattern Recognit. Lett.* **31** 651–66

[23] Jain A K and Dubes R C 1988 *Algorithms for clustering data* (Prentice-Hall, Inc.)

[24] Rendón E, Abundez I, Arizmendi A and Quiroz E M 2011 Internal versus external cluster validation indexes *Int. J. Comput. Commun.* **5** 27–34

[25] Chormunge S and Jena S 2015 Efficiency and Effectiveness of Clustering Algorithms for High Dimensional Data *Int. J. Comput. Appl.* **125**

IOP Conf. Series: Earth and Environmental Science **257** (2019) 012015 doi:10.1088/1755-1315/257/1/012015

Landscape components of Aborlan Palawan Philippines: influence on the employment and income of its populace

L. Acero[1]

1 Professor from the Department of Natural Sciences, College of Arts and Sciences, San Beda University Manila Philippines
E-mail: lilyacero1@yahoo.com

Abstract. Aborlan Palawan is known to be the sleeping town in the Southern part of Palawan Philippines As a municipality, its landscape is composed of several structures making it a municipality, home for its population. This study focuses on the use of landscape maps to describe its landscape components and how it influences the employment and income of its population. Descriptive research method and documentary analysis was used to describe the influence of landscape maps (road network, water catchment, Environmentally Critic Area Network (ECAN), soil map, land classification and land use) to the employment and income of the population Components of the landscape is described using Patch Mosaic Model Pattern Elements, specifically the patch, corridor and matrix. Result of this study revealed that the land scape of Aborlan Palawan which is composed of 57% ECAN considered as the major landscape matrix in Aborlan, influence the employment and income status of its population. The ECAN matrix, influence livelihood of its populace, since employment and income of population situated near ECAN, lag behind compared with the villages (*barangays*) in lower slopes. The employment rate and income is high in villages far from ECAN.

1. Introduction

The composition of a landscape is defined by the spatial elements that are distinguished in the map and believed to be relevant to the landscape function under consideration. Composition is mainly concerned with land cover types which are presented at courser scale [1]. Landscape ecology is a discipline that studies environmental complexity, concentrating mainly on an analysis of the importance of spatial relationships between the various components (individuals, populations, communities, and land mosaics) of the real world. When studying a landscape, it is important to be able to consider the components that are of immediate interest without losing sight of the whole [2]. In this study landscape component is define as the physical structures (land, water, roads) that composes the municipality of Aborlan Palawan Philippines. The biggest structure is express as matrix, patches as scattered mosaic of discrete structures, and corridors as linear landscape such as road networks and bodies of water. Aborlan Palawan is a first class municipality located in the Southern part of Palawan, Philippines. It is situated at the central portion of Palawan Island, 69 km southeast of Puerto Princesa City (Fig. 1). It is located at 118°32' 53" North longitude and 9°26' East latitude. It lies in a vast plain between the Sulu Sea and the mountains. Bounded on the east and west by rich and mighty bodies of water, the Sulu Sea and the West Philippine Sea facing South China Sea, and comprising part of Palawan Island's almost 2,000 kilometers of coastline. It has 19 villages/barangays including its Aborlan Poblacion. It has a population of 35,091 people. In 2014, there are also 7,310 total households with an average family size of 4 persons per household [3]. Generally, the soil of Aborlan can be classified into rough mountain soil type which is suitable for forest growth and preservation purposes

Content from this work may be used under the terms of the Creative Commons Attribution 3.0 licence. Any further distribution of this work must maintain attribution to the author(s) and the title of the work, journal citation and DOI.
Published under licence by IOP Publishing Ltd

comprising about 57% of its total land area, and clay loam which is suitable for agriculture comprising about 40%. The topography ranges from flat near the shore of the eastern and western part to rugged mountainous in the central and southwestern portion of the municipality. The total land area of 80,733 hectares is covered with different land use types, of which about, 54.8% is forest [4].

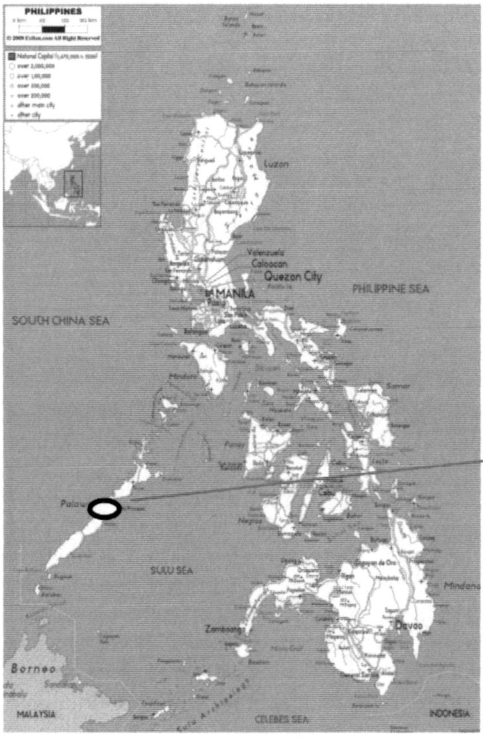

Figure 1a. Map of Philippines showing Aborlan Palawan Circled Black

Figure 1b. Map of Palawan showing Aborlan Palawan in green patch

1.1. Objectives of the study

This study describes the following landscape components of the municipality (expressed in landscape maps-road network map, water catchment, environmentally critical Area Network, soil map, land classification and land use) and its influence on the employment and income of the population.

2. Methodology

A convenient and popular model for conceptualizing and representing the elements in a categorical map pattern is known as the *patch-corridor-matrix model* (PCMM). Under this model, three major landscape elements are typically recognized, and the extent and configuration of these elements defines the pattern of the landscape. The patch mosaic model is the dominant model of landscape structure in use today and Map of land cover types is an example of using this model in studying the landscape.

This study used descriptive research method to describe landscape maps and its influence on the livelihood of the population. Documentary analysis on landscape maps, employment, and income of the population supports the results and discussion. Data on landscape map, employment, and income, were gathered from the Municipal Planning and Development Office of Aborlan Palawan (MPDO) Philippines.

In this study, PCMM was used in which landscape maps are express in mosaic (biggest structure) linear (roads and bodies of water) and patches (mosaic with discrete land structure). The employment and income are express in magnitude, proportion, and total.

3. Results and discussion

3.1. Road network

Road network map (corridors-linear landscape elements) is shown in Figure 2 in different hues of brown. Darkest brown line denotes national highway from Puerto Princesa City to Narra Palawan, and is located close to villages facing the Sulu Sea. Provincial and municipal roads (dark brown lines), connects villages which is mostly prevalent in villages facing Sulu Sea. Municipal road, also connects 2 villages (Culandanum and Apurawan) facing South China Sea. To date no road traverses villages facing South China to villages facing Sulu Sea. Thus people of Culandanum and Apurawan, use pumpboat (motored banca) to Puerto Princesa City, then travel by land via public utility vehicle (PUV) to town proper of Aborlan. Examining the road network map shows that 57% of Aborlan is ECAN, characterize by forest growth. Villages that are closer to the forest from the North are Sagpangan,

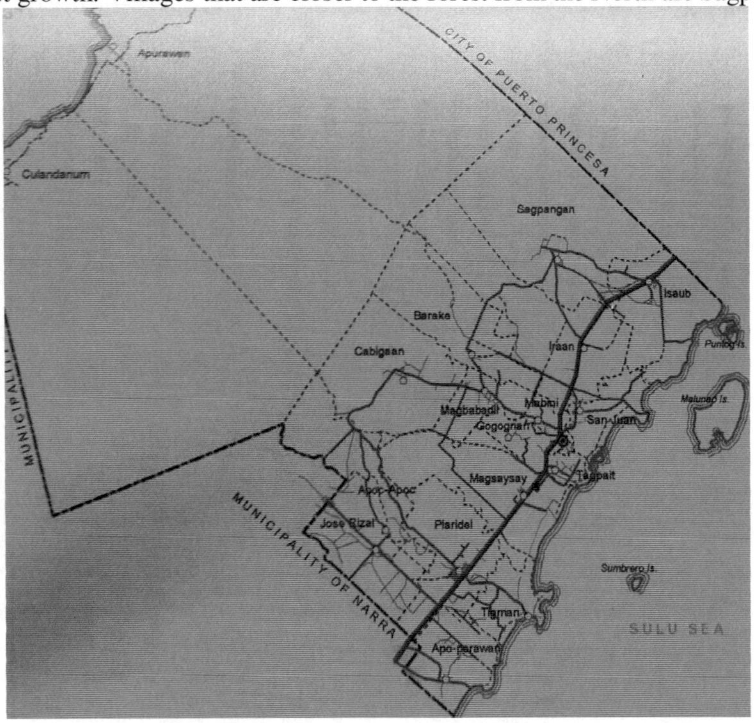

Figure 2. Road Network Map of Aborlan Palawan

Source: Aborlan MPDO Data Bank

Barake, Cabigaan and Apurawan and Culandanum facing the South China Sea Coastline is expressed in dark blue line while rivers and creeks are represented by light blue single lines which are distributed in interwoven manner, in villages facing the Sulu Sea. Most of the population inhabited areas, are in lower slopes and are accessible to road networks. Areas with higher slope are forests. However, one group of the Indigenous people in Palawan-the Tagbanuas inhabited areas near the forest such as Sagpangan, Barake and Barake facing Sulu Sea and Apurawan and Culandanum facing South China Sea [5].

3.2. Water catchment map

Water Catchment map (corridors-linear landscape elements) of Aborlan Palawan is posited in Figure 3.Water catchment is shown in 2 blue lines, which are connected to the main rivers (single blue line).Two villages facing South China Sea has more distinct and extended water catchment areas which are connected to the following rivers; Tigman, Talakaygan, Ibanu, Apurawan, Aborlan. Bubunaguan, Apurawan, Calategas and Culandanum. The rest of the vilages (facing Sulu Sea) has less extended water catchment which came from the following rivers; Isaub, Sagpangan Aborlan, Talakaygan, Tigman, Apo-Aporawan, Barake, Isaub, Aborlan and Mayligan. All of these water catchments areas terminated along the coastlines (3 blue lines) of the municipality. Villages facing South China Sea are bordered by the coastlines facing South China Sea. While villages facing Sulu Sea are bordered by coastlines facing Sulu Sea. The water catchment map influences the agricultural production in the municipality.

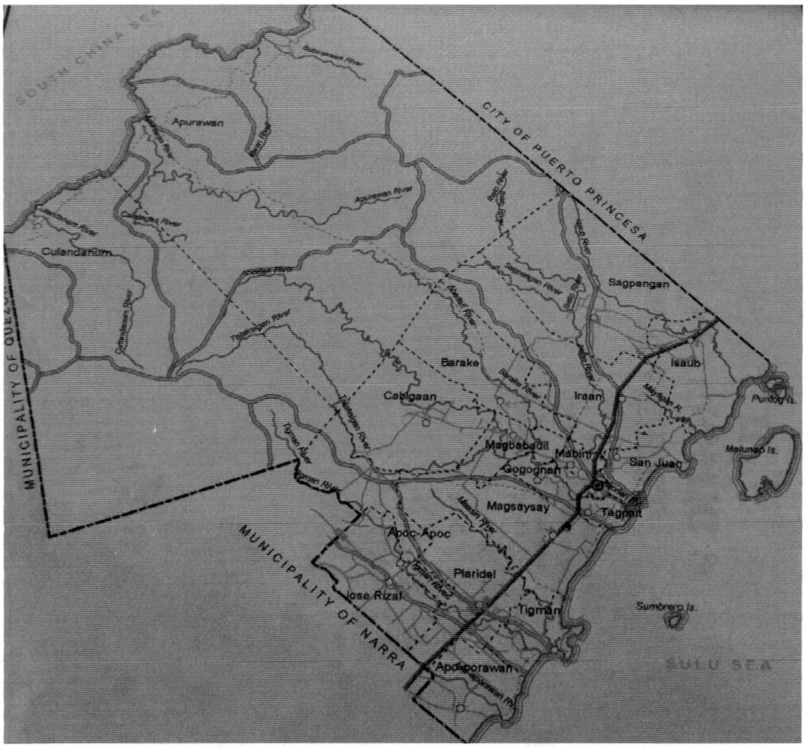

Figure 3. Water catchment Map of Aborlan Palawan

Source: Aborlan MPDO Data Bank

3.3, Environmentally critical area network map

Figure 4 shows the environmentally critical Area Network (ECAN**).** The **core** zone (red matrix) is the

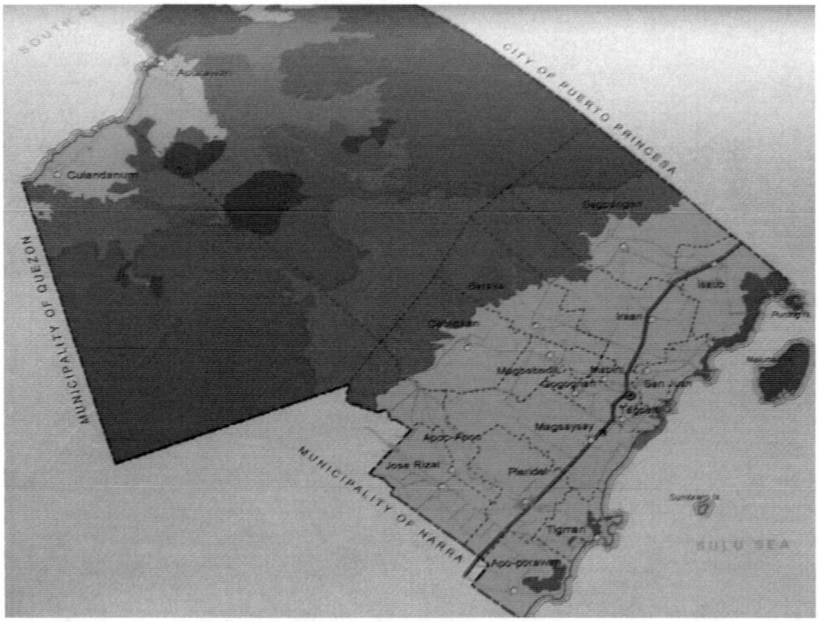

Figure 4. Environmentally Critic Area Network Map of Aborlan Palawan

Source: Aborlan MPDO Data Bank

ECAN, and are protected by Philippine Forestry Law from any human activity. It include sanctuaries of rare and endangered species in the villages of Apurawan and Culandanum. Selected coral reefs, seagrass and mangrove ecosystem reserves are found along the coastal areas in Apurawan, barangay Isaub and island Puntod facing Sulu Sea. Considering the distance of Aborlan (first town from Puerto Princesa City (going South). Cleopatras Needle in Puerto Princesa City serve as home for some endemic species of animals like the Palawan squirrel are also found in core zone in Aborlan. Restricted use zone are shown in blue patches and also found in the villages of Culandanum and Apurawan. These areas are inhabited by endemic and endangered species of flora and fauna like leopard cats. Spatial ecology study of Palawan leopard cats (*Prionailurus bengalensis heaneyi*) using live trapping, radio telemetry, and small mammal trapping from May 2013 to July 2014 in Aborlan, Palawan, Philippines, revealed the presence of the indigenous animal in the area [6]. Zone areas are posited in green patches found in villages of; Apurawan, Culandanum,Sagpangan, Barake and Cabigaan. Planting of crops are allowed for indigenous people that inhabited the areas, in order to sustain their lives. Multiple use areas are shown in yellow patches which are used for growing of agricultural crops. Most of the yellow patches are located in villages in lower slope and facing Sulu Sea. Minimal yellow patches are found in the villages facing South China Sea. Traditional use zone, (traditional agricultural practices) are presented in brown patches and also found in 2 villages facing South China Sea. Mangrove forest is displayed in purple small patch along the coastline of barangay Magsaysay facing Sulu Sea.

3.4. Soil map

Figure 5 posited the soil Map of Aborlan Palawan (patches of varying brown hues). Soil map of the municipality is described in varying hues of dark brown to light brown based on fertility of the, soil. Fertile areas facing Sulu Sea are found in villages of; San Juan, Mabini, Gogognan Magbabadel, Tigman, Apo-aporawan where main agricultural crops like rice corn, legumes are grown. Perennial crops like coconut, cashew, citrus family also thrive in these areas. Next on soil fertility are the vilages of; Isaub, Apoc-apoc, Magsaysay, Plaridel, Jose Rizal. Agricultural crops are also grown in these areas.

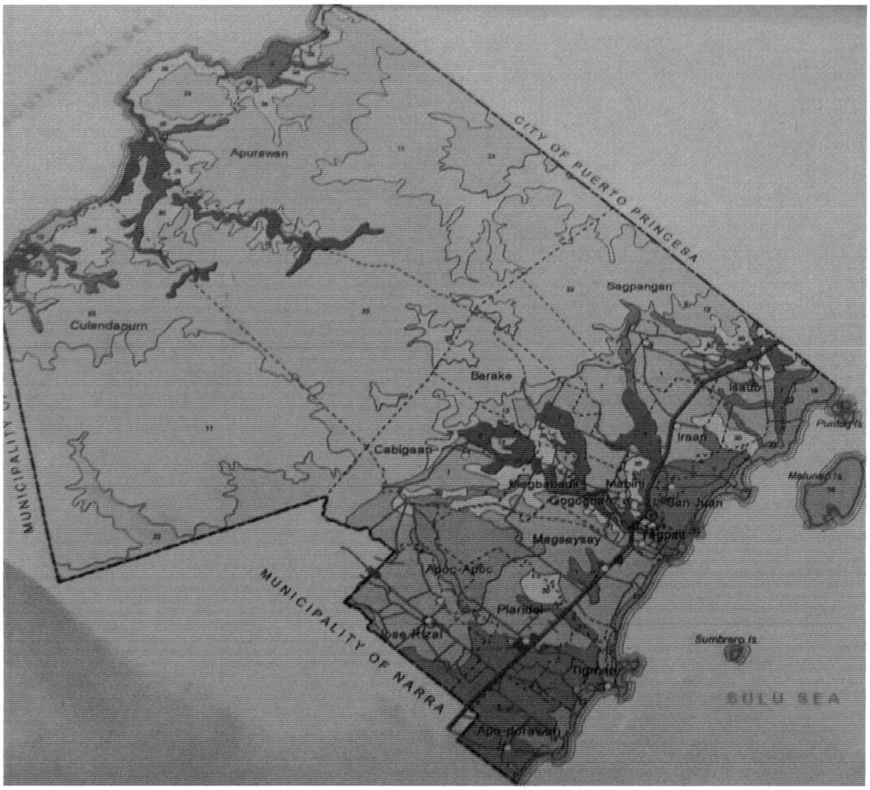

Figure 5. Soil Map of Aborlan Palawan

Source: Aborlan MPDO Data Bank

Villages facing South China is characterized by lighter color. Slope in this area is high compared with the villages facing Sulu Sea. Agricultural perennial crops like coconut thrive in dark brown to light brown patches near the coastlines, and near Apurawan river.Tigman, Apo-aporawan where main agricultural crops like rice corn, legumes are grown. Perennial crops like coconut, cashew, citrus family also thrive in these areas. Next on soil fertility are the vilages of; Isaub, Apoc-apoc, Magsaysay, Plaridel, Jose Rizal. Agricultural crops are also grown in these areas. Less fertile soils in villages facing Sulu in villages of: Sagpangan, Barake and Cabigaan,since these villages are found in higher slope. Less fertile soils in villages facing Sulu in villages of: Sagpangan, Barake and Cabigaan, since these villages are found in higher slope. Lowland crops are not suited to these villages, thus farmers in those areas, plant long variety of rice which has lower harvest than the lowland rice. Indigenous people (*Tagbanua*) mostly inhabited the three villages, where they rely much on plants in the forest like rattan gathering, honey gathering during summer months. Some species of orchids typically found in forest and some Indigenous People still sold it in lowlands.

3.5. Land classification map of Aborlan Palawan

Figure 6 shows the land classification map of Aborlan, Palawan. More than the majority (57%) of Aborlan is classified as protected land as shown in dark green matrix, where almost 60% of the villages facing South Sea China Sea and about 25 % of the land in barangays facing Sulu Sea. It covers approximately 80% of Sagpangan, 50% of Isaub, 40% of Iraan, 60% of Barake and 40% of Cabigaan These villages are inhabited by *Tagbanuas* and rely much of their livelihood on forest products. Large patch of timber/forest land, represented by light blue green patches, found mostly in villages of Culandanum and Apurawan. Small patches are found in Iraaan, Isaub and Magasaysay and Jose Rizal, where planted trees are owned by private farm owners. Large patches of arable land (light

IOP Conf. Series: Earth and Environmental Science **257** (2019) 012015 doi:10.1088/1755-1315/257/1/012015

pink) are mostly found in villages facing Sulu Sea and small patches near the coastal areas in Culandanum and Apurawan facing South China Sea

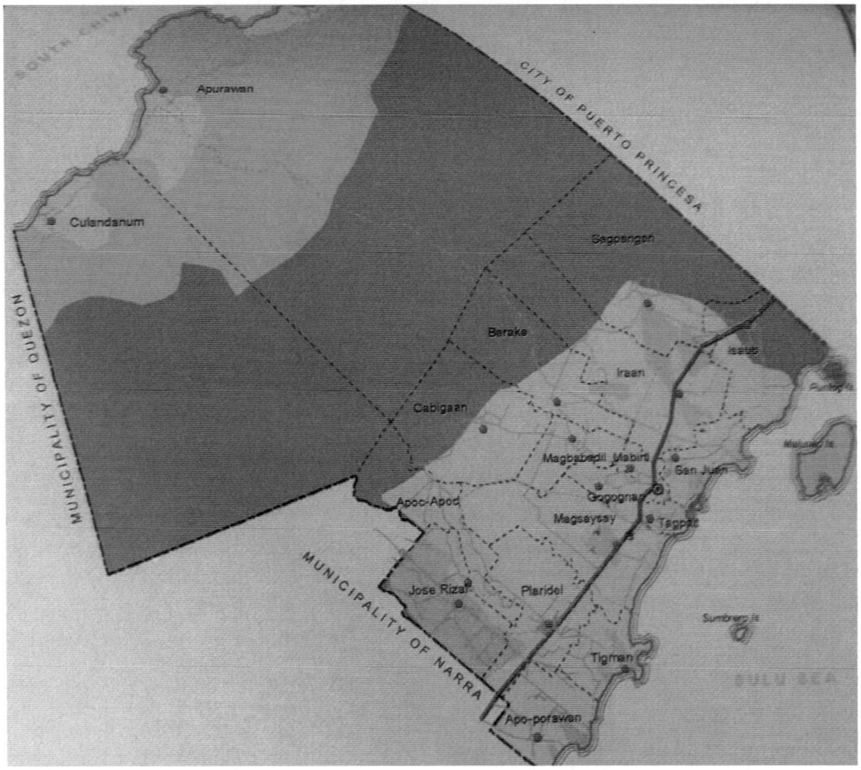

Figure 6. Land Classification Map of Aborlan Palawan
Source: Aborlan MPDO Data Bank

3.6. Existing land use map

Figure 7 illustrates the Existing general Land Use Map. Large dark green matrix is characterized by forest mostly found in 2 villages facing Sulu Sea. Light green patches characterize by grassland are mostly found in villages facing Sulu sea and small light green patches in 2 villages facing South China Sea. Agricultural lands are represented by light blue patches which are mostly found in villages facing Sulu Sea, and very small patches along the coastline of Culandanum facing South China Sea. Swamp areas (dark blue lines) are along the coastline of villages facing Sulu Sea and South China Sea. Mangroves are represented in pink patches near the coastline of barangays facing Sulu Sea and very small pink patch in barangay Culandanum. Although very minimal in Aborlan landscape, still contributed to the coastal area of the country. The coastal areas of the Philippines provide a continuous supply of goods—fish, oil, gas, minerals, salt, and construction materials— and services such as shoreline protection, sustaining biodiversity, maintaining water quality, transportation, and recreation. The direct benefits from mangroves have been estimated to be more than $600 per hectare per year in fish production and potential sustainable wood harvest. The annual direct benefits in the Philippines from mangroves are therefore at least $83 million per year [7].

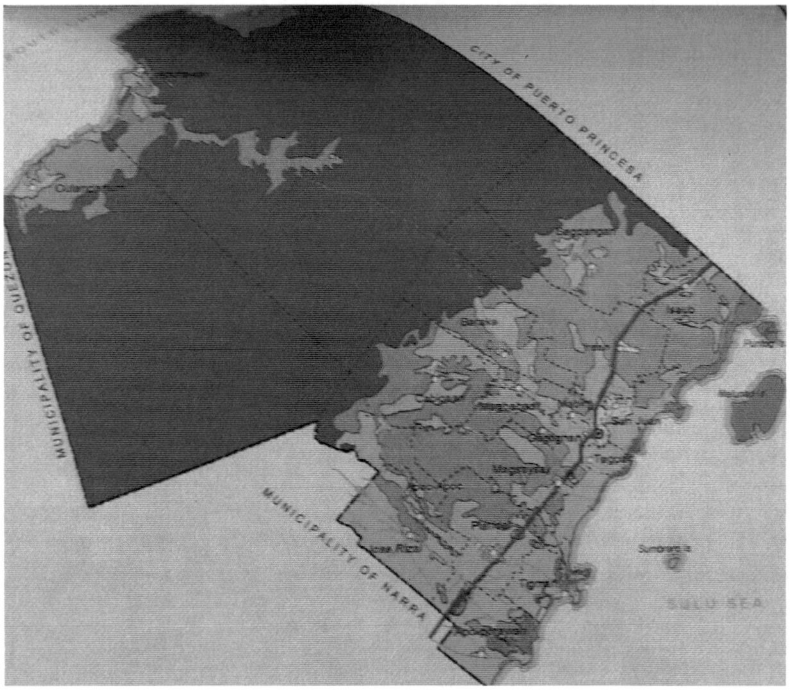

Figure 7. Existing General Land Use Map of Aborlan Palawan, Philippines
Source: Aborlan MPDO Data Bank

Table 1. Employment Rate (2014) in Aborlan Palawan

Barangay	Number of Households of the labor force	Employed members of the labor force	
		Magnitude	**Proportion**
1.Apo-Aporawan	315	293	93.02
2.Apoc-apoc	246	241	97.97
3.Apurawan	1357	1323	97.49
4.Barake	121	118	97.52
5. Cabigaan	750	727	96.93
6. Gogognan	229	220	95.07
7. Iraan	262	247	94.27
8.Isaub	333	310	93.09
9. Jose Rizal	309	297	98.12
10.Mabini	322	307	95.34
11. Magbabadil	396	377	95.2
12.Plaridel	952	899	94.43
13. Ramon Magsaysay	531	513	96.61
14. Sagpangan	481	469	97.51
15. San Juan	804	745	92.66
16. Tagpait	347	330	95.1
17. Tigman	579	557	96.2
18.Poblacion	681	651	95.59
19. Culandanum	374		
Total	9342	8942	96.72

Source: Aborlan MPDO Data Bank

3.7. Influence on employment rate

Employment rate as revealed in Table 1. Based on the report of MPDO (2014), there are 20,928 individuals with ages 10 years old and over were the source of data for employment in the municipality. Out of this number, 9,342 belongs to the labor force and 8,942 were in the workforce either employed or self –employed. Highest employment was found in the villages of Jose Rizal and Apoc-apoc (facing Sulu Sea) wherein, these areas are located near road networks and agricultural production zones. No data on employment was reported in the village of Culandanum, due to the remoteness of the village. Local facilities and employment opportunities is critical during times of need, especially for health, employment and social reasons [8].The survival of small upland farmers depends on whether they would have more scientific knowledge about farming, favourable incentives to grow environment- friendly crops such as trees and other perennials, and access to nonfarm income sources [9].Employment conditions in Philippine rural labor markets and agriculture can be characterized as casual or informal, with low skill requirements, with low productivity and returns, and a greater concentration of poverty. This is consistent with a prominent strand of development literature that posits a traditional sector, mostly located in rural areas, and highly depending on agricultural livelihood.

3.8. Influence of land structures on the income of the population

How this land components influence the income of the population, data from Aborlan MPDO (2015) such as employment and income was gathered. There are 3,421 households with income below the

Table 2. Income of the Population

Barangay	Number of Households	Households with Income below Poverty Threshold	
		Magnitude	**Proportion**
1.Apo-Aporawan	220	66	30
2.Apoc-apoc	203	91	44.83
3.Apurawan	874	433	49.54
4.Barake	176	86	48.86
5. Cabigaan	504	344	68.25
6. Gogognan	150	101	67.33
7. Iraan	370	148	40
8.Isaub	462	234	50.65
9. Jose Rizal	238	109	45.8
10.Mabini	247	124	50.2
11. Magbabadil	284	154	54.23
12.Plaridel	591	240	40.61
13. Ramon Magsaysay	569	253	44.46
14. Sagpangan	359	237	66.02
15. San Juan	596	188	31.54
16. Tagpait	237	127	53.59
17. Tigman	390	95	24.36
18.Poblacion	519	149	28.71
19. Culandanum	321	242	75.39
Total	7310	3421	46.8

Source: Aborlan MPDO Data Bank

poverty threshold or about 46.8 % of the total Households (Table 2). At the village level, Culandanum registered the highest proportion with 75.39% of the total households fall below the poverty incidence, followed by Cabigaan with 68.25% and Gogognan at 67.33%. This can attest to the land components, where areas near the ECAN cannot fully utilize the land even though it is more fertile, since it is restricted and mostly composed of forest and timber land. Close to 1.6 billion people – more than 25% of the world's population – rely on forest resources for their livelihoods and most of them (1.2 billion) use trees on farms to generate food and cash. Food Agriculture Organization, promoted the concept of better forestry, less poverty which suggests ways to prevent, mitigate, and reduce poverty through forestry-based interventions in rural areas [10]. At present, the Philippine uplands are occupied and cultivated by roughly 24 million people—one-third of the country's

IOP Conf. Series: Earth and Environmental Science **257** (2019) 012015 doi:10.1088/1755-1315/257/1/012015

population—with their lives and general wellbeing intimately linked with the forest and its resources. On these lands, a large percentage of the population suffers from extreme poverty [11].

4. Conclusion

The use of landscape maps (road network, water catchment, ECAN, soil map, land classification and land use) to assess the employment and income of the population can provide environmentalist, policy makers, researchers and laymen baseline information on whatever projects, and development plans that will be implemented in the municipality. Result of this study revealed that the landscape components of Aborlan Palawan, Philippines influenced the employment and income status of its population. It is imperative that the 57% of its total land area is ECAN, had influenced the employment and income status of the population. The lay back employment and income status of the people near ECAN could be attributed to the protected areas in the said area.

5. References

[1] Tolessa T Senbeta F and M Kidane 2016 Landscape composition and configuration in the central highlands of Ethiopia central highlands of Ethiopia. Ecology and Evolution.6(20) pp 7409–7421.
[2] Almo F 2001.Landscape and their ecological components. The Living World. (Lunigiana Museum of Natural History Aulla, Massa, Italy) 4 pp.435-444.
[3] Municipal Planning and Development Offices- Aborlan 2014. Municipality of Aborlan Palawan-Socio-economic Profile
[4] Mattson A Fors S and I Karehold 2017 Different indicators of socioeconomic status and their relative importance as determinants of health in old age. International Journal of Equity Health. 16(1) p 173
[5] Menoro M and A Tablizo 2017 The modern world: exploring the culture of tagbanua tribe in Palawan International Journal of Novel Research in Interdisciplinary Studies 4(2) pp 15-26
[6] Department of Environment and Natural Resources 2001 Slyvatrop. The Technical Journal of Philippine Ecosystem and Natural Resources. 28(1) pp 1-3
[7] White and Cruz-Trinidad 1998 .Philippine Coastal and Marine Resources: An Introduction pp.2-6
[8] Asian Development Bank. 2002 Impact of rural roads on poverty reduction: a case study- based analysis. pp.1-128
[9] Bagares I 2003 Farm income trends in the uplands: results of a nine-year study in Bukidnon, Philippines. ISPPS Working Paper 03-10, University of the Philippines Los Baños, College Laguna, Philippines.
[10] Food Agriculture Organization 2015 Forests and Poverty Reduction p 1
[11] Espiritu N Casin M and S Camacho 2010 Development pathways in the Philippine uplands impacts and influences on forest resource management and human well-being Asian Journal of Agriculture and Development. 7(2) pp 27-47

IOP Conf. Series: Earth and Environmental Science **257** (2019) 012016 doi:10.1088/1755-1315/257/1/012016

Treatment of waste emulsion using coagulation method

M. Najib Razali[1], M. A. Fattah Ahmad[1], N. Adilah Md Salehan[1] and Amira E. Anuar[1]

1 Faculty of Chemical & Natural Resources Engineering, Universiti Malaysia Pahang, Tun Razak Highway, 26300 Kuantan, Pahang, MALAYSIA
E-mail: najibrazali@ump.edu.my

Abstract. Waste emulsion is one of the major toxic wastes that are generated from the metal processing industry. Emulsions are commonly used in the metal processing production as emulsified coolant and in power plant as lubricating oil. This paper presents a research to suggest the best coagulant in treating waste emulsion from manufacturing plant via coagulation method. Chitosan and bio-solvent will represent the natural coagulants, alum and polyaluminium chloride (PAC) will represent the industrial coagulant. Jar Test apparatus was used to treat the waste emulsion. It will be conducted in a batch test with a row of six beakers. One jar will be used as a control and the other five will be of different dosages of coagulant at 150 rpm of mixing rate, 30 minutes of mixing time and at room temperature. From the results, PAC showed to be the best coagulant with 82.29%, 95.45%, 99.95%, 96.4%, and 53% removal of COD, BOD O&G, TSS, and Turbidity, respectively, at a pH of 3.7 and minimum dosage of 2wt%. In conclusion, the PAC managed to break the oil and water bonding from the waste emulsion. Thus, it was suitable to treat the industrial waste emulsion.

1. Introduction

Water is one of the most important substances on earth. All plants and creatures must have water to survive. In the event that there is no water, there would be no life on earth. It covers around 71% of the Earth's surface, and is vital for all for all known forms of living. However, just 2.5% of the Earth's water is fresh water. Economic growth and industrialisation in Malaysia has led to better living quality but at a large cost of excessive waste and pollutants. With the current population rate at 1.4%, rapid progress in urbanisation and industrialisation has become a challenge in scheduled waste management [1]. According to the Department of Environment (DOE), a total of 2,918,478.34 metric tonnes of scheduled wastes were generated in Malaysia in 2015, all of which are dross/slag/clinker/ash, heavy metal sludge, gypsum, and oil and hydrocarbon were again the main categories. Waste emulsion is one of the major toxic wastes that are generated from the metal processing industry. Emulsions are commonly used in metal processing productions as emulsified coolant and also in power plant as lubricating oil. They are also commonly referred as cutting fluid, cutting oil, coolant, or lubricant. Usually, after they have lost their efficiency, these emulsion waste will be discharged or sent for treatment [2]. It is categorised as oil and hydrocarbon type of scheduled waste. Even so, oil and hydrocarbon type of scheduled waste is only 2.69% from the total 100%. This type of scheduled waste must not be looked at lightly, where this type of scheduled waste can rise to the top of the list if improper disposal is done.

Content from this work may be used under the terms of the Creative Commons Attribution 3.0 licence. Any further distribution of this work must maintain attribution to the author(s) and the title of the work, journal citation and DOI.
Published under licence by IOP Publishing Ltd

Majority of the industrial wastes in Malaysia is sent to Kualiti Alam Sdn. Bhd. for treatment and disposal periodically. From a conducted survey, a manufacturing plant will produce around 20 tonnes of wastewater consistently in a month. This implies that 240 tonnes of wastewater is produced each year. This scheduled wastewater is send to Kualiti Alam Sdn. Bhd. at a rate of RM3,000.00 for every tonne; henceforth, each plant will spend more than RM720,000.00 every year. In Malaysia, there are more than 100 manufacturing plants operating and still developing. In this way, the plant will have to spend RM72 million yearly to treat this scheduled waste [2]. Since a huge amount of money is being spent by the industry to treat the wastewater, it is feasible to have a treatment facility that runs on an affordable budget, yet has high treatment performance. This study is going to evaluate the coagulation performance of chitosan, alum, bio-solvent and polyaluminium chloride (PAC) in handling this matter. Breakthroughs on the water treatment technologies have been developed over the years. Chemical treatment is one of the most useful and utilised techniques in treating waste emulsion [3]. Chemical treatment usually involves coagulation and flocculation process as a primary treatment, followed by a secondary stage treatment. As for the secondary stage treatment, it involves filtration, foam flotation, ion exchange, aerobic and anaerobic, electrolysis, solvent extraction, and adsorption treatment as a selection for a secondary stage treatment [4]. For this study, coagulation and filtration method will be employed by using common industrial coagulant and natural coagulant as primary treatment, followed by filtration using filter paper as the secondary stage treatment. Chitosan and bio-solvent will represent the natural coagulant, alum and polyaluminium chloride (PAC) will represent the common industrial coagulant. Coagulation method is selected because of its ability to deal with organic matter. In this case, the dissolved oil inside the waste emulsion can be broken using the coagulation method, based on the previous study [5]. Once the bond between oil and water is broken, it is easy to recover the water or oil for further usage. There are two types of coagulants selected because, according to a handful of previous studies, the common industrial coagulant used is chemical-based coagulant. Chemical coagulant may enhance the toxicity of the treated waste emulsion due to the reaction between the colloidal particles inside the waste emulsion with the chemical coagulant. This is different from the natural coagulant that will not enhance the toxicity of the treated waste emulsion because the coagulant is made from bio-material [6].

Alum or ($Al_2(SO_4)_2$) is the commonly used chemical coagulant in the industry. Since ancient times, this coagulant is popular to be used to treat suspended solid (TSS), turbidity and volatile suspended solid (VSS) in the wastewater. Alum was found to be the most effective coagulant for this application as compared to ferric chloride and lime. Alum was also the most effective coagulant in treatment of white and red wine wastewater [7]. Chitosan or poly-(1-4)-2-amino-2-deoxy-b-d-glucose is a biopolymer that can be chemically expressed as non-toxic, heterogeneous, linear, cationic, and biodegradable polysaccharide with high molecular weight produced by alkaline deacetylation of chitin [8], where chitin or poly(β -(1-4)-N-acetyl-D-glucosamine) is the major component of crustaceans shell such as crab, lobster, shrimp, and crawfish. Chitin and chitosan are both insoluble in water as well as in most organic solvents due to it hydrogen bonds between its molecules. However, it is soluble in acidic solutions due to the protonation of its amine groups. Due to that property, chitosan potentially has high affinity to adsorb pollution such as heavy metals and dye [9]. Bio-solvent is water-based dispersant that are formulated with an active ingredient called surfactants. Surfactant is design to have hydrophilic (water liking) & oleophilic (oil liking). Moreover, surfactants can reduce the surface tension of hydrocarbon molecules in oil, and the oil will spread out and be collected in droplets. Surfactant molecules possess hydrophilic head groups that associate with water molecules, while the oleophilic tails associate with oil molecules. This oil dispersant or degreaser product contains many different surfactants ingredient blended with it to solve all oil characteristic matters (EURO-CIRCUIT, 2017). PAC is a coagulant that derives from the action of its basic active constituent, namely, polyaluminium chloride. As a pre-hydrolysed coagulant, PAC is becoming increasingly popular in the coagulation-flocculation process. PAC has many benefits over the conventional hydrolysing aluminium or iron salts, such as better performance at low temperature, less aluminium residuals, less sludge volume, less effect on the pH value of raw water, and more rapid flocculation. Furthermore, there are large differences between the alum and PAC in morphology and precipitation in water [10]. This paper is a study done to suggest the best coagulant in treating waste emulsion from manufacturing plants via the coagulation method.

2. Materials and methods

2.1. Experimental materials

Approximately 50 kg of waste emulsion were collected from the CNC Machine at FTEK Lab. The collected samples were placed in a plastic container, sealed tight and labelled, before it was transported back to the laboratory. Samples were then refrigerated at about 4°C to prevent bio-activities from occurring. Analysis was done within 6 to 24 hours of collection from the plant [11]. Chitosan was purchased from Hunza Nutriceuticals Sdn. Bhd and obtained from the FKKSA lab in powder form with viscosities and DDA of about 30–3000 mPa.S (at 25°C) and 85–99%. 15 L of polyaluminum chloride in liquid form were purchased from i-Chem Solution Sdn. Bhd. 1 L of bio-solvent in liquid form was purchased from Isnas Resources Sdn. Bhd. 3 kg of alum in powder form was purchased from Train Aid Global Sdn. Bhd.

2.2. Experimental procedure

2.2.1. Preparation of chitosan. Chitosan powder was used as a starting material in this work. The procedure for the activation of chitosan was prepared by diluting it with acetic acid to produce a gel-like substance. Dilution was conducted by using a self-estimated 7:2:91 ratio of chitosan to acetic acid to distilled water to produce a certain chitosan concentration. 35 ml of acetic acid was added into 455 ml of distilled water. 10 g of chitosan powder was later added slowly into the stirring solution to improve or promote well mixing and was stirred rapidly for 15 minutes. The prepared 500 ml of chitosan coagulant was stored in a beaker for further use [2].

2.2.2. Preparation of alum. Alum powder was used as a second material in this work. The procedure for the activation of alum was prepared by diluting it with diluted water to produce alum solution. Dilution was conducted by using concentrations of alum obtained from journals. The desired concentration of alum solution is 5000 mg/L [11]. 5 g of alum powder was added slowly into 1000 ml of distilled water to improve well mixing. The solution was stirred rapidly for 15 minutes. The prepared 1000 ml of alum coagulant was stored in a beaker for further use.

2.2.3. Jar Test procedures. Jar Test apparatus was used treat the waste emulsion. It will be conducted in a batch test with a row of six beakers, each having a stirring device. One jar will be used as a control and the other five will be with different dosages of coagulant (2 wt%, 4 wt%, 6 wt%, 8 wt%, 10 wt%, and 12 wt%) at 150 rpm of mixing rate and 30 minutes of mixing time and at room temperature. Experiment was repeated by varying the coagulant types and dosages. Analysis for the desired parameter was done with treated water with all the coagulants.

2.2.4. Laboratory analysis. The analysis of the treated sample was carried out in the Central Lab of Universiti Malaysia Pahang. Parameters of pH, TSS, O&G, COD, BOD, and turbidity were analysed. These methods are based on the Examination Manual of Water and Wastewater [12]. Table 1 summarises the analysis method.

Table 1. Summarised experimental analysis method.

Parameter	Test Method	Equipment
pH	ASTM D1293-18 (ASTM, 2018)	Seven Easy pH-METTLER TOLEDO
TSS	ASTM D5907-18 (ASTM, 2018)	Laboratory apparatus
O&G	ASTM D3921 (ASTM, 2018)	Laboratory apparatus
Turbidity	ASTM D7315 - 17 (ASTM, 2018)	HACH 2100P
COD	ASTM D1252 - 06(2012)e1 (ASTM, 2018)	HACH Model DR/2400 and HACH Reactor
BOD	ASTM D 1252 - 60 (ASTM, 2018)	Yellow Springs Model 5010 and BOD Incubator

3. Results and discussion

3.1. Characterisation of coagulants

Characterising the coagulants is important to be conducted before further investigation can be performed in order to understand the nature and behaviour of the coagulants. It is also important, as it would explain the reason behind its response toward specific treatment that is adopted in this study. Hence, the study was conducted to examine the two critical parameters of the coagulant, as shown in Table 2. The table shows the conductivity and pH of the coagulants measured in the laboratory.

Table 2. Characteristic of coagulant

No	Coagulant	pH	Conductivity	Unit
1	Low molecular weight chitosan with DDA 99% (LMWC)	3.7	4.34	mS/cm
2	Medium molecular weight chitosan with DDA 99% (MMWC)	3.45	4.12	mS/cm
3	High molecular weight chitosan with DDA 99% (HMWC)	3.49	4.32	mS/cm
4	Chitosan powder with DDA 85%	3.3	2.72	mS/cm
5	Chitosan powder with DDA 75%	3.24	2.63	mS/cm
6	Poly aluminium chloride (PAC)	2.03	4.78	mS/cm
7	Bio Solvent	12.22	3.84	mS/cm
8	Alum	3.65	0.0245	mS/cm

Reasons to study both of the parameters in Table 2 are: first, is to see whether the coagulants have enough positively-charged particles or not to neutralise the negatively-charged or colloidal suspension in the waste emulsion for coagulation to occur. Second, is to select the best type of chitosan to be used in this study. From the result, it showed that low molecular weight chitosan with DDA 99% (LMWC) has a higher conductivity value compared to other types of chitosan, therefore, it will be chosen as one

IOP Conf. Series: Earth and Environmental Science **257** (2019) 012016 doi:10.1088/1755-1315/257/1/012016

of the coagulants that is going to be used in this study. Lastly, the pH value will indicate whether the coagulant can reduce or increase the waste emulsion pH value.

3.2. Characterisation of coagulants

Characterising the waste emulsion is important to be conducted before further investigation can be performed in order to understand the nature and behaviour of the coagulants. It is also important, as it would explain the reason behind its response toward specific treatment that is adopted in this study. The waste emulsion was compared with the Standard A and Standard B water quality parameter limits set by the Department of Environmental Malaysia (DOE). Discrepancy and comparison of the focused parameters are discussed in this chapter. In addition, from the comparison, it will provide a standard guideline to be compared with the treated wastewater.

Table 3. Characteristics of waste emulsion

No	Parameter	Unit	Water Quality Standards		Raw CNC Wastewater
			A	B	
1	Temperature	°C	40	40	23.3
2	pH Value	-	6.0-9.0	5.5-9.0	5.1
3	Suspended Solids	mg/l	50	100	More than 750
4	COD	mg/l	50	100	14,000
5	BOD5 at 20°C	mg/l	20	40	352
6	Oil and Grease	mg/l	1.0	10	969.0

Table 4. Physical characteristics of waste emulsion

No	Parameter	Unit	Result
1	Turbidity	NTU	100

Table 3 and Table 4 listed the values for the tested parameters of waste emulsion sample before the treatment processes. Standard A is a more stringent list compared to Standard B. This is because Standard A is applied as a regulation at the water catchment areas, which includes areas upstream of surface or above sub-surface of waterways, mainly directed to the community and for human use.

3.3. Effect of dosage on reduction of Chemical Oxygen Demand (COD)

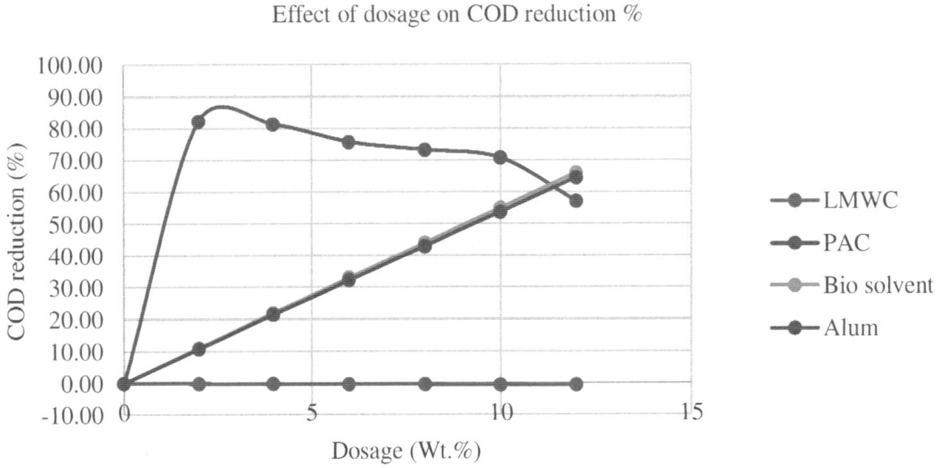

Figure 1. Effect of dosage on COD reduction (%)

From Figure 1, it can be seen that at the highest coagulant dosage (12 wt%), LMWC showed the lowest reduction of COD compared to other coagulant. Meanwhile, at the same dosage, bio-solvent showed the highest reduction of COD. PAC has 57.07% of COD reduction, while bio-solvent reduced 66.00%, LMWC reduced 0%, and alum reduced 64.50% at the maximum dosage. On the other hand, at the lowest coagulant dosage (2 wt%), PAC showed the highest reduction of COD compared to other coagulants. Meanwhile, at the same dosage, LMWC also showed the lowest reduction of COD. PAC has 82.29% of COD reduction, while bio-solvent reduced 11.00%, LMWC reduced 0% and alum reduced 10.57% at the minimum dosage. Based on the results, PAC proved to be the best coagulant in reducing COD from waste emulsion with a minimum dosage of injection.

Waste emulsion with high PAC dosage contains highly positive charged particle in exceed, that will make them repel to each other thus lowering the chance for them to attract to the negatively charged particle inside the waste emulsion. For Bio Solvent and Alum, as the dosage increase, the more the positively charged particle can settle down the negative charged, At this point, the chemical compound inside the waste emulsion used less oxygen to break the organic matter due to the organic matter is attracted to the positive charge particle. As the pH value of the waste emulsion becoming more acidic, the higher the effectiveness of coagulation, the higher the probability of organic matter to be attracted to the positive charge particle. Thus reducing the COD of the waste emulsion. As for LMWC the COD increase as the dosage increase.

3.4. Effect of dosage on reduction of Biological Oxygen Demand (BOD)

From Figure 2, it can be seen that at the highest coagulant dosage (12 wt%), PAC showed the highest reduction of BOD compared to other coagulants. Meanwhile, at the same dosage, LMWC showed the lowest reduction of BOD. PAC has 94.03% of BOD reduction, while bio-solvent reduced 75.10%, LMWC reduced 68.09% and alum reduced 71.47% at the maximum dosage. On the other hand, at the lowest coagulant dosage (2 wt%), PAC also showed the highest reduction of BOD compared to other coagulants. Meanwhile, at the same dosage, LMWC also showed the lowest reduction of BOD. PAC has 95.59% of BOD reduction, while bio-solvent reduced 19.76%, LMWC reduced 18.55% and alum reduced 19.16% at the minimum dosage. Based on the results, PAC proved to be the best coagulant in reducing BOD from waste emulsion with a minimum dosage of injection.

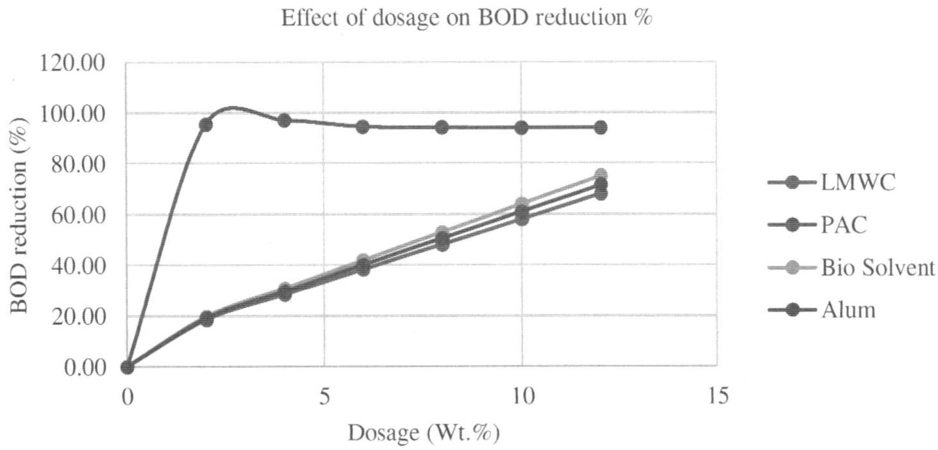

Figure 2. Effect of dosage on BOD reduction (%)

The results show that PAC have highest removal at minimum dosage due to this type of coagulant and have advantages to coagulate with a minimum number of dosage. At this point, the biological compound inside the waste emulsion used less oxygen to stabilize the organic matter. The organic matter is adsorbed by the positive charge particle as the pH value of the waste emulsion becoming more acidic. The higher the effectiveness of coagulation, the higher the probability of organic matter to be attracted to the positive charge particle thus reducing the BOD of the wastewater.

3.5. Effect of dosage on reduction of oil residue (O&G)

From Figure 3, it can be seen that at the highest coagulant dosage (12 wt%), LMWC showed the lowest reduction of O&G compared to other coagulants. Meanwhile, at the same dosage, alum showed the highest reduction of O&G. PAC has 93.29% of O&G reduction, while bio-olvent reduced 97.52%, LMWC reduced 92.98%, and alum reduced 99.48 % at the maximum dosage. On the other hand, at lowest coagulant dosage (2 wt%), PAC showed the highest reduction of O&G compared to other coagulants. Meanwhile, at the same dosage, LMWC also showed the lowest reduction of O&G. PAC has 99.90% of O&G reduction, while bio-solvent reduced 16.20%, LMWC reduced 15.48%, and alum reduced 16.62% at the minimum dosage. Based on the results, PAC proved to be the best coagulant in reducing O&G from waste emulsion with a minimum dosage of injection.

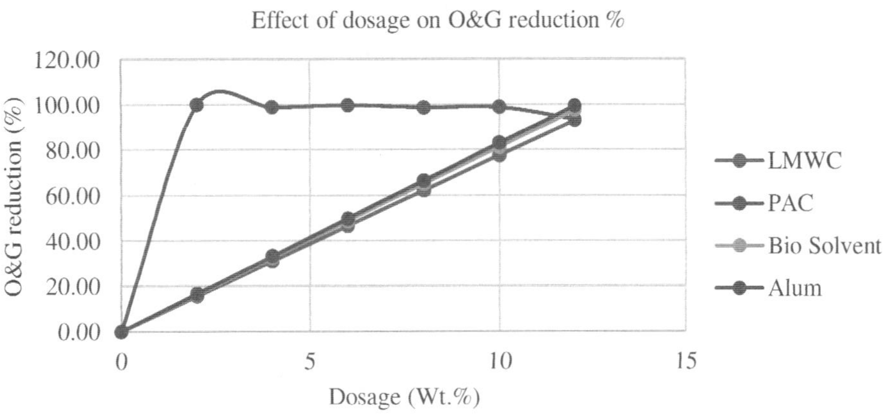

Figure 3. Effect of dosage on O&G reduction (%)

The results indicated that PAC able to break oil bond with water at a lower dosage. At this point, the dissolved solids in the oil residue in the waste emulsion are absorbed by the positively charged particle in the coagulants. Therefore the waste emulsion loss it emulsion characteristic thus reducing the O&G of the wastewater.

3.6. Effect of dosage on the reduction of TSS

From Figure 4, it can be seen that at the highest coagulant dosage (12 wt%), PAC showed the lowest reduction of TSS compared to other coagulants. Meanwhile, at the same dosage, bio-solvent showed the highest reduction of TSS. PAC has 3.07% of TSS reduction, while bio-solvent reduced 77.20%, LMWC reduced 53.20%, and alum reduced 2.40% at the maximum dosage. On the other hand, at the lowest coagulant dosage (2 wt. %), PAC showed the highest reduction of TSS compared to other coagulant. Meanwhile at the same dosage, alum showed the lowest reduction of TSS. PAC has 96.40% of TSS reduction, while bio-solvent reduced 12.80%, LMWC reduced 8.80%, and alum reduced 0.40% at the minimum dosage. Based on the results, PAC proved to be the best coagulant in reducing TSS from waste emulsion with a minimum dosage of injection.

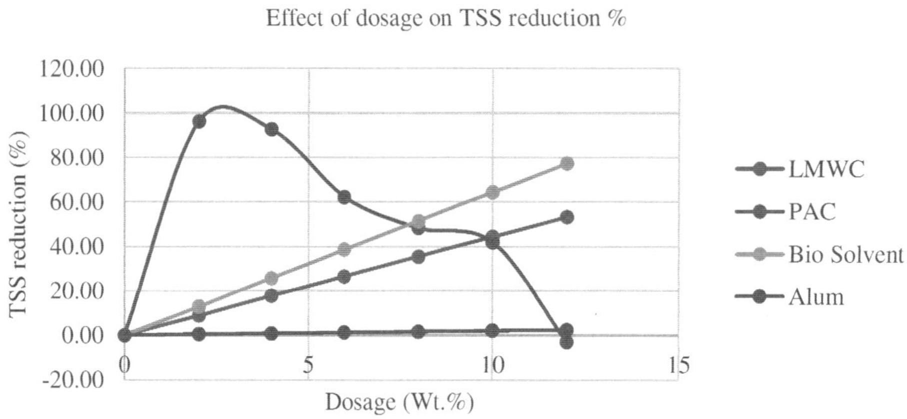

Figure 4. Effect of dosage on TSS reduction (%)

The results show that PAC showed to have an increasing and decreasing trend. Because when the waste emulsion is rich with PAC dosage, then the positively charged particle is in exceed, that will make them repel to each other thus lowering the chance for them to attract to the negatively charged particle inside the waste emulsion. At this point, the colloidal particle in the sample is absorbed by the positively charged thus reducing the TSS of the wastewater.

4. Conclusion

The present exploratory works showed that the PAC coagulant can be used for waste emulsion treatment. The optimum dosage needed for PAC, for coagulation of 500 mL of waste sample is 2wt%. The optimum operating condition for coagulation is at 150 rpm of mixing rate and 30 minutes of mixing time and at room temperature. The PAC coagulant showed satisfying results with 82.29%, 95.45%, 99.95%, 96.4%, and 53% removal of COD, BOD O&G, TSS and Turbidity, respectively, at a pH of 3.7. Based on the results, it showed that the coagulation method using PAC had managed to break the oil and water bonding from waste emulsion. Thus, it was found suitable to treat industrial waste emulsion. Therefore, it is recommended to use PAC to ensure a feasible treatment for treating emulsion waste.

IOP Conf. Series: Earth and Environmental Science **257** (2019) 012016 doi:10.1088/1755-1315/257/1/012016

5. References

[1] Che Jamin, N., & Mahmood, N. Z. (2015). *Scheduled Waste Management in Malaysia: An Overview.* Switzerland: Trans Tech Publications

[2] Hon Yvain, C. (2014). Development of emulsified wastewater treatment system. *Faculty of Chemical & Natural Resources Engineering Project Paper (Bachelor of Chemical Engineering)*

[3] Jiang, J. Q. (2015). The role of coagulation in water treatment. *Chemical Engineering*, 36–44

[4] Rajasulochana, P., & Preethy, V. (2016). Comparison on efficiency of various techniques in treatment of waste and sewage water – A comprehensive review. *Resource-Efficient Technologies*, 175–184

[5] Muhammad, I., Tahir, B., Naz, I., Naeem, A., Ruaf, A. K., & Aamir, S. (2013). The removal of COD, TSS and colour of black liquor by coagulation–flocculation process at optimized pH, settling and dosing rate. *Arabian Journal of Chemistry*, 1-12

[6] Lee, C. S., Robinson, J., & Chong, M. F. (2014). A review on application of flocculants in wastewater. *Process Safety and Environment Protection*

[7] Teh, C. Y., Budiman, P. M., Shak, K. P., & Wu, T. Y. (2016). Recent advancement of coagulation-flocculation and its application in wastewater treatmentand its application in wastewater treatment. *Ind. Eng. Chem. Res*, 1-102

[8] Vakili , M., Rafatullaha, M., Salamatinia, B., Abdullah, A. Z., Ibrahim, M. H., Tan, K. B., Amouzgar, P. (2014). Application of chitosan and its derivatives as adsorbents for dye removal from water and wastewater: A review. *Carbohydrate Polymers*, 115–130.

[9] Ngo, D. H., Vo, T. S., Ngo, D. N., Kang, K. H., Je, J. Y., Pham, H. N.-D., Kim, S. K. (2015). Biological effects of chitosan and its derivatives. *Food Hydrocolloids*, 1-77.

[10] Ning, W., Zhongguo, Z., Dan, L., Yue, W., Jun, W., & Qunhui, W. (2015). Coagulation behavior of polyaluminum chloride: Effects of pH and coagulant dosage. *Energy, Resources and Environmental Technology*, 1-16.

[11] León-Luque, A. J., Barajas, C. L., & Peña-Guzmán, C. A. (2016). Determination of the Optimal Dosage of Aluminum Sulfate in the Coagulation-Flocculation Process Using an Artificial Neural Network. International Journal of Environmental Science and Development, 346-350.

[12] APHA. (2005). *Standard Methods for the Examination of Water and Wastewater.* Washington, DC.: 21st Edn., American Public Health Association.

Acknowledgement

The authors express their gratitude and appreciation for the financial support from the Ministry of Higher Education (MOHE), Malaysia for the Fundamental Research Grant Scheme (FRGS KPT – RDU160129, Reference Number: FRGS/1/2016/TK02/UMP/03/2 entitled Rheological and Structural Characterization of Emulsified Modification Bitumen Synthesized from Industrial Wastes) and the Universiti Malaysia Pahang for the L2M Grant (UIC161003 Bio-Coagulant and Smart System for Sustainable Water Treatment). The support from the Faculty of Chemical and Natural Resources Engineering and Universiti Malaysia Pahang, Malaysia are also acknowledged.

IOP Conf. Series: Earth and Environmental Science **257** (2019) 012017 doi:10.1088/1755-1315/257/1/012017

$NaHCO_3$-promoted olivine weathering with H_2 generation and CO_2 sequestration in alkaline hydrothermal system

J Wang[1], K Nakamura[1], N Watanabe[1], A Okamoto[1] and T Komai[1]

1 Graduate School of Environmental Studies, Tohoku University, Aramaki, Aoba-ku, Sendai, Japan
E-mail: wang.jiajie.q5@dc.tohoku.ac.jp

Abstract. Hydration of Fe(II) bearing minerals, such as olivine ((Mg,Fe)SiO_4), potentially generate hydrogen (H_2). However, because of the low Fe(II) dissolution rate, the H_2 production rate is usually low. We have recently proposed a novel system to promote H_2 production and simultaneous CO_2 storage in hydrothermal conditions via $NaHCO_3$-enhanced olivine weathering. The present study reports the role of $NaHCO_3$ on both H_2 production from olivine hydration and on minerals generations on laboratory experiments in CO_2-rich (0.5 mol/L $NaHCO_3$) and CO_2-free hydrothermal conditions at pH range of 8-11. The highest H_2, HCOOH yields and carbonation rate reached at the CO_2-rich experiment with lower alkaline pH at 8.25. The addition of high concentration $NaHCO_3$ decreased pH from 10.92 to 8.25, whereas olivine and brucite dissolutions were accelerated. Thus, more Fe(II) was released from olivine and brucite, and H_2 production was accelerated. At higher pH range, olivine dissolution was promoted, but brucite dissolution was suppressed in both CO_2-rich and CO_2-free conditions. This study suggests H_2 production was promoted with the presence of HCO_3^-, but not the pH variation. The consumption of Fe(II)-bearing brucite was important in enhancing H_2 production during olivine hydration process.

1. Introduction

The rising concentration of CO_2 in the atmosphere has been identified as a cause of climate change. Control of CO_2 emission and the exploration of non-carbon-based energy sources are the leading scientific and technological challenges of the 21st century. Hydrogen (H_2) is one of the energy carriers that potentially ensure a sustainable energy future [1]. By using H_2 gas, CO_2 could be converted into valuable chemicals [2,3]. However, current H_2 production processes (e.g., steam reforming of natural gas) usually needs high cost and causes substantial CO_2 emission [1].

High-temperature hydration of ultramafic rocks often accompanied with H_2 production, owing to the oxidation of reduced Fe present in the minerals [4,5]. Olivine ($(Mg,Fe)_2SiO_4$) is one of the Fe(II) bearing minerals, which has the highest content of the ultramafic rock in the crust [6]. The hydration of olivine potentially offers significant H_2 supply. This process has been well studied in the geological environment [6-8]. However, few studies have been carried out to enhance the poor H_2 generation rate from olivine hydration.

The low H_2 generation is partially attributed to the low olivine dissolution rate, which can be enhanced by varying reaction conditions, such as elevating temperature and using acidic conditions [9,10]. We have recently proposed a novel system using the CO_2-rich hydrothermal constitution to achieve mass H_2 production and simultaneous CO_2 storage [11]. High concentration of $NaHCO_3$ is utilized to promote olivine dissolution and H_2 generation, then CO_2 was reduced to formic acid (HCOOH) or

Content from this work may be used under the terms of the Creative Commons Attribution 3.0 licence. Any further distribution of this work must maintain attribution to the author(s) and the title of the work, journal citation and DOI.
Published under licence by IOP Publishing Ltd

sequestrated in magnesite ($MgCO_3$). While the CO_2-promotion on olivine hydration was experimentally confirmed, the mechanism is less clear.

The role of $NaHCO_3$ in H_2 production from olivine hydration may be complex and multifaceted. In most previous researches concerning olivine carbonation, the improved carbonation rate with the addition of high concentration of $NaHCO_3$ was explained as a "catalyst effect" since the bicarbonate concentrations in liquid phase did not significantly change before and after the carbonation experiments [12]. The addition of $NaHCO_3$, a pH buffer, also causes pH variation. Previous research has proposed that under acidic conditions, more Mg^{2+} at the surface of olivine will readily exchange with available protons at lower pH, thus enhance olivine dissolution [10]. Olivine dissolution in the natural system commonly creates alkaline conditions, the addition of a high concentration of $NaHCO_3$ lows system pH, which potentially enhances olivine dissolution. Gadikota et al. (2014) [13] infer that $NaHCO_3$ is not a catalyst, but instead serves as a pH buffer and a source of carbonate ions during olivine carbonation. In the case of alkaline pH range, the olivine dissolution kinetics and mechanism are less clear [14, 15]. Some studies claim to have found a pH dependence of the dissolution kinetics, while others argue that the kinetics is almost independent of pH.

To investigate the separate effects of $NaHCO_3$ and pH on H_2 production via olivine hydration, experiments with and without CO_2 at pH range of 8-11 were conducted in the present study. Secondary minerals generations after the reaction was also investigated to reveal the mechanisms. Gas, fluid and mineral samples were withdrawn during the reaction and analyzed to determine the H_2 production and carbonation rate.

2. Materials and methods

2.1. Materials

Purchased olivine grains were chosen for experimental investigation, which as collected from Damaping (China). Then olivine was ball-milled (Pulverisette 6, Fritsch) to diameter <45 μm. The composition of the olivine was measured using Electron probe micro-analyzer (EPMA) and was defined as $Mg_{1.8}Fe_{0.2}SiO_4$. Solid $NaHCO_3$ was purchased from Kanto Chemical (Japan). NaOH and HCl be used for pH adjustment were obtained from Wako (Japan).

2.2. Olivine weathering experiments

Experiments were performed in a closed-batch reactor made of Hastelloy-C with an internal volume of 170 mL (Figure 1). In each experiment, 100 mL slurry prepared with 5 ± 0.01 g of olivine powders and 0.5 mol/L $NaHCO_3$ solution or Milli-Q water was poured into the reactor. pH was measured and adjusted using NaOH and HCl solution to desired ones before the experiment. Then the reactor was closed and purged with N_2 gas for 10 minutes to remove dissolved O_2 in the solution and the upper headspace. Then air outlet was closed and N_2 gas continues to be injected to reach a certain pressure. The reactor was well sealed to increase the temperature to 300 °C, with the final pressure reached 10 MPa. Gas, liquid and solid samples were withdrawn via sampling tubes. After 72 h, the experiment was stopped by reducing the temperature of the reactor using recirculated cooling water. When lowing the temperature to ambient ones, the pH of the reacted slurry was measured again. The mineral powder was filtered and dried at 50 °C for 24 h in an oven before further analysis.

2.3. Measurement and Characterization

Liquid samples were analyzed using ion chromatography (IC; 761 Compact IC, Metrohm, Switzerland) coupled with a Metrosep Organic Acids column (Metrohm, Switzerland). Gas species were analyzed using gas chromatography (GC; GC-3200, GL Science, Japan) equipped with a thermal conductivity detector (TCD). Thermogravimetric analyses (TGA) of all the weathered olivine were performed using a thermodilatometry (Thermo plus EVO TG 8120, Rigaku, Japan) to measure mineral compositions. The temperature was increased from room temperature to 1000 °C at a rate of 10 °C per minutes. The surface morphologies of the minerals were observed using scanning electron microscopy (SEM; SU-8000, Hitachi, Japan) equipped with an Energy Dispersive Spectroscopy (EDS).

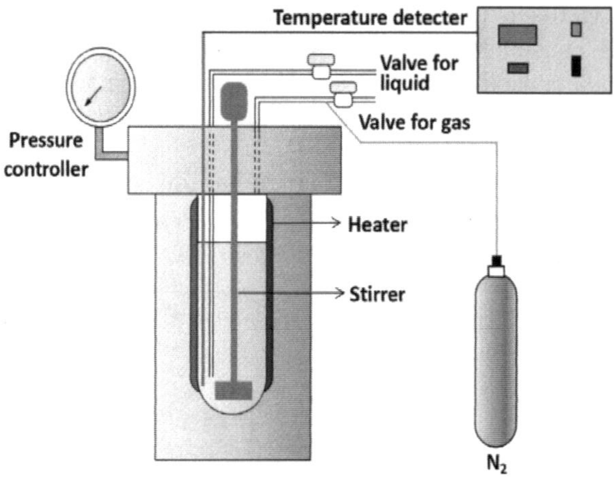

Figure 1. Schematic diagram of the experimental set-up.

3. Results and discussions

3.1. Energy production

In the present study, CO_2-rich olivine weathering experiments were conducted in 0.5 mol/L $NaHCO_3$ solution at 300 °C and 10 MPa. Before pH adjustment, the initial pH of CO_2-rich olivine slurry was 8.25. Other three slurries with different pH values of 9.36, 10.11 and 10.87 were then prepared for experiments. After 72 h reaction, the pH of the reacted slurries increased to 9.49, 9.71, 11.02 and 12.06, respectively (Table 1). The increases in pH after the reaction were attributed to the consumption of H^+ by olivine dissolution and generation of OH^- during the olivine hydration process [16].

Table 1. pH variation before and after 72 h reaction

	CO_2-rich experiments				CO_2-free experiments			
Exp.	1 (unadjusted)	2	3	4	5	6	7	8 (unadjusted)
Initial pH	8.25	9.36	10.11	10.87	8.28	9.33	10.12	10.92
Final pH	9.49	9.71	11.02	12.06	8.86	10.24	10.30	10.21

During reactions, gas and liquid samples were withdrawn at the reaction time of 72 h. Gaseous H_2 and liquid HCOOH was experimental detected as the main products in the present study. The products yields in 72 h are summarized in Figure 2. H_2 and HCOOH yields in 72 h decreased slightly with the increase in the initial system pH. When the initial pH was adjusted from 8.25 to 10.87, the H_2 generation over 72 h dropped from 196.6 to 156.3 mmol/kg of olivine, while the HCOOH yield also decreased from 54.6 to 24.5 mmol/kg of olivine. The decrease in H_2 production is attributed to the retarded Fe(II) oxidation process.

To reveal the separate effect of pH changes, CO_2-free experiments were also conducted at a pH range of 8-11. The initial pH of olivine powders (5 g) with CO_2-free Milli-Q water was 10.92. Slurries with pH values of 8.28, 9.33 and 10.12 were then prepared. After weathering at 300 °C and 10 MPa for 72 h, the pH of the slurries changed to 8.86, 10.24, 10.30 and 10.21, respectively. H_2 and HCOOH generations in 72 h were summarized in Figure 1 (dash lines). H_2 production rates kept low when compare with CO_2-rich experiments. With the increase of initial experimental pH, H_2 generation showed an upward trend. When the pH of the slurry was adjusted to 8.28, the highest H_2 generation reached 61.1 mmol/kg of olivine in 72 h. No HCOOH was detected at CO_2-free experiments.

The comparison of CO_2-contained and CO_2-free experiments at different pH infers the effect of $NaHCO_3$ on H_2 generation is multifaceted. The variation of pH caused by $NaHCO_3$ addition promoted H_2 generation. However, the pH effect is somewhat limited, further investigations of $NaHCO_3$ effect were then conducted based on minerals analysis.

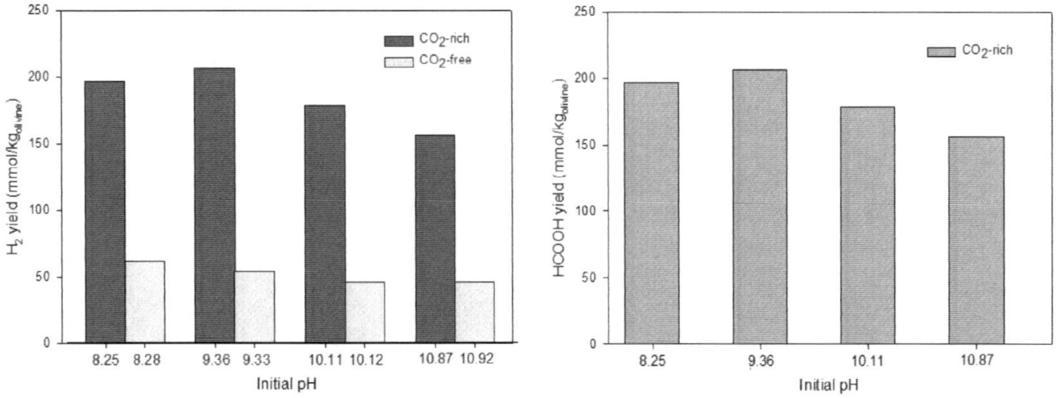

Figure 2. H_2 and HCOOH yields after 72 h olivine weathering at different pH range (8.25 - 10.92) with CO_2 (0.5 mol/L $NaHCO_3$) and without CO_2

3.2. *Minerals changes*

To investigate the pH effect on olivine weathering process, mineral particles after weathering in the hydrothermal system for 72 h were collected and analyzed using TGA, results are shown in Figure 3. In both CO_2-rich and CO_2-free experiments, olivine dissolution was promoted at higher alkaline conditions. In CO_2-rich experiments with an initial pH of 8.25, the unreacted olivine weight was 3.12 g (62.36 wt.%) after 72 h reaction. While for experiment with an initial pH of 10.87, unreacted olivine was 2.31 g (46.13 wt.%) after the reaction. Moreover, olivine dissolution rates in CO_2-rich condition were higher than that in CO_2-free condition. It indicates the addition of $NaHCO_3$ promoted olivine dissolution. However, the lower pH caused by $NaHCO_3$ addition did not play a positive role.

Serpentine (($Mg_3Si_2O_5(OH)_4$) generations at CO_2-rich and higher pH experiments showed increases, which was attributed to promoted olivine dissolution process. According to SEM observations, chrysotile was the dominant serpentine species generated during olivine weathering in all experiments. In the CO_2-free solution, the diameter of chrysotile fibers was approximately 100 nm, and the length was less than 1 μm (Fig. 4(a)). In CO_2-rich experiments with 0.5 mol/L $NaHCO_3$, chrysotile fibrous were getting longer with lengths of several micrometers (Fig. 4(b)), which is consistent with the result of Klein and McCollom[17].

In CO_2-rich condition, brucite only present in solid samples after reacting at the alkaline condition with pH of 10.11 and 10.87, as evidenced by the TGA result with an apparent mass loss at near 390 °C. SEM observations (Figure 4(c)) revealed hexagonal mineral after reaction in CO_2-rich solution with an initial pH of 10.11, which is regarded to be the typical shape of brucite. In the experiments without $NaHCO_3$ addition, although brucite amounts slightly decreased with the decrease in pH, it was observed in all experiments with the pH range of 8.28-10.92. It indicates HCO_3^- promoted brucite dissolution, the effect was much stronger than pH variation. With the increase in pH, HCO_3^- was converted to CO_3^-. According to Harrison et al. (2013), the presence of CO_3^- species will inhibit brucite dissolution [18].

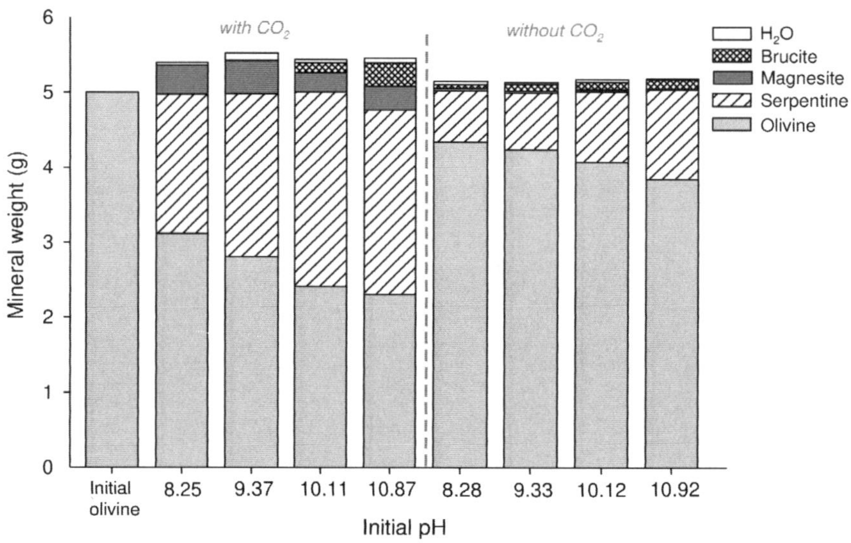

Figure 3. Mineral mass compositions after olivine weathering experiments.

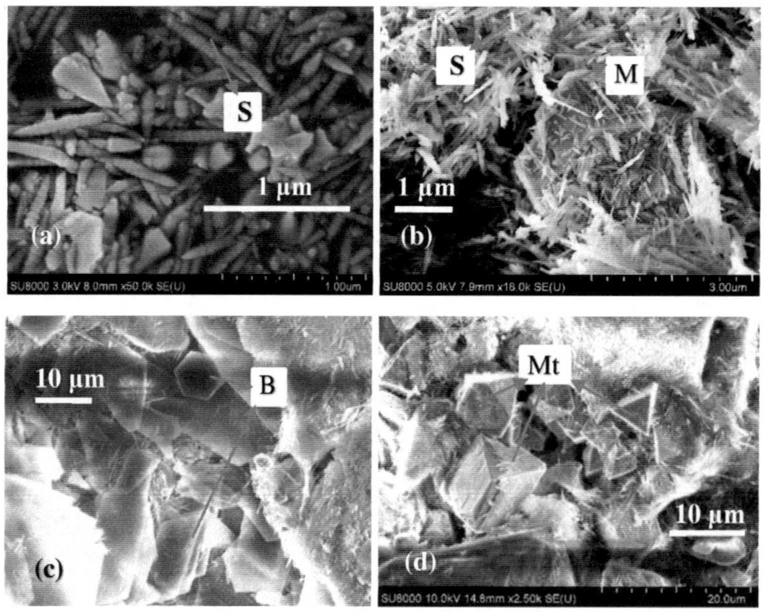

Figure 4. SEM images of minerals after reaction in CO_2-free condition with pH of 10.92 (a) and CO_2-rich conditions with a pH of 8.25 (b) and 10.11 (c, d)
S: serpentine, M: magnesite, B: brucite, Mt: magnetite.

Magnesite was the main product of olivine carbonation, which was largely generated in CO_2-rich experiments. The few magnesites generated in CO_2-free experiments was attributed to the background carbon presented in original olivine or fluid. In CO_2-rich experiments, carbonate generation was suppressed under alkaline conditions. When pH increased from 8.25 to 10.11, magnesite amount decreased from 0.38 to 0.26 g in 72 h. The suppressed carbonation process was partly attributed to the inhibited dissolution of brucite. In our study, although magnetite amount was not quantified, from SEM images (Figure 4(d)), a higher ratio of magnetite was observed in samples collected from higher alkaline experiments.

According to previous researches, Fe(II) released from olivine is ready to be incorporated in brucite and magnesite, rather than react with H_2O. Thus, H_2 generation was suppressed. To further investigate the mechanisms regarding H_2 production rate and mineral changes during olivine weathering, Fe(II) incorporation in brucite and magnesite was traced using EDS. The EDS results reveal the Mg# of the brucite generated in CO_2-rich experiments with the initial pH of 9.36, 10.11 and 10.87 were 0.946, 0.968 and 0.984, respectively (Figure 5). It indicates that more Fe(II) was incorporated into brucite during olivine hydration at lower pH. However, due to the increased brucite amount at higher pH experiment, the total amount of Fe(II) be incorporated into brucite followed an increasing trend. Mg# of the magnesite generated in different pH experiments with the CO_2-rich condition were also analyzed. Same to that of brucite, Mg# of the magnesite markedly increased at higher pH, indicates the decrease of total Fe(II) in magnesite. Fe(II) incorporation measurement in the present study proved that both brucite and magnesite have the potential to incorporate Fe(II), and 1 mol of brucite incorporates more Fe(II) than 1 mol of magnesite. Besides, partitioning of Fe(II) into brucite shows anti-correlation with H_2 yield. The results indicate the controlling of brucite generation should be one of the key points of enhancing H_2 generation.

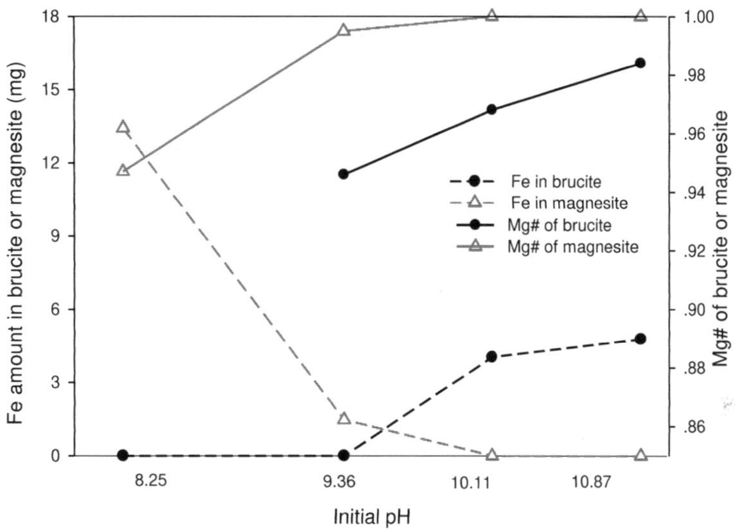

Figure 5. Mg# of brucite and magnesite and incorporated Fe(II) amount in CO_2-rich experiments

4. Conclusion

The role of $NaHCO_3$ in both H_2 production from olivine hydration and on minerals generations were investigated in the present study. The highest H_2 (196.6 mmol/kg of olivine) and HCOOH (54.6 mmol/kg of olivine) yields reached in the CO_2-rich experiment with lower alkaline pH. The addition of high concentration $NaHCO_3$ accelerated olivine and brucite dissolutions, thus promoted Fe(II) being released and consequent H_2 generation. The addition of $NaHCO_3$ caused pH variation to lower values. At lower alkaline pH range, brucite dissolution in both CO_2-rich and CO_2-free conditions was promoted, while olivine dissolution was suppressed. Fe(II) prone to be incorporated in brucite and magnesite at lower pH condition. This study suggests the addition of $NaHCO_3$ promoting olivine and brucite is crucial in H_2 production, but not the pH variation.

5. Reference

[1] Malvoisin B, Brunet F, Carlut J, Montes-Hernandez G, Findling N, Lanson M, Vidal O, Bottero JY and Goffé B 2013 *Int. J. Hydrogen Energ*. **38(18)** 7382-93.

[2] Wang W, Wang S, Ma X and Gong J 2011 *Chem. Soc. Rev*. **40(7)** 3703-27.

[3] Porosoff M D, Yan B and Chen J G 2016 *Energy Environ. Sci.* **9** 62-73.
[4] Marcaillou C, Muñoz M, Vidal O, Parra T and Harfouche M 2011 *Earth Planet. Sc. Lett.* **303** 281-90.
[5] Klein F, Bach W, Jöns N, McCollom T, Moskowitz B and Berquó T 2009 *Geochim. Cosmochim. Ac.* 7**3** 6868-93.
[6] McCollom T M and Seewald J S 2007 *Chem. Rev.* **107** 382-401.
[7] Sherwood L B, Lacrampe-Couloume G, Voglesonger K, Onstott T C, Pratt L M and Slater G F 2008 *Geochim. Cosmochim. Ac.* **72** 4778-95.
[8] Proskurowski G, Lilley M D, Seewald J S, Früh-Green G L, Olson E J, Lupton J E, Sylva S P and Kelley D S 2008 *Science* **319** 604-7.
[9] McCollom T M, Klein F, Robbins M, Moskowitz B, Berquó T S, Jöns N, Bach W and Templeton A 2016 *Geochim. Cosmochim. Ac.* **181** 175-200.
[10] Azdarpour A, Asadullah M, Mohammadian E, Hamidi H, Junin R and Karaei M A 2015 *Chem. Eng. J.* **279** 615-30.
[11] Wang J, Nakamura K, Watanabe N and Komai T 2018 *Proc. of the Twenty-eighth (2018) International Ocean and Polar Engineering Conference. Sapporo, Japan, June 10-15,* **2018**.
[12] Gerdemann S J, O'Connor W K, Dahlin D C, Penner L R and Rush H 2007 *Environ. Sci. Technol.* **41(7)** 2587-93.
[13] Gadikota G, Matter J, Kelemen P and Park A H 2014 *Phys. Chem. Chem. Phys.* **16(10)** 4679-93.
[14] Crundwell F K 2014 *Hydrometallurgy* **149** 265-75.
[15] Wogelius R A and Walther J V 1990 *Geochim. Cosmochim. Ac.* **55** 943-54.
[16] Montserrat F, Renforth P, Hartmann J, Leermakers M, Knops P and Meysman F J R 2017 *Environ. Sci. Technol.* **51** 3960-72.
[17] Klein F and McCollom T M 2013 *Earth. Planet. Sc. Lett.* **379(1)** 137-45.
[18] Harrison A L, Power I M and Dipple G M 2013 *Environ. Sci. Technol.* **47** 126-34.

Acknowledgments
The authors thank reviewers who gave helpful suggestions. This work was supported by JSPS KAKENHI Grant Number JP18J12695.

IOP Conf. Series: Earth and Environmental Science **257** (2019) 012018 doi:10.1088/1755-1315/257/1/012018

Sodium grease formulation from waste engine oil

N W Abdu Rahman[1], N S A Japar[1,2], M A A Aziz[1], A H A Razik[1] and M Y M Yunus[3]

1 Faculty of Chemical and Natural Resources Engineering, Universiti Malaysia Pahang, Kuantan, Pahang, Malaysia
2 Institute of Postgraduate Studies, Universiti Malaysia Pahang, Kuantan, Pahang, Malaysia
3 Earth Resources Sustainability Centre (*ERAS*), Universiti Malaysia Pahang, Kuantan, Pahang, Malaysia

E-mail: maizudin@ump.edu.my

Abstract. The present study was conducted to formulate sodium soap grease using waste engine oil (WEO). Greases were formulated by homogenizing sodium soap thickener in WEO for 2 hours at 150°C Different grease composition of oil-to-thickener ratio was designed which are 90/10, 80/20, 70/30 and 60/40. The formulated greases were analysed by conducting the consistency, oil bleeding, oil separation and FTIR characterization tests. The present study found that grease can be formulated using WEO and the thickener percentage have significant effect on formulated greases properties. Grease formulated with 70 – 80% of base oil and 20 – 30% of thickener was the best grease formulated as it shows desirable grease properties.

1. Introduction

Lubricating grease is a semi-solid material representing a dispersion of a thickener in a liquid lubricant [1]. It consist of three component which are oil, thickener and additives. Additives and base oil play the major roles in the grease formulation while the thickener is referred to as a sponge that hold the lubricant together. Mineral oil are often used in grease production as their fluid component. However, a better stability can be find in synthetic base oil during extreme temperature (high or low) [2]. To set the grease apart from the fluid lubricants, the thickener is used. Thickeners are molecules, polymers or particles that are partially soluble in lubricating fluid.

Earliest production of grease is from the early Egyptian or Roman eras by combining lime with oil. The product of this saponification produce calcium grease when the lime saponified the triglyceride that comprises oil. Starting from that, the development of thickeners have a big impact in grease technology. Lithium thickener has dominated approximately 70% of the grease today [3]. They searched for additives that are used to extend the capabilities of the basic product. However, this concept which is over 60 years old invented by Clarence Earle in the 1940's are the base of a current grease production. The formulator in the 1970's and 1980's target was one single grease which would be capable of a superior performance in a wide range application, temperature and environment. Unfortunately, it was too expensive to used multi-purposed grease for a basic application.

In modern days, the impact of grease production has increase environmentally and economically. As the production and consumption of oil is high every day, the disposal of the waste can give negative impact to the environment. This also bad for the environment as all the waste oil are categorized as the scheduled waste [4]. As the raw material for the grease production is a base oil, using waste oil can

Content from this work may be used under the terms of the Creative Commons Attribution 3.0 licence. Any further distribution of this work must maintain attribution to the author(s) and the title of the work, journal citation and DOI.
Published under licence by IOP Publishing Ltd

reduce the numbers of waste oil that are being disposed to the landfill. It also can help to save the production cost of the grease as the composition of grease consist of 70%-90% of base oil.
Recycling waste oil to replace the base oil can reduce the production cost as the waste oil is much cheaper in price [5]. The main purpose of this study is to design formulation of soap base grease and the production of grease by utilizing waste engine oil.

2. Methodology

2.1. Materials
In this study, waste engine oil is used as base oil. The thickener used in this experiment will be sodium soap thickener. The sodium hydroxide will undergoes saponification process with stearic acid to produces sodium soap thickener. Stearic acid is one of the material needed for saponification process that have a long-chain fatty acid consisting of 18 carbon atoms without double bonds. In nature stearic acid occurs primarily as a mixed triglyceride, or fat, with other long-chain acids and as an ester of a fatty alcohol.

2.2. Pre-treatment of waste oil
Waste oil will undergoes pre-treatment in order to remove any impurities and water in the oil. Two step involves in the treatment which are filtering process and heating process. For filtering process, the oil will be filtered by using filter paper to remove any small impurities in the oil. Then the oil will undergoes second process which is heating the oil until the temperature higher than 90°C with continuous stirring to remove water in the oil. This process will take about 1 hour with a temperature higher than 150°C.

2.3. Formulation of soap base grease
In this formulation waste oil was utilized around 60 to 90 wt% and 10 to 40% of sodium soap thickener was dispersed in the waste engine oil (WEO). The WEO was heated to > 90°C and stearic acid was melted at 60°C. 70% of the heated oil was transferred into the melted stearic acid and homogenized for about 15 minutes. 6M NaOH solution was then added into the mixture drop by drop to control excessive foaming at low speed. After the addition of NaOH, the temperature was increased to 150°C and homogenized for 30minutes. The oil balances were then added completely to the mixture and homogenized for 2 hours at high speed of 8000rpm and constant temperature. Finally, heat was removed from the sample and homogenization was continued for another 30 minutes. Grease sample was stored in an enclosed container and allowed to cool to room temperature for 2 days.

2.4. Analysis of grease

2.4.1. Consistency. Consistency test was carried out to evaluate the level of softness and firmness of the grease ranging from NLGI number 000 to 6 using SKF grease testing kit.

2.4.2. Oil bleeding. Oil bleeding test was conducted to study the oil bleeding area if the grease for fresh and used grease. In this test, greases were heated on filter paper for 2 hours at 60°C [6]. Oil stained created after the test was measured, calculated and compared between fresh and used grease. It is desirable for the grease to bleed oil within the range of +15% to -15% of oil bleed different between fresh and used grease [7]. This range shows that in an operation, the grease can still be used without changing the re-lubrication intervals. Positive values shows that bleeding is increased due to the broken of grease's thickener structure. Negative values represent the reduction of oil bled at which during the grease was used in an operation, the grease dried out before the re-lubrication interval and the surface is 'hungry' for lubrication. In this analysis, used grease is the grease that was aged in an oven for 10 days at 70°C.

2.4.3. Oil separation. Oil separation test is a test to determine the grease tendency to separate oil during storage which correlate directly with the oil separation occurs in 35-lb pails of grease during storage. This test was carried out at 25°C after the grease being stored for 2 months where oil separated on top of the grease were collected and measured. It was desired for the oil separated from the grease to be < 4% by weight [8].

2.4.4. FTIR Characterization. The base oil and the greases were characterized using Fourier-transform infrared (FTIR) spectroscopy to study the compound existed in the oil and grease.

3. Results and discussion

3.1. Consistency

Table 1 shows the consistency of each formulated greases in accordance to NLGI number ranging from 000 to 6. It was found that as the thickener percentage in grease formulation increases, the consistency of the grease becomes more firm. Commonly, the grease available in the market possess grease consistency of NLGI no. 1 to no. 3 [9]. From this test, grease of 20 – 30% of thickener content shows the most desired consistency of NLGI no. 1 to no. 3.

10% of thickener content possessed softer consistency compared to the other grease. This is due to the insufficient amount of thickener to hold the oil in its thickener system which finally led to softer consistency. This also applies towards grease with consistency more firm than NLGI no. 3 at which the amount of thickener were too excessive which cause the lack of oil content in grease. As the matter of fact, grease consistency can be altered by adjusting the oil-to-thickener ratio [10].

Table 1. Consistency number for grease formulation based on NLGI number.

Base oil (%)	Stearic acid (%)	Sodium hydroxide (%)	NLGI number
90	5	5	0
80	10	10	1 to 2
70	15	15	2 to 3
60	20	20	3 to 4
50	25	25	3 to 4

3.2. Oil separation

Oil separation usually occur when grease are being kept for a long period of times. The grease can be considered as good grease if the oil separation is under 4%. From this study, it can be seen that all formulated greases separated oil < 4% **(Table 2)**. There are no oil puddle observed for grease with thickener more than 30% where it can be concluded that the oil were fully hold by the thickener. Oil puddle is formed on top of grease formulated with less than 30% of thickener due to their soft consistency. But, oil contamination such as dust, dirt can also be the factor of oil being drew out from thickener system over time. The oil separation could be influenced by many factors such as the percentage of the base oil, the ability of the thickener to hold the oil, and the homogenize level of the finished grease.

Table 2. Oil separation of the formulated grease.

Base oil (%)	Stearic acid (%)	Sodium hydroxide (%)	Oil separation (%)
90	5	5	1.4
80	10	10	0.6
70	15	15	0.0
60	20	20	0.0
50	25	25	0.0

3.3. Oil bleeding

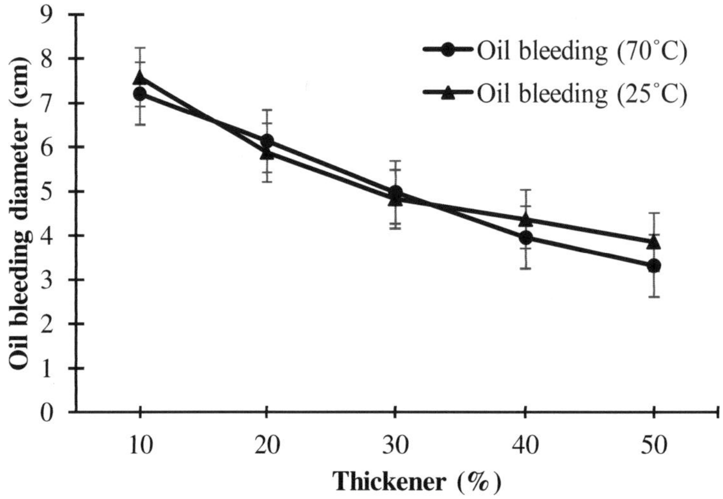

Figure 1. Graph of oil bleeding versus thickener percentage.

Figure 1 shows that as the amount of thickener had significant effects on the formulated grease's oil bleeding. In this analysis, oil bleeding was tested in two different condition which are at room condition of 25°C and at 70°C. The graph above shows that the oil bleeding are relatively linear to thickener percentage regardless of grease ageing condition. Oil bleeding happened when there is stress occur to the grease. Although this stress are extremely weak when be compared to the mechanical stress acted upon grease during usage, but it still can result in the release of a small amount of oil from the grease [11].

3.4. FTIR analysis

a)

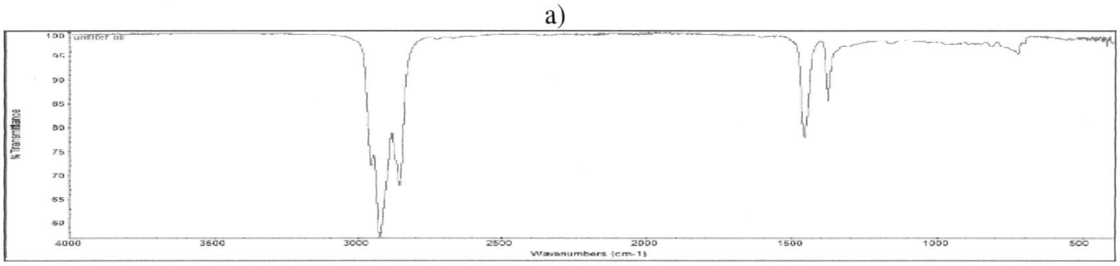

b)

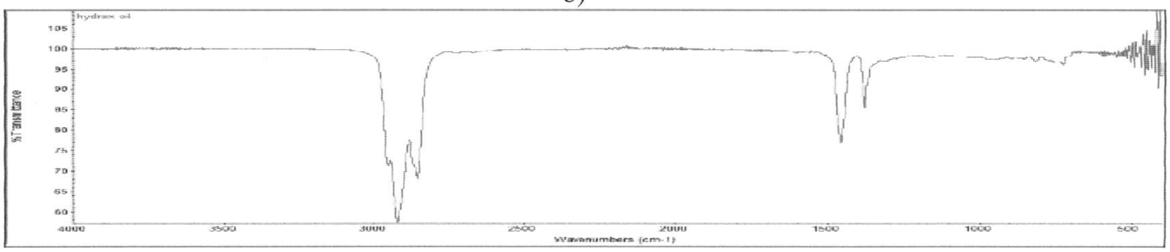

c)

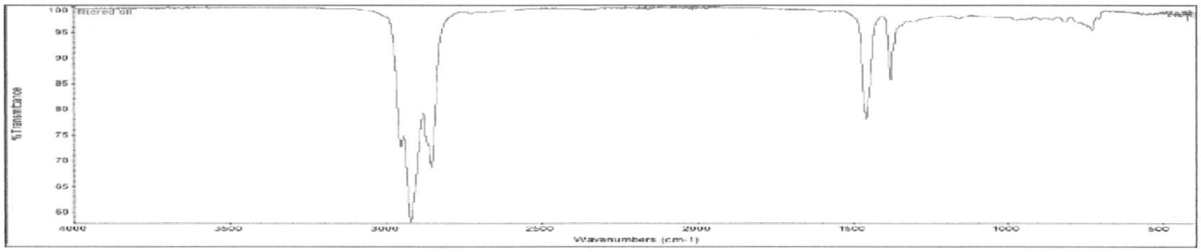

d)

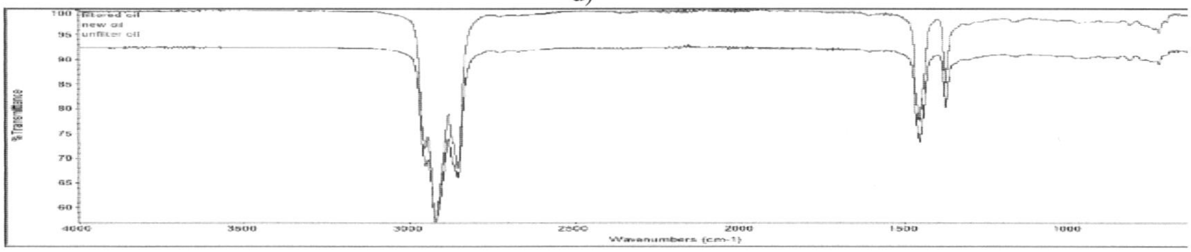

Figure 2. *FTIR analysis of (a) unfiltered waste oil, (b) new oil, (c) filtered waste oil, (d) comparison between oil.*

a)

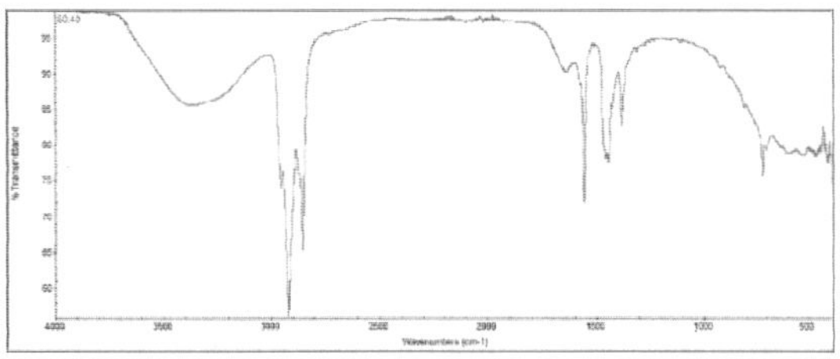

b)

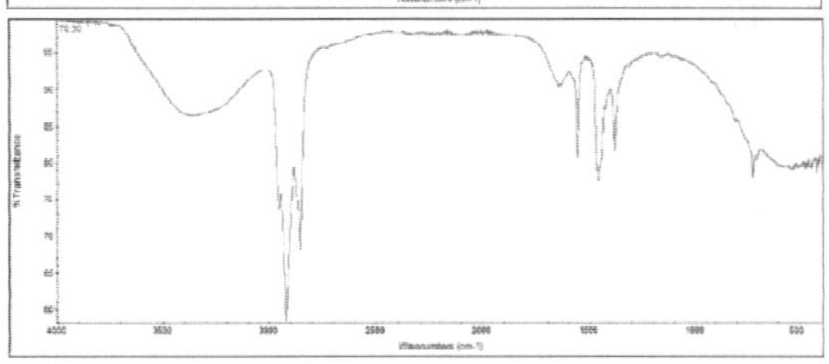

c)

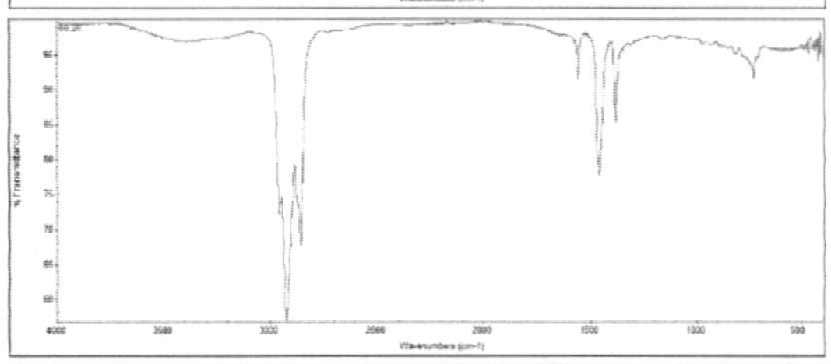

d)

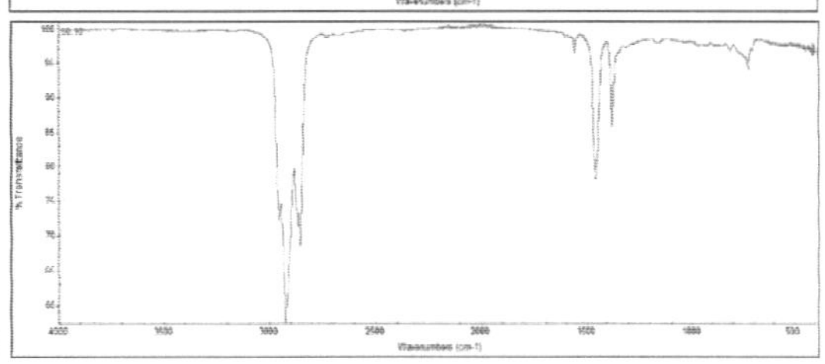

Figure 3. FTIR analysis of grease formulation: (a) 60:40, (b) 70:30, (c) 80:20, (d) 90:10

Figure 2 shows the FTIR analysis results for base oil and their comparison against each other. It was observed that there are no significant difference between new engine oil and waste engine oil (WEO) either it was filtered or unfiltered. It was also observe that there are no known contaminants presence in WEO.

IOP Conf. Series: Earth and Environmental Science **257** (2019) 012018 doi:10.1088/1755-1315/257/1/012018

Figure 3 shows the formulated greases FTIR analysis results. The FTIR spectrum for each greases are almost similar due to the same type chemicals were used in the formulation. However, the spectrum shows some variation on peaks as the thickener percentage were increased. Absorption peaks at 2973 cm^{-1} indicates the asymmetric stretching vibration of CH_3 group from the base oil. It was observed that there were visible peaks at 3400 cm^{-1} as the thickener percentage increases for which it indicates the presences of water or glycol compound in grease. This peaks was belongs to the thickener itself for which there are some traces of water produced when the soap was first formed during the reaction of NaOH and stearic acid [12,13].

4. Conclusion

Grease can be formulated from waste engine oil. FTIR spectroscopy has found that there were no significant differences between new oil, waste oil, and all formulated greases. Results of grease analysis shows that the thickener percentage have significant effect on formulated greases properties including consistency, oil bleeding and oil separation. Out of all formulation, grease formulated with 70 – 80% of base oil and 20 – 30% of thickener was the best grease formulation as it shows desirable grease properties of consistency within NLGI no.1 to no. 3, < 4% oil separation, and acceptable oil bleeding. This, it was proven that grease could be formulated using waste engine oil.

5. References

[1] Donahue C J 2006 Lubricating grease: A chemical primer *J. Chem. Educ.* **83** 862–9

[2] Wright J 2008 Grease Basics *Noria*

[3] Laurentis N De, Kadiric A, Lugt P and Cann P 2016 Tribology International The in fl uence of bearing grease composition on friction in rolling / sliding concentrated contacts *Tribiology Int.* **94** 624–32

[4] EPA 2017 Managing, Reusing, and Recycling Used Oil *United States Environ. Prot. Agency*

[5] Srivastava S P 2014 Rerefining and recycling of used lubricating oil *Development in Lubricant Technology* (Hoboken: John Wiley & Sons, Inc.) pp 299–307

[6] SKF 2009 SKF Grease Test Kit TKGT 1: Instruction Manual 1–40

[7] Ludwig L G 2012 Storing Grease to Avoid Bleed and Separation *Mach. Lubr.*

[8] Lugt P M 2013 *Grease Lubrication in Rolling Bearings* (Chichester: John Wiley & Sons, Inc.)

[9] Rizvi S Q A 2008 *A Comprehensive Review of Lubricant Chemistry, Technology, Selection, and Design* (ASTM International)

[10] Doyle D 2015 General Grease Overview and Bearing Lubrication *ALS Tribol.*

[11] ExxonMobil 2012 Grease Static Oil Bleed 1

[12] Kholijah A M S, Yeung S L C, Sazwani S and Yunus R M 2012 Production of High Temperature Grease from Waste Lubricant Sludge and Silicone Oil *J. Appl. Sci.* **12** 1171–5

[13] Núñez N, Martín-Alfonso J E, Valencia C, Sánchez M C and Franco J M 2012 Rheology of new green lubricating grease formulations containing cellulose pulp and its methylated derivative as thickener agents *Ind. Crops Prod.* **37** 500–7

Pharmaceuticals and personal care products removal from aqueous solution by nitrogen-functionalized carbon adsorbent derived from pomelo peel waste

P Prarat[1,*], K Hadsakunnee[2], L Padejepan[2], P Inlee[1], S Wongpaisan[1] and K Prasertboonyai[1]

1 Faculty of Science, Energy and Environment, King Mongkut's University of Technology North Bangkok (Rayong Campus), Rayong, Thailand
2 Department of Agro-Industrial, Food and Environmental Technology, Faculty of Applied Science, King Mongkut's University of Technology North Bangkok, Bangkok, Thailand
E-mail: panida.p@sciee.kmutnb.ac.th

Abstract. The aim of this work is to investigate the feasibility of the preparation of nitrogen-functionalized adsorbent material from waste pomelo peel by using diammonium hydrogen orthophosphate (DAP) activation for the removal of Pharmaceuticals and personal care products (PPCPs) such as carbamazepine (CBZ), clofibric acid (CFA) and oxytetracycline (OTC). Our results showed that the adsorbent prepared by chemical activation in the presence of DAP lead to mesoporous material. The adsorbent exhibits acid and basic groups at its porous surface. The kinetic data are found to follow a pseudo-second order kinetic model. The Langmuir model provided a good description of the experimental isotherms for CBZ, whereas CFA and OTC adsorption behavior rather follow the Freundlich model. Moreover, the PPCPs adsorption was found to be strongly dependent on the pH of solution as well as the pKa of both adsorbents and PPCPs. The strong adsorptive interaction between PPCPs and P-DAP was mainly attributed to the combination interactions of electrostatic interaction, hydrogen bonding and π-π interactions.

1. Introduction

In recent years, biologically active pharmaceuticals and personal care products (PPCPs) are among the most frequently detected environmental pollutants because PPCPs are extensively used in human and veterinary medicine, resulting in their continuous release to the environment. Even at low concentrations, PPCPs can cause the adverse effects on aquatic life and humans and has been suggested as an environmental micro-pollutant hazard. Therefore, the effective removal of PPCPs from water has become an increasingly important issue. Many studies have reported that most of the PPCPs are not biodegradable and often not easy to eliminate completely from water by biological and physical wastewater treatment plants due to their high polarity, aromaticity and structural complexity of these compounds [1, 2]. Consequently, other more effective treatment methods have been investigated, including chemical treatments such as ozonation [3], electrochemical degradation [4], and advanced oxidation process [5]. However, some of these intermediate compounds produces residual toxic compounds which are suspected to be more toxic than the parent compounds.

Content from this work may be used under the terms of the Creative Commons Attribution 3.0 licence. Any further distribution of this work must maintain attribution to the author(s) and the title of the work, journal citation and DOI.
Published under licence by IOP Publishing Ltd

Adsorption is now recognized as an effective technique for pollutants removal because of its flexibility in implementation, high quality treatment outcomes, and the possibility of reusing the spent adsorbent. Activated carbon (AC) is preferred adsorbent for the removal of micropollutants from wastewater because of its high surface area, well-developed porosity and rich surface chemical functional groups. However, in recent years, the industrial production of AC is facing the problem of raw material scarcity and high costs. To decrease treatment costs, attempts have been made to find inexpensive alternative precursors for carbon preparation, such as organic waste products. AC derived from organic sources have a large specific surface area, mesoporous to microporous characteristics and high adsorption performance. One of these were carbon material form pomelo peel (PP) , which has a well-designed 3D structure with large pore size [6]. This structure can provide large active surface area and fast diffusion path, so carbon adsorbent produced from pomelo peel has been used for the adsorption of organic compound from solution; such as anionic and cationic dyes [7], and antibiotic [8].

The adsorption performance of carbon material depends on not only its surface area and pore structure, but also functional groups on its pore surface. According to literature, the oxygen-containing groups on AC play a key role on the adsorption process of pollutants. As a result, chemically activation agents (such as H_3PO_4, KOH, NaOH and $ZnCl_2$) have been frequently used for chemical activation due to the rich of the oxygen-containing groups on carbon adsorbent surface. However, the amino functional groups, as well as hydroxyl functional groups, play a significant role in the field of PPCPs adsorption [9]. Nevertheless, there was rare work to prepare nitrogen-functionalized activated carbon from agricultural wastes.

The aim of this study, di-ammonium hydrogen orthophosphate; $(NH_4)_2HPO_4$, was selected as the activation agent and pomelo peel waste was used as precursor material. Ammonium phosphates release phosphoric compound and amino radicals in the process of thermal decomposition, eventually promoting the porosity and surface amino functionalization of the produced activated carbon. After the preparation and the characterization of the adsorbent prepared, the removal of representative PPCPs, three drugs named carbamazepine (CBZ), clofibric acid (CFA) and oxytetracycline (OTC), by adsorption from aqueous solution was investigated. The characteristic and chemical structure of CBZ, CFA and OTC are summarized in table 1.

2. Materials and methods

2.1. Preparation of activated carbon

Pomelo peel (precursor) used for preparation of activated carbon were obtained locally. The precursor was first washed with distilled water, cut, dried, and sieved to desired mesh size of 3–4 mm. After soaking in DAP solution at a ratio of 1:3 (g precursor/g activate reagent) for 24 h, the sample was washed with distilled water until the pH was nearly neutral, and then dried at 105 °C. After that, the sample was heated at temperature of 800 °C and hold at the carbonization temperature for 1 h. Finally, the produced carbon adsorbent (designated P-DAP) was ground, sieved and kept in a desiccator until use.

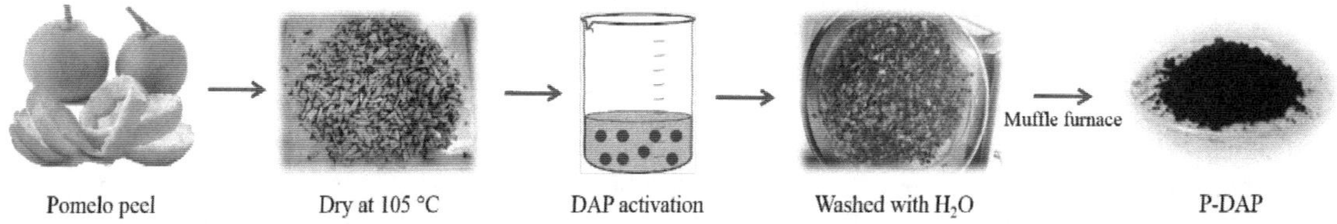

Figure 1. Schematic illustration of the N-functionalized activated carbon preparation.

2.2. Characterization of activated carbon

The physiochemical characteristics of P-DAP were characterized by several techniques. N_2 adsorption-desorption isotherms were analyzed with a V-Sorb 2800P surface area and pore volume analyzer at 77 °K. X-ray powder diffraction (XRD) patterns were obtained on a Bruker AXS D8 diffractometer using Cu-Kα radiation. The adsorbent was determined the pH at the point of zero charge (pH_{PZC}) by the acid-base titration method [10]. Fourier transform infrared (FT-IR) spectrophotometry was recorded on a Perkin Elmer Spectrum One using KBr pellets. Elemental analysis was carried out using a LECO Corporation, 628Series. Boehm's titration was carried out according to the Boehm's method [11].

2.3. Adsorption experiments

Batch adsorption experiment was studied in 125-ml Erlenmeyer flask containing 0.57 g/L of carbon absorbent. The solution pH and ionic strength was controlled by using a 10 mM phosphate buffer. The experimental samples were conducted with shaking condition at 200 rpm at room temperature, and then filtered through a glass filter (GF/C, pore size 0.45 μm). The quantity of CBZ, CFA and OTC concentration in equilibrium solutions was analyzed by a UV-Vis spectrophotometer (Shimadzu, UV 1800) at 275, 274 and 278 nm, respectively. The uptake capacity at the equilibrium q_e (mg/g) was calculated as follows:

$$q_e = \frac{(C_0 - C_e)V}{m} \quad (1)$$

where C_0 and C_e (mg/L) are the initial and equilibrium concentrations, respectively; V (L) is the volume of the solution; and m (g) is the dosage of the adsorbent.

Adsorption kinetic, the experiment was performed by fixing the initial three PPCPs concentration at 50 mg/L. Adsorption isotherm was performed at the concentration of PPCPs varied between 5 mg/L and 60 mg/L. While the effect of the pH on the adsorption capacity was investigated at pH range from 5 to 9. All the experimental samples were performed in duplicate, and the average values are reported.

2.4. Data analysis

The kinetics of the adsorption process were determined using the pseudo-first order and pseudo-second order kinetic models, as indicated in the Equation (2) and (3), respectively as follows:

$$log\,(q_e - q_t) = \log q_e - \frac{k_1}{2.303}t \quad (2)$$

$$\frac{t}{q_t} = \frac{1}{k_2(q_e{}^2)} + \frac{t}{q_e} \quad (3)$$

where q_e and q_t (mg/g) are the adsorption capacities of adsorbate at equilibrium and at time, respectively. The k_1 and k_2 are the pseudo-first-order (1/min) and pseudo-second order rate constants (g/mg.min), respectively.

Langmuir and Freundlich isotherm models were investigated the adsorption process in this work. The Langmuir and Freundlich equation can be represented by the Equation (4) and (5), respectively as follows:

Langmuir isotherm

$$\frac{1}{q_e} = \frac{1}{K_L q_m}.\frac{1}{C_e} + \frac{1}{q_m} \quad (4)$$

Freundlich isotherm

$$\ln q_e = \frac{1}{n}\ln C_e + \ln K_F \quad (5)$$

where q_m (mg/g) is a theoretical maximum adsorption capacity, K_L (L/mg) is Langmuir isotherm constant, C_e is an equilibrium concentration (mg/L), K_F (L/mg) and n are Freundlich isotherm constant and heterogeneity factor indicating the adsorption intensity of the adsorbent, respectively.

Table 1. Characteristic and chemical structure of representative PPCPs.

Name	Molecular weight (g/mol)	pKa_1	pKa_2	pKa_3	Chemical structure
Carbamazepine (CBZ)	236.27	13.90	-	-	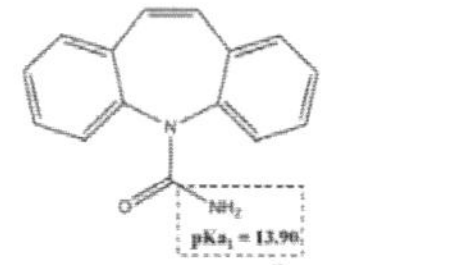
Clofibric acid (CFA)	214.65	3.61	-	-	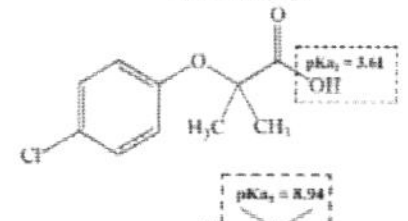
Oxytetracycline (OTC)	460.43	3.22	7.46	8.94	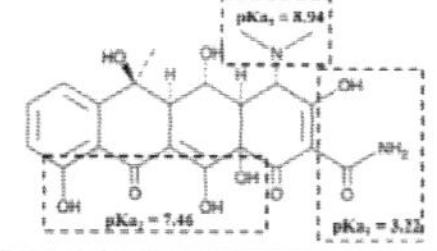

3. Results and discussion

3.1. Adsorbent Characterization

Figure 2a shows the N_2 adsorption/desorption isotherms and pore size distributions. The isotherms are mixture of type I and type IV with hysteresis loops, indicating the co-existence of micropores and mesopores. The mean pore size of P-DAP is 4.03 nm and a larger portion of mesoporous located in 20 nm (figure 2a and table 2). The BET surface area of P-DAP is 83.79 m^2/g which was lower than the other biomass activated carbons prepared under the similar activation agent. Guo et al. [12] has reported on the characteristic of carbon prepared from *Phragmites australis* by impregnating with DAP; denoted as AC-DAP. The BET surface area of AC-DAP was 495 m^2/g which was higher 5.6 times than that of P-DAP. However, the higher degree of mesoporosity is similarly present (>65%).

FT-IR spectroscopy was used to identify the major functional groups of the adsorbent (figure 2b. The peak at 3470 cm^{-1} can be assigned to the O–H stretching vibration of the hydroxyl groups, due to the existence of cellulose and lignin [13]. The peak at 2930 cm^{-1} indicates the C-H stretching in methyl and methylene groups [14].The presence of peak at 1662 cm^{-1} is characteristic of the C=O stretching in carboxyl group. The peak at 1385 cm^{-1} could be ascribed to phenolic C–O and O–H stretching modes. A small peak at 1034 cm^{-1} could result from ionized linkage P^+–O^- in phosphates and to symmetrical vibration in a chain of C–O–P [15] or C-N.

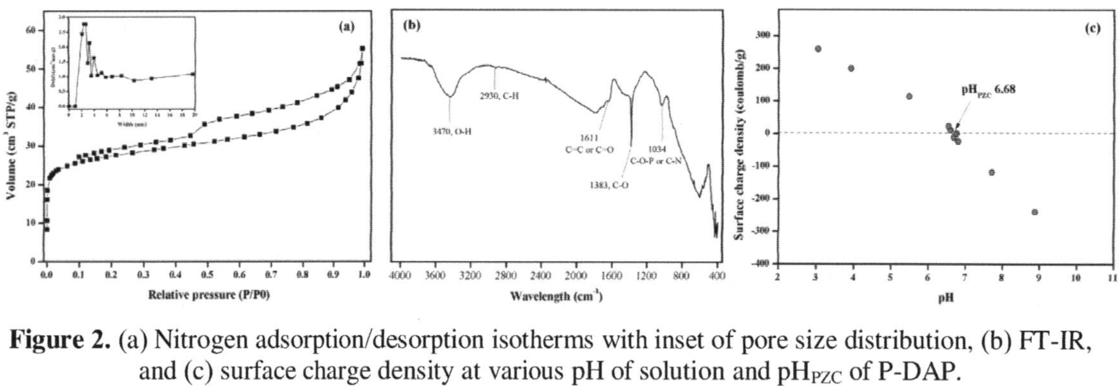

Figure 2. (a) Nitrogen adsorption/desorption isotherms with inset of pore size distribution, (b) FT-IR, and (c) surface charge density at various pH of solution and pH_{PZC} of P-DAP.

The point of zero charge (pH_{PZC}) of P-DAP is 6.68 as shown in figure 2c. This result indicates that the material is basic in nature due to the ammonium produced by DAP decomposition in activation process. Surface oxygen-containing groups of P-DAP determine its surface acidity/basicity, and act as a crucial factor affecting its adsorption capacity and selectivity. As shown in table 3, the result of Boehm titration showed that the amount of acidic and basic groups covering the surface of P-DAP. The contents of C H, N, O compositions (C, H, N, O and others) are 53.6%, 2.8%, 2.1% and 41.5%, respectively.

Table 2. Structural properties of P-DAP.

Adsorbent	S_{BET} (m^2/g)	Pore diameter (nm)	V_{total} (cm^3/g)	$V_{micropore}$ (cm^3/g)	$V_{mesopore}$ (cm^3/g)
P-DAP	83.79	4.03	0.08	0.02	0.06

Table 3. Surface functional groups and elemental composition of P-DAP.

Adsorbent	Acidic (mmol/g)	Basic (mmol/g)	Phenolic(mmol/g)	Carboxyl (mmol/g)	Lactone (mmol/g)	C (wt%)	H (wt%)	N (wt%)	O and others (wt%)
P-DAP	23.25	21.75	14.30	2.85	6.90	53.6	2.8	2.1	41.5

3.2. Adsorption kinetic

The kinetic profiles for the adsorption are presented in figure 3a and indicates that for three PPCPs adsorption took place via multi stages namely – a fast adsorption uptake that occurred within 2 h for both CBZ and CFA, followed by a slower, incremental uptake step with equilibrium reached within 8 h. Whereas with OTC adsorption, it took for long time and tended toward it at 30 h and was half maximal at ~6 h. The adsorption capacity rapidly increased in the initial stage due to the presence of number of available active sites for adsorption during the initial stages. The kinetic parameters and correlation coefficients are shown in table 4. For all three PPCPs the pseudo-second order model gave a better fit to the data (figure 3b and 3c), as defined by the higher R^2, than the pseudo-first order model. This result also exhibited that different adsorption mechanisms are involved in the adsorption process including chemical and physical adsorption, while the physical adsorption dominated in the adsorption process [16]. The adsorption mechanism on adsorbent surface should be proportional to the driving force and area. In this study, the driving force is the adsorbate concentration in the solution, and the area is the amount of active sites on the adsorbent surface.

Table 4. Kinetic parameters of CBZ, CFA and OTC.

PPCPs	$q_{e,exp}$	Pseudo-first order model			Pseudo-second order model		
		$q_{e,cal}$ (mg/g)	k_1 (min^{-1})	R^2	$q_{e,cal}$ (mg/g)	k_2 (g/mg.min)	R^2
CBZ	78.94	38.90	0.0049	0.9733	79.37	0.0126	0.9980
CFA	28.63	5.95	0.0007	0.5986	28.76	0.0025	0.9999
OTC	48.99	55.11	0.0002	0.8734	50.76	0.0197	0.9505

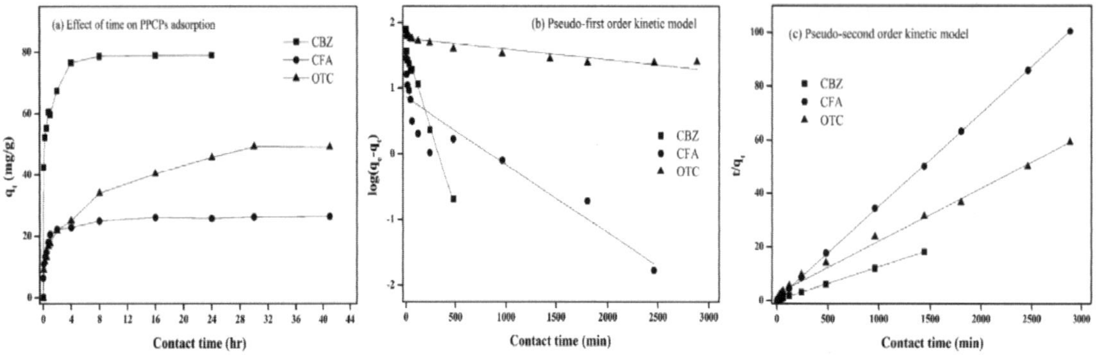

Figure 3. (a) adsorption kinetic of CBZ, CFA and OTC onto P-DAP, (b) Pseudo-first order kinetic model fitting, and (c) Pseudo-second order kinetic model fitting

3.3. Adsorption isotherms

Langmuir and Freundlich adsorption models were used to fit the experimental data. The adsorption model parameters are listed in table 5. Both the Langmuir and the Freundlich models fitted the three PPCPs isotherms well on the P-DAP, however, the exponent $1/n$ value for CFA and OTC was fairly less than half in the Freundlich models, which means that the form of the CFA and OTC adsorption equilibrium on P-DAP was more Freundlich like than Langmuir model. The results suggested that CFA and OTC adsorption behavior on P-DAP was not an ideal monolayer adsorption and the adsorption on the surface and pore of P-DAP were heterogeneous. While the value of $1/n$ of CBZ was more than 0.5, which implied that the actual data of CBZ adsorption can be described by the Langmuir model indicating that the monolayer chemisorption of CBZ on P-DAP was predominant.

Comparison on the effects of the different molecular structure of three PPCPs on adsorption capacities is shown in figure 4. P-DAP adsorbed the CBZ far more effectively than OTC and CFA. As revealed in figure 2d, the surface charge of P-DAP exhibited a nearly neutral at pH 7. Similarly, CBZ is neutral form (pK_a =13.90), whereas with CFA is anionic form (pKa = 3.61) and OTC are zwitterionic form, consequently P-DAP could adsorb CBZ via H-bonding interaction due to hydroxyl groups (pKa ~ 9.32) as well as amino groups (pKa ~ 9.6) on its surface. Moreover, CBZ can be adsorbed onto P-DAP due to the dispersive interactions between the delocalized π electrons in the P-DAP basal planes and the free electrons (aromatic rings) in the CBZ molecule. Hence, the much higher CBZ adsorption capacity should be ascribed to the combination of H-bonding and π-π interaction. With respect to CFA, its anionic form at pH 7 could contribute to repulse electrostatic interactions towards $-COO^-$ groups (pKa ~ 4.5) of P-DAP. For OTC adsorption, the zwitterion (H_2OTC and $HOTC^-$) was the major form at pH between 3.22 and 7.46, thus P-DAP could adsorb OTC by electrostatic interaction between H_2OTC and the weakly negatively charged P-DAP, including π-π interaction and H-bonding.

A comparison of the surface area and maximum PPCPs adsorption capacities per square meter (q_{max}, mg/m^2) of the P-DAP with other carbonaceous adsorbents is shown in table 6. Surprisingly, P-DAP exhibited lower surface area but obviously higher CBZ, CFA, and OTC adsorption capacities per square meter than other carbons. Accordingly, it can be mention that the excellent actual adsorption behavior due to organic functional group notably surpasses the influence of specific surface area of adsorbent as the P-DAP has a much lower surface area than the other adsorbents.

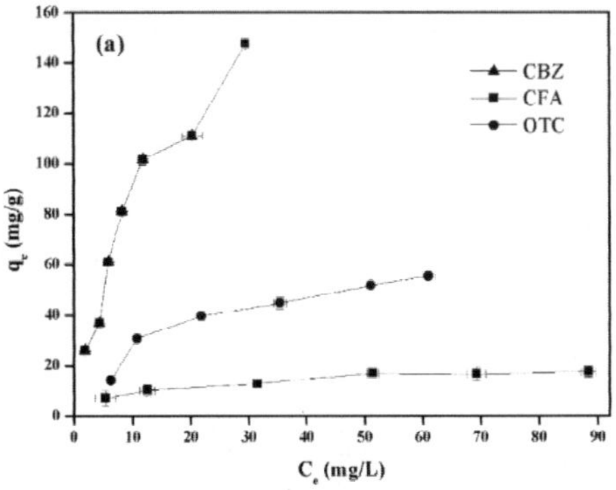

Figure 4. Adsorption isotherm of CBZ, CFA and OTC onto P-DAP at pH 7.

Table 5. Isotherm parameters of CBZ, CFA and OTC.

PPCPs	Langmuir			Fruendlich		
	q_m (mg/g)	K_L (L/mg)	R^2	K_F (L/mg)	$1/n$	R^2
CBZ	216.2	0.065	0.998	22.1	0.56	0.997
CFA	19.4	0.092	0.997	4.6	0.31	0.998
OTC	64.9	0.064	0.999	13.3	0.32	0.997

3.4. Effect of pH on PPCPs adsorption capacity

In figure 5, CBZ adsorption capacity was higher under neutral pH than that at acidic and basic pH. The pKa of CBZ is around 13.90, so it maintained neutral molecular (figure 6a) form throughout the investigation. The H-bonding and π-π interactions were responsible for the adsorption of CBZ. When the pH increase, the density of the negative surface charge at pH 9 of the P-DAP was higher than that at pH 5 and 7, leading to a lower adsorption capacity.

On the contrary, increasing the pH resulted in a lower CFA and OTC adsorption on P-DAP. As the pKa of CFA is 3.61, it exists in anionic form (figure 5b). The surface of P-DAP carbon is more positive charge at pH = 5, which is at a pH lower than its pH_{PZC}. Therefore, the higher adsorption capacity of CFA on P-DAP at low pH may be consequent on the electrostatic interaction between the positively charged surface of P-DAP and the CFA anions. With the increase in pH, the adsorption capacity decreased, which may be explained by the repulsive electrostatic interaction between P-DAP and the CFA anions. Moreover, the maximum uptake of OTC is obtained at pH = 5 and the adsorption decreased at higher pH, with reflects the unfavorable electrostatic conditions for anionic OTC molecules and negatively charged P-DAP.

Table 6. Surface area and maximum CBZ, CFA and OTC adsorption capacities per square meter of the various carbonaceous materials.

Adsorbents	BET (m^2/g)	q_{max} (mg/g)	q_{max} (mg/m^2)			Reference
			CBZ	CFA	OTC	
Commercial GO (type C)	771	CBZ = 195	0.253			[17]
Commercial GO (type M)	74	CBZ = 19	0.257			[17]
CNT	112	CBZ = 28	0.250			[17]
GAC	1181	CBZ = 297	0.251			[17]
Cotton liner fiber AC (CLAC)	2143	CBZ = 1025.2			0.478	[18]
Cotton cloth waste AC (ACC50%)	1175	CFA = 49.9 TC = 109.1		0.043	0.093	[19]
Commercial AC (CP)	907	CFA = 82.6		0.091		[20]
Commercial AC (VP)	758	CFA = 101.7		0.134		[20]
P-DAP	83.79	CBZ = 216.2	2.579			This work
P-DAP	83.79	CFA = 19.4		0.231		This work
P-DAP	83.79	OTC = 64.9			0.775	This work

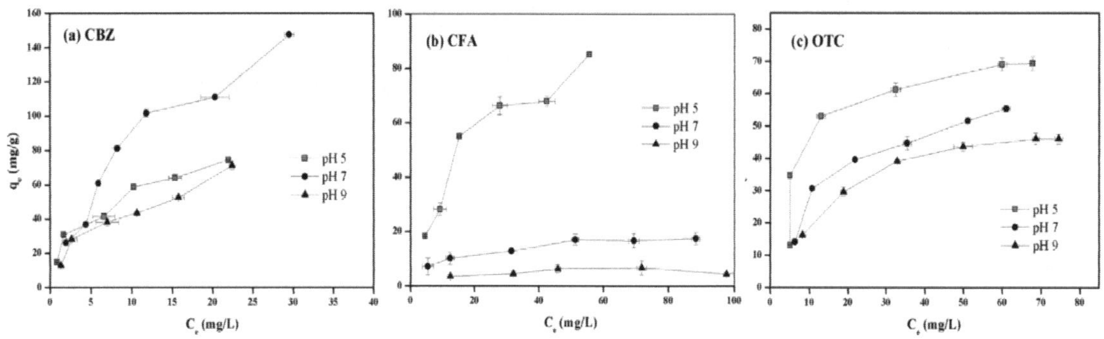

Figure 5. Effect of pH on adsorption of CBZ, CFA and OTC onto P-DAP.

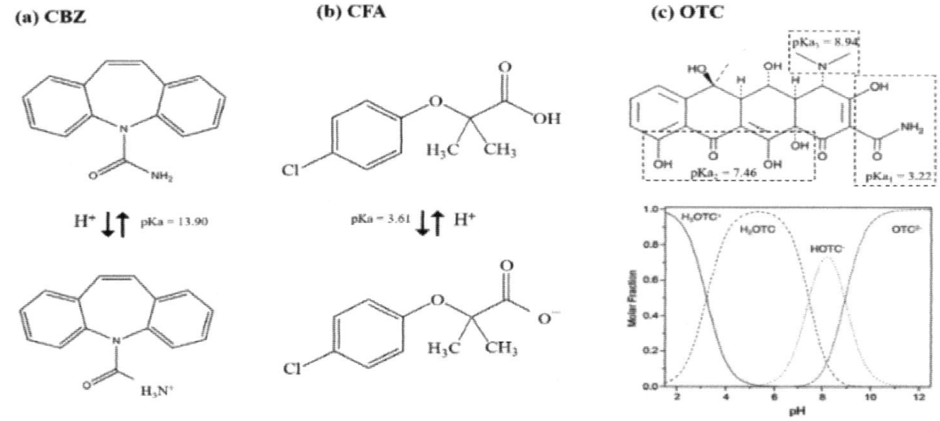

Figure 6. The pKa values of CBZ, CFA and OTC and pH-dependent speciation of OTC.

4. Conclusions

The liquid-phase adsorption of CBZ, CFA and OTC (three representative PPCPs) has been studied for the first time using nitrogen-functionalized carbon adsorbent derived from pomelo peel waste, the following conclusion can be drawn. P-DAP exhibited a high ratio of mesoporous structure (>70%), and contained the acid and basic groups at its porous surface. The Freundlich model appeared to fit the data better than the Langmuir model for describing the adsorption behavior of CFA and OTC with the exception of the equilibrium data of CBZ that is well described by the Langmuir model. The electrostatic interaction, hydrogen bonding and π-π interaction played the key role influencing the PPCPs adsorption. Operating variables including adsorbent dose, PPCPs concentration, solution of pH and coexisting of natural organic matter will be optimized using central composite design (CCD) under response surface methodology (RSM) approach in further studies.

5. References

[1] Chen Y, Vymazal J, Březinová T, Koželuh M, Kule L, Huang J and Chen Z 2016 *Sci. Total Environ.* **566-567** 1660-69

[2] Zhang D, Gersberg R M, Ng W J and Tan S K 2014 *Environ. Pollut.* **184** 620-39

[3] Coelho A D, Sans C, Agüera A, Gómez M J, Esplugas S and Dezotti M 2009 *Sci. Total Environ.* **407** 3572-78

[4] Joss A, Zabczynski S, Göbel A, Hoffmann B, Löffler D, McArdell C S, Ternes T A, Thomsen A and Siegrist H 2006 *Water Res.* **40** 1686-96

[5] Klavarioti M, Mantzavinos D and Kassinos D 2009 *Environ. Inter.* **35** 402-17

[6] Chen S, Liu Q, He G, Zhou Y, Hanif M, Peng X, Wang S and Hou H 2012 *J. Mater. Chem.* **22** 18609-13

[7] Foo K Y and Hameed B H 2011 *Chem. Eng. J.* **173** 385-390.

[8] Sun Y, Li H, Li G, Gao B, Yue Q and Li X 2016 *Bioresour. Technol.* **217** 239-44
[9] Li N, Yang S, Chen J, Gao J, He H and Sun C 2016 *Appl. Surf. Sci.* **386** 460-66
[10] Peng X, Hu X, Fu D and Lam F L Y 2014 *Appl. Surf. Sci.* **294** 71-80
[11] Boehm H P 1994 *Carbon* **32** 759-69.
[12] Guo Z, Fan J, Zhang J, Kang Y, Liu H, Jiang L and Zhang C 2016 *J. Taiwan Inst. Chem. Eng.* **58** 290-96.
[13] Liang S, Guo X and Tian Q 2011 *Desalination* **275** 212-16
[14] Zhang X, Niu J, Zhang X, Xiao R, Lu M and Cai Z 2017 *J. Chromatogr. B* **1046** 58-64
[15] Liu H, Zhang J, Bao N, Cheng C, Ren L and Zhang C 2012 *J. Hazard. Mater.* **235-236** 367-75
[16] Jung C, Heo J, Han J, Her N, Lee S J, Oh J, Ryu J and Yoon Y 2013 *Sep. Purif. Technol.* **106** 63-71
[17] Cai N, and Larese-Casanova P 2014 *J. Colloid Interface Sci.* **426** 152-61.
[18] Sun Y, Yue Q, Gao B, Li Q, Huang L, Yao F and Xu X 2012 *J. Colloid Interface Sci.* **368** 521-27
[19] Boudrahem N, Delpeux-Ouldriane S, Khenniche L, Boudrahem F, Aissani-Benissad F and Gineys M 2017 *Process Saf. Environ. Prot.* **111** 544-59
[20] Mestre A S, Nabiço A, Figueiredo P L, Pinto M L, Santos M S C S and Fonseca I 2016 *Chem. Eng. J.* **286** 538-48

Acknowledgements
This research was funded by King Mongkut's University of Technology North Bangkok; Contract no. KMUTNB-62-KNOW-33. Moreover, we would like to thank Faculty of Science, Energy and Environment, King Mongkut's University of Technology North Bangkok (KMUTNB) for the equipment.

IOP Conf. Series: Earth and Environmental Science **257** (2019) 012020 doi:10.1088/1755-1315/257/1/012020

AFFF spread over liquid surface under external radiation conditions

Baowei Wang[1,2], Shenshi Huang[1], Qize He[1], Shouxiang Lu[1], and Ruichao Wei[1]

1 State Key Laboratory of Fire Science, University of Science and Technology of China, Hefei, China
2 Fire Department of Ministry of Public Security, Beijing, China
E-mail: ustcwbw@163.com

Abstract. Aqueous film forming foam (AFFF) is effective in extinguishing oil fires. The extinguishing effect generally depends on the foam spread process and its coverage area. In this study, the foam spread process under external thermal radiation is investigated. Firstly, based on the axisymmetric foam spread model, the relationship of radiation and foam thickness is discussed. Then foam spreading experiments under pool fire condition were carried out, to simulate the actual extinguishing process of AFFF. Finally, the foam spread model considering the external radiation flux is developed which may provide theoretical guidance for fire suppression using AFFF.

1. Introduction

AFFF is one of the most effective and widely used extinguishant to put out liquid fire [1]. An important feature of the foam is that, its density is less than general oil. On the one hand, AFFF can cool down the fuel; on the other hand, the foam cuts off the contact between combustible gas and air. Therefore, it is very efficient in extinguishing oil fires.

The study of fire extinguishing process of foam has been a top topic in recent years, because it help to comprehensively evaluate the fire extinguishing efficiency of foam. Williams et al. [2] indicated that the AFFF agent shows different extinguishing effect on over different burning liquid. Although fire suppression effect is much related to the flash point of fuel, the foam coverage area is the most critical factor. Persson et al. [3,4] considered the foam spread process and the foam mass loss due to fire heat radiation independently. The basic foam spread model over liquid surface was proposed. In our previous study [5], we continued to develop the axisymmetric spread model and measured the key parameters by carrying out experiment.

In this study, the model of AFFF spread over liquid surface was investigated, with the condition of external thermal radiation. Based on the axisymmetric foam spread theory, the foam loss under heat radiation was considered. Also, the experiment was carried out to simulate the actual extinguishing process of AFFF, to give reference to the spread model. This study can finally will provide further understanding of the AFFF foam spread in fire condition and contribute to fire suppression work using AFFF.

2. Axisymmetric foam spread theory

In the previous studies [5,6], the axisymmetric foam spread theory has been developed. When putting out the oil fire, AFFF was sprayed from the hose and spread start from a point source over the whole liquid surface. Therefore the foam spread process can be illustrated as Figure 1, and the AFFF spread model without external radiation can be expressed by Equation (1).

Content from this work may be used under the terms of the Creative Commons Attribution 3.0 licence. Any further distribution of this work must maintain attribution to the author(s) and the title of the work, journal citation and DOI.
Published under licence by IOP Publishing Ltd

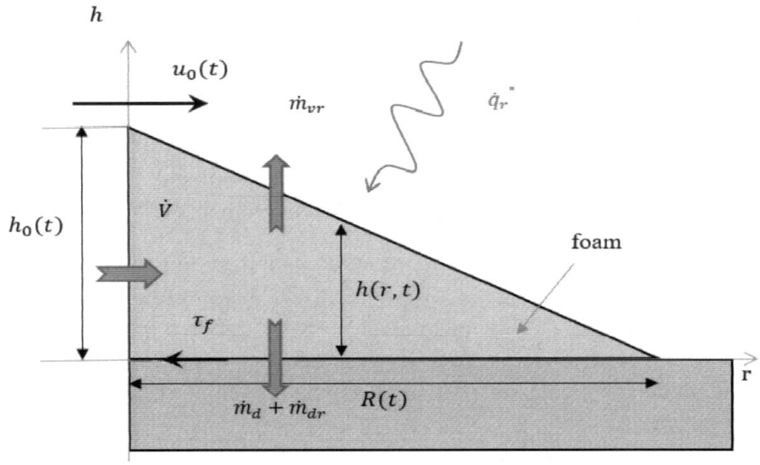

Figure 1. Illustration of foam spread process.

$$\begin{cases} h_0 = \left(\frac{27\dot{V}^3}{4\pi^3\beta^2{R_0}^2}\right)^{1/7} \cdot t^{1/7} \\ R = \left(\frac{18\beta R_0\dot{V}^2}{\pi^2}\right)^{1/7} \cdot t^{3/7} \end{cases} \tag{1}$$

Where, h_0 is the foam thickness at the source point; R is the front radius; $\dot{V}$ is the volume flow rate of foam; β is the friction parameter; and R_0 is the initial radius of foam. Therefore, both the thickness h_0 and the radius R grow as a power function over time t.

This axisymmetric foam spread model has been validated by a series of experiments in previous study. It is found that with the foam expansion about 10.0, the value of the parameter βR_0 was calculated as 2.06×10^{-5} m/s. With determined foam flow rate $\dot{V}$, the relationship foam coverage area and time can be expressed as

$$A = 0.0033\left[10^{-24/7} \times (\text{m/s})^{2/7}\right]\dot{V}^{4/7}t^{6/7} \tag{2}$$

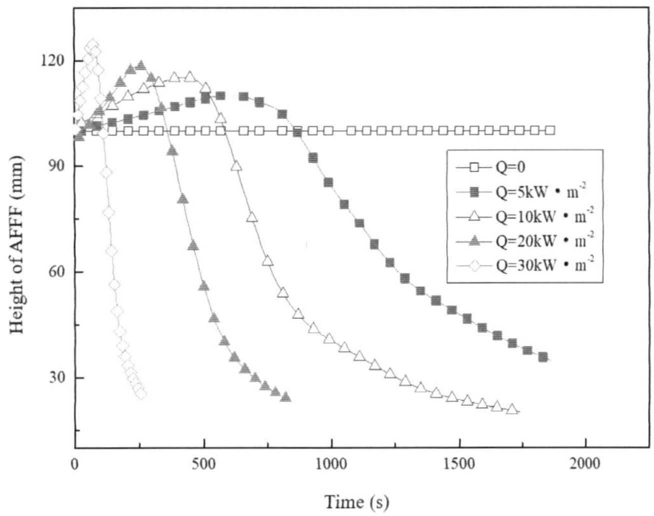

Figure 2. The height of AFFF under the condition of different external thermal radiation fluxes. (S = 13.5, h = 100 mm).

3. Radiation flux and foam thickness

In the case of external fire source radiation, the water in the foam can gradually evaporate after being heated, and the foam accelerates to collapse, thus reducing the thickness of the foam layer.

Wu et. al. studied the characteristics of typical AFFF under the condition of external thermal radiation. Figure 2 shows the change rate of AFFF height with initial foam height of 100 mm and the foam expansion of 13.5. The external thermal radiation flux was set to be 5 kW/m^2, 10 kW/m^2, 20 kW/m-, and 30 kW/m^2 respectively. It can be seen from the figure that, the decreasing rate of foam height varies greatly under different radiant fluxes.

After exposed to the thermal radiation, the foam will expand at first and the foam thickness will increase slightly. Then, due to the effect of evaporation and defoaming, the foam thickness starts to decrease rapidly. The maximum decreasing rate of foam height is obtained by differentiating the change of foam thickness, as shown in Figure 3. It is found by data fitting that, the maximum height decreasing rate of foam height presents an exponential relationship with thermal radiation flux, which can be expressed as Equation (3)

$$\left(\frac{d\delta}{dx}\right)_{max} = 0.0092 \cdot e^{0.1557\dot{q}_r} + 0.1458 \quad (3)$$

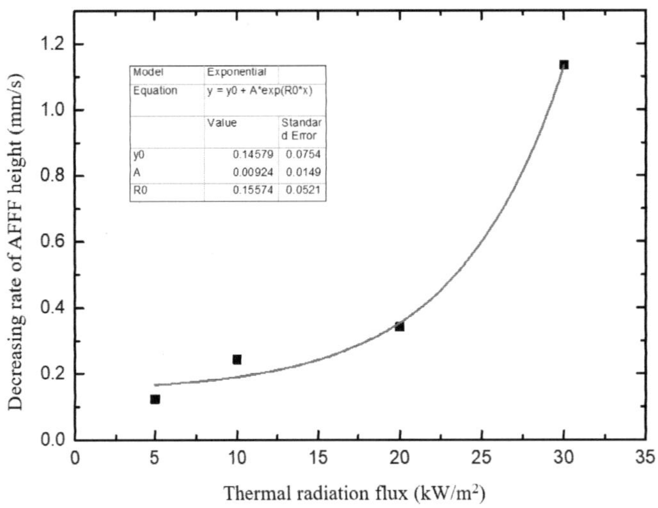

Figure 3. The relationship between decreasing rate of AFFF height and radiation flux (S = 13.5, h = 100 mm).

4. AFFF extinguishing experiment

In order to understand the actual process of AFFF extinguishing, the experiment was carried to put out a pool fire. The experiment simulated the case that foam injection from the central of the oil pan, which is similar to the real situation. In the experiment, the kerosene was used as fuel and 0.5 L heptane was used as the ignition agent. After ignition, it is ensured to have a one-minute-long pre-burning time, to generate a stable pool fire. After the pre-burning, AFFF extinguishing agent was sprayed on the oil pan.

Figure 4 shows the process of AFFF extinguishing experiment, from the time when the AFFF was applied on the fire. It can be observed from the image that the foam began to accumulate in the central of the pan gradually, while the flame shape was not significantly affected. The phenomenon was similar to the flame aggregation of the ring oil pool fire. As the foam spread over the entire surface of the pan, the flame went noticeably weaker and gradually died out. It is also found from the results that, with the external radiation, it took far longer time for the foam spread over the entire liquid surface. It can be inferred that, the heat radiation reduced the thickness of the foam, so that the volume of foam decreased, thereby the spread speed of foam was reduced.

Figure 4. The process of AFFF extinguishing experiment (S=13, $\dot{V}$=150 ml/s).

5. AFFF spread model with external radiation

In order to obtain the model of foam weight, the radiation flux should be determined first. As the relative position of foam and the fire flame has a great impact on the radiation flux, it need to be discussed in several cases, as shown in Figure 5. When the foam is inside the fire flame, the flame shape will not will be changed significantly as the experiment indicates. When the foam is outside the flame, the fire will be weakened as the foam spread and overlaps with oil surface.

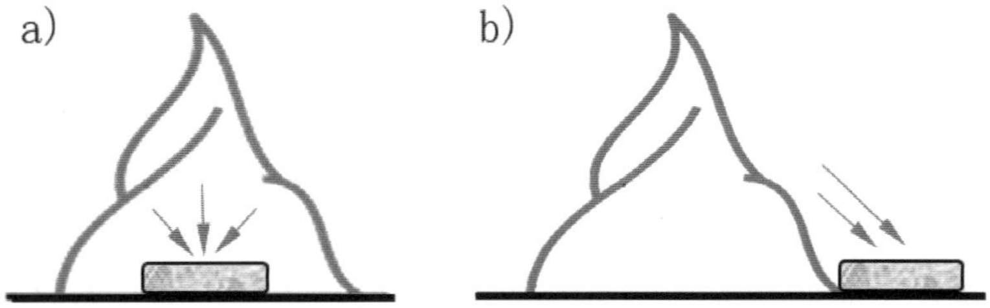

Figure 5. The relative position of fire flame and foam: a) inside; b) outside.

Using the solid flame model to calculate the fire radiation, it is calculated that view factor of fire flame to the foam surface can be expressed as Equation(4), with the cylinder flame shape assumed.

$$F_{\text{fire},H} = \frac{B-1/s}{\pi(B^2-1)^{1/2}} tan^{-1}\left(\frac{(B+1)(s-1)}{(B-1)(s+1)}\right)^{1/2} - \frac{A-1/s}{\pi(A^2-1)^{1/2}} tan^{-1}\left(\frac{(A+1)(s-1)}{(A-1)(s+1)}\right)^{1/2} \quad (4)$$

Where, $= L/R_f$, $h = H/R_f$, $A = (h^2 + s^2 + 1)/2s$, and $B = (s^2 + 1)/2s$. L is the distance between the fire flame to foam front, H is the flame height, and R_f is the radius of fire flame.

Therefore, the model of foam loss can be obtained. The spread and the evaporation of the foam contribute to the increase of the coverage, and their dynamic balance determines the foam coverage area, as described in Figure 6. The spread model determines the maximum area that the foam is able to cover. When the foam is exposed to external radiation, the thickness reduction rate can be calculated by this foam loss model. The actual foam coverage area can be obtained by subtracting the radiation loss area from the total foam area calculated from the axisymmetric spread model.

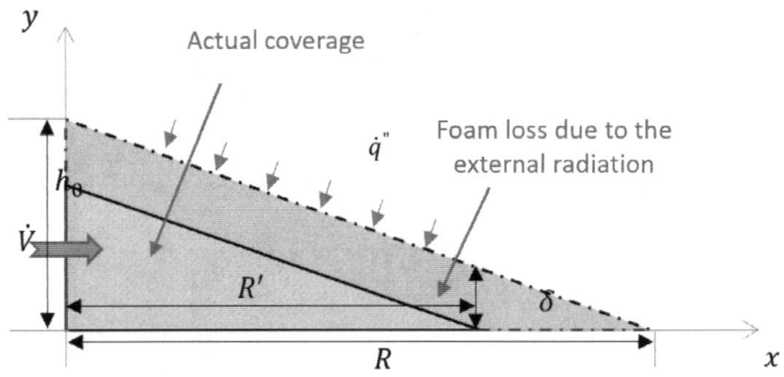

Figure 6. Illustration of foam spread process with external radiation.

It can be inferred from Figure 6 that, the key to solving the foam coverage area under external thermal radiation is to calculate the foam radius where the foam height is $\delta(t)$. And, this $\delta(t)$ is directly related to the reduction rate of foam layer thickness at this location, which can be expressed as

$$\delta(t) = (0.0092 \cdot e^{0.1557\dot{q}_r''} + 0.1458) \cdot t \tag{5}$$

For the case of both effects from foam spread and external thermal radiation, it is considered that the thermal effect can be calculated from the actual foam coverage radius $R'(t)$.

$$R'(t) = R(t) - \frac{R(t)}{h_0(t)}\delta(t) \tag{6}$$

Thus, substituting the Equation(1), the actual foam coverage radius $R'(t)$ can be written as

$$R'(t) = (\frac{18}{\pi^2}\beta R_0\dot{V}^2)^{1/7} \cdot t^{3/7} - (\frac{8}{3}\pi\beta^3{R_0}^3)^{1/7} \cdot \dot{V}^{-1/7} \cdot t^{2/7}\delta(t) \tag{7}$$

If the foam injection is stopped, the only condition will be the thermal radiation. Assuming that previous foam injection time is t_0, similarly, the $R'(t)$ can be described as

$$R'(t) = R(t_0) - \frac{R(t_0)}{h_0(t_0)}\delta(t) \tag{8}$$

Substituting the Equation(1) into Equation (8), the actual foam coverage radius $R'(t)$ can be obtained, thereby the actual foam coverage area can be calculated by $A' = \pi R'^2$.

$$R'(t) = (\frac{18}{\pi^2}\beta R_0\dot{V}^2)^{1/7} \cdot {t_0}^{3/7} - (\frac{8}{3}\pi\beta^3{R_0}^3)^{1/7} \cdot \dot{V}^{-1/7} \cdot {t_0}^{2/7}\delta(t) \tag{9}$$

6. Conclusion

The model of foam loss under external thermal radiation is focused on in this study. By analysing the experimental data of foam thickness change under thermal radiation, the relationship between foam thickness reduction rate and radiant flux is obtained. Finally, the actual AFFF spread process under external thermal radiation is modeled based on the cylindrical flame radiation theory. Taking into account both foam spread and foam loss under external thermal radiation, the fire extinguishing effect of AFFF is determined. It is believed that the results of this study can provide theoretical guidance for fire suppression using AFFF, .especially in calculating the amount of extinguishing agent in different fire scenarios.

7. Reference

[1] Kuchta, Joseph M., and Robert G. Clodfelter. Aircraft mishap fire pattern investigations. GREEN

[2] Williams, B. A., Sheinson, R. S., & Taylor, J. C. (n.d.). Regimes of Fire Spread Across an AFFF-Covered Liquid Pool, 1–10.

[3] Persson, B., & Dahlberg, M. (1994). A Simple Model For Predicting Foam Spread Over Liquids. Fire Safety Science, 4, 265–276.

[4] Persson H., "Fire Extinguishing Foams - Resistance Against Heat Radiation", SP Report 1993-54, 1993.

[5] Wang, B. W., Huang, S. S., He, Q. Z., & Lu, S. X. AFFF Spread Over Liquid Surface with Different Foam Expansion. Materials Science and Engineering, 10, 9789813226517_0160.

[6] Dahlberg M., "Foam Spread Experiments on a Liquid Surface", SP Report 1994: 16, 1994

Biomass pellets produced from filler cake as waste to energy in sugar industry

K Pajampa[1], T Wongwuttanasatian[1,2*]

1 Department of Mechanical Engineering, Faculty of Engineering, Khon Kaen University, Khon Kaen, Thailand
2 Center for Alternative Energy Research and Development, Khon Kaen University, Thailand
E-mail: *tanwon@kku.ac.th

Abstract. In the sugarcane and sugar industry, there is a waste produced from the manufacturing process called "Filter cake". It is a black solid sludge with the moisture content of 70-75%. In general, this waste is about 4% of the sugarcane mass input to the process. In every year, there are a large amount of filter cake resulting in high waste management costs. Therefore, the idea of conversion this waste to energy was proposed. The filter cake was dried off to the moisture contents of 40%, 45%, 50%, 55% and 60% and then fed into a pelletizer for the 5 different initial moistures. Then filter cake pellets were dried off by sun drying for 3 days. Dried pellets from different initial moistures were analyzed for physical properties. It was found that the pellet with initial moisture content of 40% has the best physical properties at the diameter of 4.0 mm and 68.4% of pellets felt into 30-40 mm of length. In addition, it has the highest bulk density, highest durability and lowest dust content of 222.03 kg/m^3, 96.13% and 0.54% respectively. The proximate analysis and heating value were also determined for the dried pellets. The results showed that dried pellets contained 5.33% moisture, 35.67% ash, 51.33% volatile matter, 7.67% fixed carbon and the heating value of 11.71 MJ/kg. These properties were compared to other pellets and standard and it was suggested that the filter cake pellets can be used together with bagasse as fuel for a boiler in a sugarcane industry.

1. Introduction

Currently, there are 55 sugar mills with a total capacity of 100 million tons per year in Thailand [1]. The sugarcane bagasse, molasses and filter cake are economically significant by-products in sugarcane and sugar industry. The plant can produce 13% of sugar, 52% of waste water, 28% of bagasse, 3% of molasses and 4% of filter cake as seen in Figure 1[2]. The filter cake is a by-product produced from manufacturing process, which is a black solid sludge with moisture content of about 70-75%. It is the sludge from a sugar clarifier which composes of clay sediment, contaminants from sugarcane juice and fine bagasse. The filter cake is conveyed through the conveyor belt to the truck and being dumped into a provided area far from the plant buildings. Filter cake is generally produced about 4% of total sugarcane input to the mill. Thus, with a big amount of sugarcane used in this industry, millions ton of filter cake are found to be a problem in waste management.

Content from this work may be used under the terms of the Creative Commons Attribution 3.0 licence. Any further distribution of this work must maintain attribution to the author(s) and the title of the work, journal citation and DOI.
Published under licence by IOP Publishing Ltd

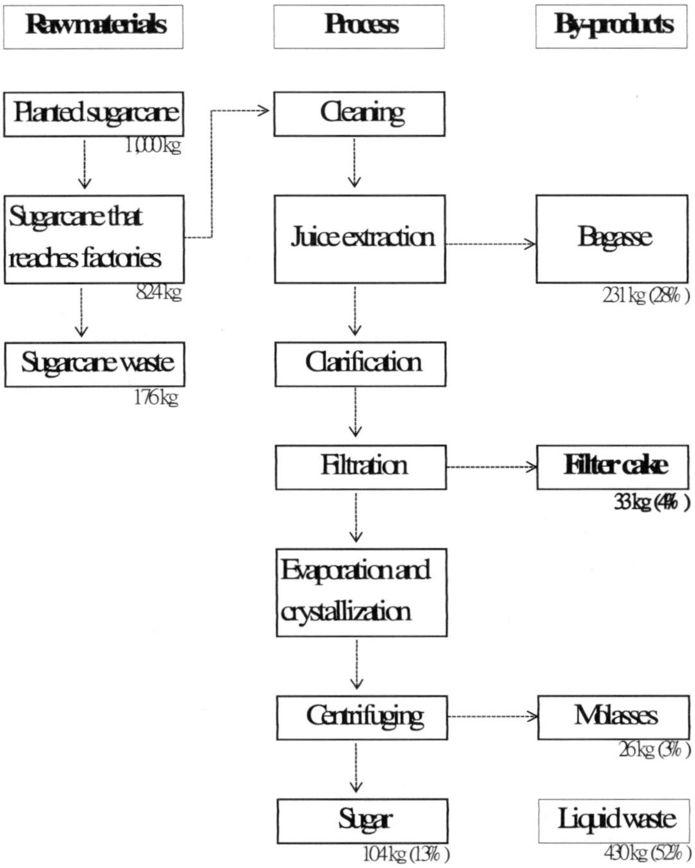

Figure 1. Manufacturing process of sugar and wastes [2].

According to the sugarcane milling information throughout Thailand in 2016-2017, there are 147.40 million tons of sugarcane fed into milling process, the total amount of filter cake is accounted for 5.89 million tons [1]. The conversion of waste to energy was proposed in this study to eliminate the cost of waste management and also to obtain a kind of biomass fuel complying with the Thai national renewable energy plan so called Alternative Energy Development Plan: AEDP2015 2015-2036. There have been many attempts to utilize filter cake, for example, the properties of compressed cement (SCC) mixed with combusted filter cake (ISFC) instead of Portland concrete (OPC) was investigated [3]. The mixing of filter cake together with soil and cow manure as ground-support material for the earthworm farming [4]. The filter cake mixed with distillery slop has also been used as a soil amendment agent for farmers [5]. In addition, the agricultural residues have been used to produce pellet fuels such as pellets from pineapple peels with cassava starch as a binder [6], pellets from olive branches [7], pellets from pine cones and reed canary grass (RCG), from timothy hay (H) and switchgrass (SW) [8], from eucalyptus bark, from mangosteen shell and papaya peel with different binders (dammar, Perseakurzii kosterm powder, and cashew nut shell liquid) [9]. Moreover, pellet fuels have been from the community waste water treatment sludge, coal powder mixed with tapioca flour as binder [10]. Therefore, the aim of the work is to produce pellet fuel from the filter cake and examine the physical properties of the pellets. It is expected that the filter cake pellets can be used with bagasse for a boiler in a sugar industry.

2. Material and methods

2.1. Raw material preparation and pelletizing process

The filter cake as raw materials were prepared with 5 different initial moisture contents of 40%, 45%, 50%, 55% and 60% and then fed into a pelletizer. The pelletizer has 4-mm holes flat die with maximum capacity of 120 kg/hr shown in Figure 2 and Figure 3. The filter cake was continuously fed through the die causing heat generated by friction in pelletizing process. Due to the different temperature and heat conditions during pelletizing process, lignin (average crude fiber analysis of bagasse is Cellulose 43.2%, Hemi cellulose 31.5%, Lignin 22.0%, Ash 3.3%) [11] inside the samples was transformed into binder helping the agglomeration of samples making the pellet to form [12]. The obtained pellets were dried for 3 days by sun drying to reduce the moisture content (below 10%) as required by the standard of pellet.

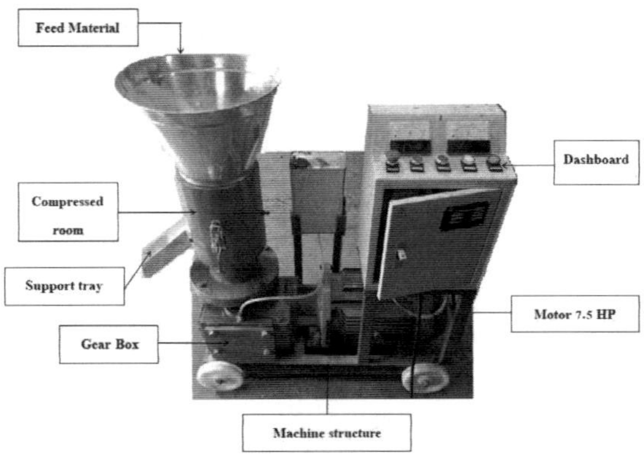

Figure 2. Pelletizer and its components.

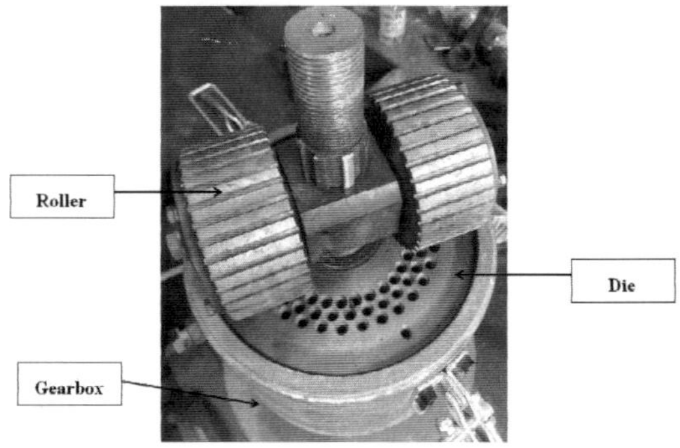

Figure 3. Die.

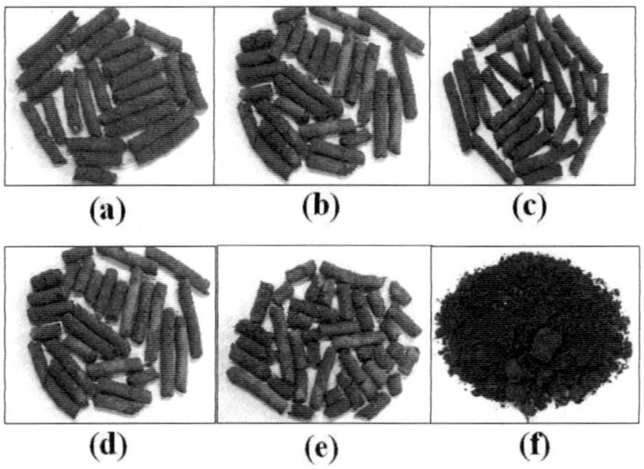

Figure 4. Pelletized filter cake pellets with various initial moisture contents at (a) 40%, (b) 45% (c) 50%, (d) 55%, (e) 60% and (f) filter cake before pelletization with 40%moisture content.

2.2. Physical properties of filter cake pellets

2.2.1. Dimension (Diameter and Length) Diameter and length of the pellets were randomly measured by a vernier caliper and mass was measured using a balance with a resolution of 0.1 g. Sample of 500 g of each pellet types (a-e) seen in Figure 4 were measured. The diameters and lengths were then averaged. The lengths of pellets were classified into 3 different ranges, i.e., 30-40 mm, 20-30 mm and below 20 mm. Then each pellet group was weighted to determine percentage of mass.

2.2.2. Bulk density Bulk density of 5 types of pellets was examined according to EN 15103 standard. The bulk density of each type of pellets was averaged from 3 samples.

2.2.3. Durability and Fines Durability testing was carried out according to EN 15210-1 standard. Each type of pellets were tested for 2 samples and averaged.

$$\%Durability = \frac{mass\ of\ pellets \vee crumbles\ after\ tumbling}{mass\ of\ pellets \vee crubles\ before\ tumbling} x100$$

The durability indicated the strength of the pellets that cannot be broken under a tested condition. While, Fines was tested according to EN 15149-1 standard. Fines is an index that indicates quantity of small pieces broken during the test. The pellets with higher fines indicated that the pellets were brittle, breakable and crumby. Similarly, 2 samples of each type of pellets were tested and averaged. The fines can be calculated by $\%Fines = \frac{W-W_1}{W} x100$
Where F = Fines content (%mass), W = original mass (g) and W_1 = remaining screened mass (g)

2.3. Proximate Analysis and Heating value
The proximate analysis of dried pellets after 3 days of sun drying was determined based on ASTM D3172. Moisture content (%M), Volatile Matter (%VM) and ash (%A) were determined and then Fixed carbon (FC) could be obtained by FC % = 100 - (M % + VM % + A %)
The heating value was determined according to ASTM D 240 standard using IKA C5000 Bomb calorimeter.

3. Experimental results and discussions

3.1. Physical properties of filter cake pellets

3.1.1. Diameter and Length It was found that the diameter of filter cake pellets was 4.0 mm according to the 4-mm holes die. The percentage of mass of pellets for 3 classified lengths depended on the initial moisture content as shown in Figure 5. With initial moisture content of 40%, 68.40% of mass felt into 30-40 mm meaning that perfect shape of pellets was most obtained, and they were not easily broken. On the other hand, by increasing initial moisture, perfect pellets were less formed. By comparing to the EN 14961-1 standard, length of filter cake pellets agreed well with the standard. But the diameter of 4 mm is less than wood pellet standard. However, if 6-mm holes die is used, the desired diameter will be obtained.

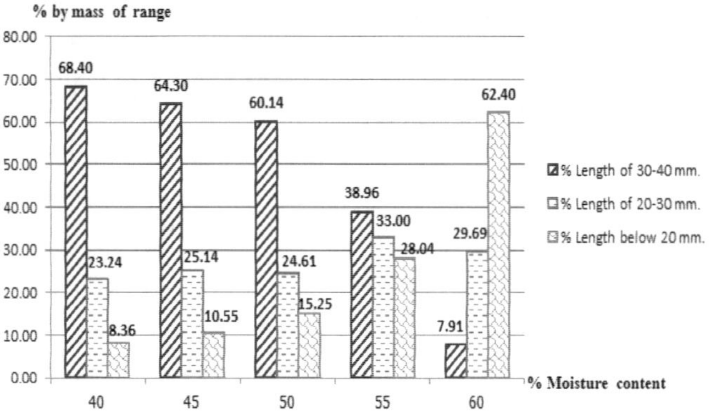

Figure 5. Percentage of mass for 3 classified lengths.

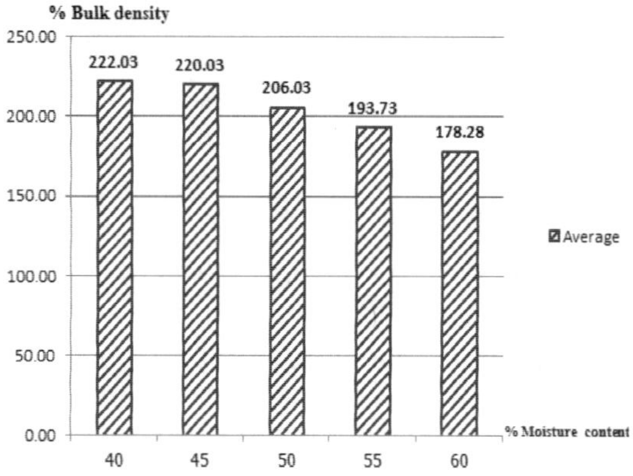

Figure 6. Bulk density.

3.1.2. Bulk density The bulk densities of pellets with 5 initial moisture contents were slightly different. The filter cake pellets with 40% initial moisture content had the highest bulk density of 222.03 kg/m^3 and the density decreased when initial moisture increased as shown in Figure 6. The bulk density value of biomass fuel based on EN 14961-1 standard should be greater than 600 kg/m^3 [13]. Thus, it is suggested that pelletizing process must be altered to increase the density. It is suggested that for current bulk density, filter cake pellets must be used on site instead of being transported.

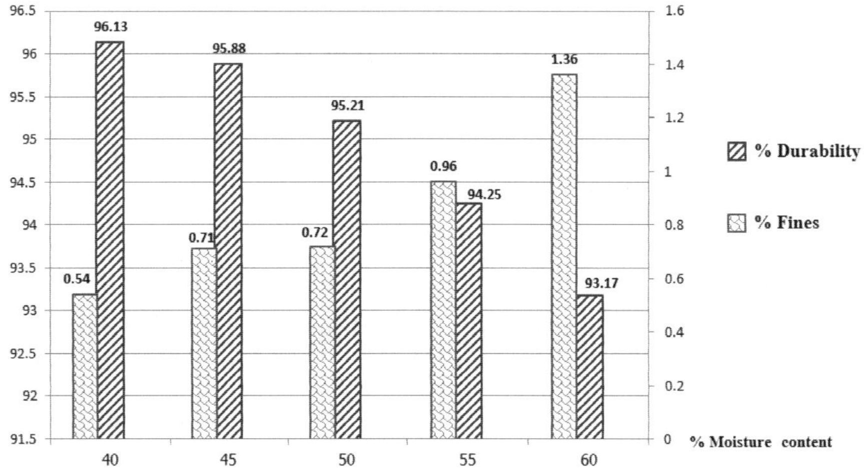

Figure 7. %Durability and %Fines of filter cake pellets.

3.1.3. Durability and fines analysis For 5 types of pellets seen in Figure 7, averaged durabilities were 96.13%, 95.88% 95.21%, 94.25% and 93.17% respectively. It was found that only at 40% initial moisture, the durability of filter cake pellets was acceptable according to the standard depicted in Table 1. From the fines analysis, the pellets made from 5 initial moisture contents indicated in the fragility of pellets. It was shown that with less initial moisture, the pellets could be less breakable. Similarly, with 40% initial moisture content, the fines value of pellets agreed well standard. The Durability and fines were found to have inversely relation.

Table 1. Physical properties of filter cake pellets and olive trees pellets.

Parameter	Filter cake pellets					Pellets made from olive trees [7]			Biomass pellets
	(a)	(b)	(c)	(d)	(e)	OL	OP	OW	Standard EN 14961-1 [13]
Diameter : mm	4.00	4.00	4.00	4.00	4.00	6 or 8	6 or 8	6 or 8	(6 ± 1mm)
Length : mm	30-40 (68.4%)	30-40 (64.3%)	30-40 (60.14%)	30-40 (38.96%)	< 20 (62.40%)	12.29	-	28.74	(3.15-40 mm)
Bulk density : kg/m^3	222.03	220.03	206.03	193.73	178.23	<1,000	<1,000	<1,000	≥ 600.00
Durability : %	96.13	95.88	95.21	94.25	93.17	0	42.3	91.7	DU ≥ 96.5
Fines : %	0.54	0.71	0.72	0.96	1.36	-	-	-	F≤ 1.0

(a): initial moisture content 40%, (b): initial moisture content 45%, (c): initial moisture content 50%, (d): initial moisture content 55%, (e): initial moisture content 60%, OL: Olive leaves, OP: Leaves and small branches, OW: Olive wood

From Table 1, it can be concluded that the physical properties of filter cake pellets were within the standard. On the other hand, the bulk density was too low compared with wood pellet standard. This suggested that the filter cake pellets are suitable for onsite use and have a disadvantage of logistic issue.

3.2. Proximate analysis and heating value
From proximate analysis shown in Table 2, filter cake pellets were compared with some biomass pellets and briquettes. It can be said that filter cake pellets having heating value of 11.71 MJ/kg can be used as fuel. Filter cake pellets was found to have higher heating value than bagasse since pellets contained much less moisture than bagasse. Although, the fixed carbon is quite low and ash content is too high, the pellets could be fired with bagasse as an additional fuel. Because of sun drying, the moisture of filter cake pellets was considerably low resulting in higher heating value compared with bagasse.

Table 2. Proximate analysis and higher heating value.

Parameters	Filter cake pellets	Bagasse [11]	Briquettes made from Different materials				Pellets made from olive trees [7]			Wood pellet Standard EN 14961-1 [13]
	FP		PP[6]	MP[14]	DP[14]	SW[15]	OL	OP	OW	
Moisture: M %	5.33	7.0-32.0	17.80	5.65	6.68	7.20	11.02	10.89	7.42	M≤ 10
Volatile matter: VM %	51.33	69.4-81.7	62.90	86.55	88.37	76.80	14.17	5.50	1.43	-
Ash: A %	35.67	3.3-7.21	3.30	5.03	4.57	5.30	63.74	79.80	90.15	A≤ 0.7
Fixed carbon: FC %	7.67	12.0-16.0	15.90	2.77	0.37	10.70	11.07	3.81	1.00	-
Heating value: (MJ/kg)	11.71	5.57-14.0	13.90	18.20	16.33	17.34	19.64	18.20	17.53	Q ≥16.5

FP = Filter cake pellet, PP = Pineapple peel, MP = Mangosteen peel, DP = Durian peel, SW = Swamp
OL = Olive leaves, OP = Leaves and small branches, OW = Olive wood

4. Conclusion
Filter cake considered as waste in sugar industry was pelletized with 5 different initial moisture contents. Filter cake pellets obtained from those initial conditions were then analyzed for their physical properties. After sun drying of the pellets, proximate analysis and heating value were determined. The results showed that with 40% initial moisture content, the pellets gave better physical properties, i.e., rigid form, less fragile, high bulk density and low fines. These properties were compared to wood pellets standard and agreed well within the standard. The filter cake pellets also gave similar heating value compared with bagasse. However, high ash content of the pellets was a drawback. In addition, with low bulk density of the pellets, it was suggested that filter cake pellets should be used onsite and not suitable to transport as the logistic cost may be too high. Finally, it is recommended that filter cake pellets should be used as co-firing with bagasse for boilers in sugar industries.

5.References
[1] Office of the Cane and Sugar Board 2014 *http://www.ocsb.go.th/th/home/index.php*
[2] Makul N and Sua-iam G 2016 Characteristics and utilization of sugarcane filter cake waste in the production of lightweight foamed concrete. *Journal of cleaner production.* **126** p 118-133
[3] Sua-Iam G and Makul N 2017 Effect of incinerated sugarcane filter cake on the properties of self-compacting concrete. *Construction and Building Materials.* **130** p 32-40
[4] Jala A and Chakkrawan S January 2016 Bedding material mixed with different ratio of filter cake from sugar cane which effected on growth of 2 types of earthworm. *Thai Journal of Science and Technology.* p 43-55

[5] Prado R, Caione G and Campos C 2013 Filter Cake and Venessa as Fertilizers Contributing to Conservation Agriculture. *Applied and Environmental Soil Science.* **2013** 8 pages
[6] Tantisatayakul Th, Sayda K, Phusongsri S and Ngernreuangroj S 2015 Feasibility study of biomass briquettes production from pineapple peel. *Journal of Science and Technology.* **23 No.5** *(Special Issue)* p 754-773
[7] Garcia-Maraver A, Rodriguez M L, Serrano-Bernardo F, Diaz L F, Zamoranoa M 2015 Factors affecting the quality of pellets made from residual biomass of olive trees. *Fuel Processing Technology.* **129** p 1–7
[8] Haruna N Y and Afzalb M T 2016 Effect of Particle Size on Mechanical Properties of Pellets Made from Biomass Blends. *Procedia Engineering.* **148** p 93 – 99
[9] Jamradbedluk J and Lertsatitthanakorn Ch 2017 Influences of Mixing Ratios and Binder Types on Properties of Biomass Pellets. *Energy Procedia.* **138** p 1147–1152
[10] Sungkapan A and Eksangsri T 2018 Co-pelletized briquettes production from municipal wastewater sludge; Case study of pattaya municipal wastewater treatment plant. *The 9th Thailand national and international conference, 20 March 2018 Bangkok.* p 1523-1533
[11] Omoniyi, T.E., Olorunnisola A.O 2014 Experimental Characterization of Bagasse Biomass Material for Energy Production. *International Journal of Engineering and Technology* **4 No.10** p 582-589
[12] Torsagun S, Suparat J and Phanasdee N 2015 Design and construct the machine for compressing the biomass fuel pellets. *The 8th Thailand renewable energy for community conference* p 91-94
[13] Alakangas, Eija, (VTT) 2011 New European Pellet Standard - EN14961-1 coordinator of EUBIONET III. Finland. p 1-9
[14] Asavrukulchai A, Semsayan Sh, Prapakdee N, Piamsuwansiri N and Choochart N 2011 The durian and mangosteen rind are used in the form of solid fuel. *The 49th Kasetsart University Annual Conference Bangkok Thailand* p 162-168
[15] Tantisatayakul Th, Phongkasem S, Phuya P and Taipanguy P 2015 Community Renewable Energy from Compost *Science and Technology.* **23 (3)** p 418-429

IOP Conf. Series: Earth and Environmental Science **257** (2019) 012022 doi:10.1088/1755-1315/257/1/012022

Solar energy potential in Bangka belitung islands, Indonesia

Yuant Tiandho[1,2], Irwan Dinata[1,3], Wahri Sunanda[1,3], Rika Favoria Gusa[1,3] and Dwi Novitasari[4]

[1]Research Center for Energy and Information Technology, Universitas Bangka Belitung, Indonesia
[2]Department of Physics, Universitas Bangka Belitung, Indonesia
[3]Department of Electrical Engineering, Universitas Bangka Belitung, Indonesia
[4]Center for Energy Studies, Universitas Gadjah Mada, Indonesia
E-mail: yuant@ubb.ac.id

Abstract. Bangka Belitung Islands is one of the provinces in Indonesia which has hundreds of small islands. Because the islands are separated from each other it is almost impossible to distribute electricity from the main electricity source in the Bangka Island to all the islands. Because it is located in the tropic country, solar cells can provide an alternative solution to the fulfillment of electricity rather than create a distribution cable across the sea. The solar energy potential in Bangka Belitung Islands can be calculated based on climatological, meteorological, and geographical data. Various solar radiation transmittance coefficients have a higher value in the dry season. This condition indicates that in the dry season the total solar radiation is higher than in the rainy season. Due to the high value of total solar irradiance in the Bangka Belitung Islands, it reaches 4.95 kWh/m^2 day, it is interesting to utilize this potential to provide electrical energy in the Bangka Belitung Islands.

1. Introduction

In a modern civilization the need for electrical energy has become a basic necessity. Electricity will support people in various activities: social, economic, technology, and others. Most traditional electric power generation use oil and its derivatives to produce energy. But the issue of limited sources of fossil energy and the impact of pollution has led many countries around the world develop electricity conversion technologies from a renewable energy [1, 2]. The photovoltaic panel has been widely used because it provides various advantages in terms of both economy and environment [3]. Additionally, because the installation process is easy and has good durability, the solar cells can support electricity supply for people who do not have access to the main electricity source [4]. Solar cells have been widely applied in various geographical areas such as cities, countryside, and deserts [5].

Generally, tropical countries have abundant solar energy. Even tropical countries can receive about 300 sunny days with 8 hours sunshine in a year [6]. Therefore, the use of solar energy as a source of electrical energy in the tropics has a promising prospect.

Bangka Belitung Islands is a province in Indonesia that has many small islands. There are about 1015 islands in the province [7]. Therefore, the effort to distribute electricity from the main electricity source in Bangka Island will have economic difficulties. Alternatively, the use of solar cells to provide electrical energy in small islands in the Bangka Belitung Islands is interesting to be developed because the province is located in the tropics.

In this paper we analyze the potential of solar radiation energy in Bangka Belitung Islands based on climatology, meteorology, and geographical data. The model of solar irradiance potential is

Content from this work may be used under the terms of the Creative Commons Attribution 3.0 licence. Any further distribution of this work must maintain attribution to the author(s) and the title of the work, journal citation and DOI.
Published under licence by IOP Publishing Ltd

constructed based on Bird-Hulstrom [8] and Villicana-Ortiz [9] model. It is expected that the calculations in this paper will be taken into consideration in the effort of electrification of small islands in Bangka Belitung Islands.

2. Solar Irradiance Physical Model

The solar energy that reaches the earth is partly of a electromagnetic radiation emitted by the sun. The radiation travels from the solar core and is distributed uniformly and isotropically throughout the universe. When it reaches the Earth's atmosphere the radiation undergoes several interactions such as scattering, absorption, and reflection. The various interactions occurs because the atmosphere composed by various molecules such as water vapor, CO_2 and O_2, ozone, and various types of aerosols [9]. In the model of Bird-Hulstrom, total radiation (I_T) is the amount of direct radiation (I_d) with diffuse radiance (I_{sa}) and it corrected with albedo parameters: ground (r_g) and sky/atmospheric (r_s) [8],

$$I_T = \frac{\left(I_d + I_{as}\right)}{1 - r_g r_s} \tag{1}$$

Since it is difficult to obtain an explicit value of r_s then in the model the parameter r_s is proposed as [8]

$$r_s = 0.0685 + \left(1 - B_a\right)\left(1 - \tau_{as}\right) \tag{2}$$

where B_a is a forward-scattering factor ratio and it is related to asymmetry factor in Mie theory. But by looking at the asymmetry factor data for different types of aerosols, Bird-Hulstrom suggested the value of B_a=0.84. The τ_{as} is the transmittance coefficient due to absorbance and scattering by dry air in the Watt model [8],

$$\tau_{as} = 10^{-0.045\left(\frac{P}{101325} m_{rel}\right)^{0.7}} \tag{3}$$

where P is atmospheric pressure whose value depends on altitude, z, $P = 101325\exp(-0.0001184z)$ and m_{rel} is relative air mass. Solar direct irradiance is determined through the relationship [9],

$$I_d = 0.9662 C_r \tau_s \tau_o \tau_g \tau_w \tau_a \cos\theta \tag{4}$$

where C_r is daily solar constant, τ_s is transmission coefficient by scattering due to air molecules, τ_o is transmission coefficient due to ozone absorption, τ_g is transmission coefficient due to absorption by uniform gases mixture, τ_w is transmission coefficient due to absorption of water vapor, τ_a is transmission coefficient due to absorption and scattering by the presence of aerosols, and θ is solar zenith angle. Explicitly, the parameters in solar direct irradiance defined as in Table 1. In Table 1, C is normalized solar constant (C = 1367 W/m^2), n is the Julian day, m_a is optical air mass, L_o is thickness of ozone layer, Ww is amount of water capable of precipitate in the vertical direction over the place, H is relative humidity, T is absolute temperature, α and β is parameter that related to aerosol particle size and turbidity level, respectively.

In the other hand, the model of diffuse solar irradiance is defined as [8],

$$I_{as} = 0.79 C_r \tau_o \tau_g \tau_w \tau_{aa} 0.5 \frac{1 - \tau_s + B_a\left(1 - \tau_{AS}\right)}{1 - m_a + m_a^{1.02}} \cos\theta \tag{5}$$

where τ_{AS} is transmission coefficient due to aerosols scattering and τ_{aa} is transmission coefficient due to absorption by aerosols. Mathematically, the both coefficient defined as (15) and (16). While ω_0 is a parameter called as single scattering albedo [8],

$$\tau_{aa} = 1 - \left(1 - \omega_o\right)\left(1 - m_a + m_a^{1.06}\right)\left(1 - \tau_a\right) \tag{6}$$

$$\tau_{AS} = \frac{\tau_a}{\tau_{aa}} \tag{7}$$

Table 1. Explicitly definition of parameters in the direct solar irradiance [8,9].

Parameters	Eq. no.
$C_r = C\left(1+0.33\cos\frac{360n}{360}\right)$	(8)
$\tau_s = \exp\left(-0.0903 m_a^{0.84}\right)\left(1+m_a-m_a^{1.01}\right)$	(9)
$m_a = \exp\left(-0.0001184z\right) m_{rel}$	(10)
$m_{rel} = \frac{1}{\cos\theta + 0.15\left(93.885-\theta\right)^{-1.253}}$	(11)
$\tau_o = 1-\left(0.1611\left(L_o m_{rel}\left(1+139.48 L_o m_{rel}\right)^{-0.3035}\right)+\frac{0.002715 L_o m_{rel}}{1+0.044 L_o m_{rel}+0.003\left(L_o m_{rel}\right)^2}\right)$	(12)
$\tau_g = \exp\left(-0.0127 m_a^{0.26}\right)$	(13)
$\tau_w = 1-\frac{2.4959\left(Ww\, m_{rel}\right)}{\left(1+79.034\left(Ww\, m_{rel}\right)\right)^{0.628}+6.385\left(Ww\, m_{rel}\right)}$	(14)
$Ww = 0.493\left(\frac{H}{T}\right)\exp\left(26.23-5416/T\right)$	(15)
$\tau_a = 0.12445\alpha - 0.0162 + \left(1.003-0.125\alpha\right)\exp\left(-\beta m_a\left(1.089\alpha-0.1523\right)\right)$	(16)

3. Results and Discussion

The solar energy availability model proposed by Bird-Hulstorm requires several geographical, climatological, and meteorological parameters. To evaluate the solar irradiance it is necessary to use climate data available in existing weather stations, i.e. Badan Meteorologi, Klimatologi, dan Geofisika (BMKG) - Depati Amir Meteorology Station, Pangkalpinang from January 1, 2016 to January 1, 2018. In Fig. 1, the temperature and humidity patterns for this time period are presented. Through temperature and humidity patterns it appears that the values of temperature and humidity in Bangka Belitung Islands have a relationship that is inversely proportional. The temperature range is 24°C to 30°C. While the moisture range is 69% to 99%. The Bangka Belitung Islands has two seasons: the rainy season that occurs in October - May and the dry season that occurs in June - September [10].

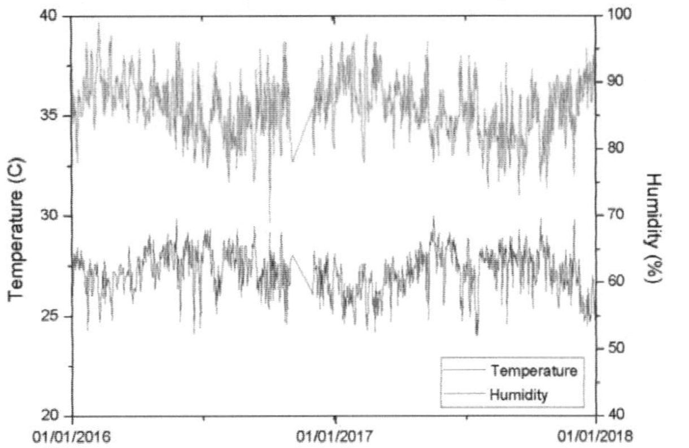

Figure1. Temperature and humidity patterns in Bangka Belitung Island since 2016 to 2018.

To analyze solar direct irradiance then required coefficients as stated in eq. (4). The first coefficient is the transmission coefficient by scattering due to air molecules, τ_s. Based on eqs. (6)-(8), the transmission coefficient is influenced by altitude and solar zenith angle. According to ref. [11] it is stated that some areas of the Bangka Belitung Islands lies in altitude 0-699 masl. Although for the Pangkalpinang area itself is only in altitude 0-70 masl. Based on the pattern of relationship between transmission coefficient by scattering due to air molecules with altitude and solar zenith angle as shown in Fig. 2(a) it appears that the dominant factor that changes the value of τ_s is solar zenith angle. Above 1 rad the value of τ_s decreases significantly. This condition can be understood as a result of the dependence of τ_s to the relative air mass which also changes rapidly above 1 rad.

The second transmission coefficient is the transmission coefficient due to ozone absorption, τ_o. Based on eq. (9), τ_o is influenced by solar zenith angle and the ozone layer thickness. However, the thickness of ozone layer in Indonesia in the recent decades is relatively stable. Since 2003, the thickness is in the range 246.26 DU to 249.46 DU [12]. Thus, it is safe to say that the average ozone thickness in Indonesia is 247.82 DU or about 0.11 mol/m^2. By using eq. (9) can be obtained the relationship between τ_o with solar zenith angle as shown in Fig. 2(b). It appears that the value of τ_o changes rapidly for solar zenith angle above 1 rad.

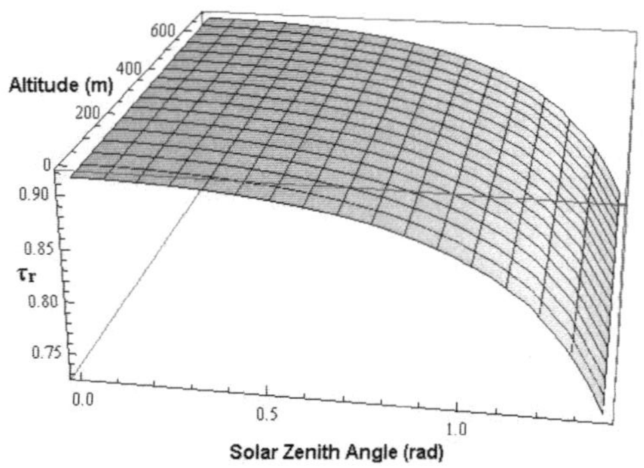

(a)

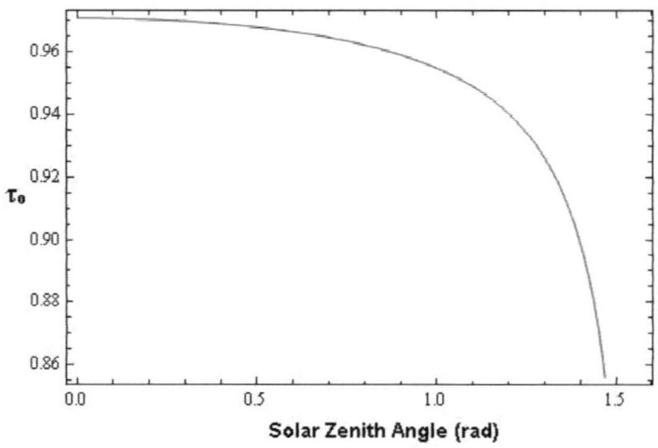

(b)

Figure 2. The relationship between: (a) τ_s and (b) τ_o with the solar zenith angle.

According to eq. (10) transmission coefficient due to absorption of gas mixtures, τ_g, influenced by altitude and solar zenith angle. In Fig. 3 it appears that τ_g is also changing rapidly due to changes in solar zenith angle above 1 rad. However, the change in τ_g is not significant and the value remains at around 0.98.

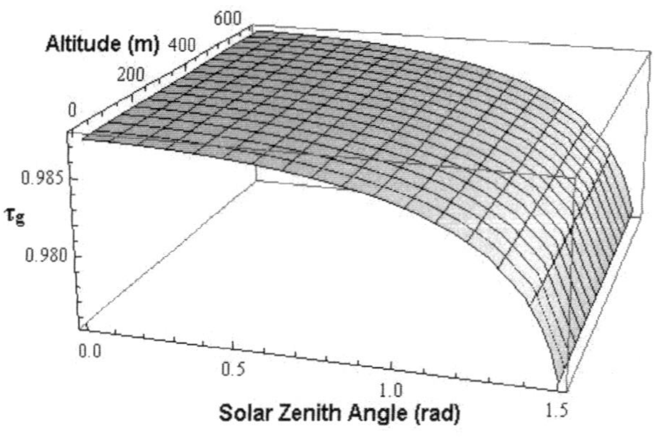

Figure 3. The relationship between τ_g with the solar zenith angle and altitude.

To calculate the transmission coefficient due to the absorption of water vapor, τ_w, then the required amount of *Ww* as indicated in eq. (12). Based on the temperature (absolute) and humidity data as expressed in Fig. 2 then the daily *Ww* within the period 2016 to 2018 can be calculated as in Fig. 4(a). The pattern of fluctuations in *Ww* almost related with the season that occurred Bangka Belitung Islands. According to the type of season it can be calculated the average value of *Ww* for the dry season is 508.7 while in the rainy season its value is 515.4. The high value of *Ww* in the rainy season because the amount of water vapor that may become water has a greater probability than when the dry season. Thus by using the result it can be analyzed the value of τ_w for both seasons as shown in Fig. 4(b). It appears that the transmission coefficient τ_w in the dry season is higher than the rainy season because in the dry season the amount of water vapor that absorbs solar radiation is more slight. However, the differences in both conditions are not so significant.

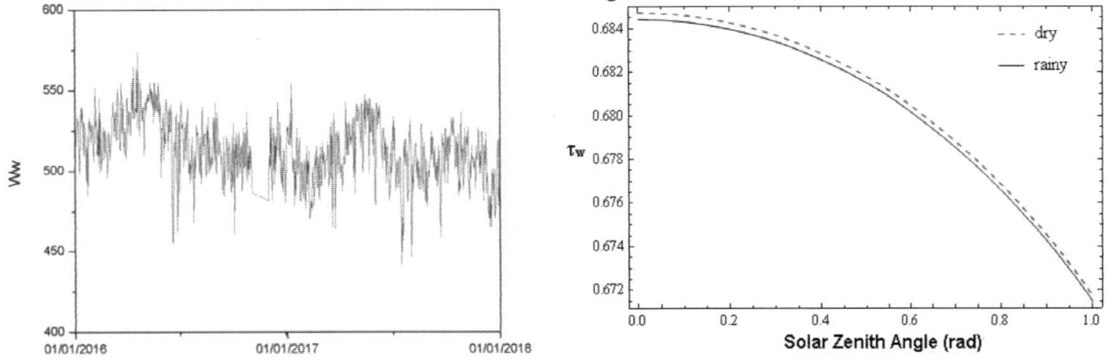

(a) (b)

Figure 4. (a) The value of *Ww* since 2016 until 2018, (b) the relationship between τ_w for dry season and rainy season with the solar zenith angle.

The last transmission coefficient on direct solar irradiance is the transmission coefficient due to absorption and scattering by aerosols. In eq. (13) it has been stated that the transmission coefficient is affected by the size of the aerosol particles (α) with the mean size being 1.3 μm and the level of atmospheric turbidity (β). In this article it is assumed that there are only two conditions of atmospheric turbidity: (i) β=0.1 for conditions with cloud-free sky that occur during the dry season and (ii) β=0.4 for cloudy sky conditions that occur in the rainy season. According to the assumption the value of coefficient τ_a shows in Fig. 5. Based on the calculation it appears that the value of τ_a for the dry season is greater than the value of τ_a for the rainy season.

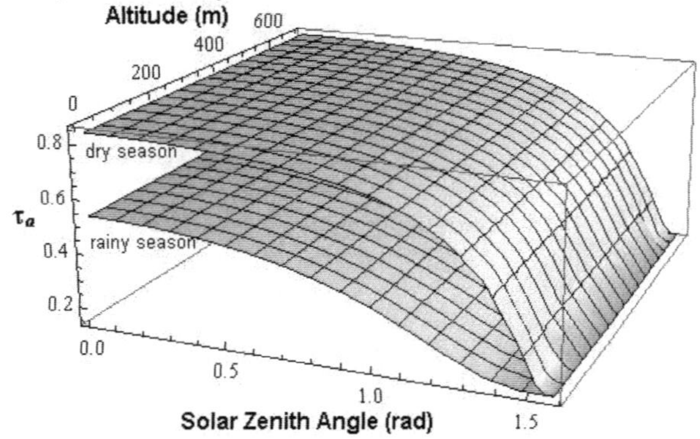

Figure 5. The relationship between τ_a with the altitude and solar zenith angle for dry season and rainy season.

In addition to direct solar irradiance, total solar irradiance is also contributed by diffuse solar irradiance on horizontal surfaces from atmospheric scattering. In the Bird-Hulstrom model, the ratio of B_a for aerosols in atmospheres is 0.84 whereas for albedo single scattering, ω_0, is 0.9. But before calculate the solar irradiance we require information related to transmission coefficient due to aerosol absorption as in eq. (15). By substituting the related parameters, we obtain the transmission coefficient τ_a and τ_{AS} for the rainy and dry seasons as shown in Fig. 6(a) and (b). It appears that the both τ_{aa} and τ_{AS} transmission coefficients for the dry season are higher than the rainy season.

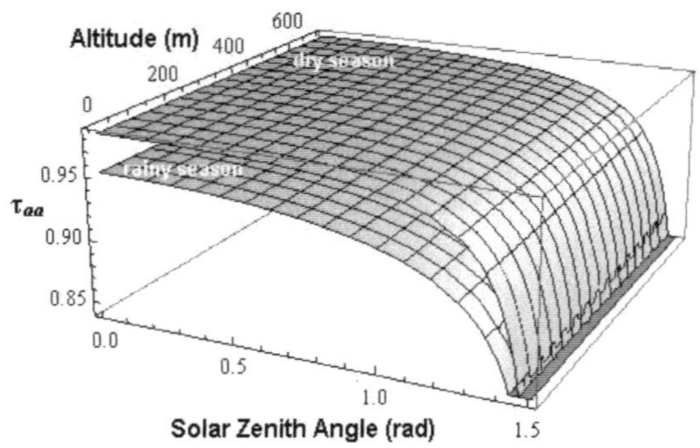

(a)

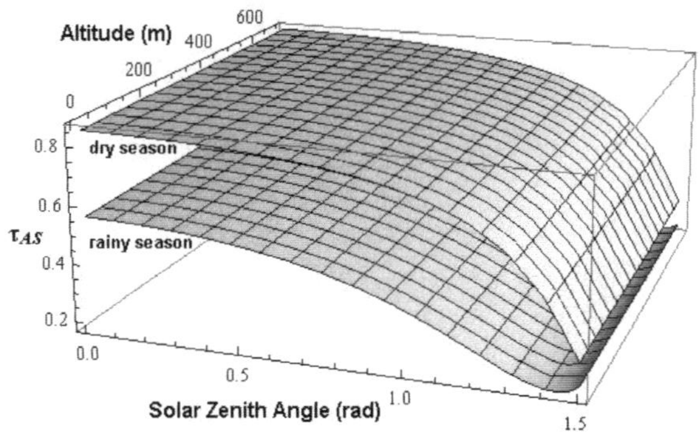

(b)

Figure 6. The relationship between solar zenith angle and altitude with: (a) τ_{aa} and (b) τ_{AS} for dry season and rainy season.

Through the results obtained above, we can calculate the total solar irradiance in Bangka Belitung Islands. The total solar irradiance in one day can be determined through multiplication of solar irradiance power with the average daylight hours for 2016 to 2018 (solar power above 120 W/m^2; Depati Amir Station). To simplify the conditions then the analysis will be divided based on the seasons: the dry season and the rainy season and it assumed the solar zenith angle in 0° and the altitude, z=33 masl. The average duration of daylight hours for the dry season is 6.19 hours/day while for the rainy season is 4.14 hours/day. The albedo ground conditions are assumed as bare soil (r_g=0.17), green grass (r_g=0.25), and concreate (r_g=0.55) [13]. The calculations are presented in Table 2. Generally, total daily solar radiance in the dry season is higher than in the rainy season. In addition, based on the results it can be stated that the Bangka Belitung Island has the high potential of solar energy for the development of renewable energy technology.

Table 2. Total, direct, and diffuse irradiance for rainy season and dry season

	Dry season (kWh/m^2 day)	**Rainy season (kWh/m^2 day)**
Solar direct irradiance	4.26	1.81
Solar Diffuse irradiance	0.47	0.66
Total (bare soil)	**4.79**	**2.51**
Total (green grass)	**4.82**	**2.52**
Total (concreate)	**4.95**	**2.59**

4. Conclusions

Climatological, meteorological, and geographical factors determine the level of attenuation of solar irradiance and the amount of solar energy abundance. In addition to the zenith solar angle, an important factor that significantly changes the value of the transmission coefficient is the type of season. Various transmittance coefficients such as τ_w, τ_a, τ_{aa}, and τ_{AS} have a greater value when the dry season. This condition also has implications on the total solar irradiance. In the dry season the solar potential in Bangka Belitung Islands tends to be higher than in the rainy season. In the dry season the total solar irradiance can reach 4.95 kWh/m^2 day so it can be utilized as a source of electrical energy.

5. Acknowledgement

We gratefully acknowledge the funding from USAID through the SHERA programme – Centre for Development of Sustainable Region (CDSR).

6. References

[1] Tobnaghi D, Madatov R and Farhadi P 2013 *Electric Power Engineering & Control Systems*, . November 90
[2] Razykov T, Ferekides C, Morel M, Stefanakos E and Ullal H 2011 Solar Energy 85 1580
[3] Kazem H 2015 International Journal of Applied Engineering Research 10 43572

[4] Khemariya M, Mittal A, Baredar P, and Singh A 2017 *Journal of Energy Storage,* 14 62.
[5] Al-Shafeey M and Harb A 2018 *9th International Renewable Energy Congress (IREC)* Hammamet 1
[6] Mahtta R, Joshi P and Jindal A 2014 *Renewable Energy* 71 255
[7] Soyusiawaty D, Umar R and Mantofani R 2007 *Seminar Nasional Aplikasi Teknologi Informasi 2007 (SNATI 2007)* Yogyakarta K-17.
[8] Bird R and Hulstrom R 1981 *A simplified clear sky model for direct and diffuse insolation on horizontal surface* (USA: U.S. Department of Energy)
[9] Villicana-Ortiz E, Gutierrez-Trashorras A, Paredes-Sanchez J and Xiberta-Bernat J 2015 *Renewable Energy* 81 534
[10] Fadholi A 2013 *Jurnal Geografi* 10 112
[11] Ngadenin, Syaeful H, Widana K and Nurdin M 2014 *Eksplorium* 35 69
[12] Fajar M 2011 *Pengukuran Ketebalan Lapisan Ozon Menggunakan Data MODIS sebagai Bagian dari Sistem Pemantauan Bencana* (Depok: Universitas Indonesia)
[13] Markvart T and Castaner L 2003 *Practical Handbook of Photovoltaics: Fundamentals and Applications* (UK: Elsevier)

The effect of electric field on glycerine sedimentation and reaction acceleration using multiple high voltage electrodes designed for biodiesel production

Tharin Ratanabuntha[1], Tanakorn Wongwuttanasatian[2] and Amnart Suksri[2]

1 Department of Electrical Engineering, Khon Kaen University, Khon Kaen, Thailand
2 Center of Alternative Energy Research and Development, Khon Kaen University, Khon Kaen, Thailand
E-mail: tharin_rat@kkumail.com, samnar@kku.ac.th

Abstract. Biodiesel is produced from fatty acids, using palm oil that is widely used in food industry as a raw material. This research aims to introduce an innovative method of acceleration in biodiesel production process. The biodiesel production was prepared by 1 liter substrate and conducted in the reactor chamber. The chamber was designed and built with coaxial cylindrical electrode consisted of multiple inner rod electrode and outer electrode. Supply high voltage level was at 10 kV and exposure time was 30 minutes. It is found that strong electric field can accelerate the rate of reaction on biodiesel production. The rate of glycerine sedimentation resulting from electric field process using single rod, double rod, triple rod and quadruple rod calculated from the results are 3.6, 4.2, 5.9 and 9.1 mL/min respectively. Electric field enhancement can be obtained by utilizing multiple rod electrodes so that the biodiesel reaction is confined within controlled environment. The fastest reaction process is by utilizing quadruple rod inner electrode. Glycerine yield separation was obtained at 100 milliliter within 11 minutes. This is 2.5 times faster than single rod configuration.

1. Introduction

Biodiesel is derived from vegetable oil or animal fats through chemical processes. The final product of the process is fuel which is consisted of Mono-alkyl esters of fatty acids (biodiesel). Presently, biodiesel production utilizes chemical reaction process called trans-esterification. Vegetable oil or animal fat reacts with methyl alcohol and using catalyst to develop reaction with conventional procedure. Methanol is commonly used as a substitute for alcohol in biodiesel production due to its low cost and can react with fats easily. It is also a good solvent for base catalysts such as sodium hydroxide or potassium hydroxide, etc. The trans-esterification process can be accelerated further by introduce heating effect during reaction time. Reheating of reaction is generally heated by boiling. The reaction time is roughly 2 hours. Then the oil is transferred to the glycerine separation process later. In normal glycerine separation process, sedimentation has been done by gravity that will take more than 2 hours for 1 liter substrate [1, 2]. Another method may be the use of centrifugation process by decanter centrifuge [3,4] which is expensive and consumes lots of energy. Nevertheless, chemical are used for reducing the reaction time in biodiesel process such as deep eutectic solvent (DES) [5] or activated charcoal [6]. Chemical contamination will still be in the water after washing process.

Content from this work may be used under the terms of the Creative Commons Attribution 3.0 licence. Any further distribution of this work must maintain attribution to the author(s) and the title of the work, journal citation and DOI.
Published under licence by IOP Publishing Ltd

In order to shorten the production time of biodiesel and reducing environmental problems, there are many technologies to assist in the process of production [7, 8, 9]. It has been found that using high voltage electrical field helps accelerate the reaction time of biodiesel production from used vegetable oil [10]. By using 100 milliliter of vegetable oil substrate and expose them to the supply voltage using ring and cylinder electrode at 10 kV. It was reported that, the process under an applied electric field was lasted for 2 minutes, there is 99.8 per cent yield of biodiesel. The glycerin and ester layer is clearly separated.

A 2-pole square electrode [11] was used with 100 milliliter of soy bean oil using an applied voltage of 14 kV. It was found that with spacing of the 5 cm electrode could cause sedimentation of glycerine. This is the fastest time that was achieved with 53 seconds. But when electrodes were spaced apart, it will takes more time to precipitate. The study of trans-esterification by electrostatic field electrodes have been studied recently [12]. Each electrode consisted of double rod, point-to-point, double plate and wiring of 100 milliliters of vegetable oil, using 13 kV. It was found that double plate electrode configuration with a spacing of 5 cm apart was able to deliver the lowest possible separation time about 20 seconds. Biodiesel reaction can also be accelerated by using high-voltage electric fields utilizing various vegetable oils and electrodes [2, 13, 14]. However, it is necessary to investigate the effectiveness of high voltage electric field and electrodes design that can be applied to the Biodiesel production. Soy oil and Palm oil will be used as the substrate and characteristics of electrodes will be explored.

2. Effect of electric field to reaction

Improvement of biodiesel production process for faster production requires a catalytic reaction between natural oils and alcohols. The process can be done by providing input energy to the substrate during the reaction. In the biodiesel production process, the feedstock is typically heated with conventional heating method. Oil is nonpolar molecule that contained balance positive and negative charges. Under electric field, oil molecule will be emitted by moving electron and can be separated into free charges. The positive charges will move in the same path of e-field and collide with negative charge that moves in an opposite direction, which directly affects the convection coefficient as shown in Figure 1. [6, 15]

Likewise, alcohol was used as a solvent in the reaction, it is also polar molecules. Under the electric field, molecules of alcohol move and rearrange themself according to the direction of the electric field. If electric fields were present, these charges will try to move in the direction of the polarity in electric field. The dipole rotation causes the molecules to rotate as shown in Figure 2.

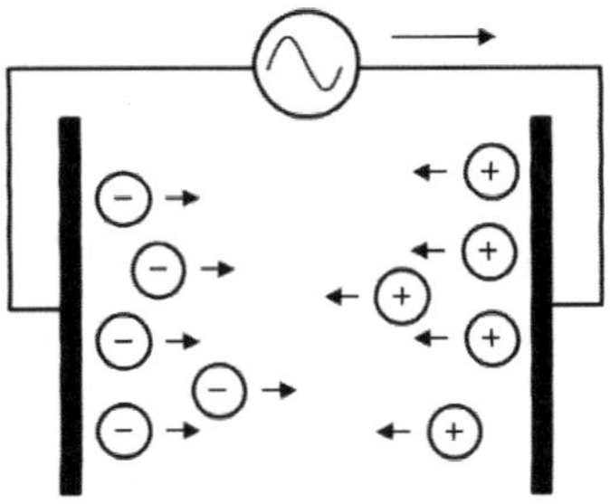

Figure 1. Effect of electric field to charges movement

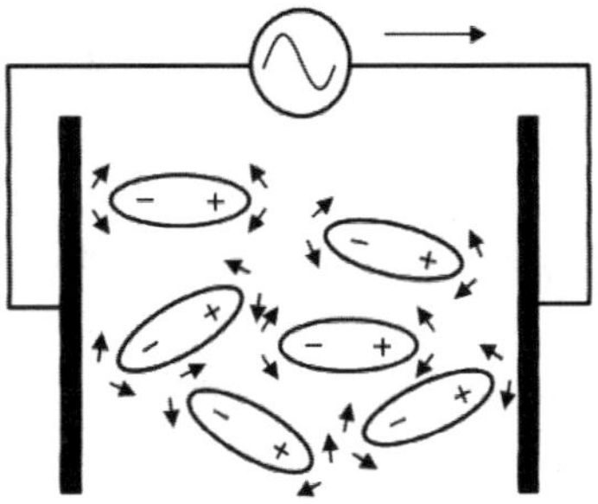

Figure 2. Effect of electric field to polar molecules movement

The rate of reaction depends on the frequency and direction of the collision in the reaction. So if there is more collisions, the more likely it is to accelerate the reaction. Based on the theory of the electric field, the electric field causes particles to move, it will increases the chance of collisions. So, the electric field can pass on energy to the molecular level. These effects can shorten the biodiesel reaction time.

3. Design of reactor chamber

The configuration of electrodes electric field design is based on a coaxial electrode, which is an electrode that has the ability to widely distribute an electric field as shown in Figure 3 [16]. The chamber consisted of a cylindrical internal electrode, which is exposed to the substrate. The insulation is a container for the substrate. There is also an outside electrode surrounded serving as a return electrode. The design and modification of internal electrodes will be optimized and investigated for biodiesel production.

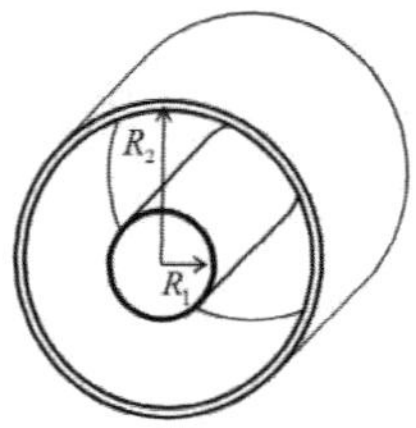

Figure 3. Coaxial cylinder electrode.

The electric field strength between the cylindrical electrodes and internal electrode can be calculated by (1).

$$E = \frac{V}{R \ln \frac{R_2}{R_1}} \tag{1}$$

E	=	maximum electric field strength (V/m)
V	=	supply voltage (V)
R	=	distance from center to notice point (m)
R_1	=	radius of internal electrode (m)
R_2	=	radius of external electrode (m)

For constant supply voltage, there are 2 ways to increase the electric field in the chamber. First, external electrode radius must be decreased and second, inner electrode can be increased. The decreasing of external radius will also decrease the volume of chamber. On the other hand, increasing of internal radius will increases the weight of electrode. Bigger electrode will result to higher cost and difficult to build. To reduce the cost of electrode, authors were able to increase the area of reaction by using bundled electrode design as shown the figure 4. Electric field contour derived from using different number of rod electrode are shown in Figure 5. The electric field was calculated by COMSOL Multiphysics software.

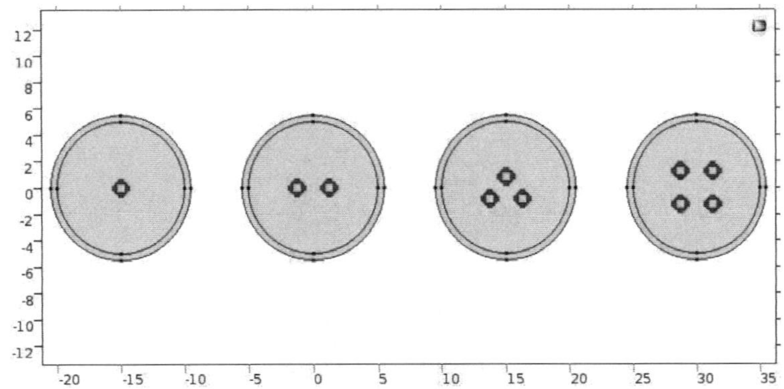

Figure 4. Bundled electrode rod.

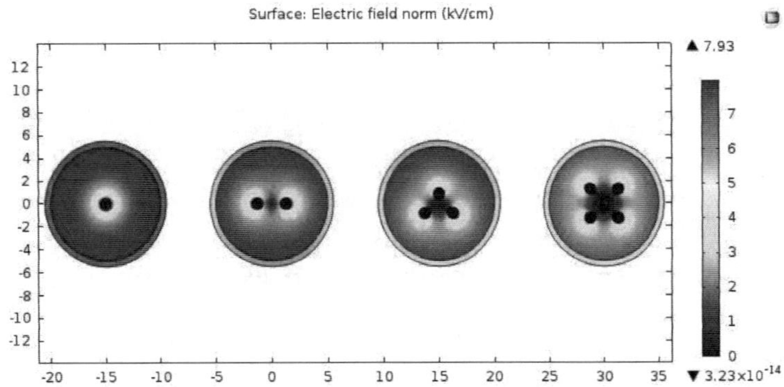

Figure 5. Coaxial cylinder electrode.

The average electric field strength in reaction area was calculated by integration method in the software was shown in table 1.

Table 1. Number of rod and average electric field strength.

Number of rod	Average electric field strength (kV/cm)
1	1.4376
2	1.8627
3	2.0121
4	2.2110

The electric field stresses at some sharp edge on electrode surface are main causes for partial discharge and resulting to a power loss. To reduce this loss, author design a 1 liter chamber with top electrode cover and external toroid shape serve as a return electrode as shown in Figure 6.

IOP Conf. Series: Earth and Environmental Science **257** (2019) 012023 doi:10.1088/1755-1315/257/1/012023

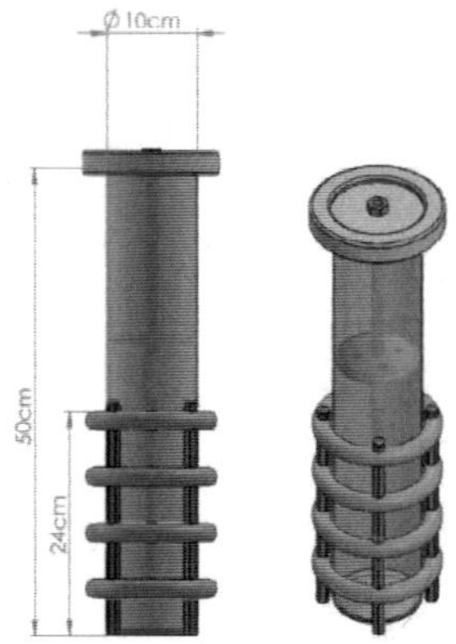

Figure 6. Designed reactor chamber.

4. Experiment

The main raw material for biodiesel production was palm oil. Methyl alcohol is used as a solvent with a molar ratio of 1:6. Also, using potassium hydroxide (KOH) as the catalyst at 1 wt. %[17]. For substrate preparation, palm oil 1 liter was warmed up with temperature of 60 °C. At the same time the methanol was mixed with KOH until a homogeneous liquid phase in the glass bottle is achieved. When the oil was warmed, it was mixed by methanol with catalyst. Stirred for 1 min and filled in reactor chamber as shown in Figure 7.

Figure 7. Reactor chamber.

High voltage electric field generator includes autotransformer for controlling the voltage supplied to the main transformer. The voltage from the main transformer that supply to the test chamber reactor, is measured by a voltage divider that is connected to the voltmeter. The supply voltage to chamber was fixed at 10 kV for 25 minutes and recording the glycerine sedimentation. Also, the control set of biodiesel oil (without supplied voltage) was prepared to compare the result. The configuration of the test rig is shown in Figure 8.

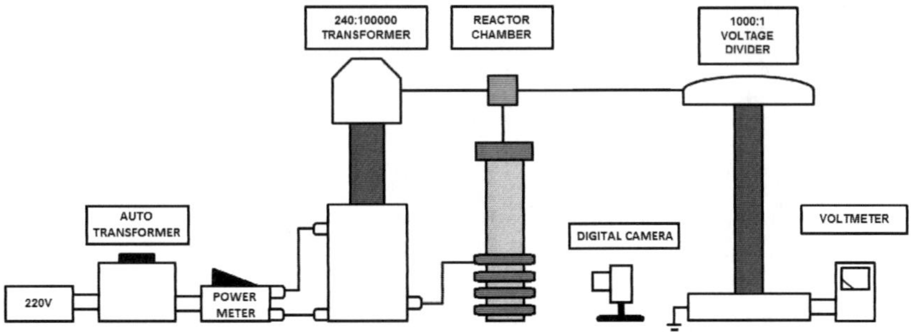

Figure 8. Equipment configuration.

In order to compare the rate of reaction in each period, researchers have measured and compared the volume of glycerine by-products from the reaction. In some cases, the reaction is not completed and the substrate was not formed as biodiesel but the experiment will have to terminate at 30 minutes time.

5. Results

The reactor was energized with 10 kV at the top cover electrode of chamber for 30 minutes. During the process, sedimentation of glycerine was recorded by digital camera. Power consumption throughout the experiment was recorded by energy meter. The timer for process duration counting start when circuit breaker was on and stop when the volume of glycerine is reached to 100 milliliter which is the saturation point of reaction. The results from each of electrode cases are shown in table 2.

Table 2. Number of rod, process time and energy consumption.

Number of rod	Process Duration (min)	Energy Consumption (W.hr)
1	28	4.8
2	24	4.8
3	17	3.1
4	11	2.3

Results from table 1 and table 2 showed that, the number of electrodes used has affected the process time duration. The increasing number of rod electrode will result to higher average electric field strength. The energy consumption is depending on process duration. The fastest process is quadruple rod that complete the glycerine sedimentation within 11 minutes and consume energy of 2.3 W.hr. The slowest one is using single rod processes that complete the sedimentation of 28 minutes using 4.8 W.hr. Moreover, the ester layer from the quadruple rod process is more transparency than others as shown in Figure 9.

Figure 9. Glycerine sedimentation after process.

6. Conclusion

The acceleration rate of the glycerine sedimentation on biodiesel production process by electric field is depending upon average electric field strength in the chamber area. Increasing the number of internal rod will result to an increase of electrical field strength inside chamber area. The rate of glycerine sedimentation resulting from electric field process using single rod, double rod, triple rod and quadruple rod calculated from the results from table 2 are 3.6, 4.2, 5.9 and 9.1 mL/min respectively. It can be concluded that electric field enhancement can be obtained by utilizing multiple rod electrodes so that the biodiesel reaction is confined within a tight environment. Strong electric field causes particles to move and increases the chance of collisions. Therefore electric field passes on energy to break up the molecular bond level. These effects can shorten the biodiesel reaction time as well as glycerine separation. The fastest process is using quadruple rod, glycerine yield separation was obtained at 100 mL within 11 minutes. This is faster than single rod configuration by 2.5 times.

7. References

[1] Rungthip C, Chakrit T, Kulchanat P 2013 Biodiesel production from used frying oil using sodium methoxide as catalyst in continuous process. *The 3rd International Thai Chemical Engineering and Applied Chemistry Conference* (Khon Kaen, Thailand)

[2] Shakinaz A 2010 Production of biodiesel using the microwave technique. *Journal of Advanced Research* pp 309-314

[3] Kraai GN 2009 Novel highly integrated biodiesel production technology in centrifugal contactor separator device, *Chemical Engineering Journal* pp 384-389.

[4] Pedro F 2006 Production of biodiesel from waste frying oils. *Waste Management* pp 487-494.

[5] Gu L, Huang W, Tang S, Tian S and Zhang X 2015 A Novel Deep Eutectic Solvent for Biodiesel Preparation Using a Homogeneous Base Catalyst. *Chem Eng Journal* pp 647-652.

[6] Abdelrahman BF, Mohammed MD, Abdul-Qader YA 2012 Purification of biodiesel using activated carbons produced from spent tea waste. *Journal of the Association of Arab Universities for Basic and Applied Sciences* pp 45-49.

[7] Chen KS, Lin YC, Hsu KH and Wang HK 2012 Improving biodiesel yield form waste cooking oil by using sodiummethoxide and a microwave heating system, *Energy* pp 151-156.

[8] Cubas ALV 2015 Biodiesel production using fatty acids from food industry waste using corona discharge plasma technology. *Waste Management* pp 149-154.

[9] Hamed M 2010 Ultrasonic-assisted biodiesel production process from palm oil using alkaline earth metal oxides as the heterogeneous catalysts. *Fuel* pp 1818-1825.

[10] Noppadol Sirirat 2011 Bio-diesel production from transesterification using electric field. *Master of Engineering Thesis* (Department of Energy Engineering Graduate School, Chiang Mai University, Thailand)

[11] Komol Siriwatanakul 2012 Sedimentation technique of biodiesel using oil to flow through high voltage. *Independent Study Master Degree* (Engineering Department of Energy Engineering College, Khonkaen University, Thailand)
[12] Ahmad A., Barat G and Gholamhassan N 2014 Electrostatic coagulation for separation of crude glycerine from biodiesel. *Advances in Environmental Biology* pp 321-324.
[13] Barmwal BK and Sharman MP 2005 Prospects of biodiesel production from vegetable oils in India. *Renewable and Sustainable Energy Reviews* pp 363-378.
[14] Guan G 2009 Synthesis of biodiesel fuel using an electrolysis method. *Chemical Engineer Journal* pp 159-164.
[15] Tanongkiat K 1999 The effect of electric field in increasing heat transfer in dielectric fluid, *Journal of Research and Development* (KMUTT, Thailand) pp 93-106.
[16] Samruay S 2012 High Voltage Engineering. Electric field stress and insulation durability. (Bangkok. Thailand)
[17] Luque R and Melero J 2012 Advances in biodiesel production: processes and technologies (Woodhead Pub Ltd.)

Acknowledgments
This research project is fully supported by Center for Alternative Energy Research and Development, Khon Kaen University, Thailand.

IOP Conf. Series: Earth and Environmental Science **257** (2019) 012024 doi:10.1088/1755-1315/257/1/012024

Empowering remote island communities with renewable energy : a preliminary study of Talaud Island

Meita Rumbayan[1], Sherwin Sompie[1] and Yosuke Nakanishi[2]

1 Lecturer and Researcher, Faculty of Engineering, Sam Ratulangi University, Manado, Indonesia
2 Professor, Graduate School of Environment and Energy Engineering, Waseda University, Tokyo, Japan
E-mail:meitarumbayan@unsrat.ac.id

Abstract. This paper describes a preliminary study of the empowering remote island communities with renewable energy. This research is a case study for Talaud island which is located in the border island between Indonesia and Philipines. It is located in Kiama village Talaud Island regency of North Sulawesi province of Indonesia. Method of implementation of this program consists of several stages, namely: (1) Implementation of small-scale solar home system and a wind turbine; (2) Through technology transfer in the capacity building activities to the selected community, in order to provide the skill and knowledge to island community in Kiama Village, Talaud Islands of Indonesia. It is expected that technology implementation will be maintained in the future and sustainable. The life cycle cost analysis is done by comparing two system of pilot projects, namely solar home system and solar home system with a wind turbine. The cost of generating electrical energy with renewable energy is relatively high then it is necessary to provide strategies for the development of power infrastructure for the sake of the energy security of island communities in Indonesia. Further both monitoring and evaluation of the renewable energy system implementation should be conducted for the detail report and analysis.

1. Introduction

The eastern region of Indonesia, especially the island which is the outermost island that becomes the front border requires special attention in the case of energy problems. Dependence on the supply of fossil fuels from the island to the island that have high transportation costs and produce greenhouse gases need to be minimized. For this reason the utilization of renewable energy as an alternative energy source for energy generation needs to be studied.

Talaud island was chosen as the location to be studied because of its specificity located on the border between Indonesia and the Philippines is the outermost island which became the front porch of Indonesian territory. Geograpphically, Talaud island is located between 125° 9'28" – 125° 24'25" E and 02° 4'13" – 02° 52'47" N. Talaud island is situated in regency of Talaud Island. It takes 10 hours by boat or 1 hour by flight from Manado, the capital city of North Sulawesi provience. The island of of Talaud islands is shown in Figure 1.

Content from this work may be used under the terms of the Creative Commons Attribution 3.0 licence. Any further distribution of this work must maintain attribution to the author(s) and the title of the work, journal citation and DOI.
Published under licence by IOP Publishing Ltd

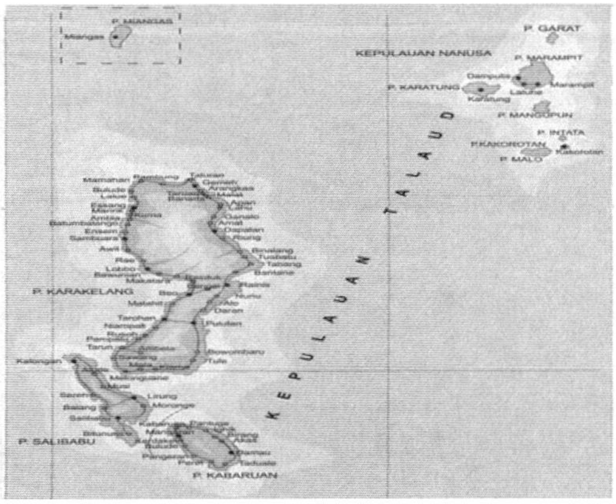

Figure 1. Map of Talaud Islands

Talaud islands are facing the high cost of diesel fuel due to remoteness from Sulawesi mainland. This paper presents a preliminary study of renewable energy utilization based on local availability of renewable energy resources in the remote island of Indonesia for an independent system electrical power on a rural house scale.

2. Literature Reviews

Analysis of solar energy potential has been done through a research [1]. The implementation of solar home system in rural Bangladesh [2] has also been reviewed as a reference for the implementation of solar power plant systems in rural Indonesia.

Many small islands in every region in the world use or propose to use renewable energy: the Azores and Canary islands in the North Atlantic, Gotland and Samsoe in the Baltic, Sardinia and Sicily in the Mediterranean, Mauritius and Reunion in the Indian Ocean, Fiji and the Hawaiian islands in the Pacific, as well as Dominica and the Guadeloupe islands in the Caribbean. Many small islands have achieved their goal of transitioning to renewable energy [3].

References [3] stated that around the world a few islands already have taken the decision to become a Renewable Energy Island (REI) in the short or medium term. Samsoe (Denmark), Pellworm (Germany), Aeroe (Denmark), Gotland (Sweden), El Hierro (Spain), Dominica and St. Lucia have an explicit target of becoming 100% selfsufficient from renewable energy sources. Nearly 70% of the islands in the overview that are utilising renewables for electricity generation are producing between 0.7-25% of their electricity from renewable energy sources.

References [4] states "Life cycle cost (LCC) analysis is a tool used to compare the ultimate delivered costs of technologies with differenr cost structures". The function of LCC analysis is to calculate the cost of delivering over the life of the projects, is is not only calculate the initial costs or operating cost.

The life cycle cost (LCC) analysis which is consist of capital cost, recurring maintenance cost, fuel cost and replacement cost has been used for economic evaluation of a stand alone residential of photovoltaic for remote area in Bangladesh [5]. The LCC was analyzed by considering 20 years of life cycle, discount rate as 7%, general inflation as 3% and the estimated of the load power supply as 640 Wh per day. In this analysis, the result indicated the LCC to be economically feasible compare to conventional electricity using diesel generator.

The cost of generating electrical energy with renewable energy is relatively high then there is a need for policies and strategies for the implementation of renewable energy in the archipelago, among others: cooperation with parties that have been successful with the implementation of renewable energy technologies, increased priority implementation of renewable energy-based energy infrastructure for potential locations. The adoption of a pro-island energy policy in the foremost island

IOP Conf. Series: Earth and Environmental Science 257 (2019) 012024 doi:10.1088/1755-1315/257/1/012024

as a terrace that needs to be enriched for the sake of security, welfare and beauty as an added value in Indonesia's border region [6].

3. Methods

The data collected for the input, in the form of population data, the existing condition of electricity in Miangas island are given in Table 1.

Table 1. Communities Data in Talaud Island

Number of People	775
Number of Households	324
Daily Load Average	500 kWh
Installed Capacity of Diesel Powerplant	250 kW

Feasibility of solar energy system mainly depends on solar radiation available at the specific location [7]. Data of solar energy sources in term of solar irradiation in Talaud island have been taken from NASA (*National Aeronautics and Space Administration*) website by input the latitude and longitude of the location. The average annual solar radiation equals to 5.53 kWh/m^2/day. NASA also gives the wind speed data for Talaud island. The data of solar radiation and wind speed in Talaud island are presented in Table 2.

Table 2. Average Monthly Solar Irradiation and Wind Speed Data in Talaud island

Month	Average Solar Irradiation (KWh/m^2/day)	Average Wind Speed (m/s)
January	5.01	2.57
February	5.43	2.57
March	6.05	2.57
April	6.40	2.57
May	5.73	2.57
June	5.00	3.08
July	5.32	3.59
August	5.50	4.11
September	5.84	3.08
October	5.71	3.08
November	5.38	2.05
December	5.01	2.57

The implementation of solar home system technology in coastal communities of Talaud island provides measurable outcomes that affects the application of appropriate technology to the community and improves the welfare and convenience of the community in Kiama village as described in Figure 2.

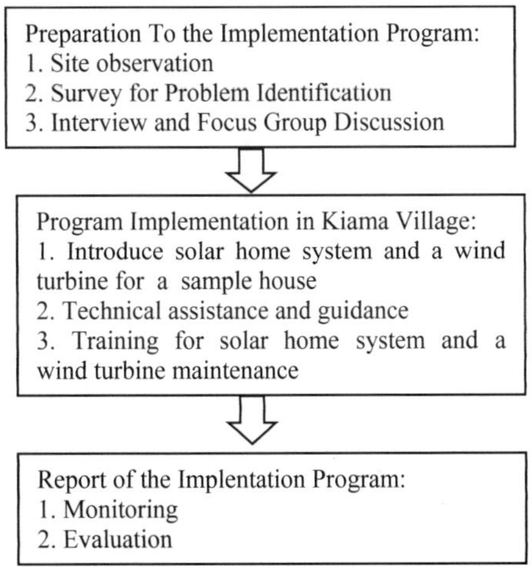

Figure 2. Implementation Method of the Hybrid Home System Program Activities

Implementation of solar home system technology in Kiama village has been done in several stages as described below:

3.1. Introduces small-scale solar wind power plant technology.
Implementation of solar home system as a small pilot in a home which is located in Kiama village in Talaud island has been done by using a solar panel 50 Wp, batterey, inverter and system controller. The total capital cost of equipments and installation are counted about Rp. 5.000.000,-. The cost of wind turbine with capacity 50 Watt and the simple tower are counted about Rp. 5.000.000,-.

3.2. Technical assistance and guidance on small-scale solar power plant operation.
Assistance on how to operate an installed solar power plant at sample houses has been done by the team who has knowledge of this technology. A standard operating guide has been written and taught to small home-scale solar power plant users. Assistance was carried out after the installation of the solar power plant at the sample house was done.

First, the approach methods are survey, interview and field visit. Based on preliminary survey, site observation, and data collection in Kiama village, identificate the problem by Implementation Team.

Second, at this stage the approach method is to conduct a focus group discussion between implemention teams and partners in Kiama village for finding solutions to the problems of coastal communities in the demand of energy.

Third, the approach method is the application of appropriate technology through the design of the installation of solar home system as a solution in solving communities's problems, mentoring and technical guidance, as well as counseling for increasing the knowledge and skills of the community on technology and given solutions.

From the results of observation and interviews, it is known that this village has a problem of limited access of houses to the electricity grid. In the initial discussion, implementedon team offered a solution that introduces the community to small-scale solar power plant technology. Through the introduction of small scale home system technology, the problem of limitation of electricity accsess in the village could be solved.

Since Hybrid Home System technology is still relatively expensive, the team decided to install the system at the sample house located in Kiama village, which is located in Talaud island, one of the remote island in North Sulawesi provience of Indonesia.

Installation of a small-scale SHS system in a Kiama village in Talaud island is followed by mentoring and technical guidance on the operation of the system and training on how to treat the SHS and wind turbine for rural and remote communities.

4. Results

The proposed of hybrid power generation model based on PV and wind energy for a rural house in Talaud island as shown in Figure 3.

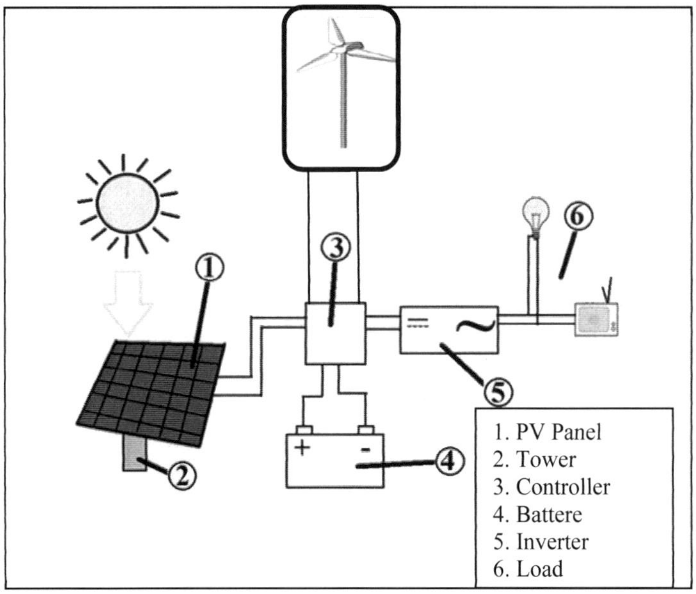

Figure 3. The Proposed of Hybrid Home System Model for A Rural House in Talaud Island

The installation results of system model are shown Figure 4.

Figure 4. The result of Implementation for the Model of Wind Solar Power System in Talaud Island

Monitoring and evaluation of program implementation of solar wind power model for a rural house in Kiama village are done by observation and economic evaluation through life cycle cost analysis. The sustainability of this program can be sustained through the transfer of knowledge and skills regarding the operation and maintenance of technologies that already introduced to the coastal communities in Kiama village, Talaud island.

The results of economic evaluation of a stand-alone solar home system in Talaud island are presented in Table 3.

Table 3. Economic Evaluation of a SHS in Talaud Island

	Daily Energy Generated (Wh)	Total Energy in 20 years (kWh)	LCC of SHS only (Rp/ kWh)	LCC of SHS with wind turbine (Rp/ kWh)
1	200	1460	3424	6848
2	300	2190	2283	4566
3	400	2920	1712	3424
4	500	3650	1369	2738
5	600	4380	1141	2282

The economic evaluation analysis of a Solar Home System in Kiama Village which is located in Talaud island has been presented in Table 3. It is found that the Life Cycle Cost of a SHS will be varied depend on the daily energy generated to supply energy for a home. The more energy generated for this system installed, the more total energy could be generated for 20 years of the system life. The life cost for this SHS will be achieved good price in the below of 2000 Rp/ kWh as compare to the electricity price from the grid of PLN. The LCC analysis shows double cost of Solar Home System with a wind turbine installation to power a rural house in Kiama village. Through direct observation of wind turbine indicates frequently work due to the wind has no occur in the area of installation.

5. Conclusion

The problem of limited access of electricity for the coastal community in Kiama village, which is located in Talaud island has been searched to find the solution. Implementation of small scale solar home system technology has been introduced in a house as a small pilot project. In the future, detail data will be collected from the implementation in order to monitor and evaluation of the practices.

The implementation of SHS in Kiama village which is located in Talaud island have been done through the introduction and capacity building of SHS technology and a wind turbine in a sample house as a small pilot.

The outcomes the implementations of SHS system in coastal communities are as follows: design and installation of SHS technology for a small pilot home as well as improvement of the community's knowledge and skills about SHS technology by a training.

6. References

[1] Rumbayan M and Nagasaka K 2010 *Resources and economic assessement of solar irradiation in an island community*, Proceeding of Indonesia - Japan Joint Scientific Symposium (IJJSS) , pp.59

[2] Khan S and Azad A. K 2014 *Social impact of solar home system in rural Bangladesh*, IAFOR Journal of Sustainability, Energy and the Environment, International Academic Forum, Japan.

[3] Jensen Thomas L 2000 *Renewable Energy on Small Islands*, Forum for Energy and Development (FED). Sandia National laboratories (1991), Stand alone photovoltaic power system, A Handbook of Recommended Design Practices.

[4] Sandia National laboratories 1991 *Stand alone photovoltaic power system*, A Handbook of Recommended Design Practices.

[5] Bhuiyan M.M. Ali Asgar, Mazumder R. K, and Hussain M 2000 *Economic evaluation of a stand alone residential photovoltaic power system in Bangladesh*, Renewable Energy 21, 403-410.

[6] Rumbayan, M and K Nagasaka 2018 *Techno Economical Study of PV-Diesel Power System for a Remote Island in Indonesia : A Case Study of Miangas Island*, IOP Conf. Ser.: Earth Environ. Sci. 150

[7] Chmiel Z and Bhattacharya SC 2015 *Analysis of off-grid electricity system at Isle of Eigg (Scotland): Lessons for developing countries*, Renewable Energy 81; pp 578-588.

Acknowledgement
This program has been funded by Ministry of Research, Technology and Higher Education of Indonesia.

IOP Conf. Series: Earth and Environmental Science **257** (2019) 012025 doi:10.1088/1755-1315/257/1/012025

Green ethanol production from cotton stalk

K Shahzad[1], M Sohail[2] and A Hamid[2]

1Institute of Chemical Engineering and Technology, University of the Punjab, Lahore-54000, Pakistan

2 Department of Environmental Sciences, Kinnaird College for Women, Lahore-54000, Pakistan

Abstract. This study was initiated to produce green ethanol from cotton stalk lignocelluloses using separate hydrolysis and fermentation (SHF) process. The influence of sodium hydroxide (2%) pretreatment at different residence times was also investigated. The effect of incubation period, enzyme concentration, pH and substrate concentration during enzymatic saccharification on sugar yield and ultimately on the ethanol production was studied. The results show that pretreatment with 2% NaOH for 90 min at 121°C was most effective and gave the highest percentage of cellulose (78.2±1.64%) along with highest delignification rate (63.9%). During the enzymatic saccharification, under the optimized conditions, maximum concentration of reducing sugars of the hydrolyzate was 67.83 ± 0.35 g/L with the maximum saccharification efficiency of 78.06%. The total sugars were 77.48 ± 2.37 g/L and total phenols were found to be 1.6 mm. Fermentation was carried out by using the enzymatically saccharified hydrolyzate and monoculture of *Saccharomyces cerevisiae*. Highest ethanol concentration of 22.93 ± 1.74 g/L with 0.36 g/g ethanol yield was achieved after 48 h of incubation at 30°C, pH 5 and 3% inoculum size.

1. Introduction

Energy demand is on the rise with escalating population of the world. According to the latest statistics, Global energy demand will rise to about 30% by 2035 with an average growth of 1.3% annually, while in developing countries it is estimated that the energy demand will increase by almost 30% by 2040 [1]. Presently, the renewable energy sources like solar, wind, biomass etc provide an attractive alternative to meet the growing energy demands as well as protecting the environment. In many countries, government policies ad incentives are promoting use of renewable energy sources. Renewable energy is the world's fastest growing source of energy, at an average rate of 2.6 % annually. According to an estimate, by 2040, $44 trillion is going to be invested in the energy supply and there is a decline in the share going to the fossil fuels. It is believed that by year 2040 the renewable power generation capacity rise to 60% [1].

Globally, biomass fuel is turning out to be an attractive and reasonable alternate for conventional fossil fuels because of growing interest for green energy approaches and declining fuel reserves. It not only decreases reliance on oil import from other countries but also sustains rural economies by creating employment opportunities and provides resources for income [1].The most common renewable fuel today is bioethanol; being the most widely used biofuel in over 64 countries [2]. There is a large variety of biomass sources which can be used for bioethanol production including some aquatic plants, woody plants and herbaceous plants/grasses [3]. Agricultural residues are considered to be one of the most suitable choices for ethanol production worldwide as these residues can serve as a replacement of grains for ethanol production with no threat of food scarcity [4].

Corn has been extensively used for bioethanol production; almost all of the ethanol in USA is produced from corn [5, 6]. Maize, corn stover, wheat and rice straw have been investigated for high ethanol yield by many researchers [7-13]. The lignocellulosic biomass is a favorable resource among all for bio-fuels production [14, 15]. These renewable raw materials have the potential to create green products [16, 17].

Content from this work may be used under the terms of the Creative Commons Attribution 3.0 licence. Any further distribution of this work must maintain attribution to the author(s) and the title of the work, journal citation and DOI.
Published under licence by IOP Publishing Ltd

Many different technologies are being developed and many are underway/developmental stage to produce bioethanol from lignocellulosic biomass applicable to different stages of production process of bio-ethanol; pretreatment, hydrolysis/saccharification and fermentation [18].

Cotton stalk is one such lignocellulosic biomass. It is the residue of plant after harvesting of cotton. Cotton is produced in over 80 states in the World, which generates the huge quantity of cotton stalk [19].Cotton stalk is a potential raw material for conversion to ethanol because it is rich in cellulose (32–46%) and hemicellulose (20–28%) [19]. Bio-ethanol from cotton stalk was produced utilizing two-stage dilute acid hydrolysis and fermentation of detoxified hydrolysate a [20]. In Pakistan, during 2010-2011, approximately 1,474,693 metric tons of cotton residue was produced. This residue of the cotton crop has the potential of power generation around 3,071 GWh [21].

Cotton stalk has high holocellulose content which makes it a potential feedstock for bio-ethanol production [17]. In a study done on cotton stalk, cellulose in the range of 40.10 ± 1.55%, hemicelluloses 13.60 ± 0.64% and lignin 29.40 ± 0.95% was found [22]. As cotton stalk/residue is extensively produced biomass in Pakistan, the present study was designed to study the potential of producing bioethanol from cotton stalk which not only offers a cleaner fuel but also will help in recycling waste and help in waste management. Hence, this study was initiated to produce bioethanol from cotton stalk and also to optimize its process to achieve maximum ethanol production.

2. Materials and method

In practical work of the current study, analytical grade chemicals and reagents were used. The Cotton stalk *(Gossypium arboretum)*biomass was acquired from the local market. After the careful inspection of the biomass, the undesirable dust particles were removed and after that it was dried in an oven for 48 hours (h) at 100°C. Physical pretreatment was carried out by reducing the size of cotton stalk to 2 mm by crushing in a hammer beater mill, followed by chemical pretreatment using sodium hydroxide (NaOH). Cotton stalk (100 g) were soaked for one hour in 1000 ml (2%) NaOH solution in 3 flasks and autoclaved at the residence times of 30, 60 and 90 min at constant temperature of 121°C with 15 psi pressure. The solid liquid ratio was taken as 1:10.

2.1. *Enzymatic saccharification*

Enzymatic saccharification of pretreated cotton stalk was performed with 0.05M citrate buffer (100 ml) with 0.02% sodium azide and added in pretreated substrate for incubation in saccharification process. The commercial enzyme cellulase was added in it. It was then incubated at 50°C, 120 rpm for 92 h and was removed from the incubator at 24 h intervals for sugar analysis. The enzymatic conversion of cotton stalk was enhanced by optimizing the concern parameters.

2.2. *Fermentation*

The microorganism strains of *Saccharomyces cerevisiae* was used during experimental work. The strains were maintained on PDA (Potato Dextrose Agar) slants. pH of medium was maintained at 6.0 and sterilized at 121°C for 15 mins. The inoculum was prepared by adding YPD (Yeast, Peptone and Dextrose) media and strains of *Saccharomyces cerevisiae* for 48 h at 30°C. The prepared cultures of *Saccharomyces cerevisiae* were used as inoculum in fermentation.

Fermentation was the 3rd step in bioethanol production from cotton stalk. First, the inoculum size was optimized followed by fermentation by using all pre-optimized parameters and inoculum size to achieve the maximum ethanol concentration. To determine optimum inoculum size of *Saccharomyces cerevisiae* for fermentation, different inoculum sizes (1-4%) were studied. The enzymatically saccharified hydrolysate (100ml) with the highest reducing sugars were added in the fermentation medium consisting of 1.5 g yeast extract, 1 g each of peptone and $(NH_4)_2SO_4$, 0.5 g each of K_2HPO_4, $MgSO_4.H_2O$ and

$MnSO_4$at pH 5. The medium was sterilized for 15 min at 121°C. After cooling the media, different inoculum sizes of 1-4% was added into 4 flasks containing sterilized media. Fermentation was carried out for 96 hours at 30°C. Initially, aerobic conditions were provided for 24 hours followed by anaerobic conditions for 48 hours. The samples were collected at 24h intervals throughout the fermentation process and analyzed for ethanol content and reducing sugars.

2.3. Analytical methods

The percentage of cellulose, hemicellulose and lignin in cotton stalk samples was estimated by method of [23-25].The total phenol and ethanol estimation was carried out by using method adopted [26,27] respectively. The ethanol yield was calculated according to the equation given in a study done by [28].

3. Results

3.1. Chemical composition of raw cotton stalk

The results revealed that the cellulose content in the raw (untreated) cotton stalk was 43.7 ± 0.45 % and hemicellulose was 12.53 ± 0.40 %;whereas, the lignin content was found to be 28.6 ± 0.45 %.Cellulose and hemicelluloses content in a defined combination makes the holocellulose, which was found to be 55.8% in the present study on dry weight basis.

3.2. Results of naoh pretreatment of cotton stalk

High lignin content (28.6 ± 0.45%) was found in the present study, so delignification with NaOH pretreatment constituted an important step. The results after physiochemical treatment indicated that when cotton stalk was soaked in 2% NaOH at 121°C in an autoclave with the residence times of 30, 60, and 90 min, the cellulose exposure increased with time. Table 1 summarizes the results of chemical composition of cotton stalk before and after pretreatment, along with the holocellulose and delignification percentages. Cotton stalks autoclaved for 90 min gave the highest percentage of cellulose (78.2±1.64) and also highest delignification rate (63.9%). 30 min residence time in an autoclave was not very effective and the cellulose was increased by only 8% from the untreated cotton stalks.

3.3. Enzymatic saccharification of naoh treated cotton stalk

Table 1. Chemical Composition of Cotton Stalk

Autoclave residence time (min)	Cellulose (%)	Hemicellulose (%)	Holocellulose (%) (cellulose + hemicellulose)	Lignin (%)	Delignification (%)
Untreated	43.7 ± 0.45	12.5 ± 0.40	56.2	28.6 ± 0.45	------
30 min	52.5 ± 1.64	11.2 ± 0.43	63.7	19.8 ± 0.83	30.7
60 min	65.4 ± 1.63	6.3 ± 1.52	71.7	13.4 ± 0.95	53.1
90 min	78.2 ± 1.64	4.9 ± 0.32	83.1	10.3 ± 1.23	63.9

All results are represented as mean value ± standard deviation among triplicates

In enzymatic saccharification, the pretreated cotton stalk with 78.2 ± 1.64 % cellulose and 63% delignification was selected for enzymatic saccharification. Pretreated cotton stalk was incubated with 0.05M citrate buffer containing 0.02% sodium azide (100 ml) as the production medium for 96 h and were removed from the incubator for sugar analysis at 24 h intervals. Figure 1 shows the optimization of incubation period. It was found out that hydrolysis effect increased at first 72 h but started to decrease after 96 h.

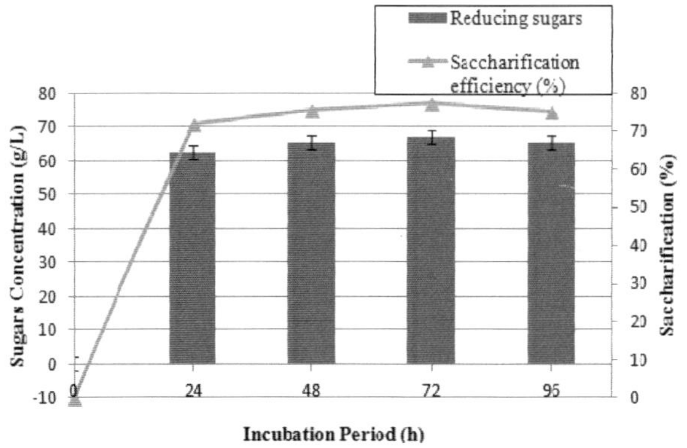

Figure 1. The Effect of Incubation Period on Reducing Sugar Concentration and Saccharification Efficiency

The effect of cellulase concentration on the enzymatic saccharification was carried out by adding cellulase enzyme at different concentrations of 200 IU/ml, 400 IU/ml, 600 IU/ml and 800 IU/ml. According to the results of the effect of enzyme concentration on the reducing sugar yield, the reducing sugars increased at

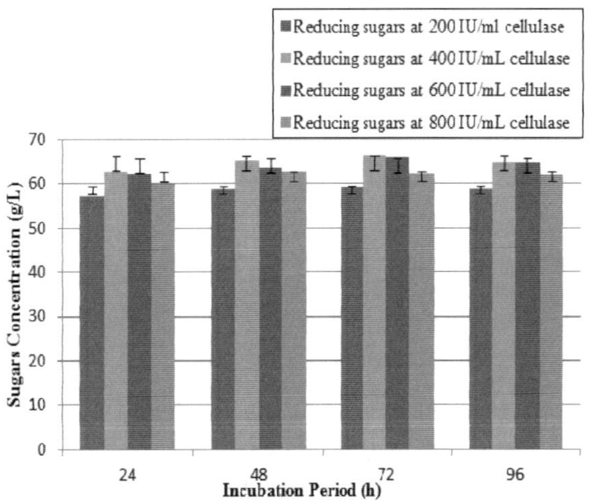

Figure 2. Effect of Incubation Period and Enzyme Concentration on Reducing Sugar Concentration

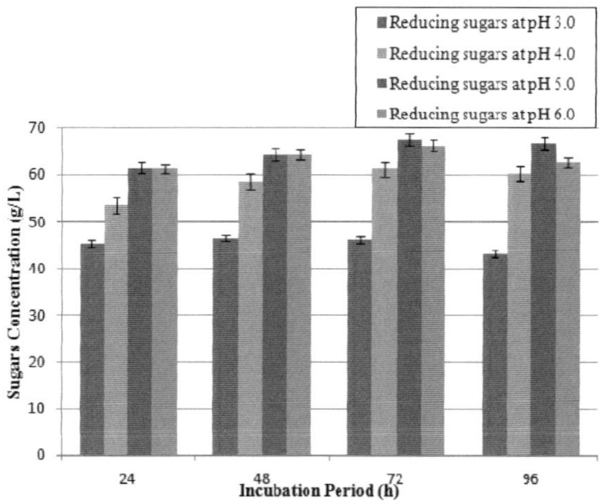

Figure 3. Effect of Incubation Period and pH on Reducing Sugar Concentration

first 72 h of saccharification but started to decrease after 96 h at all enzyme concentrations. The results of enzyme optimization at all concentrations are shown in figure 2. The effect of pH on the enzymatic saccharification was studied at various loadings of pH ranging from 3.0 to 7.0. Figure 3 shows the results of pH optimization. Initially, the reducing sugars increased with the increase in pH till 5.0. Further increase in pH decreased the sugars.

The substrate concentration on the enzymatic saccharification was studied by using 0.5 g, 2.5g, 5g and 7.5g of substrate per 100mL of citrate buffer. The effect of substrate concentration on reducing sugars (figure 4) shows that saccharification efficiency decreased when the substrate taken was increased from 5g/100mL buffer. When all conditions of the enzymatic saccharification were optimized, an experiment was conducted again with all the pre-optimized conditions i.e. 400 IU/ml cellulase enzyme was used, which was added in 5g substrate in 100mL citrate buffer with pH 5 at 50°C temperature for 72 h incubation period. The total sugars, reducing sugars and total phenols of the hydrolysate obtained after 72 h of optimized conditions were taken.

The reducing sugars of the hydroyzate were found to be 67.83 ± 0.35 g/L with the maximum saccharification efficiency of 78.06%. The total sugars of the enzymatically saccharified hydrolysate were 77.48 ± 2.37 g/L and the total phenols were found to be 1.6 mm.

3.4. Fermentation of enzymatically saccharified cotton stalk

After the enzymatic saccharification, the next task was to determine optimum inoculum (*Saccharomyces cerevisiae*) size for maximum ethanol yield. The enzymatically saccharified hydrolysate attained after all the optimized parameters with the highest reducing sugars (67.83 ± 0.35 g/L) were utilized in the optimization of inoculum (*Saccharomyces cerevisiae*) size for fermentation. The experiment was conducted for 96 h at 30°C with pH adjusted at 5.0 and varying the inoculum size from 1% to 4%.

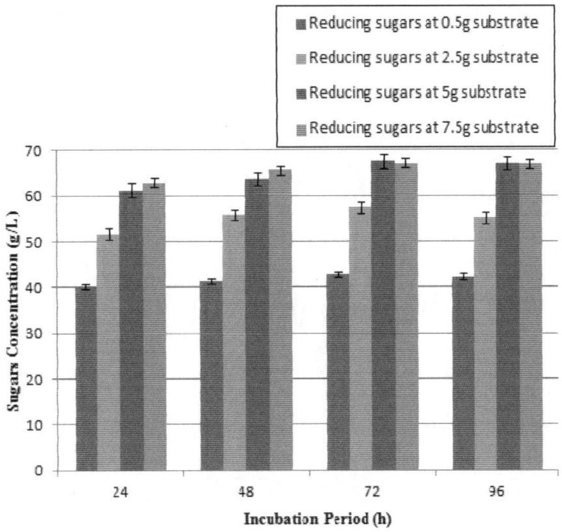

Figure 4. The Effect of Incubation Period and Substrate Concentration on Reducing Sugar Concentration

Maximum ethanol concentration and reducing sugar consumption rate was found at 48 h of incubation period in fermentation with ethanol concentration increasing with increase in inoculum size till 3% (v/v), after which it started to decrease with further increase in inoculum size. The reducing sugars also decreased with the increase in inoculum size as shown in Table 2. The inoculum size of 3% showed the maximum ethanol concentration of 22.79 ± 0.75 g/L with the ethanol yield of 0.36 g/g after 48 h of incubation at 30°C with pH 5.

Table 2. The Effect of Inoculum Size of *Saccharomyces cerevisiae* on Reducing Sugars, Ethanol Concentration and Ethanol Yield after 48 h of Fermentation

Inoculum size % (v/v)	Reducing sugars g/L	Ethanol concentration g/L	Ethanol yield g/g
1	7.33 ± 2.35	17.85 ± 1.52	0.29
2	6.26 ± 1.76	20.58 ± 0.38	0.33
3	5.47 ± 1.84	22.79 ± 0.75	0.36
4	5.10 ± 2.61	21.21± 0.53	0.33

All results are represented as mean value ± standard deviation among triplicates

After inoculum size optimization, fermentation of enzymatically saccharified hydrolysate was conducted at 30°Cat pH 5 by using inoculum size (3%) of *Saccharomyces cerevisiae*.The reducing sugars in the hydrolysate were 67.39 ± 0.35 g/L before inoculum addition (at 0 h). According to the results given in Table 3 and Figure 5, ethanol concentration and yield increased at first 48 h and started to decrease after 48 h. The highest ethanol concentration was 22.93 ± 1.74 g/L with 0.36 g/g ethanol yield at 62.2 % reducing sugars consumption rate.

Table 3. The Effect of Incubation Period on Reducing sugars, Ethanol concentration and Ethanol yield

Incubation Period (h)	Reducing sugars g/L	Ethanol concentration g/L	Ethanol yield g/g
24 h	18.85 ± 0.59	17.23 ± 1.32	0.35
48 h	5.23 ± 1.61	22.93 ± 1.74	0.36
72 h	5.11 ± 1.58	21.64 ± 1.35	0.34
96 h	4.62 ± 1.74	21.43 ± 1.86	0.33

All results are represented as mean value ± standard deviation among triplicates

The highest ethanol concentration of 22.93 ± 1.74 g/L with 0.36 g/g ethanol yield was achieved after 48 h of incubation at 30°C, pH 5 and 3% inoculum size.

4. Discussion

The similar results of high delignification with 2% NaOH treatment was found in another

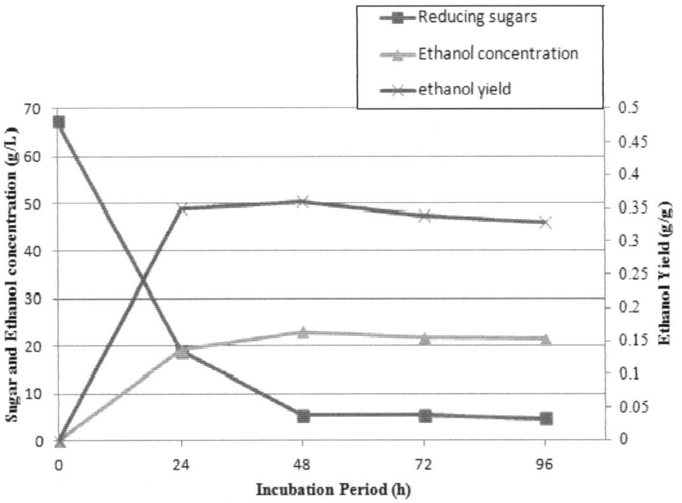

Figure 5: The Effect of Time on Reducing Sugars, Ethanol Concentration and Ethanol Yield

study done by [29]. Alkali pretreatment, especially with NaOH is very effective for delignification as it breaks the crosslinking ester bonds of lignin, also breaks the hetromatrix of carbohydrate fraction. This effect causes the porosity of biomass to increase and decrease the crystallinity of cellulose [30]. The maximum holocellulose content also changed significantly from 55.8% (untreated) to 83% (2% NaOH, 90 min) on dry weight basis. In a current study, the maximum hollocellulose content conversion is slightly higher than the hollocellulose conversion found in a study done by [22] which is 53.70 ± 0.58% (untreated) to 76.20 ± 1.21% (3% NaOH treatment) on dry weight basis.

In enzymatic saccharification, the pretreated cotton stalk with 78.2 ± 1.64 % cellulose and 63% delignification was selected for enzymatic saccharification. Cirtrate buffer (0.05 M) was used as the production medium for enzymatic saccharification. Citrate buffer (0.05 M) for the hydrolysis of cotton stalk was reported by [31] in a study done on cotton stalks and also in another study done on sorghum straw by [32]. Cellulase enzyme was added during saccharification and one factor was kept constant at a time to optimize the activity of cellulase enzyme. Temperature was kept constant at 50°C during all experiments. The optimum temperature for the cellulase enzyme was reported to be 40-50°C [33]. The optimum temperature range provided by *Trichoderma koningii*is also in the range of 40-55°C. According to a study, the parameters which were optimized during the enzymatic saccharification played an important role in the reducing sugar yields during enzymatic hydrolysis. It was reported that for the enzymatic hydrolysis of the lignocellulosic biomass, substrate concentration, cellulase activity and conditions such are temperature and pH are the main factors for the hydrolysis rate [33].

During the optimization of incubation period, the hydrolysis effect increased at first 72 h but started to decrease after 96 h. The highest reducing sugar values were 67.25 ± 1.62 g/L obtained after 72 h of hydrolysis with the saccharification efficiency of 77.39 %. The maximum released sugars were obtained at short period of 72 hours as the commercial cellulase enzyme was used in the current study. The maximum saccharification efficiency after 72 h of cotton stalk saccharification was also found in another study done [20].

The ratio of enzyme to substrate used is an important factor in the efficient hydrolysis rate [33].Optimization of the commercial cellulase enzyme is important to find its optimum range, as the increase in cellulase levels would also increase the cost of the process. The effect of enzyme concentration on the reducing sugar yield showed that the reducing sugars increased at first 72 hours of saccharification but started to decrease after 96 hours at all enzyme concentrations. The maximum sugars were obtained at 400 IU/ml (2g) of enzyme concentration at 72 h with 66.17 ± 0.68 g/L of reducing sugars with the saccharification efficiency of 76.15%. At higher concentrations of cellulase i.e. 600 IU/ml and 800 IU/ml, the sugars and hydrolysis efficiency started to decrease. It was found in another study that the increase in enzyme loading limit the hydrolysis yields. This happens due to the improper bindings and product inhibition effect [32].

Enzymatic saccharification of cellulose by the cellulase enzymes works best under specific, mild conditions i.e. pH 4.5-5 and temperature 40-50°C. This reduces the corrosion problems; utility consumption is low and also the low toxicity of the hydrolyzates [33]. In the present study, the maximum sugars were obtained at pH 5.0 after 72 h. Initially, the reducing sugars increased with the increase in pH till 5.0. Further increase in pH decreased the sugars.

The saccharification efficiency decreased when the substrate taken was increased from 5g/100mL buffer. It was reported that substrate concentration is one of the main factors in the yield of sugars in the initial rate of enzymatic hydrolysis. Inhibition can be caused by high substrate concentration which can cause low hydrolysis rate [33]. The maximum sugars of 67.63 ± 0.35 g/L were obtained with 5g substrate with the saccharification efficiency of 78.06% after 72 h. Further increase in the substrate level resulted in the downfall of sugars, which indicates that 5g substrate is the optimum concentration of substrate for the maximum hydrolysis rate. The increase in the substrate from the optimum concentration of substrate results in the decrease of sugars shows the fact that the hydrolysis of biomass is linked with the characteristics of the biomass feedstock [32]. The highest reducing sugars of the hydrolysate after the enzymatic saccharification at all optimized conditions had the reducing sugars of 67.83 ± 0.35 g/L with the maximum saccharification efficiency of 78.06%. The total sugars of the enzymatically saccharified hydrolysate were 77.48 ± 2.37 g/L and the total phenols were found to be 1.6 mm. Similar results of 68.20 ±1.16 g/L of reducing sugars with different conditions were obtained after the hydrolysis of alkaline extraction of steam exploded cotton stalk in another study done by [20]. The maximum ethanol concentration and reducing sugar consumption rate was found at 48 h of incubation period in fermentation. The inoculum size of 3% showed the maximum ethanol concentration of 22.79 ± 0.75 g/L with the ethanol yield of 0.36 g/g after 48 h of incubation at 30°C with pH 5.The inoculum size (v/v) of *Saccharomyces cerevisiae* at 3%was also optimized in another study done by [28].

The ethanol concentration and yield increased at first 48 h and started to decrease after 48 h. The highest ethanol concentration was 22.93 ± 1.74 g/L with 0.36 g/g ethanol yield at 62.2 % reducing sugars consumption rate. It was reported in a study that the decline in the ethanol production after 48 h can be due to the simultaneous consumption of accumulated ethanol and sugars in the medium by the adapted yeast. The results of the present study are in harmony with the results found in a study done by [22]; they reported the ethanol concentration of 23.17±0.84 g/L with a yield of 0.44 g/g with the monoculture of *Saccharomyces cerevisiae* with different conditions. The results obtained by [34] were slightly higher than the present study. The ethanol concentration was 24.4 g/L with 0.44 g/g ethanol yield [34].

5. Conclusions

Cotton stalk is a promising feedstock for bioethanol production as it has high cellulose content. The pretreatment with 2% NaOH solution at 121°C for 90 min had the most significant effect on cellulose exposure and delignification. Sugar yield was affected by the variety of factors, including the incubation period, enzyme concentration, pH and substrate concentration during enzymatic saccharification. By using the enzymatically saccharified hydrolysate and monoculture of Saccharomyces cerevisiae, the highest ethanol concentration of 22.93 ± 1.74 g/L with 0.36 g/g ethanol yield was achieved after 48 h of incubation at 30°C, pH 5 and 3% inoculum size. However, further research is needed on making the process more feasible and improving the ethanol yields to make the bio-ethanol from cotton stalk more economically viable, before making it on the commercial-scale.

6. References

[1] World Energy Outlook 2016. World Energy Outlook. 2016; 10.1787/weo-2016-en

[2] Arshad M. Bioethanol: A Sustainable and Environment Friendly Solution for Pakistan. Science Vision 2010; 16 and 17.

[3] Patni N, Pillai SG, Dwivedi AH. Wheat as a Promising Substitute of Corn for Bioethanol Production. Procedia Engineering 2013; 51:355–62. Doi:10.1016/j.proeng.2013.01.049

[4] Zhao X-Q. Bioethanol from Lignocellulosic Biomass. Biotechnology in China III: Biofuels and Bioenergy Advances in Biochemical Engineering/Biotechnology 2011:25–51.Doi: 10.1007/10_2011_129.

[5] Mohanty B, Abdullahi II. Bioethanol Production from Lignocellulosic Waste-A Review. Biosciences, Biotechnology Research Asia 2016; 13:1153–61. Doi:10.13005/bbra/2146.

[6] Wong YC, Fikri AM. Synthesis of Bio-ethanol from Corn Stalk by Fermentation Process. Oriental Journal of Chemistry. 2014; 30(2):637–42.

[7] Oleskowicz-Popiel P, Lesiecki P, Holm-Neilson J, Bjerri A, Thomsen N. Ethanol production from maize silage as lignocellulosic biomass in anaerobically digested and wet-oxidized manure. Bioresource Technology. 2008; 99(13):5327–34.

[8] Zuo Z, Linga VR, Sing KP. Soaking Pretreatment of Corn Stover for Bioethanol Production Followed by Anaerobic Digestion Process. Applied Biochemistry and Biotechnology. 2012Jun; 167(7):2088–102.

[9] Smuga-Kogut MC, Wnuk A, Zgorska K, Kubiak M, Wojdalski J, Kupczyk A, Szlachta J, Luberanski A. Production of ethanol from wheat straw. Polish Journal of Chemical Technology. 2015Jan; 17(3).

[10] Koti S, Govumoni S, Gentela J, Rao L. Enhanced bioethanol production from wheat straw hemicellulose by mutant strains of pentose fermenting organisms Pichia stipitis and Candida shehatae. Springerplus. 2016; 5(1).

[11] Aditiya HB, Sing KP, Hanif M, Mahlia TM. Effect of Acid Pretreatment on Enzymatic Hydrolysis in Bioethanol Production from Rice Straw. International Journal of Technology. 2015;6(1):3

[12] Wu X, Zhang J, Xu E, Liu Y, Cheng Y, Addy M, Zhou J, Griffith R, Chen P, Ruan R. Microbial hydrolysis and fermentation of rice straw for ethanol production. Fuel. 2016Apr27; 679–86.

[13] Nachaiwieng W. Potential in bioethanol production from various ethanol fermenting microorganisms using rice husk as substrate. Biodiversitas, Journal of Biological Diversity. 2015Jan; 16(2):320–6.

[14] Achinas S, Euverink GJW. Consolidated briefing of biochemical ethanol production from lignocellulosic biomass. Electronic Journal of Biotechnology. 23:44–53. Doi:10.1016/j.ejbt.2016.07.006

[15] Anwar Z, Gulfraz M, Irshad M. Agro-industrial lignocellulosic biomass a key to unlock the future bio-energy: A brief review. Journal of Radiation Research and Applied Sciences 2014; 7:163–73. Doi:10.1016/j.jrras.2014.02.003.

[16] Belal EB. Bioethanol production from rice straw residues. Brazilian Journal of Microbiology 2013; 44:225–34. Doi: 10.1590/s1517-83822013000100033.

[17] Mussatto SI, Teixeira JA. Lignocellulose as raw material in fermentation processes. Applied Microbiology and Microbial Biotechnology. 2010; 4710–057.

[18] Abril D, Abril A. Ethanol from lignocellulosic biomass. Ciencia e investigaciónagraria 2009; 36. Doi: 10.4067/s0718-16202009000200003.

[19] Wang M, Zhou D, Wang Y, Wei S. Bioethanol production from cotton stalk: A comparative study of various pretreatments. Fuel 2016; 184:527–32. Doi:10.1016/j.fuel.2016.07.061

[20] Keshav PK, Naseeruddin S, Rao LV. Improved enzymatic saccharification of steam exploded cotton stalk using alkaline extraction and fermentation of cellulosic sugars into ethanol. Bioresource Technology. 2016; 214:363–70
[21] Aziz N. Pakistan's Potential to Produce from Biomass. Biomass and Bioenergy 2017; 58:379–89. Doi:10.1016/j.biombioe.2017.08.003
[22] Keshav PK, Shaik N, Koti S, Linga VR. Bioconversion of alkali delignified cotton stalk using two-stage dilute acid hydrolysis and fermentation of detoxified hydrolysate into ethanol. Industrial Crops and Products. 2016; 91:323–31
[23] Gopal, K. and Ranjhan, S.K. Laboratory manual for nutrition research. Roland Press (India) Private Ltd. New Dehli, India. 1980.
[24] Milagres, A.M.F. Production of xylanases from Penecilliumjanthinellum and its use in recovery of cellulosic textile fibers. Enz. Microbiol. Technol. 1994; 16(7): 627-632.
[25] Van'soest, P.J. Determination of lignin and cellulose in acid-detergent fiber with permanganate. J. Assoc. Off. Anal. Chem. 1976; 51:780-785.
[26] Carralero Sanz, Luz Mena M, Gonzalez-Costes A. Anal Chim Acta. 2005; 528, 1–8.
[27] Caputi, A., J.M. Ueda and T. Brown. Spectrophotometric determination of chromic comples formed during oxidation of alcohol. Am. J. Enol. Vitic. 1968; 19: 160-165.
[28] Nadeem M, Aftab M, Irfan M, Mushtaq M. Production of ethanol from alkali-pretreated sugarcane bagasse under the influence of different process parameters. Frontiers in Life Science. 2015; 8(4):358–62.
[29] Lackner KS. Chapter 1. Comparative Impacts of Fossil Fuels and Alternative Energy Sources. Issues in Environmental Science and Technology Carbon Capture. 2009;:1–40
[30] Silverstein RA, Chen Y, Sharma-Shivapa RR, Boyette MD, Osborne J. A Comparison of chemical pretreatment methods for improving saccharification of cotton stalks, Bioresour. Technol. 2007; 98, 3000-3011.
[31] Miller GL. Use of dinitro salicylic acid reagent for determination of reducing sugar. Anal Chem. 1959; 31:426–428.
[32] Rodhe AV, Sateesh L, Sridevi J, Venkateswarlu B, Rao LV. Enzymatic hydrolysis of sorghum straw using native cellulase produced by T. Reesei NCIM 992 under solid state fermentation using rice straw. Biotech. 2011;
[33] Taherzadeh MJ, Karimi K. Enzyme based hydrolysis processes for ethanol from lignocellulosic materials: a review. . Bioresources 2007; 2:707–38.
[34] Govumoni SP, Koti S, Kothangouni SY, Venkateshwar S, Linga VR. Evaluation of pretreatment methods for enzymatic saccharification of wheat straw for bioethanol production. Carbohydrate Polymers. 2013; 91(2):646–50.

Acknowledgment.
We are very grateful to Dr. Rubina Nelofer and Pakistan Council of Scientific and Industrial Research (PCSIR) for facilitation during this work.

IOP Conf. Series: Earth and Environmental Science **257** (2019) 012026 doi:10.1088/1755-1315/257/1/012026

The gasification efficiency improving by self-steam gasifier using RDF from municipal solid waste

T Khosasaeng and R Suntivarakorn

Department of Mechanical Engineering, Faculty of Engineering, Khon Kaen University, Khon Kaen, 40002, Thailand
E-mail: ratchaphon@kku.ac.th

Abstract. This research aims to improve the efficiency of the gasification by self-steam gasifier stove (SSG) which has been designed to produce steam for using gasification system.The waste heat is generated around the heat exchanger with water for steam production. Then the experiment was conducted to produce syngas from an air gasification and self-steam gasificaion using Refuse-derived fuels (RDF) from municipal solid waste. The RDF with density of 930 kg/m^3 and feed rate of 10 kg/hr was used as a fuel for gasification. The ER was varied from 0.15-0.50 in order to study the production of the producer gas and evaluate optimum point of heating value and cold gas efficiency for Comparison between air gasification and Self-steam gasification. From the experiment, it was found that The air gasification had optimum ER of 0.35,with maximum syngas heating value of 5.87 MJ/Nm^3 and The Self-steam gasification had optimum ER of 0.30 and steam feed rate of 0.96 kg/hr, with maximum syngas heating value of 6.56 MJ/Nm^3. The self-steam gasifier shows higher cold-gas efficiency than the conventional gasifier of 13. 87%

1. Introduction

Presently, municipal solid wastes (MSW) have continually increased, bringing with them environmental pollution problems. Municipal solid wastes contain multiple items such as metals, paper, textiles, plastics, wood and food garbage [1,] which consist carbon, hydrogen, nitrogen and oxygen[2]. Some of the components in waste can be used to produce energy, i.e., carbon and hydrogen.Gasification is a chemical process that changes the solid organic components of wastes into gases such as O_2, CO_2, CO, H_2, and CH_4 [3]. The important gases affecting the heating value of syngas and cold gas efficiency of gasification are CO, H_2 and CH_4. There are many types of gasification, for example, fluidized bed and fixed bed, which can be used with down-draft, up-draft, or cross-draft gasifiers [4]. The down-draft gasifier has certain advantages, namely, it is easier to control than the fluidized bed type, and releases less dust and tar than the up-draft and cross-draft gasifiers [5]. Therefore, it has frequently been used for real engineering manufacturing. Moreover, the gasification technology which is installed with a single throat is advantageous since the single throat enables RDF flowing in the combustion zone to compress well and hence making RDF receive thorough heat and burn well. RDF is thus appropriate for utilization as a renewable energy source because of the ease of transportation, low pollution, and small storage space requirement [6]. Municipal solid wastes are therefore popular to be produced and used as renewable energy.

The efficiency of the gasification process can be improved by means of the use of steam and air as the reactant in the gasifier. The resulting combustible gases will contain a high proportion of methane and hydrogen [7], which means an increase of the thermal value of the gases and higher efficiency of the overall system[8]. However, the steam used in the process has to be produced and brought in from an

Content from this work may be used under the terms of the Creative Commons Attribution 3.0 licence. Any further distribution of this work must maintain attribution to the author(s) and the title of the work, journal citation and DOI.
Published under licence by IOP Publishing Ltd

external source. This increases the cost of fuel gas production. A Self-Steam gasifier (SSG) that can produce its own vapor by the exchange of internal heat with the fixed water coil pipe will reduce the cost since there is no need to import steam.

This article compares the appropriate air feed, fuel components, thermal value of fuel gas and the efficiency of gasification between the conventional gasifier that relies on air and the self-steam gasifier (SSG), which uses air and steam as the reactant. The appropriate temperature of the gasifier was also determined, which will yield steam with the best fuel thermal values and highest gasification efficiency.

2. Meterials and Experimental Methodology

2.1. Material of Feed stock

Refuse-derived fuels (RDF) used in the experiment were obtained from municipal wastes that had been separated. Most parts were plastic materials. Then the wastes were shredded into tiny pieces and pressed into small pellet with a diameter of roughly 8 mm, a length of 25 mm and density of 930 kg/m^3. The moisture content was not allowed to exceed 4%. Next, an ultimate analysis was performed using Perkin elmer, Series ll CHNS/O Analyzer 2400. A proximate analysis was performed with Perkin Elmer instrument, Pyris Diamond TG/DTA model. Heating values were analyzed with Bomb Calorimeter, IKA brand, C5003 control Model. The details of RDF are given in Table 1.

Table 1. The Characteristics of RDF.

Proximate analysis (%wt)		Ultimate analysis (%wt)	
Moisture	4.00	C	58.16
Volatile matter	81.47	H	9.80
Fixed carbon	9.73	N	0.73
Ash	4.80	O	31.31
High heating value (kJ/kg dry basis)		29,410	
Density (kg/m^3)		930	

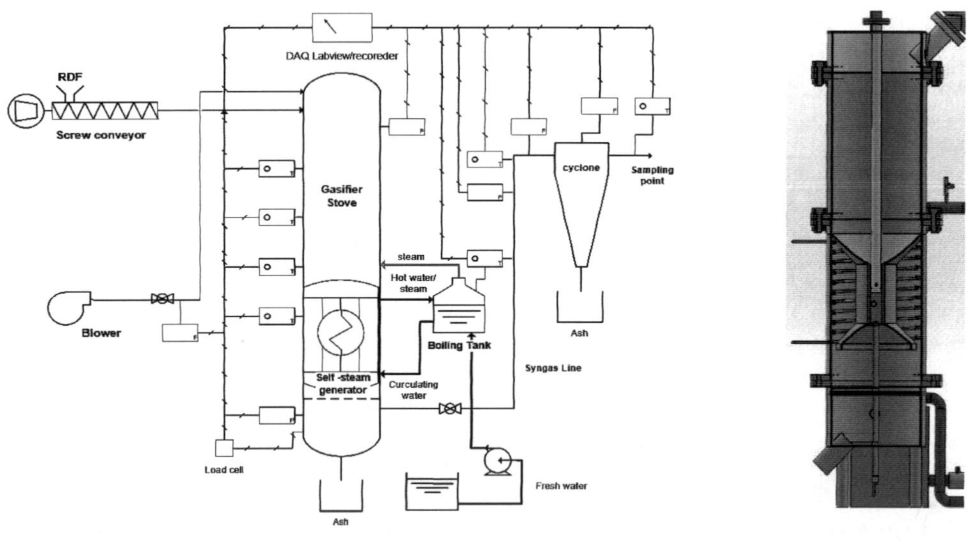

(a) (b)

Figure 1. (a) The schematic of Self-Steam gasifier experimental setup , (b) Section View of SSG.

2.2. Experimental setup

The experiment was conducted in the single-throat downdraft gasifier and Self-Steam gasifier (SSG) using RDF. The operation was carried atmospheric air as gasification medium. The experimentation plan started with the design of a single-throat down-draft gasifier with a capacity of 30 kw. The gasifier was 1.70 m high, with a diameter of 0.50 m and the single throat tilting at 45°. Four air ducts were attached to one side of the gasifier. At the bottom, a mesh and hand-stirred to get rid of ash were installed. Water was used to seal the ash exit and prevent gas leakage. The gasification system itself was composed of a gasifier, cyclone, screw converyor, blower and sensors as illustrated in Figure1. (a)

The experiment started by feeding 10 kg of RDF into the top of the gasifier which was placed over burning charcoal to ignite the system, Then the gasifier cover was closed and the blower opened to draw air in. The blower used was a 1 hp, 3-phase Hitachi blower, VB-007-DN Model, having a maximum capacity of 2.1 m^3/min. Air was fed in through the four air pipes at the side of the gasifier. The air flow rates were measured by the Testo 340 and calculated for the air feed rate. Next, the temperatures in the gasifier were monitored using the Thermocouple type k and were measured by the NI-DIQmx temperature sensor. All of the data obtained were stored in the computer.

The self-steam gasifier has been designed to produce steam at 1.5 kg/hr. Water will flow in 23 copper coils of 8mm diameter placed in the space between the bottleneck and gasifier wall. Water enters from the bottom of the copper coils and flows out from the top for another heat exchange. The steam produced is used in the gasification and the hot water is fed into the copper coil, again for heat exchange. Steam is produced 15 minutes after the system starts to operate. The temperature of the gasifier is between 850-750 °C, which results in the transfer of heat to the coils and steam is produced at 108-100 °C. At this point, the valve is closed in order for the steam to be fed into the gasifier.show in figure 1(b)

The experiment was performed to determine the suitable air feed rate for the conventional gasifier and self-steam gasifier. Controls were made over the fuel feed rate of 10 kg/hr, steam feed rate, steam temperatures, and gasifier temperature. The variables for the fuel feed rate)ER(were set within the range 0.1-0.5 to compare the efficiency and thermal values between the conventional gasifier and self-steam gasifier.

2.3. Self-Steam gasifier stove design

The self-steam gasifier has been designed to produce steam Water will flow in copper coils placed in the space between the bottleneck and gasifier wall the Temperature inside the copper wall is calculated by (1) [9]

$$T_4 = T_1 + Q\left(\frac{\ln(\frac{r_2}{r_1})}{2\pi K_a L} + \frac{1}{h_a A} + \frac{\ln(\frac{r_4}{r_3})}{2\pi K_b L} \right) \quad (1)$$

where T_1 is temperature inside SSG stove (°C) r_1,r_2,r_3 and r_4 are distance from the center of the Stove (m) K_a and K_b are thermal conductivity (w/m^2.k) ha is Heat transfer coefficient of air (w/m.k) L is The length of the SSG stove wall (m) and A is area of heat convection (m^2)

Temperature of steam is calculate by (2) and (3)

$$T_b = (T_i - T_w)\exp\left(-\frac{hA}{mc_p} \right) + T_w \quad (2)$$

$$m = \rho VA \quad (3)$$

where T_i and T_w are temperature of water inlet and temperature inside the cooer wall (°C) h is heat transfer coefficient of water (w/m.k) A is area of copper tube (m^2) m is mass of water (kg) c_p is specific heat capacity of water (J/kg.c) ρ is density of water (kg/m^3) and V is velocity of steam (m/s)

2.4. Cold gas efficiency

The air flow rate was effects the quality of syngas production. The air flow rate (M_f:kg/hr) is calculate by (4) [10]

$$M_{fa} = \left(\frac{32 \cdot C}{12} + 8 \times H + S - O \right) \times ER \times M_{fRDF} \tag{4}$$

where C H N S and O are mass fraction of Carbon, Hydrogen, Nitrogen, Sulfur and Oxygen, respectively by ultimate analysis, ER is equivalent ratio and M_{fRDF} is RDF feed rate (kg/hr).
The Volumetric composition of the production gas affect to lower heating value (LHV) and quality of syngas. The lower heating value of syngas (LHV_{sg}; kcal/Nm^3) is calculated by (5)

$$LHV_{sg} = (h_{co} \times x_{co}) + (h_{H_2} \times x_{H_2}) + (h_{CH_4} \times x_{CH_4}) \tag{5}$$

where h_{CO}, h_{H2} and h_{CH4} are lower heating value (kcal/Nm^3) of Carbon monoxide, Hydrogen and Methane and x_{co}, x_{H2} and x_{CH4} are Volumetric composition of Carbon monoxide, Hydrogen and Methane, respectively.
The cold gas efficiency (η_{cg}) is the energy input over the potential energy output that is calculated by (6) [11]

$$\eta_{cg} = \frac{Q_{sg} \times LHV_{sg}}{M_{fRDF} \times LHV_{fRDF}} \times 100 \tag{6}$$

where Q_{sg} (Nm^3/hr) is Volumetric of syngas and LHV_{fRDF} is lower heating value of RDF.

3. Result and Discussion

3.1. The effect of the Equivalent ratio (ER) on syngas concentration

The experiment revealed the control of air for the reaction in the self-steam gasifier (SSG) as follows: CH_4 9.52% at ER 0.25, CO 15.80 % at ER0.30 and H_2 11.96% at ER 0.25. This is higher than the air gasifier (single-throat downdraft gasifier) which had the following results: CH_4 8.76% at 0.25, CO 14.72 at ER 0.35 and H_2 8.82% at ER 0.25. The results are consistent with the reaction:

$C + H_2O = CO + H_2$ water gas reactor

$CO + H_2O = CO_2 + H_2$ water gas shift reaction

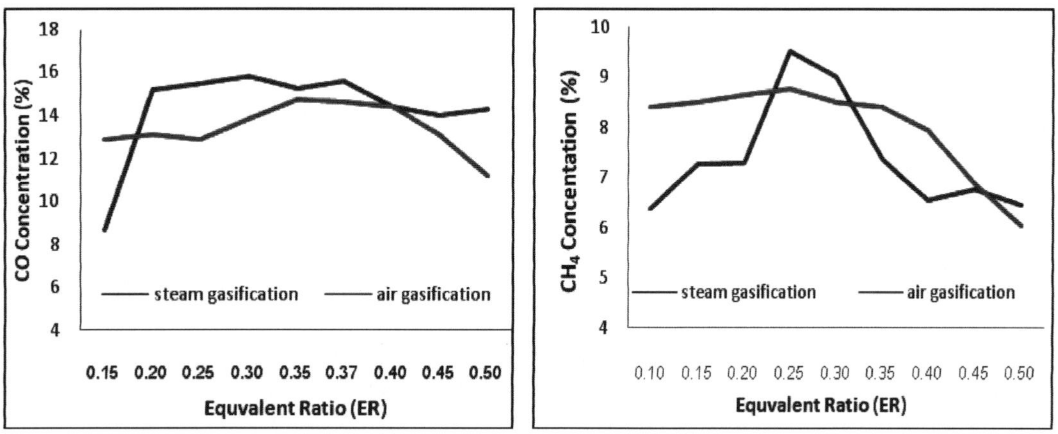

(a) (b)

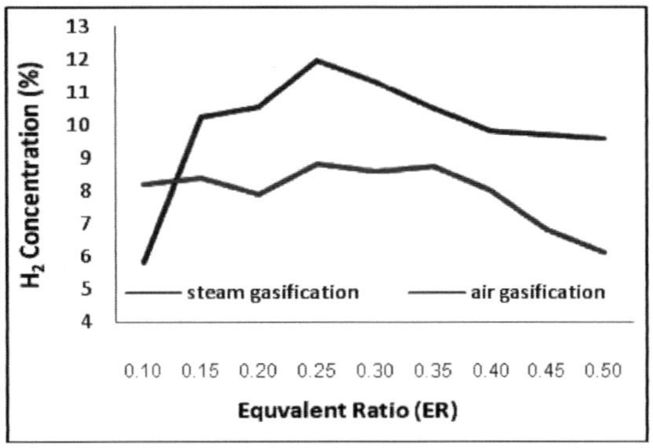

(c)

Figure 2. (a) The effect of ER on CO concentration , (b) The effect of ER on CO concentration CH_4 concentration and (c) The effect of ER on CO concentration H_2 concentration.

Both equations demonstrate that steam underwent reaction until CO and H_2 increased, as shown in Figure 2. (a-c). `In the steam control, the valve was opened when the temperature in the boiler reached 100 °C. The air fed into the gasifier was controlled with the appropriate ER for self-steam gasifier being 0.25-0.35 or an amount of 25.60-18.28 m^3/hr of 10 kg of RDF and the steam feed rate of 0.95 kg/hr.

3.2. The effect of the Equivalent ratio (ER) on efficiency

The efficiency of gasification in the self-steam gasifier is higher than the conventional gasifier both in terms of system efficiency and the thermal value of the fuel gas. The trend of the graph is increasing at the range of ER 0.15-0.35 and decreasing at the range of ER 0.50-0.35. This is consistent with the graph showing the gas components in Figure 2. Higher ER means the system approaches complete combustion, making the syngas concentration less dense with less efficiency. In Figure 3(a) and (b), the self-steam gasifier at ER 0.35 will yield the maximum thermal values of fuel gas at 6.56 MJ/Nm^3 and the maximum efficiency of 83.22%. The self-steam gasifier shows higher cold-gas efficiency than the conventional gasifier of 13.87% at ER 0.35. This is due to the fact that the self-steam gasifier uses the heat produced to the fullest extent by means of the heat exchange with the water coils to produce steam instead of wasting the heat. The steam produced also increase the density of hydrogen and carbon monoxide, which are the fuel gases and hence the overall efficiency of the system becomes high.

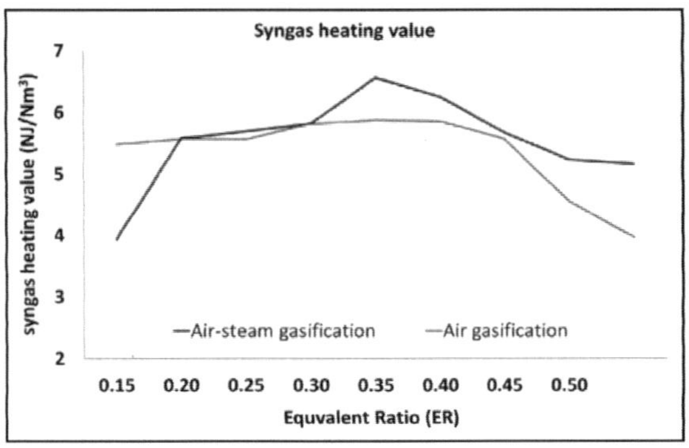

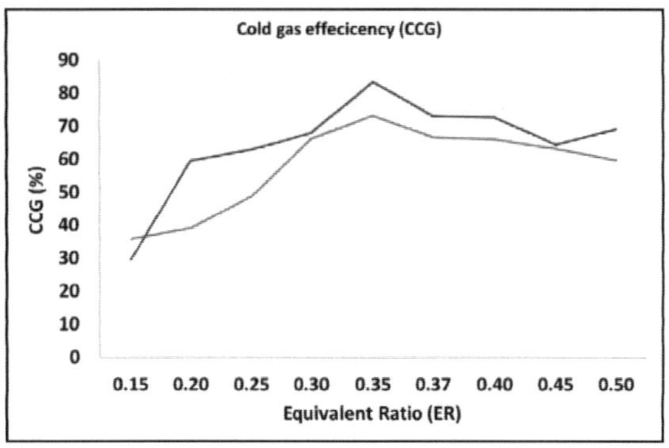

(a) (b)

Figure 3. (a) The effect of ER on syngas heating value and (b) The effect of ER on syngas heating value

4. Conclusion

The study shows that the self-steam gasifier designed in this study can increase the efficiency of the gasification system to 83.22%, which is 13.87% higher than the conventional gasifier that uses only air. This study used the compact refuse RDF as the fuel with the feed rate of **10**kg/hr. The air feed rate and the fuel feed rate (ER) were between 0.10-0.50. The system could produce steam too at the temperatures between 100-105 ^{o}C with the steam fee rate of 0.96 kg/hr. The experiment shows that the self-steam gasifier can produce fuel gas with a higher density of CO, CH_4 and H_2 than the conventional air gasifier. The appropriate ER values are in the range of 0.30-0.35 and the biomass gases obtained show the syngas heating value of 6.56 MJ/Nm^3, which is 11.68% higher than the conventional air gasifier. The self-steam gasifier shows higher cold-gas efficiency than the conventional gasifier of 13.87% at ER 0.35

5. Acknowledgments

The authors would like to express their gratitude to Energy Conservation Promotion Fund, Energy Policy and Planning, Ministry of Energy for financial support, and Farm Engineering and Automation Technology Research Group (FEAT), Khon Kean University for tool and equipment supported in the study.

6. References

[1] Srimanta P, Hovichit P 2013 Evaluation of greenhouse gas reduction from improvement of solid waste management of Khon Kaen University. KKU Engineering Journal 40(2);p.203-212.
[2] Arafat H A, Jijakli K 2013 Modeling and comparative assessment of municipal solid waste.Waste Management 33:p.1704-1713.
[3] Minhaj U M, Azrina A A, Risky A K, Abu Y 2018 *Gasification of lignocellulosic biomass to produce syngas in a 50 kW downdraft reactor*. Biomass and Bioenergy 119; 335-345
[4] Guangul F M, Sulaiman S A, Ramli A 2012 Gasifier selection, design and gasification of oil palm fronds with preheated and unheated gasifying air. Bioresource Technology 126;p.224-232.
[5] Yoon S J, Son Y I, Kim Y K, Lee J G 2012 Gasification and power generation characteristics of rice husk and rice husk pellet using a downdraft fixed-bed gasifier. Renewable Energy 2012;42;p.163-167.
[6] Chalermcharoenrat S, Laohalidanond K, Kerdsuwan S 2015 Optimization of Combustion Behavior and Producer Gas Quality from Reclaimed Landfill Through Highly Densify RDF-Gasification. Energy Procedia 79;p.321-326.
[7] E. Shayana, V. Zare, I. Mirzaee 2018 Hydrogen production from biomass gasification; a theoretical comparison of using different gasification agents. Energy Conversion and Management 159; P.30-41
[8] Nadia C, Francesco Z, Luca C, Mauro P, Massimo O C, Vito V 2017 *Air-steam and oxy-steam gasification of hydrolytic residues from biorefinery*. Fuel Processing Technology 167 ;P 451-461
[9] John H L IV, John H L V 2001 *A heat transfer textbook Third Edition*
[10] Basu P 2010 Biomass Gasifiacation and Pyrolysis Practical Design and Theory.Elsevier
[11] Uisung L,J.N. Chung, Herbert A I 2014 High-Temperature Steam Gasification of Municipal Solid Waste Rubber, Plastic and Wood . Energy&fuel 28;4573-4587

Modelling and optimization of biomass-based cogeneration plant

Abdul Razak N.A.A[1] and Abdulhalim Abdulrazik[1*]

1 Faculty of Chemical and Natural Resources Engineering, Universiti Malaysia Pahang, Kuantan, Pahang, Malaysia.
*E-mail: abdhalim@ump.edu.my

Abstract. High energy demand and energy availability lead to the increasing of unpleasant energy situation. It is exacerbated if fossil fuels are the only energy source as it contributes to climate change and global warming. Thus, governments are focusing on the usage of natural resources as fossil fuel substitution. This study focused on the modelling and optimization of biomass-based cogeneration plant. The objective was to model, simulate and optimize the cogeneration plant which used torrefied EFB pellet as fuel in Aspen Plus simulator. Firstly, suitable biomass resources in Malaysia was identified. Next, a typical process flow diagram of cogeneration from the published literature was referred. Then, parametric and structural optimization were conducted. From the simulation, 1.764 MW of power was generated and it was observed that pellet flow rate, water flow rate, air flow rate and boiler temperature influenced the power generated. Five options with different structural designs were formulated in GAMS. It was found that the plant should focused on producing power and medium pressure steam to optimize the profit. The findings concluded that biomass-based cogeneration plant was technically feasible to be deployed. However, further refinements in the optimization aspect should be developed to obtain a more accurate optimal results.

1. Introduction

Nowadays, developing countries such as Malaysia faces energy crisis including power due to rapid growth of energy demand compared to developed countries [1]. Meanwhile, fossil fuels such as coal and natural gas which are the main energy sources in Malaysia, cannot be preserved anymore due to the lack of reserves and environmental impacts such as climate change and global warming [2]. Besides, the abundance of wastages production is eventually occurring [3]. Thus, exploration of new technologies and alternative energy sources must be increased in order to support the drastic population growth and energy demand. The Malaysian government had initiated an alternative solution by announcing a Small Renewable Energy Power (SREP) program to encourage private sectors to invest in small power generation projects utilizing renewable energy sources such as biomass and solar energy [4]. This program has raised the power industry's interest towards cogeneration plant. Cogeneration has become an option as it produces industrial energy need with more economically and efficiently compared to the conventional plant due to its effective use of wasted thermal energy [5]. Furthermore, Malaysia being one of the world's primary palm oil producers begins to encourage the use of renewable energy especially in palm oil due to its availability, continuity and capacity for renewable energy solution [6]. However, there are only few references available regarding biomass-based cogeneration plant especially in the modelling and simulation perspective. This research revised a more in depth of modelling and optimization of cogeneration plant using biomass in Malaysia. Major issue for this potential is the modelling and process analysis as it is

Content from this work may be used under the terms of the Creative Commons Attribution 3.0 licence. Any further distribution of this work must maintain attribution to the author(s) and the title of the work, journal citation and DOI.
Published under licence by IOP Publishing Ltd

better to model, simulate and improve the system through optimization before constructing it. To simulate the process, Aspen Plus was chosen for this study as it is a powerful simulation tool for chemical engineering in various fields including oil and gas production, refining and power generation [7]. Meanwhile, General Algebraic Modeling System (GAMS) was used as an optimization tool to solve mixed integer optimization problem [8]. Hence, Aspen Plus modelling and usage of GAMS for optimization tools of cogeneration plant by using biomass would be a contribution for this work. The objectives were to model and simulate the cogeneration plant which uses biomass as fuel in Aspen Plus and to improve the biomass-based cogeneration plant by optimization. The scopes of study for this research covered the modelling, simulation and optimization of biomass-based cogeneration plant where treated biomass in the form of torrefied empty fruit bunches (EFB) pellet was used as a fuel and the targeted power produced was 1.3 MW.

2. Materials and methods

Figure 1 shows the sequence in order to achieve the objective. Explanation of each of the block will be explained in the following sections.

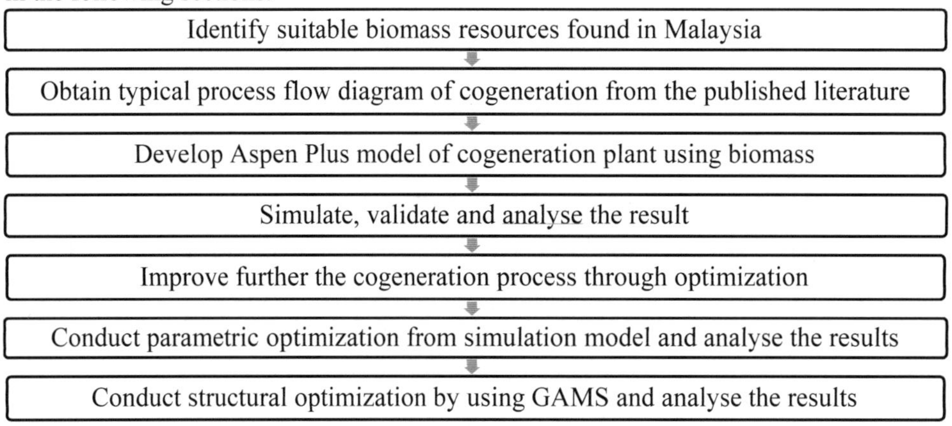

Figure 1. Process flow diagram of the study

2.1. Identify suitable biomass resources found in Malaysia

Findings and observations regarding biomass in Malaysia were carried out where torrefied empty fruit bunches (EFB) pellet was chosen for this study.

2.2. Obtain typical process flow diagram of cogeneration from the published literature

A research on process flow diagram (PFD) of cogeneration plant from published literature was carried out where it was used as a reference for modelling and result validation purposes. Figure 2 shows the chosen PFD from Wu et.al. (2017) as a base case study where oil palm waste as fuel.

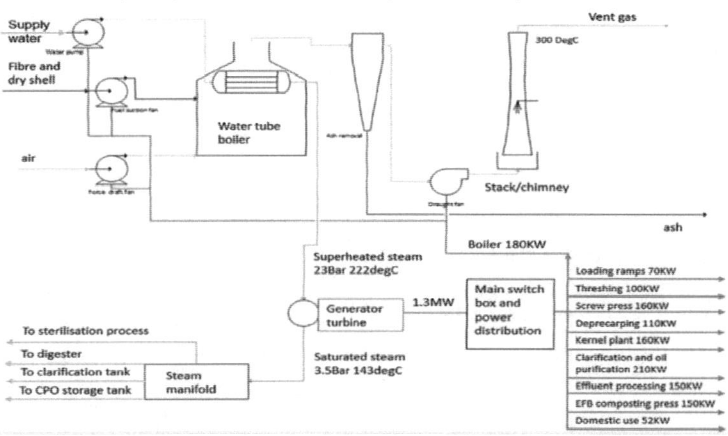

Figure 2. Process flow diagram of a biomass-based cogeneration plant [9]

IOP Conf. Series: Earth and Environmental Science **257** (2019) 012027 doi:10.1088/1755-1315/257/1/012027

2.3. Develop Aspen Plus model of cogeneration plant using biomass

Figure 3 shows the model of biomass-based cogeneration plant developed by using Aspen Plus where the PFD and operating conditions were referred to the published journal by Qibai Wu et.al.

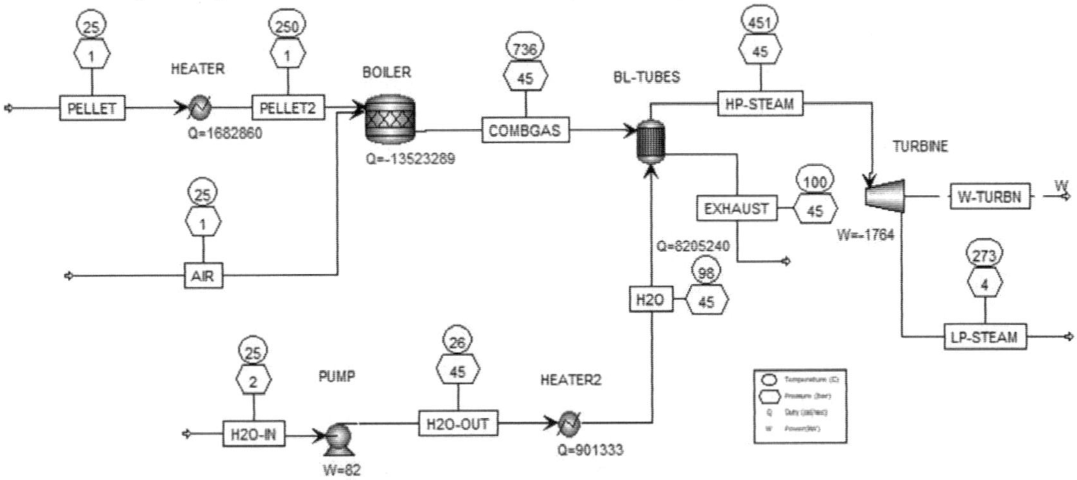

Figure 3. Flowsheet of biomass-based cogeneration plant in Aspen Plus

The components involved were carbon graphite, hydrogen, nitrogen, oxygen, sulphur, water, carbon dioxide, nitric oxide and nitrogen dioxide. The operating conditions for the inlet streams were shown in Table 1.

Table 1. Table of operating conditions for the inlet streams [9]

Stream	Operating conditions	
PELLET	i.	Temperature: 25°C
	ii.	Pressure: 1 bar
	iii.	Mass flowrate: 5 kg/s
	iv.	Valid phases: Vapor-liquid
AIR	i.	Temperature: 25°C
	ii.	Pressure: 1 bar
	iii.	Mass flowrate: 37.5 kg/s
	iv.	Mole fraction: 0.79 (Nitrogen) and 0.21 (Oxygen)
	v.	Valid phases: Vapor-liquid
H2O-IN	i.	Temperature: 25°C
	ii.	Pressure: 2 bar
	iii.	Mass flowrate: 11.5 kg/s
	iv.	Valid phases: Vapor-liquid
	v.	Mole fraction: 1

2.4. Simulate, validate and analyse the result

Firstly, the components involved were defined in the properties section of Aspen Plus by using Peng Robinson equation of state as property method. Based on Figure 3, the process was divided into two routes before entering the heat exchanger which were from a boiler and a heater. The combination of the boiler and heat exchanger represent a water tube boiler in Aspen Plus as there was no direct water tube boiler system in the software. In the first route, there were two inlet streams entering the boiler which were torrefied EFB pellet as fuel and air for oxygen supply in the combustion reaction of the pellet. The pellet stream entered at 25°C and 1 bar before heated to 250°C with a heater. The composition of elemental component of torrefied EFB was defined based on its ultimate analysis composition at 250°C as shown in Table 2.

Table 2. Composition of torrefied EFB pellet (wt%) [10]

Sample	C	H	N	S	O
Torr 250	49.47	5.91	0.52	0.54	43.56

Next, the pellet and air entered a boiler which in this case, a stoichiometric reactor was used (BOILER). The reactor operated at 736°C and 45 bar with vapor-liquid as its valid phases to produce a superheated steam of 45 bar at 451°C which was the same as in the base case study. The combustion equations for feed components such as carbon, hydrogen, sulphur, nitrogen and nitric oxide were listed below respectively and the reactions were defined under the section "Reactions" of stoichiometric reactor. It was assumed that the conversion for carbon and hydrogen were the same which was 80% while sulphur, nitrogen and nitric oxide had a conversion of only 10%.

$$C + O_2 \rightarrow CO_2 \quad (1)$$
$$2H_2 + O_2 \rightarrow 2H_2O \quad (2)$$
$$S + O_2 \rightarrow SO_2 \quad (3)$$
$$N_2 + O_2 \rightarrow 2NO \quad (4)$$
$$2NO + O_2 \rightarrow 2NO_2 \quad (5)$$

Then, the combusted gas produced entered a heat exchanger (BL-TUBES) which represent water tubes in water tube boiler and the combusted gas was connected to the hot inlet of the heat exchanger. A shortcut, counter current heat exchanger was used with a minimum temperature approach of 1°C while the hot stream outlet temperature was at 100°C. In the pressure drop section, outlet pressure of 45 bar was assumed in order to produce a high pressure steam. Simultaneously, a water inlet stream operated at 25°C and 2 bar was pumped to a pressure of 45 bar which was then connected to a heater (HEATER2) to be heated to 98°C before entering the heat exchanger as its cold inlet. The exhaust gas from the hot outlet stream of the heat exchanger was released. Meanwhile, the cold outlet stream of the heat exchanger which was labelled as SH-STEAM became a superheated steam which was then connected to a back-pressure steam turbine for power and heat production. The steam turbine (TURBINE) was assumed to be an isentropic turbine. According to Qibai Wu et.al, the steam turbine operated at 3.5 bar with 50% efficiency. The electricity generated was transferred to the main switch box for power distribution while exhaust gas was sent to a steam manifold for the plant's heating purposes. Once the simulation converged, the results were analysed and compared with the power output of the chosen published journal to validate the developed process. Further discussions were also made.

2.5. Improve further the cogeneration process through optimization
Optimization methods were analysed and selected by considering feasibility aspect of the available data. Two types of optimization method were selected and conducted to improve the model by applying parametric optimization and structural optimization.

2.6. Conduct parametric optimization from simulation model and analyse the results
Parametric optimization was conducted to measure the performance of the model and also to optimize the selected parameters such as pellet flow rate, water flow rate, air flow rate and boiler temperature. Comparison between the parameters and power generated were made and analysed.

2.7. Conduct structural optimization by using GAMS and analyse the results
Structural optimization was carried out in GAMS in order to choose the optimum structural design of the model by optimizing the correct combination of structural options. Five options with different structural design were modelled in Aspen Plus beforehand with each of the options had its own strength and weakness. The options are as follow:

- Option 1: Cogeneration process with installed air preheater and low pressure steam production
- Option 2: Cogeneration process with steam turbine stages and low pressure steam production

- Option 3: Cogeneration process with installed air preheater and medium pressure steam production
- Option 4: Cogeneration process with steam turbine stages and medium pressure production
- Option 5: Cogeneration process without installed steam turbine and only focused in medium pressure steam production

Next, the processing cost of each options were calculated by using equation (6).

$$COM = 0.18FCI + 2.73C_{OL} + 1.23(C_{UT} + C_{WT} + C_{RM}) \tag{6}$$

Where C_{OM} is cost of manufacturing, F_{CI} is fixed capital investment, C_{OL} is cost of operating labor, C_{WT} is cost of waste treatment and C_{RM} is cost of raw material [5]. Before formulation was conducted, few parameters, decision variables, constraints and objective function were identified.

2.7.1. Parameters. The parameters identified were medium pressure steam price, low pressure steam price, raw material cost, processing cost to produce 1 kg of steam and processing cost to produce 1 kWh of power.

2.7.2. Decision variables. Below are the list of decision variables:

- x_1 : Power production from option 1
- x_2 : Low pressure steam flowrate production from option 1
- x_3 : Power production from option 2
- x_4 : Low pressure steam flowrate production from option 2
- x_5 : Power production from option 3
- x_6 : Medium pressure steam flowrate production from option 3
- x_7 : Power production from option 4
- x_8 : Medium pressure steam flowrate production from option 4
- x_9 : Medium pressure steam flowrate production from option 5

Binary variables were termed as "0-1" where 0 indicates rejection of the variable while 1 indicates the acceptance of the variable [11]. Below are the list of binary variables:

- y_1 : Binary for power production from option 1
- y_2 : Binary for low pressure steam flowrate production from option 1
- y_3 : Binary for power production from option 2
- y_4 : Binary for low pressure steam flowrate production from option 2
- y_5 : Binary for power production from option 3
- y_6 : Binary for medium pressure steam flowrate production from option 3
- y_7 : Binary for power production from option 4
- y_8 : Binary for medium pressure steam flowrate production from option 4
- y_9 : Binary for medium pressure steam flowrate production from option 5

2.7.3. Constraints. The power production was assumed to be 1400 kW and steam flowrate production for both low pressure and medium pressure steam were assumed to be 41400 kg/h. This value constraints were based on the minimum output value simulated from previous Aspen Plus model. The constraints are as below:

$$x_1 + x_3 + x_5 + x_7 = 1400 \tag{7}$$

$$x_2 + x_4 + x_6 + x_8 + x_9 = 4140 \tag{8}$$

$$y_1 + y_3 + y_5 + y_7 = 1 \tag{9}$$

$$y_2 + y_4 + y_6 + y_8 + y_9 = 1 \tag{10}$$

2.7.4. Objective function. Objective function of the study was to maximise the profit of the plant per year where profit can be calculated as in equation (11).

$$Profit_i = Revenue_i - Processing\ Cost_i \tag{11}$$

Where i can be stated as power, medium pressure steam or low pressure steam. Equation (12) is obtained after calculating the profit for each options.

$$125872.14x_1y_1 + 98127.59x_2y_2 + 135018.37x_3y_3 + 98126.4x_4y_4 + 44102.88x_5y_5 + 103012.4x_6y_6 + 74361.22x_7y_7 + 103011.2x_8y_8 + 103013.6x_9y_9 = profit \quad (12)$$

2.7.5. Optimization method. Mixed integer nonlinear programming (MINLP) was formulated where it refers to optimization problem involving continuous and discrete variables and also nonlinear functions in the objective function or the constraints. MINLP can be used to simultaneously optimize the system structure which is a discrete variable and parameter which is a continuous variable [4]. Discrete and Continuous Optimizer (DICOPT) was used as a solver and the data obtained from GAMS was analysed.

3. Results and discussions

3.1. Aspen Plus model of biomass-based cogeneration plant

From the simulation, 1.764 MW of power and a low pressure steam of 4 bar were generated from the turbine which was lower compared to the base case study where 2.2 MW of power generated by using ECLIPSE software. This differences may be due to optimization was carried out beforehand for the base case study while Aspen Plus model had not yet undergone optimization. However, 1.764 MW of power generated from Aspen Plus still exceeded the actual need of electricity for the mill which was 1.3MW. Thus, the plant can produce more than enough electricity and steam for the production process when torrefied EFB pellet was used as fuel. The additional electricity can be sold to the national power grid. Thus, the old and conventional fossil fuel plants can be replaced by renewable and environmental friendly plants which use torrefied EFB as fuel. Besides, the dependency and demand towards fossil fuels can be reduced and minimized the environmental degradation.

3.2. Parametric optimization from Aspen Plus model

The results of the modifications from parametric optimization are shown in Table 3 until Table 7.

Table 3. The value of power generated at certain pellet flow rate

Pellet Flow Rate (kg/s)	2	3	4	5
Power Generated (MW)	1.347	1.458	1.599	1.764

Table 4. The value of power generated at certain water flow rate

Water Flow Rate (kg/s)	11.5	12.0	12.5	13.0	13.5
Power Generated (MW)	1.764	1.689	1.631	1.591	1.569

Table 5. The value of power generated at certain air flow rate

Air Flow Rate (kg/s)	34.5	35.5	36.5	37.5	38.5	39.5
Power Generated (MW)	1.558	1.622	1.691	1.764	1.841	1.920

Table 6. The value of power generated at certain boiler temperature (Continued)

Temperature (^{O}C)	660	680	700	720	740	760	780	800	820
Power Generated (MW)	1.387	1.469	1.564	1.671	1.789	1.916	2.043	2.170	2.296

Table 7. The value of power generated at certain boiler temperature (Continued)

Temperature (^{O}C)	840	860	880	900	920	940	960	980
Power Generated (MW)	2.421	2.545	2.668	2.790	2.912	3.033	3.152	3.271

The results show that the initial pellet flow rate and water flow rate should be maintained at 5 kg/s and 11.5 kg/s respectively to generate maximum power. Furthermore, 37.5 kg/s of initial air flow rate can be increased to 39.5 kg/s and the temperature of the boiler should be increased for a maximum power generation. Meanwhile, the pattern of each modifications can be seen in Figure 4 until Figure 7.

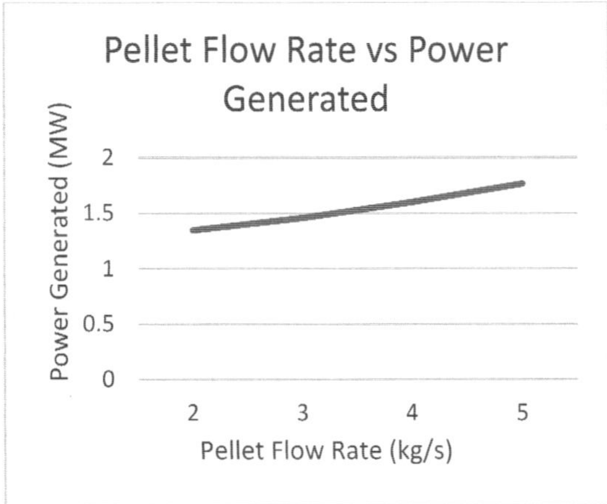

Figure 4. Graph of pellet flow rate against power generated

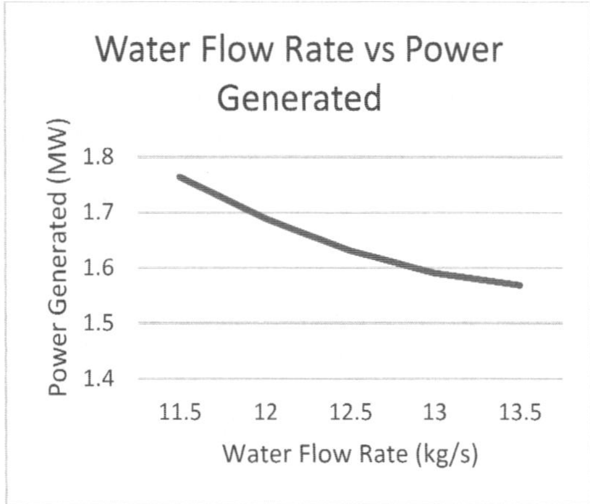

Figure 5. Graph of water flow rate against power generated

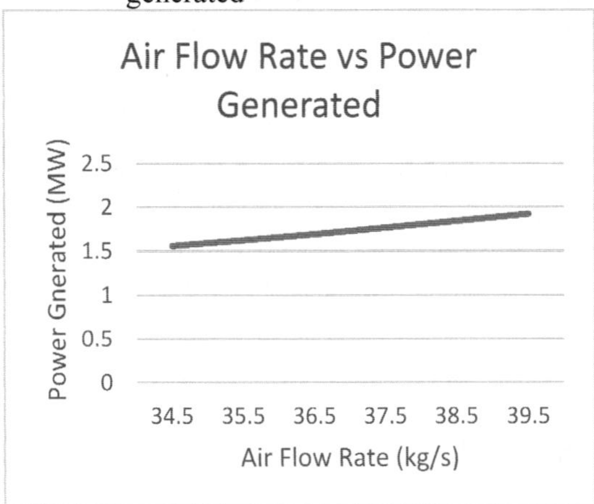

Figure 6. Graph of air flow rate against power generated

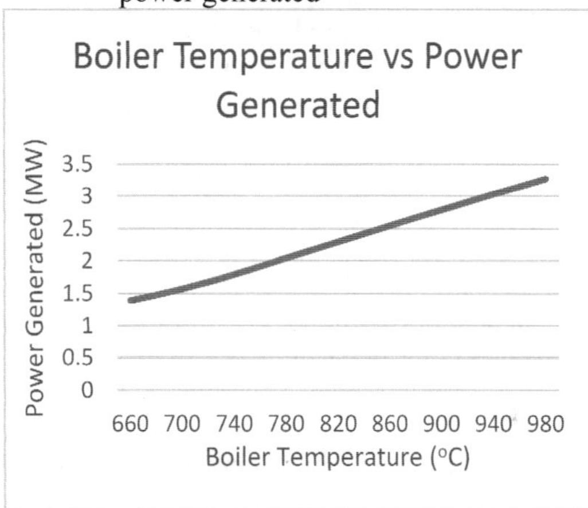

Figure 7. Graph of boiler temperature against power generated

It can be seen that power generated increases as pellet flow rate increases, water flow rate decreases, air flow rate increases and boiler temperature increases. This may be due to the factor of reaction rate where higher temperature will lead to higher kinetic energy and larger amount of molecules possess the required activation energy thus, an increase in the effective collisions between particles [12]. Besides, more fuel to be acted upon the air for combustion reaction increases the reaction rate. However, beyond 5 kg/s of EFB pellet resulted in infeasibility of the model as air became its limiting reactant. Furthermore, reduction in water flow rate was favourable to increase power generated as more heat can be supplied to heat the water which increases the temperature of steam produced and simultaneously increases the work done by the turbine. Moreover, higher air flow rate lead to a near complete combustion reaction and low production of incomplete combustion gases as amount of oxygen supply increases [13].

3.3. Structural optimization results from GAMS

The optimization results from GAMS accepted the binary variable of y_3 and y_6 while others were rejected. It was also shown that a profit of RM4,453,739,078.00 can be achieved if the plant operates to produce power and medium pressure steam at 1400 kW and 41400 kg/h respectively. However, the profit obtained was too high and might be unrealistic to achieve in real case study. This may be due to the range of constraints applied were too large for the process and it was not enough to reduce the boundary of steam and power production as reduction in the boundary values can lead to a more accurate and realistic data. Thus, further refinement should be made in the future in order to reduce the scope of the constraints and to apply more restriction towards the process.

4. Conclusions and recommendations

It can be concluded that the simulation of cogeneration plant by using torrefied EFB pellet as fuel in Aspen Plus was able to produce power of greater than 1.3 MW and optimization procedures helped in deciding the optimum value for parameters and also in the selection of production where power and medium pressure steam were favourable. Thus, biomass-based cogeneration plant has a great potential to replace the current technologies which is mostly using conventional fossil fuels. This findings are significant as it could reduce the carbon emission which can lead to a greener and healthier environment for electrical power generation sector. This research can be further improved by taking into consideration of few parameters such as raw materials' availability and products' demand in order to achieve more accurate and reliable optimal results.

5. References

[1] S. Mekhilef, R. Saidur, A. Safari, and W. E. S. B. Mustaffa, "Biomass energy in Malaysia: Current state and prospects," *Renew. Sustain. Energy Rev.*, vol. 15, no. 7, pp. 3360–3370, 2011.

[2] A. Darabadi Zareh, R. Khoshbakhti Saray, S. Mirmasoumi, and K. Bahlouli, "Extensive thermodynamic and economic analysis of the cogeneration of heat and power system fueled by the blend of natural gas and biogas," *Energy Convers. Manag.*, vol. 164, no. March, pp. 329–343, 2018.

[3] T. H. Oh, M. Hasanuzzaman, J. Selvaraj, S. C. Teo, and S. C. Chua, "Energy policy and alternative energy in Malaysia: Issues and challenges for sustainable growth – An update," *Renew. Sustain. Energy Rev.*, vol. 81, no. July 2017, pp. 3021–3031, 2018.

[4] L. S. L. Thomas F. Edgar, David M. Himmelblau, Optimization of Chemical Process. 2001.

[5] T. and Richard and W. B, *Analysis, Synthesis, and Design of Chemical Process*, Third Edit. 2009.

[6] N. Abdullah and F. Sulaim, "The Oil Palm Wastes in Malaysia," *Biomass Now - Sustain. Growth Use*, no. December, 2013.

[7] C. Y. Ket, "Optimization on Acrylic Acid Plant by Using Aspen Plus," University of Malaysia Pahang, 2013.

[8] G. D. Corporation, "GAMS – Documentation," no. October, 2018.

[9] Q. Wu, T. C. Qiang, G. Zeng, H. Zhang, Y. Huang, and Y. Wang, "Sustainable and renewable energy from biomass wastes in palm oil industry: A case study in Malaysia," *Int. J. Hydrogen Energy*, vol. 42, no. 37, pp. 23871–23877, 2017.

[10] B. Bevan, A. Ahmad, A. Johari, T. Amran, and T. Abdullah, "Torrefaction of Pelletized Oil Palm Empty Fruit Bunches," *21st Int. Symp. Alcohol Fuels – 21st ISAF*, pp. 15–19, 2015.

[11] H. Ausen, "A Study in MINLP-class Optimization Problems for Simulated Petroleum Production," no. June, 2012.

[12] W. L. Masterton, "Rate of Reaction," in *Chemistry: Principles and Reactions*, 8th ed., Belmont, CA Brooks Cole, Cengage Learning 2009, 2015, p. 274.

[13] H. Biteau, A. Fuentes, G. Marlair, and J. L. Torero, "The influence of oxygen concentration on the combustion of a fuel/oxidizer mixture," *Exp. Therm. Fluid Sci.*, vol. 34, no. 3, pp. 282–289, 2010.

Acknowledgements

The authors would like to acknowledge Universiti Malaysia Pahang (UMP) under the research grant of RDU 1703170 and the Faculty of Chemical and Natural Resources Engineering for the financial support and opportunity in conducting this research.

IOP Conf. Series: Earth and Environmental Science **257** (2019) 012028 doi:10.1088/1755-1315/257/1/012028

Evaluating the environmental and economic sustainability of energy efficiency measures in buildings

Yemane W. Weldu[1] and Sami G. Al-Ghamdi[1,*]

1 Division of Sustainable Development, College of Science and Engineering, Hamad Bin Khalifa University, Qatar Foundation, Doha, Qatar
* Correspondence: Sami G. Al-Ghamdi
Email: salghamdi@hbku.edu.qa

Abstract. It is estimated that buildings in extreme weather regions waste large amounts of money each year in thermal comfort costs, and that if addressed, significant amounts of environmental impacts would be averted. Retrofitting is considered the most immediate, pressing, and effective mechanism for reducing energy consumption and greenhouse gas emissions in the building sector. An environmental life cycle costing was applied to evaluate the sustainability of retrofitting existing buildings. A portfolio of environmental and economic impact results was developed for a baseline building and five alternative retrofitted building scenarios based on insulating the building wall. The results demonstrated a reduction in the environmental impacts and economic impacts by retrofitting the building. Improving the thermal performance of the building resulted in a reduction of environmental life cycle cost of 5%. Externality, which is primarily caused by energy usage, was the key source of the environmental life cycle cost. People in hot climate zones spend much of their time indoors, therefore, it is critical to keep comfortable thermal environment in buildings. This study provides strategically important evaluations for transforming existing buildings into more cost-effective and environmentally friendly sustainable buildings.

1. Introduction

It is known that buildings waste a large amount of money each year in energy costs, and that if addressed significant amount of greenhouse gases (GHG) can be reduced. Massive energy efficiency measure in the building sector represents a key factor for reducing GHG emissions [1,2]. Retrofitting is considered as the pressing mechanism to reduce energy consumption and carbon emissions in the building construction sector. Thermal insulation is a proven green strategy for energy reduction through improvement in the thermal transmittance (U-value) of building envelope [3,4]. The building envelope regulates the heat transfer between indoor and outdoor, and reduces the energy demand of buildings [5]. Thermal insulation materials influence both the environmental impacts of building construction and the use phase of the building [6–8]. The thermo-physical properties of the insulation material determines the thermal behavior of the building wall [1]. Enhancing the building envelope can reduce energy costs and improve thermal comfort for building occupants [5,9]. Study demonstrated that the environmental impact mainly results due to the use of the product life cycle [5]. To this effect, the operating phase significantly determines the economics of a building due to the associated operational expenditure [1,10].

Addressing the energy reducing solutions is very important; however, it is not the only parameter that should be accounted when designing a building envelope [1]. It is important to assess the economic sustainability of a building envelope design. Most life cycle assessment studies of insulation materials

Content from this work may be used under the terms of the Creative Commons Attribution 3.0 licence. Any further distribution of this work must maintain attribution to the author(s) and the title of the work, journal citation and DOI.
Published under licence by IOP Publishing Ltd

examined the integrated energy and environmental impacts of insulation materials. It is critical that further studies complete these results with the economic implications of retrofitting buildings [11]. Very few studies have investigated the integrated environmental and energy performance of insulation materials [5]. However, the economic sustainability dimension of building design has been limitedly integrated to the life cycle environmental assessment.

The building sector represents a great potential for sustainability intervention due to its large amount of energy consumption and associated environmental impacts. Current energy efficiency policies and Green Building rating systems encourage retrofitting over construction of new buildings [12]. Environmental Life Cycle Costing (ELCC) results can inform designers and construction developers with the cost-effective ways of retrofitting existing buildings when designing a building. This research determines the economic sustainability of retrofitting existing buildings, examines the integrated environmental and cost effectiveness of differing insulating materials for a specified u-value, and identifies the hot spot to achieving economic sustainable buildings.

2. Methodology

2.1 Environmental modelling

A life cycle assessment (LCA) method is applied to analyze the environmental impacts of alternative thermal insulation materials for improved energy efficiency. The studied products were modelled using GaBi software [13]. This study focuses on retrofitting the building wall by applying thermal insulation materials to reduce energy consumption. Hotels are the most sensitive environments with high energy consumption; therefore, a hypothetical nine storey hotel building in the education city of Qatar was modeled as case to analyze the sustainability of retrofitting existing building external walls [3]. Using a functional unit of 1m2K/W thermal resistance, this study compares the environmental impact of the most common insulation materials, namely stone wool (SW), glass wool (GW), expanded polystyrene (EPS), extruded polystyrene (XPS) and polyurethane (PU) that represent 90% of the global market with a baseline building with no insulation [11]. It is assumed that the opaque part of the retrofitted building wall will have the same U value and R-value, yet different thickness and mass value across scenarios.

External Thermal Insulation Composite System (ETICS) is the most recommended façade system for warmer climate zones like Qatar [11]. This study considered ETICS for retrofitting existing building wall. ETICS are a build-up system for envelopes composed of multiple layers, including the insulation material, wall, adhesive, and reinforcing coating. All of the life cycle activities from the production, transportation to the building site, the installation in the building, and the use and maintenance phase, to the end-of life of the insulation materials were included from a cradle to grave approach (Figure 1).

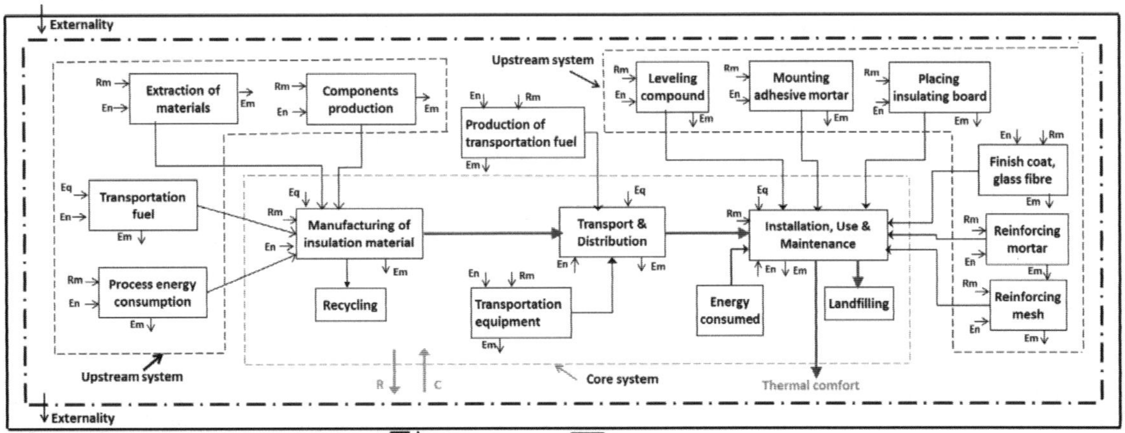

Figure 1. System boundary of building retrofitting.

Note: R: Revenue; C: Cost; En: Energy; Rm: Raw material; Em: Emission; Eq: Equipment

Environmental life cycle data were compiled for the baseline building scenario and for the retrofitted building wall scenarios based on the proposed improved U-value specifications. Energy modeling was carried out using Revit Autodesk software [14]. The amount of insulation material was quantified by relating the functional unit to the heat transferred through its surface area (A) [11]. The thermal resistance R represents the heat transfer through an insulation material. R is related to the thermal conductivity (λ) of a material and its thickness (t). Thus, R corresponds to the density (ρ) and weight of the insulation material. The life cycle reference flow was quantified per functional unit of the same R-value for all insulation materials for building refurbishment. When the U-value of the baseline building is improved from 1.532 W/m2K (i.e., R=0.65 m2K/W) to 0.34 W/m2K (i.e., R=2.94 m2K/W), the R-value for the insulating layer will be 2.29 m2K/W. As such, the material required to achieve a thermal resistance of 2.29 m2 K/W for a 1m2 panel was quantified.

2.2 Economic modelling

The environmental modeling was followed by a life cycle costing method to analyze the economic sustainability of the case study. A LCC model analyzes all costs associated with the building scenario. A computer-based Excel spreadsheet programs have been developed to quantify the life cycle cost of alternative investments. The LCC modeling was specified with the same system boundary as for the LCA to value the economic sustainability per functional unit of 1m2K/W thermal resistance (Figure 1). In this study, the LCC was analyzed based on a study period of 50 years of economic life span through 2018 to 2068. The cost per functional unit of 1m2K/W thermal resistance was quantified by linking the input/output flows from the LCA model to the LCC model. The LCC for building scenarios was quantified using the present-value method. Costs can be incurred before, during, or at the end of the service period. Future costs that arise during the system's service period, including salvage values were discounted from the end of the year in which they occur to the base date [15]. The constant dollar method at 0.01% real discount rate was considered to estimate future cash flows. The life cycle cost factors for the building scenario can be categorized into investment, operating, and externality costs. Internalized costs of GHG emissions, which will be internalized in the near future, were included as externality cost.

3. Results and discussion

3.1. Environmental and economic comparison of improving the U-value

The life cycle environmental impacts per 1 m2K/W functional unit of the various building wall systems were compared using LCA. Retrofitting the building wall reduced the impact for all environmental categories when compared to the base building. On the other hand, the net present values for the manufacturing, transport, installation, use & maintenance, and end-of-life LC stages were summed up to determine the economic impact. Improving the thermal performance of the building resulted in reduced economic impact.

3.2. Integrated environmental and economic assessment

A portfolio of the integrated LCC and LCA performances per FU were presented on a two dimensional axis to identify the sustainability balance of retrofitting an existing building (Figure 2). The results demonstrated a reduction in the environmental impacts and economic impacts of retrofitting a building. It is more cost effective and environmentally friendly when the thermal performance of a building wall is improved from a life cycle perspective.

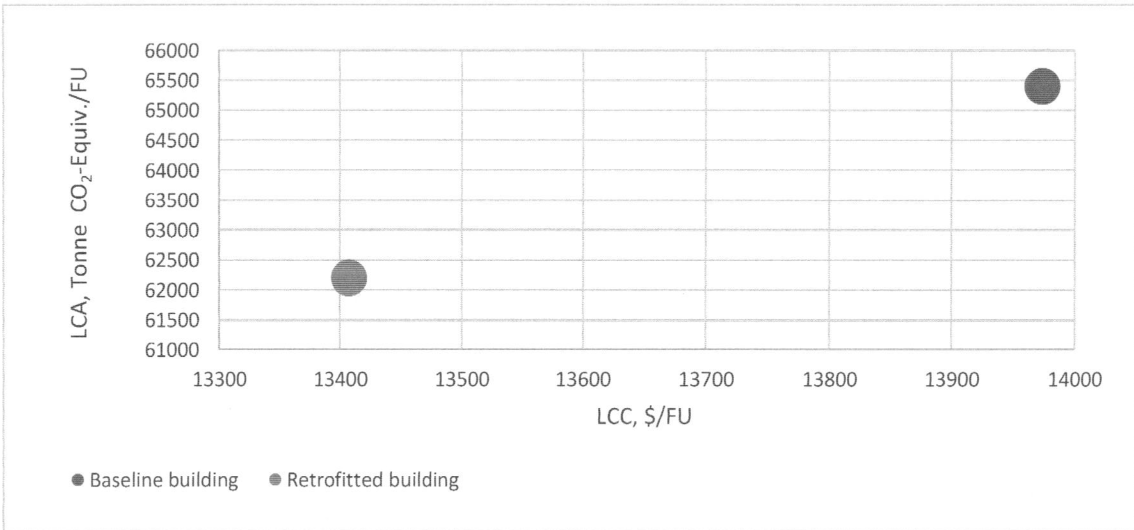

Figure 2. Integrated LCA and LCC impacts per FU of thermal comfort ($: dollar)

4. Conclusion

Retrofitting is considered as the most immediate, pressing and effective mechanism to reduce energy consumption and carbon emissions in the building sector. This research applied an Environmental life cycle costing (ELCC) approach to evaluate the sustainability of retrofitting existing buildings. A portfolio of environmental life cycle assessment (LCA) and economic impact results were presented for a baseline building and five alternative retrofitted building scenarios based on insulating the building wall. Retrofitting the building wall reduced the impact for all environmental categories, whereas the life cycle impacts among the different insulation systems was insignificant overall. The reduction in global warming potential represented much higher when compared to the rest of environmental impacts. Improving the thermal performance in the external walls should be a consideration for climate change conscious building stakeholders. Improving the thermal performance of the building resulted in reduced environmental life cycle cost by 5%. Thus, economic sustainable buildings can be achieved by retrofitting existing buildings from a life cycle perspective. The results demonstrated a reduction in the environmental impacts and economic impacts of retrofitting a building.

All stakeholders including policy-makers, developers, and architects should account for energy efficiency measures to achieve a return on investment and to address environmental concerns. Externality, which is primarily caused by energy usage, was the major contributor to the total environmental life cycle cost. As such, insulating building walls will become more cost-effective as energy costs increase in the future from the current low prices in the state of Qatar. All building scenarios achieved their least economic impact when the C price was decreased. Decreasing and increasing the build-up cost had little effect on the total outcome in absolute values. Therefore, local standards and policy should target the processes that cause massive energy consumption in order to address environmental issues. The ELCC methodology has improved the significance of the LCA methodology for decision making by integrating economic sustainability impacts and by demonstrating the benefits of thermal improvement measures in the building construction sector. The results found in this research are valid and consistent with previous results for environmental impacts [2,3], energy consumption [1,8,10], and economic impacts [5,9].

Author Contributions: Y.W.W. designed the research, collected the data, modelled the LCA & LCC, analyzed and interpreted the results, and wrote the manuscript. S.A.G modelled the energy use, contributed to the ELCC analysis, reviewed the manuscript, and helped with manuscript improvement and data interpretation.
Funding: This research received no external funding.
Acknowledgement: The authors would like to thank Qatar Foundation.
Conflicts of Interest: The authors declare that they have no competing interests.

5. References

[1] S. Schiavoni, F. D'Alessandro, F. Bianchi, F. Asdrubali, Insulation materials for the building sector: A review and comparative analysis, Renew. Sustain. Energy Rev. (2016). doi:10.1016/j.rser.2016.05.045.
[2] G.M. Lo Giudice, F. Asdrubali, A. Rotili, Influence of new fac tors on global energy prospects in the medium term: Compar ison among the 2010, 2011 and 2012 editions of the iea's world energy outlook reports, Econ. Policy Energy Environ. (2013).
[3] H.M. Taleb, Enhancing the skin performance of hospital buildings in the UAE, J. Build. Eng. (2016). doi:10.1016/j.jobe.2016.07.006.
[4] J. Coma, G. Pérez, C. Solé, A. Castell, L.F. Cabeza, Thermal assessment of extensive green roofs as passive tool for energy savings in buildings, Renew. Energy. (2016). doi:10.1016/j.renene.2015.07.074.
[5] R. Azari, Integrated energy and environmental life cycle assessment of office building envelopes, Energy Build. (2014). doi:10.1016/j.enbuild.2014.06.041.
[6] B.P. Jelle, Traditional, state-of-the-art and future thermal building insulation materials and solutions - Properties, requirements and possibilities, Energy Build. (2011). doi:10.1016/j.enbuild.2011.05.015.
[7] N. Pargana, M.D.M.D. Pinheiro, J.D.J.D. Silvestre, J. de Brito, Comparative environmental life cycle assessment of thermal insulation materials of buildings, Energy Build. (2014). doi:10.1016/j.enbuild.2014.05.057.
[8] L.F. Cabeza, A. Castell, M. Medrano, I. Martorell, G. Pérez, I. Fernández, Experimental study on the performance of insulation materials in Mediterranean construction, Energy Build. (2010). doi:10.1016/j.enbuild.2009.10.033.
[9] R. Dylewski, J. Adamczyk, Economic and ecological indicators for thermal insulating building investments, Energy Build. (2012). doi:10.1016/j.enbuild.2012.07.021.
[10] F. Asdrubali, F. D'Alessandro, G. Baldinelli, F. Bianchi, Evaluating in situ thermal transmittance of green buildings masonries: A case study, Case Stud. Constr. Mater. (2014). doi:10.1016/j.cscm.2014.04.004.

[11] J. Sierra-Pérez, J. Boschmonart-Rives, X. Gabarrell, Environmental assessment of façade-building systems and thermal insulation materials for different climatic conditions, J. Clean. Prod. (2016). doi:10.1016/j.jclepro.2015.11.090.
[12] F. Asdrubali, G. Baldinelli, F. Bianchi, S. Sambuco, A comparison between environmental sustainability rating systems LEED and ITACA for residential buildings, Build. Environ. (2015). doi:10.1016/j.buildenv.2015.01.001.
[13] thinkstep, GaBi Database & Modelling Principles, 2016.
[14] Autodesk Inc., Revit Server, Www.autodesk.co.uk. (2016).
[15] Y.W. Weldu, G. Assefa, The search for most cost-effective way of achieving environmental sustainability status in electricity generation: Environmental life cycle cost analysis of energy scenarios, J. Clean. Prod. (2017). doi:10.1016/j.jclepro.2016.11.047.

Feasibility analysis on building air-condition system renovation: Thailand case study

Bancha Sreewirote[1], Atthapol Ngaopitakkul[2]

1 Department of Electrical Engineering, Faculty of Engineering, Thonburi University, Bangkok, Thailand
2 Faculty of Engineering, King Mongkut's Institute of Technology Ladkrabang, Bangkok, Thailand
E-mail: Atthapol.ng@kmitl.ac.th

Abstract. Energy efficiency improvement in existing buildings has been an interesting issue for researcher especially in heating, ventilation, and air conditioning (HVAC) system that appear to take up the large proportion of energy consumption. So, this paper aims to present an analysis on renovation measure to improve energy efficiency in building air-condition system. The study has been done using multi-purpose building in Thailand as a case study. The analysis of proposed renovation measurement will be done in both energy performance and economic perspective. Standard from Thailand building energy code is also taken into consideration when selecting equipment for replacement. The obtained result has indicated the suitability of proposed measure application in building with significant energy efficiency improvement and attractive economic parameter.

1. Introduction

Energy consumption in Thailand's urban area has been rapidly increasing due to significant growth in economic and population that lead to attain high-rise building with significant energy demand. The commercial and residential sections, that most buildings are based on these two categories, have taken up a significant proportion of overall energy consumption up to 22% according to 2016 Thailand energy statistic [1]. Thus, the government has set 20-Year Energy Efficiency Development that aims to reduce energy intensity (EI) by 20% (30,000 ktoe) from business-as-usual (BAU) case in 2030 with 17.6% share of energy saving from building in residential and commercial sectors [2]. One of the steps that government is implementing to achieve the target is elected "Building Energy Code (BEC)" as a law in 1992 for both newly constructed and renovated existing building with total area over 2,000 m^2 to be complied with this code prior to submission for approval [3]. This standard regulated building energy performance in six main perspectives that have significant impact on overall energy consumption; building envelope, air-conditioning, lighting, hot water, renewable energy and whole building energy usage.

In Thailand, heating, ventilation, and air conditioning (HVAC) system especially air-condition has taken up to 60% of overall energy consumption in building due to Thailand location in tropical climate zone. This results in average hot and humidity weather that requires a lot of cooling capacity in order to achieve comfort zone for building occupant [4]. So, air-condition system will be main focus for researcher to improve energy efficiency by various measures, methodologies and technologies.

Literature review in field of energy efficiency improvement measure, impact and case studies especially in air-condition system has been done in this research [5]-[9]. The energy efficiency

Content from this work may be used under the terms of the Creative Commons Attribution 3.0 licence. Any further distribution of this work must maintain attribution to the author(s) and the title of the work, journal citation and DOI.
Published under licence by IOP Publishing Ltd

technologies and measures for non-residential building in European Union have been presented [5]. Research by Jin has presented the model and simulation to evaluate the air-condition usage pattern based on occupancy behaviour [6]. The field study on air-condition in office building in China has been presented [7]. The life cycle and economic analysis on air condition has been studied to evaluate impact on environment and economic perspective [8]. The energy saving measure in air-condition in case of Thailand has been also presented [9]. However, the measure presented in this paper only focuses on large size air-condition system.

From the literature review, it can be seen that HVAC system, especially air-conditioning system in case of building located in tropical climate such as Thailand, has taken up significant percentage of overall energy consumption in building, thus research on measure to improve energy efficiency in air-condition system with feasibility for implementing those measures in terms of economic perspective must be evaluated. Hence, this paper presents an analysis on air-condition renovation measure using multi-purposed building located in Bangkok, Thailand as a case study. The performance analysis on proposed measure has been done in both energy performance and economic perspective. The result and methodology from case study can be applied to other projects to improve energy efficiency in both existing and newly constructed building in the future.

2. Case study building

The case study in this research is 12-floor multi-purposed building at Faculty of Engineering, King's Mongkut's Institute of Technology Ladkrabang (KMITL) located in Bangkok, Thailand. The overall exterior of the simulated building is illustrated in Fig. 1. The floor plan of the building, location of installed outdoor and indoor air-condition unit in the building are depicted in Fig. 2. The figure shows an example of 9th floor layout. However, the building has similar floor layout throughout 1st to 12th floors with minor adjustment on room section on each floor, thus the area of room with air condition is very similar on each floor. The estimated total floor area of the building is 25,000 m^2 and floor area of the room with air-conditioning installed is 13,900 m^2.

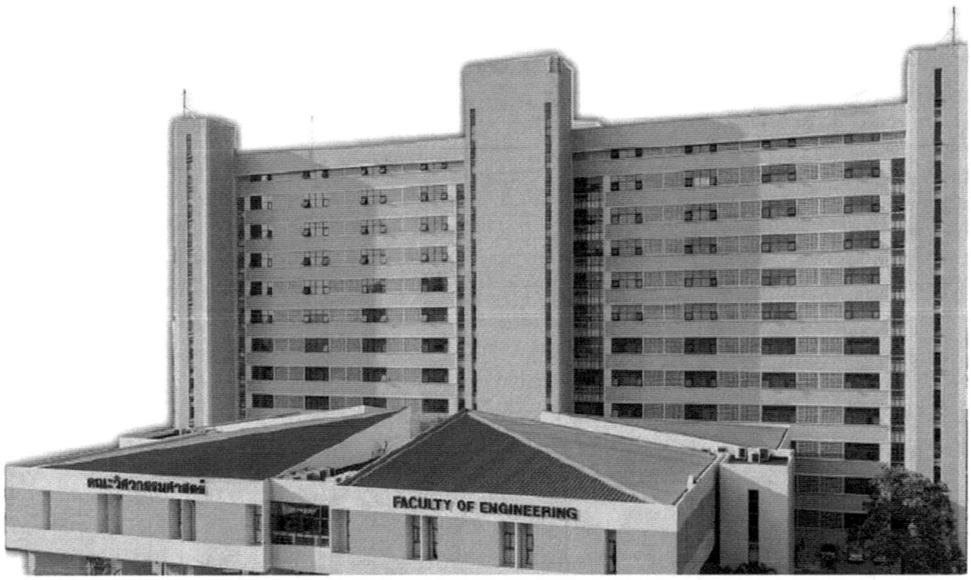

Figure 1. Exterior of the building in case study.

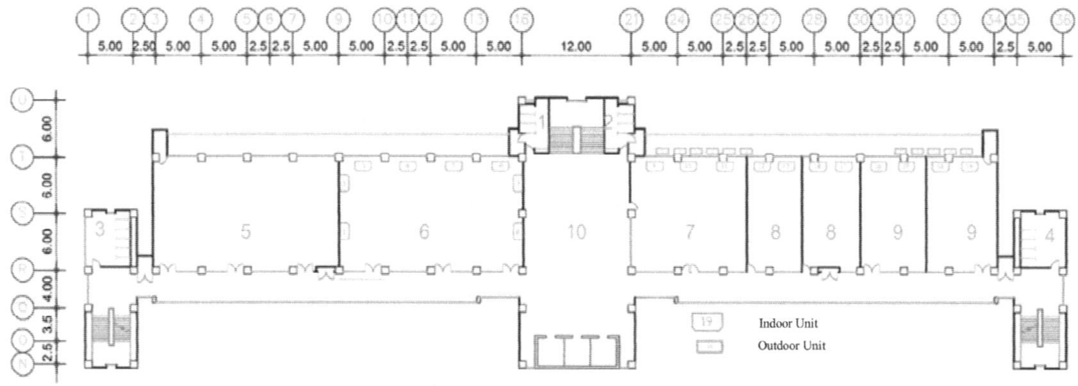

Figure 2. Floor plan of case study building (9th Floor).

According to BEC standard, the code only used as Coefficient of performance (COP) or Energy efficiency ratio (EER) as parameter to evaluate energy performance of split types air-condition, the standard has categorized air-condition into three categories according to its sizing as presented in Table. 1 [3]. However, standard for spilt types air-condition only covers for sizing up to 17.6 kW. For sizing larger than 17.6 kW, COP and EER value will be using 2.56 and 8.74 respectively. This value has been used as a reference for choosing air-condition size and types that will be installed in the building.

Table 1. Energy performance requirements for split types air-condition

Type and Size	COP (EER)
Less than 3,500 watts (0.995 RFT)	2.82 (9.62)
Between 3,500 watts and 17,600 watts (5.00 RFT)	2.82 (9.62)
Greater than 17,600 watts (5.00 RFT)	2.56 (8.74)

3. Analysis and result

3.1. Energy Performance

Energy performance of currently used air-condition has been measured from actual equipment in case study building. These obtained data will then be used to calculate EER value. The EER value of air-condition can be calculated from COP value using power consumption and cooling capacity as presented in equation (1) and (2).

$$COP = \frac{Q_L}{E} \tag{1}$$

$$EER = 3.412 \times COP \tag{2}$$

When

Q_L is Cooling Capacity

E is Energy Consumption

Air-condition system currently used in case study building is split types air-condition. The total number of currently installed air-condition in the building is 291 units with 20,000 and 30,000 BTU sizing capacity. The specification of currently used and replacement air-condition in terms of EER value, energy and power consumption is shown in Table. 2 and 3 respectively.

Table 2. Specification of current equipment in air-condition system.

Size (BTU)	Installation year	Power (kW)	Energy (kWh/year)	Energy Efficiency Ratio (EER)
20,000	1994	3.53	6,77.6	5.07
	2008	1.99	2,865.6	8.53
30,000	1994	4.44	8,524.8	6.7
	2008	3.53	6,624.0	9.77

Table 3. Specification of newly renovate equipment in air-condition system.

Size (BTU)	Power (kW)	Energy (kWh/year)	Energy Efficiency Ratio (EER)
20,000	1.3	3,120	15.4
30,000	1.94	4,656	15.25

According to the Table, air-condition replacement has significant improvement in terms of EER value and requires less energy consumption at the same BTU size. This is because currently used air-conditions in the building are aging with more than 20 years of installation and the efficiency value is lower than standard limit for the education types building according to BEC.

The energy performance of air-condition replacement measure can be summarized in Table 4. From the table, it can be seen that power consumption in case of air-condition replacement can be reduced up to 317 kW and results in saving of 647 kWh per year.

Table 4. Energy performance of air-condition renovation measure.

Number of Air-Condition (unit)	291.00
Current Power consumption (kW)	920.25
Replacement Power consumption (kW)	647.95
Power Reduction (kW)	317.43
Annual Energy Saving (kWh/yrs.)	647.95

3.2. Economic Analysis

The economic analysis for air-conditioning renovation in this research will be done using two important parameters that can assert the feasibility of the project; discounted payback period and internal rate of return (IRR). Discounted Payback can be calculated using equation (3) and (4).

$$\text{Discounted Cash Inflow} = \frac{\text{Actual Cash Inflow}}{(1+i)^n} \tag{3}$$

$$\text{Discounted Payback Period} = A + \frac{B}{C} \tag{4}$$

When
i is Discounted rate
n is Period of cash flow (years)
A is Negative discounted cumulative cash flow
B is Discounted cumulative cash flow at the end of the period A
C is Discounted cash flow during the period after A

For the IRR value, it can be calculated using equation (5)

$$I_0 = \sum_{t=1}^{n} \frac{ES_t}{(1+IRR)^t} = 0 \tag{5}$$

When
I_0 is Initial investment
n is Project life (years)
ES_t is Annual energy cost saving from year 1 to n
IRR is Internal rate of return

Initial cost of the project including equipment and installation cost for all air-condition replacement in building is approximately 606,748 USD.
The calculation of energy consumption is based on current electric Thailand electric price at 3.98 THB/kWh or 0.121 USD/kWh (1 USD = 32.75 THB) and inflation rate at 2% per year according to Bank of Thailand inflation forecast. The result from economic analysis is shown in Table 5.
From the table, it shows that investment in building air-condition system replacement measure can achieve discounted payback period within 7.43 years with IRR value 14.06%.

Table 5. Payback period for air-condition renovation measure.

Year	Installation Cost	Energy Cost Reduction	Maintainance Cost	Net Present Value	Balance
0	606,748.40	0.00	0.00	-606,748.40	-606,748.40
1	-	92,126.91	4,442.75	85,964.86	-520,783.53
2	-	91,214.77	4,442.75	83,402.51	-437,380.97
3	-	90,311.65	4,442.75	80,916.14	-356,464.79
4	-	89,417.48	4,442.75	78,503.29	-277,961.28
5	-	88,532.15	4,442.75	76,162.34	-201,798.91
6	-	87,655.60	4,442.75	73,890.62	-127,908.28
7	-	86,787.72	4,442.75	71,686.28	-56,222.08
8	-	85,928.44	4,442.75	69,547.25	13,325.22
9	-	85,077.66	4,442.75	67,471.68	80,796.90

4. Conclusion

This paper proposed energy and economic analysis on air-condition renovated measure using building in Thailand as case study. The replacement of air-condition has been selected using COR and EER value within BEC standard. In terms of energy performance, air-condition replacement can reduce annually energy consumption by 647.95 kWh.

In terms of economic perspective, the renovation measure proposed in this study can achieve discounted payback period around 7.43 years with 14.06% IRR. This value is attractive for investor to invest into renovation project.

It can be seen that air-condition system renovation in existing building can achieve long-term benefit in perspective of improving overall energy efficiency and reducing annual energy cost with attractive economic parameter. The renovation methodology and result proposed in this research can be applied to other renovating projects to evaluate the feasibility and the renovation in low efficiency system.

5. References

[1] Energy Policy and Planning office (EPPO) 2016 *Energy Statistics of Thailand 2016* Energy Policy and Planning Office, Ministry of Energy of Thailand

[2] Ministry of Energy 2011 *Thailand 20-Year Energy Efficiency Development Plan (2011-2030)* Energy Policy and Planning Office, Ministry of Energy of Thailand

[3] *The Energy Conservation Promotion Act B.E. 2535,* 1992

[4] Department of Alternative Energy Development and Efficiency (DEDE) 2012 *Research and Development in the field of Energy Conservation and Renewable Energy in Thailand* Department of Alternative Energy Development and Efficiency, Ministry of Energy

[5] D. D'Agostino, B. Cuniberti, and P. Bertoldi 2017 Energy consumption and efficiency technology measures in European non-residential buildings *Energy and Buildings* vol 153 pp 72-86

[6] J. Yao 2018 Modelling and simulating occupant behaviour on air conditioning in residential buildings *Energy and Buildings* vol 175 pp 1-10

[7] T. Wu, B. Cao and Y.n Zhu 2018 A field study on thermal comfort and air-conditioning energy use in an office building in Guangzhou *Energy and Buildings* vol 168 pp 428-437

[8] K. Alutairi, G. Thoma, J. Burek, S. Algarni, and D. Nutter 2015 Life cycle assessment and economic analysis of residential air conditioning in Saudi Arabia *Energy and Buildings* vol 102 pp 370-379

[9] W. Thanuanram, N. Auppapong, P. Nupteotrong and D. Buayorm, 2015 The Energy Saving in Air Condition System of Thailand Building and Factories *Energy Procedia* vol 79 pp 859-864

Acknowledgements

The authors wish to gratefully acknowledge financial support for this research from Faculty of Engineering, King Mongkut's Institute of Technology Ladkrabang Research fund (No. 2559-01-01023), Thailand.

IOP Conf. Series: Earth and Environmental Science **257** (2019) 012030 doi:10.1088/1755-1315/257/1/012030

Reducing energy usage in multi-family housing

A Ameen* and M Cehlin

Department of Building, Energy and Environmental Engineering, University of Gävle, Kungsbäcksvägen 47, 801 76 Gävle, Sweden
E-mail: arman.ameen@hig.se

Abstract. The energy usage in residential sector have been around 22% of the total energy use in the world and increasing due to the population growth and higher living standards. The energy sources for this are made up primarily of non-renewable energy resources which generates a large amount of global greenhouse gases. A lot of countries have implemented various regulations and rules to reduce the energy usage in buildings and promoting the use of renewable energy technologies. This paper presents a parametric study of a typical multi-family building in its pre-design stage. The climate location used is Sweden (Gothenburg) and Japan (Osaka). The aim of the study is to compare various configurations and to examine how they affect the energy use. The most interesting configurations are the use of heat pump and solar cells. Other configurations that are examined are infiltration levels, pressure coefficients, wind impact, ventilation with heat recovery, ventilation scheduling, building orientation and finally changing U-values in the building material. Results of this study show that the energy saving, by utilizing a heat pump and solar panels, can reduce the total energy use by 34.9% for Gothenburg and 32% for Osaka. The results also show that the difference in total energy use between the two cities reduce substantially (3% difference) when utilizing a heat pump in combination with solar panels.

1. Introduction

The energy usage in residential buildings have been around 22% of the total energy use in the world for 2016 and increasing due to the population growth and higher living standards. The energy sources for this are made up primarily of non-renewable energy resources which generates a large amount of global greenhouse gases. Although there have been an increase in several renewable energy sources such as wind and solar power, they are only covering a small portion of the world energy demand which is dominated by fossil fuels [1]. A recent report from the UN Intergovernmental Panel on Climate Change (IPCC) indicates that the carbon pollution have to be cut by 45% by 2030 compared to previous targets for limiting global warming to 1.5°C [2]. The European Union has already implemented regulations and rules to promote the use of renewable energy in buildings and this will further intensify if more reports come out that shows that we have less time to stop and reverse the climate change [3]. The rapid increase in cutting carbon pollution requires change in thinking when designing new buildings.

One of the tools that are used to evaluate the energy usage in buildings is building performance simulation (BPS) software. With BPS software one can predict and evaluate the various aspects of a buildings. BPS software are mainly used in the pre-design stages of a building or when carrying out a retrofit for an existing building [4-6]. Another area they are used in are for case studies such as this paper. There have been numerous studies carried out with the help of BPS software [7-10]. The primary use of BPS software in those studies have been to evaluate energy demand, heating load and thermal comfort. The energy usage in a building is dependent on many factors that interact which each

Content from this work may be used under the terms of the Creative Commons Attribution 3.0 licence. Any further distribution of this work must maintain attribution to the author(s) and the title of the work, journal citation and DOI.
Published under licence by IOP Publishing Ltd

IOP Conf. Series: Earth and Environmental Science **257** (2019) 012030 doi:10.1088/1755-1315/257/1/012030

other [11]. In this study a typical multi-family building (regarding Sweden standards) in its pre-design stage is evaluated based on the following factors:

- Site geographical location, orientation and climate file
- Wind profile
- Infiltration
- Pressure coefficients
- Ventilations system
    - AHU with supply and exhaust (includes heat recovery)
    - Return air only (with no heat recovery)
- Heating delivery system (HDS)
    - Electricity only
    - Ideal heat pump
- Electricity from solar panels
- Building material

The aim of this study is to evaluate what parameter, or a combination of parameters, have the highest influence on energy performance of the building in Gothenburg (Sweden) and Osaka (Japan). The BPS software that is used in this study is Indoor Climate and Energy (IDA ICE) 4.8 which is a dynamic simulation software that supports multiple zones to evaluate energy usage, energy loads and thermal comfort. This can be done for room level or for the entire building [12, 13]. The accuracy of the software has been investigated and shown good correlation against real measurements [14]. Another study made by Travesi J et al. (2001), who examined the accuracy of the software, carried out an empirical validation study related to the thermal performance of buildings and the results showed good agreement between the measured data and the simulated cases [15]. In 2001 IDA ICE was validated according to European Strandard prEN 13791 [14].

2. Method

2.1. Baseline model

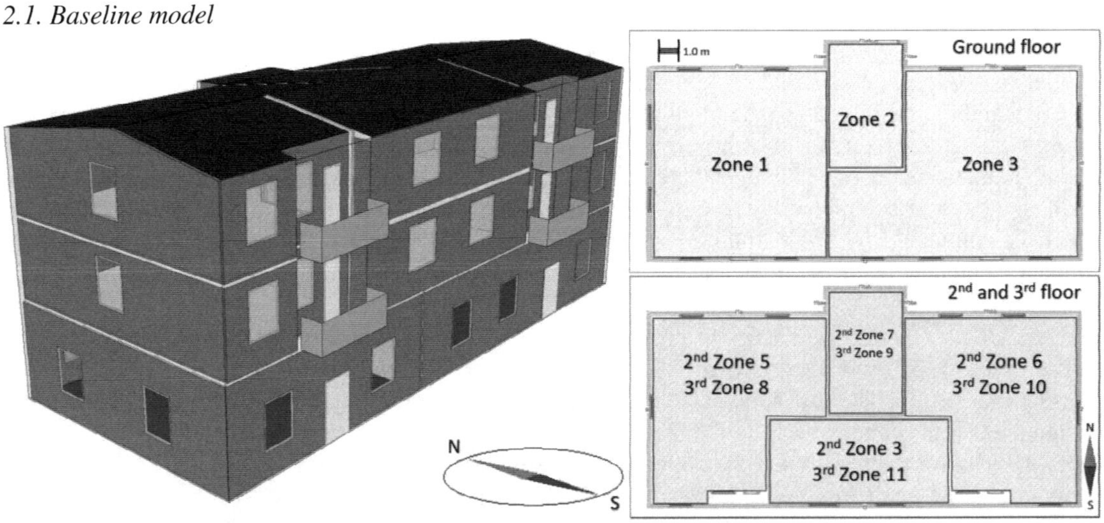

Figure 1. 3D view and floorplan overview of the building.

At start, a baseline model was created based on a building drawing obtained from plans for a future building construction in the city of Gothenburg. Next, parameters were implemented according to Swedish standards for inputs such as lighting, equipment, number of occupants, hot water consumption, temperature setpoints and U-values for building materials [16-18]. A 3D-layout as well as a floor plan of the building is show in Figure 1. The building has three floors in total. The ground floor is planned to host two small businesses, while the 2nd and 3rd floor are used for apartments. The layout for 2nd and 3rd floor is identical. In total, there are 2 areas reserved for small businesses (Zone 1 & 3) and 6 areas are reserved for apartments (Zone 4-6, 8 and 10-11). Zone 2, 7 & 9 are reserved for stairwell. The U-values for the building components are shown in Table 1.

Table 1. U-values of building construction parts.

U-values for building components	U-value [W/(m^{3}K)]
External walls	0.18
External roof	0.13
External floor	0.15
Windows	1.19
External doors	1.09

The climate locations chosen are Sweden (Gothenburg) and Japan (Osaka). These climate files are obtained from ASHRAE [19]. The lighting and equipment for the apartment zones are set to 2000 kWh/year + 800 kWh/occupant/year and for the small business zones it is set to 180.7 kWh/m^2 and year. 70% of this energy is utilized as internal heat gain. The lighting and equipment scheduling are the same as occupancy for all the zones. The number of occupants in the apartments, the clothing and the schedules are set according to Table 2 and the activity level is set to 1.0 MET.

Table 2. Number of occupants, clothing and schedules times

Zone	Clothing (CLO)	Number of occupants	Schedule
Zone 3 & 11 (1 room and kitchen)	0.75 ±0.25	1.42	17-07 Every day
Zone 5, 6, 8, 10 (2 room and kitchen)	0.75 ±0.25	1.63	17-07 Every day
Zone 1 & 3 (local store)	0.85 ±0.25	5	07-18 Every day

Table 3. Input parameters in IDA ICE baseline model

Parameters	Value
Model volume	1 826.1 m^3
Model envelope area	990.9 m^2
Window to envelope ratio	6.9%
Heating set point[1]	21°C
Cooling set point[1]	25°C
AHU (FTX) heat recovery efficiency[1]	70%
AHU (FTX) Specific Fan Power (SFP)[1]	1.8 [kW/(m^3/s)]
Average hot water use[1]	12 534 kWh/year
Airing energy loss[2]	2 155 kWh/year
Infiltration[1]	0.6 ACH @ 50 Pa
Pressure-coefficients[3]	Exposed
Wind profile[3]	Ocean (ASHRAE 1993)

[1] The value is obtained from Swedish building regulation [16]
[2] The value is obtained from SVEBY guidelines [18]
[3] The values is obtained from ASHRAE [19, 20], see also IDA ICE help section for more in depth details

Other important simulation parameters are presented in Table 3. The total floor area and the volume of the building is 214.6 m^2 and 1962.0 m^3. The floor area of the business zones (1 & 3) are 83.4 m^2 and

100.2 m^2. The floor area of the apartment zones varies between 24.2 m^2 and 70.5 m^2. The floor area of the stairwell zones is 24.2 m^2. The size of the windows are 1.2 x 1.5 m and the external doors 1.2 x 2.0 m. The ceiling height is 3.3 m for the ground floor and 2.6 m for the rest. The heating system is based on electric radiators. There are two air handling units (AHU) for the building, one for the apartment zones and the other for the business zones. All the cooling capacities of the AHU is disabled. There are however separate cooling units in the business zones (1 & 3) which have COP 3. Solar radiation level at which integrated shadings are drawn is set to 100 W/m^2. The ventilation flowrate for the business zones is set to 3.0 l/(s/m^2) for supply air and 3.24 l/(s/m^2) for exhaust air. For the apartments it is set to 0.35 l/(s/m^2) for supply air and 0.38 l/(s/m^2) for exhaust air.

2.2. Parametric setup
The parametric setup is carried out by making some changes to the baseline model and save this as a specific model [21]. Table 4 shows all the model configurations and what has been changed compared to the baseline model. After the creation of these models a simulation is carried out for all these configurations to evaluate the energy usage of each model. The simulation type is dynamic, and the period of the simulation is one year (normalizing weather data is used).

Table 4. Model configurations for parametric study in IDA ICE.

Parametric model	Change
Model 1	Infiltration – 0.2 ACH @ 50 Pa
Model 2	Pressure-coefficients – sheltered
Model 3	Wind profile – suburban (ASHRAE 1993)
Model 4	AHU – exhaust air only
Model 5	AHU – change the scheduling for apartment zones
Model 6	Change orientation by +90 °
Model 7	Change the U-value for the building components (higher)
Model 8	Installed solar panels (86 m^2 solar panels (30° tilt))
Model 9	Cooling enabled in all zones
Model 10	Using an ideal heat pump with COP 3
Model 11	Using ground heat pump with COP 3 + 86 m^2 solar panels (30° tilt)

In Model 4 both the AHUs are replaced with exhaust air only. In Model 5 the flowrate to the apartments are decreased by 70% between 07-19 (when it's assumed that the occupants are not at home). In Model 7 the U-values are increased for the external walls (0.24 W/(m^3K)), roof (0.17 W/(m^3K)) and windows (1.55 W/(m^3K)). In Model 8 solar panels are installed on the roof side facing south. The total area of the solar panels is 86 m^2, the tilting 30° and the overall efficiency is 15%. In Model 9 the cooling is enabled which makes it possible to remove any access heat during the hot periods. The COP value for the cooling units are 3. In Model 10 an ideal heat pump with a COP 3 is used for the heating system. In Model 11 the solar panels in Model 8 is combined with the heat pump (COP 3) in Model 10.

3. Result and discussion
This section is divided into four parts. The first part covers the clime difference between the two cities. The second and third part will cover the total energy use of the baseline model and the parametric models in Osaka and Gothenburg and the fourth will cover the difference between the two cities.

3.1. Climate analysis
Since a comparison is made between two different climate region it's important to evaluate the difference in weather condition between the two regions. Figure 2 shows the average temperature for both Osaka and Gothenburg over the simulation period. The average temperature in Osaka is higher

than in Gothenburg which leads to lower heating demand in Osaka. On the other hand, the cooling demand will be higher in Osaka during the warmer periods.

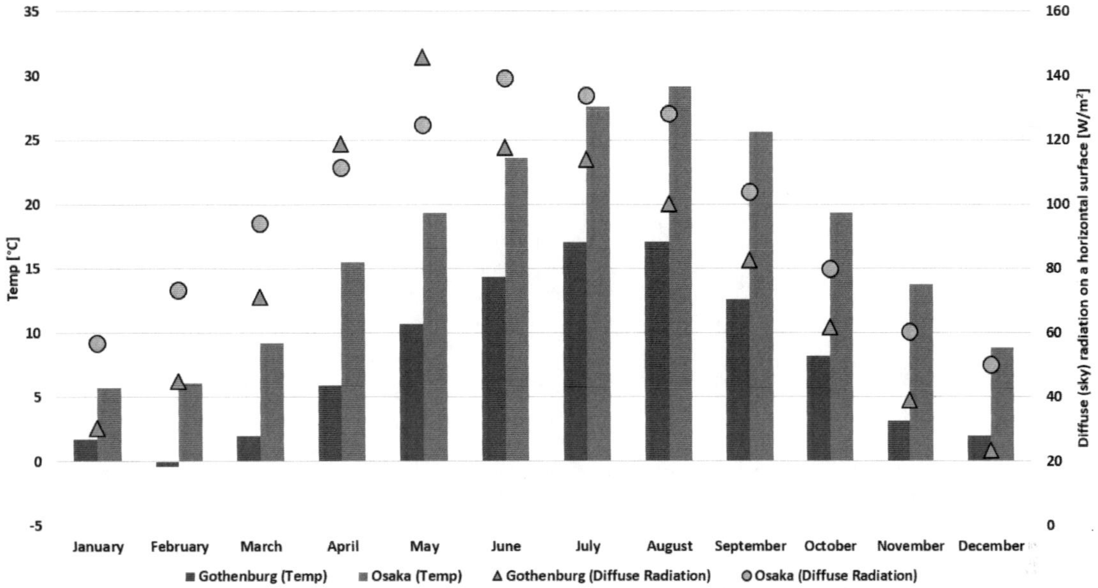

Figure 2. Average monthly temperature and diffuse radiation on a horizontal surface for Gothenburg and Osaka.

The graph also shows that the largest temperature difference between the two locations are during the summer time. From this it's possible to assume that the heating requirement will be higher for Sweden and the cooling requirement will be higher for Japan. Another important aspect is the average diffuse radiation on a horizontal surface which is higher in Osaka during most of the months in the year.

3.2. Total energy use for Osaka location

The total energy use in kWh/m^2 for Osaka is shown in Figure 3. The result shows that Model 1, which was the change in infiltration, has very low impact on the total energy usage in the building. There is a slight reduction in the total energy use, around 0.4%. In Model 2 and 3 a similar pattern is shown; the total energy use is reduced by only 0.4%. In Model 4, which is the change of the AHU to exhaust air only, the total energy usage has increased by 7.6% due to removal of the possibility of heat recovery from the exhaust air. There has been a substantial reduction on the HVAC auxiliary power (47.6%) but that reduction is not enough to offset the increase in heating energy usage. A slight reduction (3.1%) is also noted for the cooling requirement. Model 5 reduces the ventilation requirements for the apartments when the occupants are not present. This generates a reduction of 3.2% in the total energy use. In Model 6 the building is rotated + 90° which means that the windows facing south will face west and the north windows will face east. This change increases the total energy use by 0.7% which is provided primarily by an increase in heating energy use (3.1%). In Model 7 the U-value of several building components are increased which leads to a decrease in the insulation level of the building which in turn leads to an increase in the heating energy use (13.0%) and a small decrease in the cooling. The total energy use is increased by 2.2%. Model 8 introduces solar panels which allows the building to generate its own electricity and a reduction of total energy use by 19.6%. This electricity is distributed throughout the building and covers some part of the electricity requirements of each area. Total electricity production is 16 643 kWh/year. There is also a slight overcapacity (202 kWh/year) during some periods when the electricity can be sold to the grid or stored. In Model 9 cooling units are installed in the apartments. This leads to an increase in the cooling energy use (38.3%) which at the end leads to an increase in the total energy use by 3.5%. In Model 10 an ideal heat pump with COP 3 is replacing the electric heating system which operated at COP 1. This change creates a reduction in heating energy usage by 66.8% resulting in a total reduction of 12.6%. Finally, a combination of Model 10 and Model 8 is created as Model 11. This configuration takes advantage of both the heat pump and the solar panels. This combination reduces the energy usage across almost all energy sections and results in a total reduction of 32.0 % in energy usage and a small electricity overproduction of 455 kWh/year.

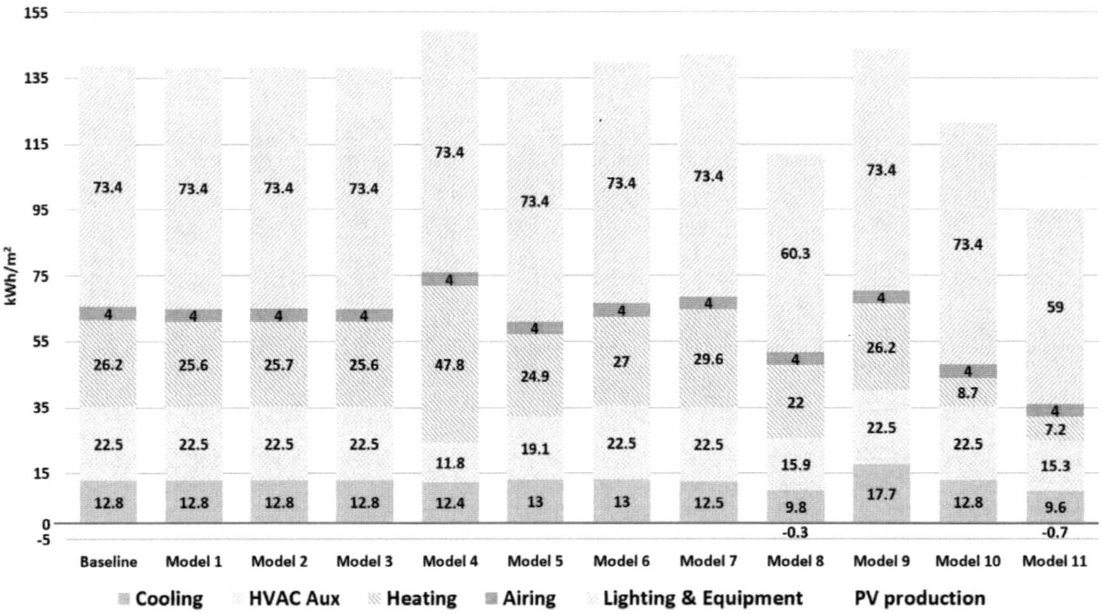

Figure 3. Total energy usage for Osaka baseline and parametric models.

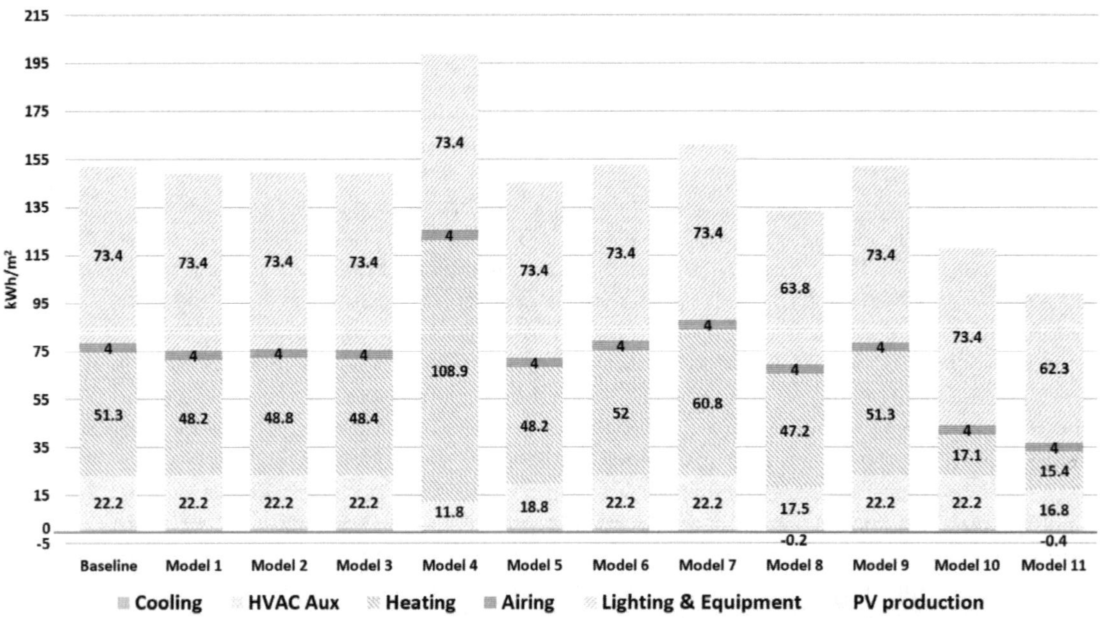

Figure 4. Total energy usage for Gothenburg baseline and parametric models.

3.3. Total energy use for Gothenburg location

The total energy use in kWh/m^2 for Gothenburg is shown in Figure 4. The cooling energy usage is very low across all models, between 0.7 and 1.1 kWh/m^2. Because of these low levels they have not been highlighted in the figure. Similar to the Osaka simulation, Model 1-3 show minor change in the total energy use, a reduction by 1.6 - 2.0%. In Model 4 the heating energy use increases by 112.3% due to the removal of heat exchanger and no supply air. There is also a 46.8% reduction in HVAC auxiliary energy use. In Model 5 there is a small reduction in the energy use for heating and HVAC auxiliary. In Model 6 a slight increase occurs in the heating energy use. Model 7 creates a higher energy use for heating which leads to a total increase of energy use by 6.1%. In Model 8, the solar panels electricity generation reduces the total energy use by 12.4% with a small overproduction of 129 kWh/year. In Model 9 the enabling of cooling units had almost no effect on the total energy use. In Model 10 the ideal heat pump reduces the heating energy use by almost 2/3 and in Model 11 the combination of heat pump and solar panels creates a total energy reduction by 34.9%.

3.4. Comparison between Osaka and Gothenburg

The major variable that effects the simulation between these two cities is the climate. The results show that cooling energy use is substantially higher in Osaka across all models compared to Gothenburg. In contrast the heating energy use in Gothenburg is almost twice as that of Osaka. In terms of total energy use Gothenburg's energy use for Baseline model and Model 1-3, 5-6 is between 7.7 and 9.4% higher than Osaka. Interestingly there is a huge difference for Model 4 where Gothenburg requires 33.1% more energy use, the reason being a large increase of heating energy needed due to the loss of heat recovery potential. In Model 7 the difference is 13.5% between the two cities. The lower insulation values have greater impact in a cold climate such as Sweden compared to Japan. In Model 8 the difference increases to 19.2% due to much higher solar radiation levels in Japan compared to Sweden which is utilized by the solar panels. In Model 9 the total energy use in Gothenburg is 5.7% higher than in Osaka. The reason for this is that the cooling requirements are much higher in Osaka. This will decrease the gap in energy use compared to the Baseline model, but Osaka still have a lower total energy use. Surprisingly, the total energy use in Model 10 is 3% lower in Gothenburg compared to Osaka. This is the only model that showed a lower total energy use for Gothenburg. The reason for this is that the cooling units in all models are already set to COP 3. Since the cooling requirement for Osaka is much higher than in Gothenburg there is no change in that section compared to the Baseline model. On the other hand, the heating energy use is reduced by almost 2/3 for both cities but it is more beneficial to Gothenburg since it has twice the amount of energy use for heating compared to Osaka. Finally, in Model 11 the difference tip in favor for Osaka since the electricity generation of the solar panels are higher there than in Gothenburg. Gothenburg have 4.8 % higher total energy use than Osaka. Another interesting aspect is that Model 10 makes the difference between the cities very small in terms of total energy use.

4. Conclusion

The models that had the lowest impact on the total energy use in both cities are 1-3 and 5. The model that reduced the total energy most was 11 which reduced the energy use for Gothenburg and Osaka by 34.9 and 32% respectively. Once a heat pump for heating is installed, the difference in total energy use between the two cities reduce substantially (-3%) despite the large difference in yearly average temperature and solar radiation. It's also worth mentioning that the heat pump is an ideal one that operates at a constant COP 3 when it is utilized. This also apply for the cooling units.

Nomenclature

AHU	=	Air Handling Unit
BPS	=	Building Performance Simulation
COP	=	Coefficient of Performance
HDS	=	Heating Delivery System
IPCC	=	UN Intergovernmental Panel on Climate Change

5. References

[1] IEA, *World Energy Balances 2018: Overview*. 2018. p. 24.
[2] IPCC, *Global Warming of 1.5°C*. 2018.

[3] EC. *Buildings*. 2018; Available from: https://ec.europa.eu/energy/en/topics/energy-efficiency/buildings.
[4] Foucquier, A., et al., *State of the art in building modelling and energy performances prediction: A review.* Renewable and Sustainable Energy Reviews, 2013. **23**: p. 272-288.
[5] Hensen, J.L., *Towards more effective use of building performance simulation in design.* 2004.
[6] Kalamees, T., *IDA ICE: the simulation tool for making the whole building energy and HAM analysis.* Annex, 2004. **41**: p. 12-14.
[7] Karlsson, F., P. Rohdin, and M.-L. Persson, *Measured and predicted energy demand of a low energy building: important aspects when using building energy simulation.* Building services engineering research and technology, 2007. **28**(3): p. 223-235.
[8] Liu, L., et al., *Comprehensive investigation on energy retrofits in eleven multi-family buildings in Sweden.* Energy and Buildings, 2014. **84**: p. 704-715.
[9] Baranova, D., et al., *Correlation of energy efficiency and thermal comfort depending on the ventilation strategy.* Procedia Engineering, 2017. **205**: p. 503-510.
[10] Luc, K.M., M. Kotol, and T. Lading, *Energy-efficient building in Greenland: investigation of the energy consumption and indoor climate.* Procedia Engineering, 2016. **146**: p. 166-173.
[11] Zhao, H.-x. and F. Magoulès, *A review on the prediction of building energy consumption.* Renewable and Sustainable Energy Reviews, 2012. **16**(6): p. 3586-3592.
[12] Björsell, N., et al. *IDA indoor climate and energy*. in *Proc. of the 6-th IBPSA Conference*. 1999.
[13] Crawley, D.B., et al., *Contrasting the capabilities of building energy performance simulation programs.* Building and environment, 2008. **43**(4): p. 661-673.
[14] Kropf, S. and G. Zweifel, *Validation of the Building Simulation Program IDA-ICE According to CEN 13791 "Thermal Performance of Buildings–Calculation of Internal Temperatures of a Room in Summer Without Mechanical Cooling–General Criteria and Validation Procedures".* Hochschule Technik+ Architektur Luzern. HLK Engineering, 2001.
[15] Travesi, J., et al., *Empirical validation of Iowa energy resource station building energy analysis simulation models.* IEA Task, 2001. **22**.
[16] Boverket. *Konsoliderad version av Boverkets byggregler*. 2018; Available from: https://www.boverket.se/contentassets/a9a584aa0e564c8998d079d752f6b76d/konsoliderad_bbr_2011-6.pdf.
[17] Sveby, *Brukarindata kontor Version 1.1.* 2013, Sveby.org. p. 42.
[18] Sveby, *Brukarindata bostäder Version 1.0.* 2012, Sveby.org. p. 37.
[19] ASHRAE, *2017 ASHRAE Handbook -- Fundamentals (SI).* ASHRAE Handbook, Fundamentals. 2017.
[20] ASHRAE, *ASHRAE Handbook -- Fundamentals. 1993.* American Society of Heating, Refrigerating, and Air Conditioning Engineers, Atlanta, 1993.
[21] Poirazis, H., Å. Blomsterberg, and M. Wall, *Energy simulations for glazed office buildings in Sweden.* Energy and Buildings, 2008. **40**(7): p. 1161-1170.

IOP Conf. Series: Earth and Environmental Science **257** (2019) 012031 doi:10.1088/1755-1315/257/1/012031

Building envelope renovation for net zero energy building in hot humid climate

C Boonyaputthipong[1]

[1] Assistant Professor, Faculty of Architecture, Khon Kaen University, Thailand
E-mail: bchumn@kku.ac.th

Abstract. Net Zero Energy building has become a topic that has drawn much attention from architects worldwide. Its concept is that buildings can meet all of their energy requirements from renewable sources. In Thailand, the renewable energy that is most suitable for integration into the idea of net zero energy building is photovoltiacs. With government funding, the communication affairs division building, Khon Kaen University, is selected for the renovation existing building to be net zero energy building. It is successfully proved that, after one year record, the energy produces by the PVs, 22,944 kWh is more than energy consumption, 20,932 kWh. This paper focuses on envelope renovation because it is an important part of the building that can apply to other renovation project in the future. It found that the envelope renovation for Net Zero Energy Building can be separated to be two part, for energy saving and for renewable energy installation. For photovoltaic installation, it can simply install on the metal sheet roof by using metal frame with special clip. The research and renovation on the envelope from this project can apply for the future net zero energy building in Thailand.

1. Introduction

Net Zero Energy building has become a topic that has drawn much attention from architects worldwide. The idea is to design a building to generate energy for its own use. It is proposed that in Thailand, the renewable energy that is most suitable for integration into the idea of a net zero energy building is photovoltiacs. Best practices suggest that the creation of a net zero energy building should start in the early stages of the design process to allow the building to utilize both energy-efficient building systems and to generate renewable energy. However, in Thailand, there is a number of existing government buildings that have been renovated for a different function other than they were originally designed. To renovate these existing buildings to become net zero buildings will be a more complicated process than to design new net zero buildings. In Thailand's hot-humid climate, it is proposed that the existing buildings' envelopes are the main element that have to be renovated for energy savings. Photovataic installations are proposed to be the most efficient method to achieve net zero energy building status. The feasibility of renovating existing government buildings for net zero energy building in Thailand need to be proven. The existing Communication Affairs Division (CAD) building, located at Khon Kaen University, was selected for the current research project. The data collected from a CAD building survey, conducted as part of this research project, was used to inform the recent CAD building renovation. This paper focuses on the building envelope renovation. For energy savings in Thailand's hot humid climate, the envelope needs to be designed to protect the interior spaces from outside heat, while providing enough day-lighting for the building occupants visual tasks. Building envelope renovations are proposed to be viable interventions that can also be applied to other project in Thailand and elsewhere in the future.

Content from this work may be used under the terms of the Creative Commons Attribution 3.0 licence. Any further distribution of this work must maintain attribution to the author(s) and the title of the work, journal citation and DOI.
Published under licence by IOP Publishing Ltd

IOP Conf. Series: Earth and Environmental Science **257** (2019) 012031 doi:10.1088/1755-1315/257/1/012031

2. Net Zero Energy Building

Ministry of Energy created Thailand's 20 years Energy Efficiency Development Plan (2011-2030) which was aimed at increasing the energy efficiency of government and private buildings beyond the Building Energy Code (BEC) toward Net Zero Energy Building (NZEB) by 2030. The Zero Energy Building concept is the idea that buildings can meet all of their energy requirements from low-cost, locally available, non-polluting, renewable sources [1]. Medium and large scale, cost-effective, zero energy buildings are expected to be viable in the next 15-20 years. However, rapid building technology development and determined building owners and designers have made medium and large-scale buildings successful today [2].

The Communication Affairs Division building, at Khon Kaen University, was selected to be a case study for this research project.(Figure 1-2) This building currently include a cafeteria on the first floor and offices on the second floor. From the building survey conducted as a part of this research project, it was found that the building consumed 98.60 kWh/day or 35,987.70 kWh/year. To renovate this building to become a net zero energy building, the process was to reduce energy consumption in the building by 30 – 40% and then provide renewable energy for the balance of the building's energy needs. The process for reducing the energy used in building included:

- Replaced office equipment
- Replaced air-conditioning
- Installed cooling pads for condensing units
- Renovated the building's envelope and roof
- Replaced the lighting systems
- Installed light pipe
- Installed photovoltiacs on the roof
- Initiating a growing energy awareness for the employees in the workplace

The renewable energy source for this renovation project was photovoltaics installed on the rooftop. The building survey had showed that the energy required for a whole year after reducing energy use in the building was 21,000 kWh/year. So, the 78 panels of 260 watt PVs modules were installed on the building's roof which generated 26,000 kWh/year in electricity.

Figure 1. The communication affairs division building before the renovation.

Figure 2. The communication affairs division building after the renovation.

3. Building Envelope Renovation for Net Zero Energy Building

3.1. For energy saving

The building envelope is a critical element in the energy performance of any building, but it is absolutely vital to the performance of net zero energy buildings [3]. In the hot humid climate of Thailand, the building envelope is the key component to provide thermal comfort for the building. The building envelope works as both the heat protection and the heat release for the building's interior spaces. Proper design of the building envelope can result in a dramatic reduction in energy consumption for cooling. Often, the energy used for air conditioning is the biggest part of the energy consumption for buildings in Thailand.

The transparent quality of the existing building envelope's glass allowed thermal energy (in form of solar radiation) into the building, which impacted energy consumption for cooling. Previously, the building had a glazing area equaling approximately 40% of the total wall area, mostly faced to the south. Even though the building wall was shaded by an asbestos cement sunscreen the heat passing through the glazing area was still high. The renovation set a goal to reduce the glazing area from 146 m2 to 78 m2 and to increase the opaque area from 222 m2 to 290 m2. (Table 1)

Table 1. Comparison of the walls before and after the renovation.

	Before			After		
Orientation	Opaque(m^2)	Glazing(m^2)	Total (m^2)	Opaque(m^2)	Glazing(m^2)	Total (m^2)
NE	78	35	112	99	13	112
SE	56	16	72	59	13	72
SW	47	65	112	89	24	112
NW	41	31	72	44	28	72
Total (m^2)	222	146	368	290	78	368
%	60%	40%	100%	79%	21%	100%

The study of F. S. Westphal, M. A. Yamakawa, L. T. de Castro, 2011 [4], showed that in a hot climate the energy consumption for cooling can be decreased through the application of thermal insulation to the exterior walls. For this building, the three inches fiberglass insulation was used for the opaque walls. By decreasing the glazing area and adding three inches of insulation, the OTTV of the building was reduced from 39.79 W/m2 to 18.95 W/m2. The heat transfer through the interior space was reduced by 52% while the window-to-wall ratio (WWR) of the building was measured as 0.21.

3.2. For photovoltiacs installation

PV installation in a building can occur on a building's wall or roof. For this project, the photovoltaics were designed to be installed on the rooftop. Before the renovation, the roof of the building consisted of asbestos cement on a corrugated hip form. The PV installation in this location required a roof slope of approximately 15 degrees facing to the south. However, the roof had to be tilted along the length of the building making the roof too high to meet the 15 degree slope. Therefore, by design the roof slope has been reduced to 10 degrees facing the south-east side. (Figure 3-4)

Figure 3. The building roof before the renovation

Figure 4. The building roof after the renovation

Sheet metal panels with 4 inches fiberglass insulation were used for the roof material. The 78 photovoltaic panels were installed on the sheet metal roof by using metal clips with steel frames. The 0.60 m. walking space between the panels is for service and maintenance purposes. (Figure 5)

IOP Conf. Series: Earth and Environmental Science **257** (2019) 012031 doi:10.1088/1755-1315/257/1/012031

Figure 5. The photovoltaic installation on the metal sheet roof

4. Conclusion

After the renovation, the CAD building became, officially, the first net zero energy among the government buildings in Thailand. The records for one year have successfully proven that the energy produced by the PVs, 22,944 kWh is more than its energy consumption, 20,932 kWh.

The building envelope renovation for a Net Zero Energy Building can be separated into two parts, for energy savings and for renewable energy installation. For energy savings, the envelope materials have to be reviewed. In Thailand's hot humid climate, the envelope needs to be designed to protect the interior spaces from outside heat, while providing enough day-lighting. The renewable energy installation can be accomplished on the building's roof or walls. For a low-rise building in this climate, it is better to install the photovoltaic on the roof as the roof is the part of the building that receives solar radiation most of the time. Recently, the photovoltaic panel was simply installed on the sheet metal roof by using a metal frame with special clips.

This research project shows that the envelope renovation for net zero energy buildings can integrate the knowledge of architecture and engineering to achieve success. It is suggested that best practices to renovate existing buildings into net zero buildings would include discussion and knowledge sharing beginning during the early stages of design and continue through the construction process.

5. Acknowledgments

The paper is part of the Net Zero Energy Building research funded by the Energy Policy and Planning office (EPPO), Ministry of Energy, Thailand. The research team are Asst.Prof.Denpong Sudpakdee, Research Leader, and Assoc.Prof.Tanakorn Wongwuttanasatian, Co-Researcher.

6. References

[1] P. Torcellini, S. Pless, and M. Deru, 2006, Zero Energy Buildings: A Critical Look at the Definition, ACEEE Summer StudyPacific Grove, California, 14-18 August 2006.

[2] C. Yimprayoon, 2016, Review Article: Zero Energy Building, Journal of Architectural /Planning Research and Studies (JARS) 13(2), Thammasat University, Bangkok, Thailand.

[3] H., Thomas, 2013, Net Zero Energy Design: A Guide for Commercial Architecture, John Wiley & Sons, Inc., Hoboken, New Jersey.

[4] F. S. Westphal, M. A. Yamakawa, L. T. de Castro, 2011, Thermal Insulation of Building Envelope Toward Zero Energy Design in Hot-Humid Climate, Proceedings of Building Simulation 2011: 12th Conference of International Building Performance Simulation Association, Sydney, 14-16 November 2011

IOP Conf. Series: Earth and Environmental Science **257** (2019) 012032 doi:10.1088/1755-1315/257/1/012032

Forecasting annual energy consumption using machine learnings: Case of Indonesia

Robi Kurniawan[*) 1, 2] **and Shunsuke Managi**[3]

1 Graduate School of Environmental Studies, Tohoku University, Japan
2 Ministry of Energy and Mineral Resources, Indonesia
3 Department of Urban and Environmental Engineering, Kyushu University, Japan
[*)] Corresponding author email: kurniawan.tohoku@gmail.com

Abstract. To understand the future trajectory of energy consumption, we propose to utilize two different machine learning algorithm, artificial neural networks (ANN) and a model tree. Taking Indonesia as a case, the annual gross energy consumption was estimated by modelling a function of urbanization, real GDP per capita proxy for affluence (economic growth), and real capital use per capita. Utilizing the time period of 1971–2014, we train and test the model. Utilizing the root mean square error and the mean absolute error for model selection, we found the tree-based model has a better performance rather than the ANN. Having more superior performance, the tree-based model was then used to forecast the annual energy consumption for the future years. Using specific scenario, the energy consumption is predicted will increase from 883 kg per capita in 2014 to become 1243 kg per capita in 2040. Providing better accuracy, the approach applied in this study can easily be replicated for other countries. Furthermore, it also can be considered in simulating energy demand and environmental consequence in the future.

1. Introduction

Considered as the world's 10th largest economy, Indonesia has been enjoying high rates of economic development. As the fourth most populous country with fast urbanization growth, the natural capital extraction, including energy, has increased significantly in the few decades [1]. It is highly linkage to Indonesia's sustainable development trajectory. Increasing economic and urbanization has led to increasing energy demand in the country. Having accurate forecasting has many benefits for fulfilling future energy demand and environmental consequence to the country.

Several works found the strong contribution of some socioeconomic variables such as urbanization, economic development, and capital use to energy consumption. Urbanization which takes place in most of developing countries potentially will generate more energy consumption. Urbanization is one of the main determinants of energy consumption in some ASEAN countries including Indonesia [2]. Urbanization, a shifting from agriculture sector toward industries and services sector, is also able to structure out the economic patterns of resource use [3]. Several studies also confirmed the linkage between energy consumption and economic growth. Using Saudi Arabia as a case, Alshehry and Belloumi [4] found a long-run relationship between energy consumption and economic growth in the country. Modifying the Stochastic Impacts by Regression on Population, Affluence and Technology, it is found that the capital use per capita has a significant contribution to energy consumption [5].

On the other hand, many different estimation methods have been developed to forecast the future energy demand. For instance, Bianco, Manca [6] employ parametric approach, multiple and single regression models, to predict electricity consumption in Italia using the socioeconomic data. To

Content from this work may be used under the terms of the Creative Commons Attribution 3.0 licence. Any further distribution of this work must maintain attribution to the author(s) and the title of the work, journal citation and DOI.
Published under licence by IOP Publishing Ltd

IOP Conf. Series: Earth and Environmental Science 257 (2019) 012032 doi:10.1088/1755-1315/257/1/012032

estimate the future trajectory of China's primary energy consumption, Yuan, Liu [7] developing two univariate models. Combining multiple linear regression and artificial neural networks, Günay [8] conduct forecasting the annual electricity demand in Turkey case. Even though after mentioned forecasting approach are developed, selection of the most appropriate technique is of a vital issue. The parametric approach such as linear regression has several drawbacks such as developed from strong assumptions of the data and the model's specification has to be specified by the user.
For forecasting purposes, instead of using the parametric models, in this study we utilize two different non-parametric machine learning methods known as model tree and ANN. Providing the technical literature on forecasting energy consumptions is the main contribution of this study. Furthermore, the understanding of the future trajectory of energy consumption is crucial especially for Indonesia, the world's 10th largest economy and the fourth most populous country. The future energy of the country also highly linkage to the environment since primary energy will be fulfilled by fossil fuel such as oil, coal and natural gas [9].
This study is organized as follows. Section 2 provides a methodology and characteristic of the datasets. The results and discussion are presented in section 3. Conclusions and policy implications are highlighted in section 4.

2. Datasets and methods

2.1. Datasets

In this study, annual data of GDP per capita, urbanization, and capital use per capita, are treated as an explanatory variable of the energy consumption per capita in Indonesia. The data for GDP per capita and capital use per capita have been converted in US$ 2010 constant prices. Summary of the descriptive statistics utilized in this study is presented in table 1.

Table 1. Descriptive statistics

Variable	energy consumption	GDP per capita	urbanization	capital use
Mean	578	1941	34	753
Maximum	884	3693	53	1296
Minimum	297	804	17	201
Std. Dev.	201	797	11	291
Unit	KgOE	US$ 2010	%	US$ 2010
Observations	44	44	44	44

Historical data for GDP per capita, urbanization, capital use per capita, and the energy demand per capita has been taken from the World Banks world development indicators. The data period of this work is 1971-2014 as shown in figure 1.

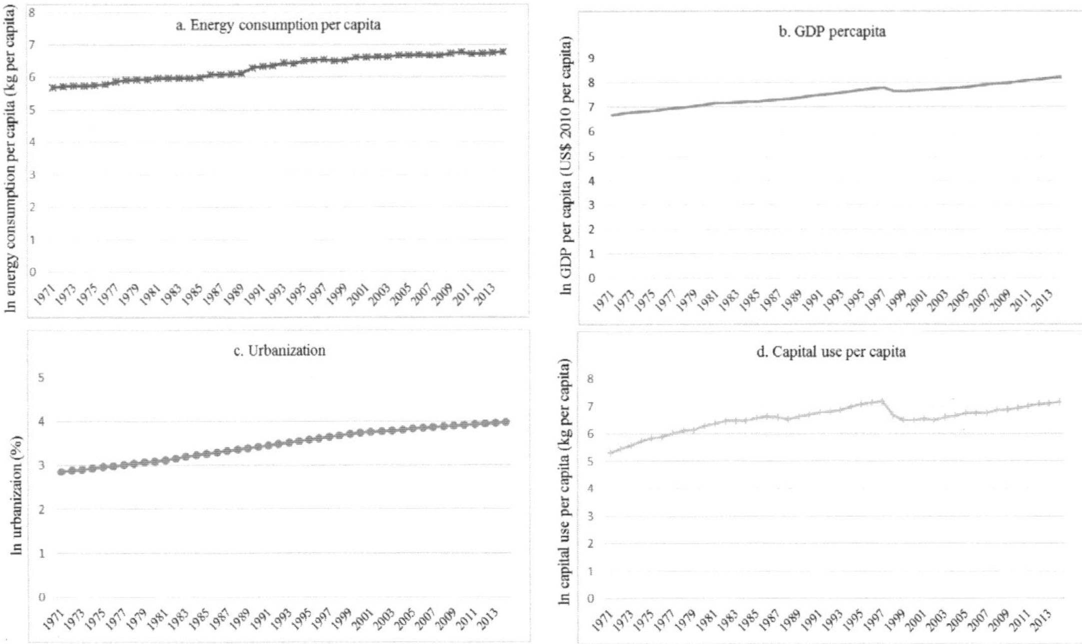

Figure 1. Data used in this study: (a) energy consumption per capita, (b) GDP per capita, (c) urbanization, (d) capital use per capita

2.2. Methods

We employ two different non-parametric machine learning approach, ANN and model tree. Capturing some unknown information which hidden in the data is considered as the primary advantage of ANN approach [10]. ANN capture the linkage between dependents and independent variable using a model derived from the biological brain responds to the stimuli from the input sensor. While the brain employs a neuron, an interconnected cell within the brain, the ANN use the artificial neuron to learn and solve the problem. For estimating the ANN model, we employ the neuralnet package (R environment package) developed by Günther and Fritsch [11].

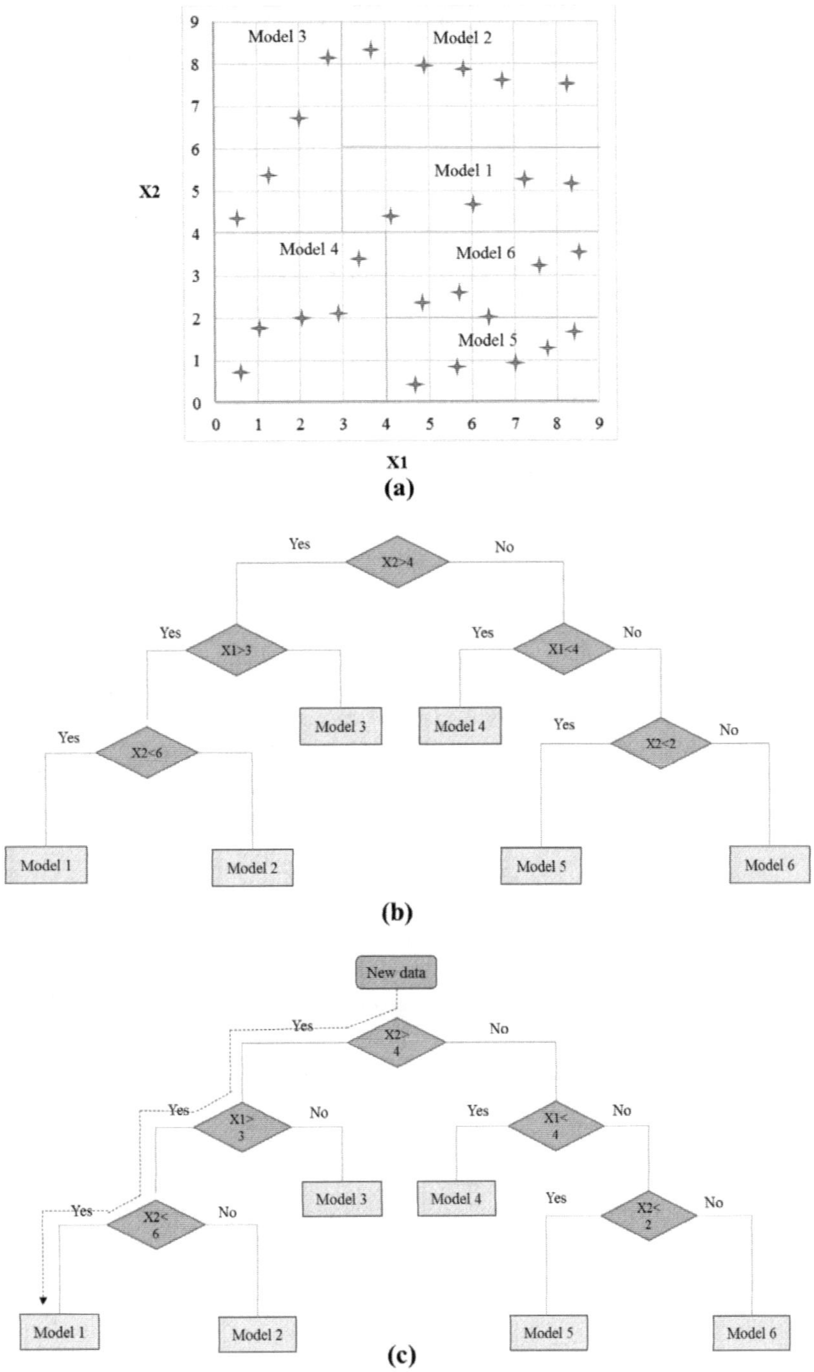

Figure 2. M5P-model tree algorithm schematic (a) splitting datasets (b) tree development (c) new input data for prediction.

In the other side, we employ a model tree which replaces the single value of leaf nodes with a linear regression model in order to improve the model performance of the regression tree [12]. Doing this approach instead of improving the accuracy, it enables us to maintain models simplicity [13]. To conduct the approach, we utilize the M5P algorithm (see Hornik, Buchta [14]) which provided in the RWeka package of R environment. Figure 1 presents the schematic of the model.

To get higher accuracy, we select the model using the RMSE and the MAE which are regularly employed in prediction model evaluation studies [15]. Without considering the direction of the errors, MAE measures the average magnitude of them in the set of predictions. On the other hand, RMSE is calculated by a quadratic scoring rule that also measures the average magnitude of the error. The indicator is a function of the square root of the average of squared differences between prediction and actual measurement.

3. Result and discussion

This section consists of three main parts. First, we split the datasets randomly into training and testing. Training data, which consist of 70% of the data, is used to build the model. Then, we assess the model accuracy using the remaining 30% of the data. For modelling Indonesia's energy consumption over the three decades, we apply an ANN and tree model. ANN and tree model were developed to simulate the predicted values of these variables to forecast the consumption using testing model. The results were compared with the real energy consumption in training datasets. This way, the success and the reliability of the forecasting approach applied in this work was validated.

Table 2. Predictive performance of the model

Indicator/Model	ANN	M5P
Root mean squared error	0.066565	0.0222
Mean absolute error	0.056363	0.022967

Table 2 presents the performance of ANN and M5P model. We utilized MAE and RMSE which capture average model prediction error in units of the variable to evaluate the models. The indicators are considered as negatively-oriented scores. According to this approach, lower values of the score are better. From the table, we can derive that the M5P model tree is more superior compared to the ANN approach.

Second, after selecting the appropriate model, we utilize the approach for the out of sample forecasting (2015-2040). Concerning population, we consider the world population prospect of the United Nations for Indonesia projected population. We adapted the socioeconomic projection of shared socioeconomic pathways 1 (SSP1). Considering the scenario, Indonesia's annual urbanization grows at 1.38 % on average. Furthermore, the economy grows at a constant rate of 4.4 % per annum, we assume the capital also grow at the same rate. Figure 3 presents the historical and forecasting of energy consumption in Indonesia using after mentioned assumption.

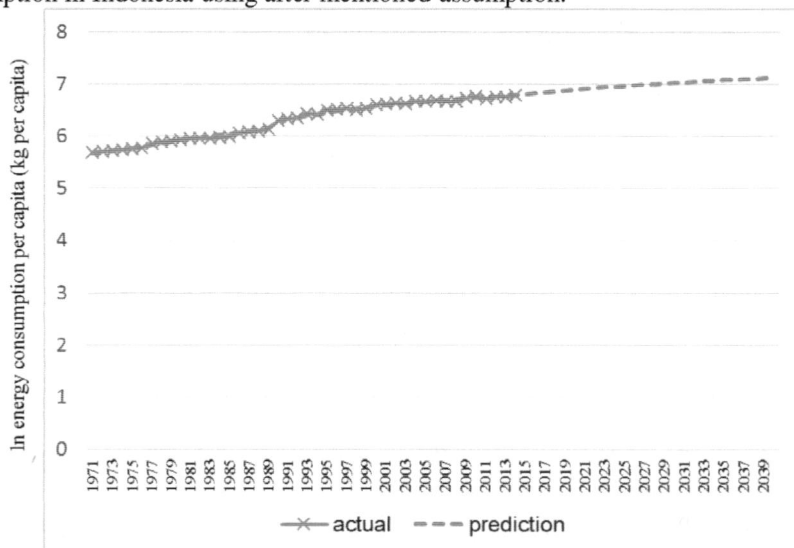

Figure 3. Actual and forecasting of the energy consumption using model tree

According to historical datasets, the average annual growth rate of Indonesia's energy consumption per capita reaches 2.63 %. Using the scenario of economic and urbanization projection, the growth rate of consumption per capita will decrease by 1.3 %. Translating the result into kg per capita, energy consumption in the country will increase from 883 kg per capita in 2014 to become 1243 kg per capita in 2040.

Third, we discuss the result of the study with another study which applies machine learning method. ANN is considered as the most widely used approach among the machine learning-based model, which has been applied in the field of energy. For instance, Kankal et al. utilized the model to forecast Turkey's energy consumption [16]. Kialashaki and Reisel also employed the method for modelling the energy demand in the industrial sector of the United States[17]. Comparing the ANN and model tree, our study revealed the model tree outperformed the ANN approach in the energy sector.

On the other hand, focusing on the causality of energy and CO2 emission in Indonesia, Hwang, & Yoo found the bi-directional causality existence among the variables [18]. According to their finding, an increase in energy consumption in the country directly linkage to CO2 emissions, while CO2 emissions also stimulate more energy consumption. It is related to the high share of coal in electricity generation which led to increasing indirect emissions in Indonesia [19]. Some socioeconomic variable such as urbanization economic growth and coal consumption are cointegrated over long-term in Indonesia [20]. The trend of Indonesia's energy consumption in the next decades should be considered in Indonesia's CO2 emission mitigation.

4. Conclusion

This study aims to provide the alternative to the technical literature on forecasting energy consumptions. Covering the time period of 1971–2014 datasets, we build two different machine learning approach, artificial neural networks and model tree. We utilized historical data of urbanization, real GDP per capita, and real capital use per capita to capture energy consumption. Employing the RMSE and MAE for model selection, we found the tree model outperformed the ANN which widely utilized for forecasting energy management.

Using a model tree, we used to forecast the annual energy consumption for the future years using the specific scenario of the independent variable. Accordingly, the energy demand will increase from 883 kg per capita in 2014 to become 1243 kg per capita in 2040.

The result brings several implications. First, having strong accuracy, the approach applied in this study can easily be replicated for other countries to make better forecasting of energy consumption for the future. Second forecasting result of the study can be considered to the stakeholder in simulating energy demand and environmental consequence in the future.

5. References

[1] Kurniawan, R. and S. Managi, *Economic Growth and Sustainable Development in Indonesia: An Assessment.* Bulletin of Indonesian Economic Studies, 2018(just-accepted): p. 1-31.

[2] Azam, M., et al., *Factors determining energy consumption: Evidence from Indonesia, Malaysia and Thailand.* Renewable and Sustainable Energy Reviews, 2015. **42**: p. 1123-1131.

[3] Parikh, J. and V. Shukla, Urbanization, energy use and greenhouse effects in economic development: Results from a cross-national study of developing countries. Global Environmental Change, 1995. **5**(2): p. 87-103.

[4] Alshehry, A.S. and M. Belloumi, *Energy consumption, carbon dioxide emissions and economic growth: The case of Saudi Arabia.* Renewable and Sustainable Energy Reviews, 2015. **41**: p. 237-247.

[5] Shahbaz, M., et al., The effect of urbanization, affluence and trade openness on energy consumption: A time series analysis in Malaysia. Renewable and Sustainable Energy Reviews, 2015. **47**: p. 683-693.

[6] Bianco, V., O. Manca, and S. Nardini, *Linear regression models to forecast electricity consumption in Italy.* Energy Sources, Part B: Economics, Planning, and Policy, 2013. **8**(1): p. 86-93.
[7] Yuan, C., S. Liu, and Z. Fang, Comparison of China's primary energy consumption forecasting by using ARIMA (the autoregressive integrated moving average) model and GM (1, 1) model. Energy, 2016. **100**: p. 384-390.
[8] Günay, M.E., Forecasting annual gross electricity demand by artificial neural networks using predicted values of socio-economic indicators and climatic conditions: Case of Turkey. Energy Policy, 2016. **90**: p. 92-101.
[9] Hasan, M.H., et al., A review on the pattern of electricity generation and emission in Indonesia from 1987 to 2009. Renewable and Sustainable Energy Reviews, 2012. **16**(5): p. 3206-3219.
[10] Zhang, Z., Artificial neural network, in Multivariate Time Series Analysis in Climate and Environmental Research. 2018, Springer. p. 1-35.
[11] Günther, F. and S. Fritsch, *neuralnet: Training of neural networks.* The R journal, 2010. **2**(1): p. 30-38.
[12] Wang, Y. and I.H. Witten, Induction of model trees for predicting continuous classes. 1996.
[13] Lantz, B., *Machine learning with R.* 2013, Birmingham-UK: Packt Publishing Ltd.
[14] Hornik, K., et al., *Rweka: R/weka interface.* 2016.
[15] Chai, T. and R.R. Draxler, Root mean square error (RMSE) or mean absolute error (MAE)?–Arguments against avoiding RMSE in the literature. Geoscientific model development, 2014. **7**(3): p. 1247-1250.
[16] Kankal, M., et al., Modeling and forecasting of Turkey's energy consumption using socio-economic and demographic variables. Applied Energy, 2011. **88**(5): p. 1927-1939.
[17] Kialashaki, A. and J.R. Reisel, Development and validation of artificial neural network models of the energy demand in the industrial sector of the United States. Energy, 2014. **76**: p. 749-760.
[18] Hwang, J.-H. and S.-H. Yoo, Energy consumption, CO 2 emissions, and economic growth: evidence from Indonesia. Quality & Quantity, 2014. **48**(1): p. 63-73.
[19] Kurniawan, R., Y. Sugiawan, and S. Managi, *Cleaner energy conversion and household emission decomposition analysis in Indonesia.* Journal of Cleaner Production, 2018. **201**: p. 334-342.
[20] Kurniawan, R. and S. Managi, Coal consumption, urbanization, and trade openness linkage in Indonesia. Energy Policy, 2018. **121**: p. 576-583.

Acknowledgements
This paper was supported by Grant-in-Aid for Specially Promoted Research (26000001) by the Japan Society for the Promotion of Science. Any opinions, findings, conclusions, and recommendations expressed in this study are those of the authors and do not necessarily reflect the views of the institutions and funding agencies. We acknowledge comments from the reviewer and the editor of International Conference on Future Environment and Energy 2019.

Heating energy use in China: the current situation, challenges, and possibilities

Qiaoran Wang[1, 2], Xianming Yang[1] and Na Ma[3]

1 Institute of Development Studies, Yunnan University, Kunming, 650091, China
2 Department of Planning, Aalborg University, Aalborg, 9000, Denmark
3 Department of Economics, Kunming University of Science and Technology, Kunming, 650091, China
E-mail: wqr891006@foxmail.com

Abstract. Heating energy consumption in China is continually increasing because of the rapid urbanization and accompanied rising living standards. A panoptic description of China's current heating status is still in need of developing effective heating policies. In this paper, heating policy in China was introduced and a brief summary of the status of heating in different parts of China was described, the key features were characterized. Then, current and foreseen challenges for China's heating energy use were analyzed. Finally, related policy suggestions were proposed and the future possibilities of heating energy-conservation in China were discussed.

1. Introduction

Driven by rapid economic growth and fast-paced urbanization, heating energy consumption in China has become a significant part of the total energy consumption. Though China's heating energy use is significantly lower than that of developed countries[1], it is growing fast and accounted for 24.14% of the total building energy consumption in 2015, with the specific amount reached 208.53 Mtce, almost double the consumption of 2001. In the northern part of China, district heating has supplied more than 80% of the total urban building areas and covered almost all cities by 2015, with the building area served by district heating increased from 5 billion m^2 in 2001 to 13.2 billion m^2 in 2015[2]. In southern China, distributed heating increased rapidly, the total residential heating energy consumption in hot summer and cold winter zone reached 17.53 Mtce in 2015, representing an average annual increase of 18.4% in 15 years[3]. As a consequence of urbanization and improvement in people's living standards, heating will continue to be a key energy end-user in China.

In response to this growing trend, the Chinese government has implemented a series of policies, aiming at limiting the increase in heating energy use through technical advances, while the impacts of economic factors like urbanization and accompanying improvement of living standard, as well as shift of consumption patterns on China's heating sector, is still not clear.

Intensive studies have focused on the district heating energy use in northern cities of China[4,5,6]. Some studies have also discussed whether district heating should be provided in the hot summer cold winter zone[7,8,9]. These studies have provided plenty of useful information on heating energy use in northern China, but most existing studies failed to quantify the heating in southern China because of the lack of constant nationwide data.

Therefore, an overview of China's heating energy use is still in need for developing effective heating policies. Nevertheless, the impact of economic factors like urbanization on heating energy use should be taken into account when talking about the future heating energy use. A comprehensive

Content from this work may be used under the terms of the Creative Commons Attribution 3.0 licence. Any further distribution of this work must maintain attribution to the author(s) and the title of the work, journal citation and DOI.
Published under licence by IOP Publishing Ltd

understanding of the present heating policy and the heating situation in different regions in China thus should, therefore, be enunciated, along with a detailed simulation of heating energy consumption under possible urbanization rates. In line with this, the challenges of urbanization and a possible pathway of heat development could be discussed, which is the aim of this paper.

Section 2 provided a brief summary of the status and key characteristics of district heating in northern China and individual heating in southern China. Section 3 discussed the challenges of heating caused by urbanization, different scenarios were set to discover the influence of urbanization on heating demand and predict the heating energy consumption trend. In Section 4, possibilities of energy conservation were proposed, potential ways to cut back the rising heating energy consumption were identified, policy recommendations were also given in this section. Finally, conclusions were drawn in Section 5 of the paper.

2. Status quo for space heating in China

2.1. Heating policies in China

China is a large country with a complex climate. China's national standard Thermal Design Code for Civil Building (GB50176-93) defines five climate zones as severe cold (SC), cold (C), hot summer and cold winter (HSCW), hot summer and warm winter (HSWW), and temperate (T) zones [10]. The zoning criteria are mainly based on the average temperatures in the coldest and hottest months of the year.

The varying climate among different regions in China leads to unique characteristics in regional energy consumption patterns. In addition, the current heating policy has given rise to the significant differences. In the 1950s, the central government set district heating zones in China according to the "Qin Mountains-Huai River" boundary, named the North-South heating line. According to this policy, district heating is only available at the northern part of the North-South heating line, which covers most urban areas in severe cold and cold zones, including 15 provinces and occupying more than 70% of the total mainland area[2]. In contrast, the district heating is not available in the southern part of the North-South heating line. While some parts of the HSCW zone in this area has a significant requirement for heating since it is at least as cold in winter as in some district heating areas in the North, in some cases even colder as far as humidity is concerned[7]. As a result, individual heating facilities have been widely utilized in the HSCW zone.

2.2. Data on heating energy in China

So far, the lack of consistent heating data from official statistics makes it difficult to get a comprehensive view of China's heating situation and to understand the underlying changes that affect energy consumption in heating. In order to get more information on key characteristics of China's heating sector, some researchers have carried out detailed regional or city-level surveys[11,12,13,14]. In addition, some researchers and institutions conducted nationwide surveys, among which, the China Residential Energy Consumption Survey (abbr. CRECS) administrated by the Department of Energy Economics at Renmin University of China and the China Building Energy Model (abbr. CBEM) built by Building Energy Research Center of Tsinghua University (abbr. BERC) are most influential.

The CRECS started in 2012 and included a sector of space heating and cooling. While only the data for the year 2012 has been open to the public by now. The CBEM was based on a series of surveys which included key information on China's heating energy use. Since 2007, Annual Report on China Building Energy Efficiency based on the data of CBEM has been published by BERC. According to these reports, building energy use in China was divided into four subsectors, namely, Space Heating in Northern Urban China, Commercial and Public buildings (excl. SHNUC), Urban Residential buildings (excl. SHNUC) and Rural Residential buildings. Heating in the HSCW zone was included in the Urban Residential buildings subsector. This paper used the data from CBEM to draw a picture of the heating energy use situation in both north and south China.

2.3. Status quo for space heating in different regions in China

IOP Conf. Series: Earth and Environmental Science **257** (2019) 012033 doi:10.1088/1755-1315/257/1/012033

2.3.1. Heating energy use, energy intensity, and heating floor area.
Figure 1 shows the energy use and energy intensity for heating in the SC&C zone and the HSCW zone. The amount of heating energy use in both regions increased rapidly from 2001 to 2015. In the SC&C zone, heating energy use rose from 115 Mtce to 191 Mtce, with an annual increase of 3.69% and accounting for 22% of total China's building commercial energy use by 2015[3]. While the heating energy use in the HSCW zone grew around 10 times from about 1.64 Mtce in 2001 to 17.53 Mtce in 2015. The annual growth rate of heating energy use in the HSCW area was 18.4%, almost 5 times that of the SC&C zone. Figure 1 also indicates the huge gap between the SC&C zone and the HSCW zone in total heating energy use, which of the SC&C zone was 69 times larger than that of the HSCW zone in 2015. One of the reasons for the increasing trend of heating energy use is the growing heating floor area. China is experiencing rapid urbanization, to satisfy the housing demand of new urban residents, urban residential floor area is increasing rapidly. In the SC&C zone, floor area served by heating rose from 5 billion m^2 to 13.2 billion m^2, with an annual increase of 7.15%. While that of the HSCW zone increased from 3.8 billion m^2 to 9.5 billion m^2, with an annual growth rate of 6.71%. Heating floor area in the SC&C zone was 1.38 times of that in the HSCW zone in 2015, it is one of the reasons for the gap in heating energy consumption in these two regions.
As for the energy use intensity, trends were different in the SC&C zone and the HSCW zone. As shown in Figure 1, the average heating energy use per unit of the floor area of the SC&C zone declined by 36%, from 22.8 kgce/ m^2 in 2001 to 14.5 kgce/ m^2 in 2015, showing the great progress of energy conservation. In contrast, that of the HSCW zone rapidly increased from 0.43 kgce/ m^2 in 2001 to 1.84 kgce/ m^2 in 2015, with an average annual growth rate of 10.94%. Though it increased rapidly in the past 15 years, the energy intensity of heating in the HSCW zone is still relatively low compared with that of the SC&C zone (which is about one-eighth of that in the SC&C zone).

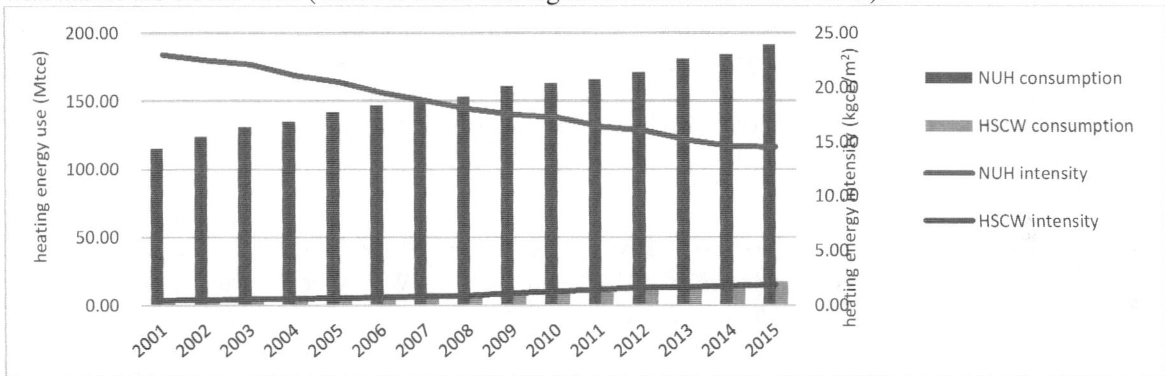

Figure 1. Heating energy use and energy intensity in the SC&C zone and the HSCW zone

2.3.2. Heating devices, usage mode, heating season and temperature.
Large differences in heating energy use and intensity of these two regions are not caused solely by climate conditions or heating floor area but are also significantly influenced by heating devices, usage mode and heating season.
(1) Heating devices. Heating devices vary significantly in the two regions. In the SC&C zone where district heating is widely used, the main heating devices are radiant floor heating and radiant heating systems. More than 60% of urban residential households in the SC&C zone utilized municipal centralized heating systems and another 14% used a gas boiler, while only 4% of households used other heating devices or did not have a heating device[14]. In contrast, most households in the HSCW zone used a decentralized heating device. The most popular heating devices in this region were air-source heat pumps and electrical heaters, used by 29% and 27% of households, respectively. In recent years, centralized heating systems expanded rapidly, from less than 3% in 2009 to about 10% in 2015. Only 29% of residents indicated that they use other heating devices or do not have a heating device.
Different heating devices have led to dissimilar intensities of heating energy consumption. For households using air-source heat pumps, the average consumption per unit of floor area was about 3.6

kWh/ m^2 (1.15 kgce/ m^2). For people using gas boilers, that was around 6 to 12 kgce/ m^2; for district heating, it was 15 to 20 kgce/ m^2. Different heating devices and heat sources are the main reasons for the huge gap between the energy consumption of the SC&C zone and the HSCW zone.

(2) Usage mode. In the SC&C zone, the district heating network keeps the heating system on without adjustment of the temperature for individual residences throughout the whole heating season. While in the HSCW zone, heating equipment is implemented separately by each household, consequently, people can adjust the indoor environment and use heating only when they feel cold in the room. The heating facilities in the HSCW zone are turned off during the most time during the heating season[13]. In a word, the district heating in the SC&C zone is operated full-time-full-space while the individual heating in the HSCW zone is operated part-time-part-space[15]. The relatively low heating energy use of the HSCW zone would increase if residents in this region use heating for a longer time and for more rooms.

(3) Heating season. According to China's heating policy, the heating period for district heating in the SC&C zone is determined by standard design, usually from mid-November to mid-March the following year in the cold zone and from late-October to mid-April the following year in the severe cold zone[16]. While in the HSCW zone, people can decide the length of the heating season, therefore, it varies among different households. According to BERC, the heating season in this region varies from less than one month to more than four months[7]. While in general, the heating season in the HSCW zone is shorter than that in the SC&C zone.

(4) Temperature. District heating helps to maintain 20°C indoors in the SC&C zone during the heating season. While in the HSCW zone, the indoor temperatures vary among different households. For most of them, the temperatures of the heated room are around 11 to 18℃, however, in some cases, the rooms have temperatures of less than 5℃. By comparing the temperatures to the American Society of Heating, Refrigerating, and Air Conditioning Engineers (abbr. ASHRAE) comfort zone limits, it is obvious that indoor thermal environments are relatively low in the HSCW zone[1,13,17,18]. If residents in this region want to get the same level of comfort as residents in the SC&C zone, they need to consume more energy. In other words, the gap in heating energy use between these two regions will be narrowed if we take comfort into account.

2.4. Key characteristics of heating in China

Based on the facts described above, several key characteristics of heating can be drawn.

Firstly, heating energy use in both the SC&C zone and the HSCW zone increased sharply from 2001 to 2015. Secondly, the trend in heating energy intensity is opposite in these two regions. The heating energy intensity in the HSCW zone has increased in the past years, while that of the SC&C zone has shown a great decrease, implying the possibility of energy conservation in the heating sector. Thirdly, the current heating energy use and energy intensity in the HSCW zone are relatively low compared with those of the SC&C zone, which is a result of the warmer climate, flexible heating devices, shorter heating season and lower indoor temperature. However, the indoor environments in the HSCW zone need to be improved, which indicates the huge growth potential of heating energy use in this region.

3. Challenges of China's space heating

A lot of researches have indicated that heating energy use is a function of climate, heating floor area, building performance, heating equipment, heating season and resident behaviors[1]. It seems most of these factors will increase heating energy use in China as the strong expectations for urbanization and improving living standard. Urbanization would result in great growth potential in heating floor area, besides, there is still sustained improvement potential in living standard caused by urbanization, which will lead to more energy use and higher energy intensity, especially in the HSCW zone. All of these expectations have put forward big challenges for heating energy conservation in China.

3.1. Challenge of increasing heating floor area caused by urbanization

China is experiencing rapid urbanization, as shown in Figure 2, from 2001 to 2015, the urbanization rate of China increased rapidly with an annual growth rate of 2.89%. To satisfy the housing demand of new urban residents, urban residential floor area has maintained a steady growth, showing an annual

increase of 6.96% during this period. Figure 2 visually demonstrate the relationship between heating floor area and urbanization rate, with a correlation coefficient of 0.99.
Considering the current urbanization level, China's urbanization process is unlikely to end soon, indicating more newly completed urban buildings and higher district heating demand in the future.

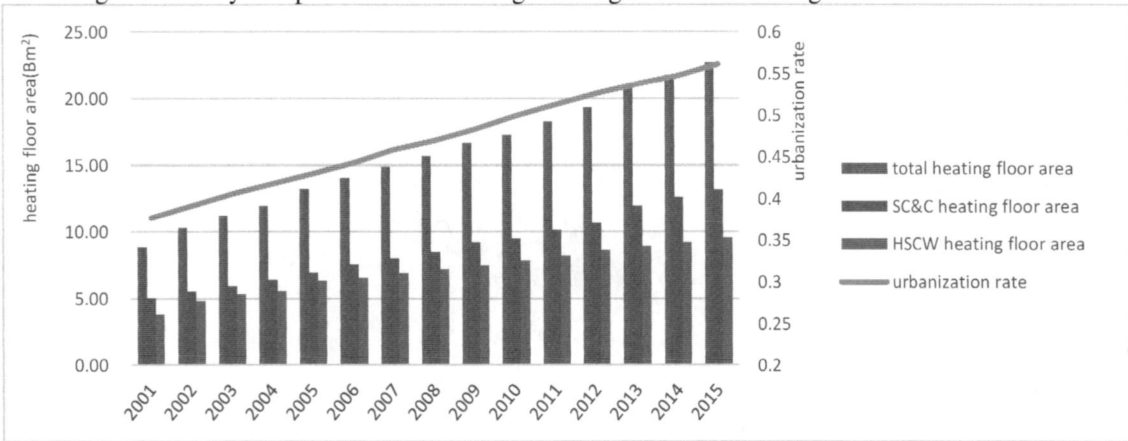

Figure 2. Urbanization rate and heating floor area (2001-2015)

3.2. Challenge of improving living standard caused by urbanization

Urbanization could also result in an increase in heating energy use by improving residents' income and living standard. The indoor comfort is far lower in the HSCW zone than that in the SC&C zone when heating season and temperature are taken into account. With the rapidly increasing urbanization rate and the accompanying growing household income and improvement of living standards, the residents in the HSCW zone will use heating equipment for a longer time to heat more rooms, they will also use larger capacity heating equipment to improve the indoor environment, which will result in more heating energy use in the HSCW zone[19].

3.3. Scenario analysis of heating energy demand in China

In 2016, the State Council of China released the National Population Development Strategy (2016-2030) [20], in which the rapid urbanization is expected to continue until 2030. As stated in this report, the total population of China will be 1.42 billion in 2020 and 1.45 billion in 2030, with the urbanization rate of 60% and 70%, respectively. That is, China's urban population will be 0.852 billion in 2020 and 1.015 billion in 2030.
In Scenario 1, the impact of urbanization on heating energy use was analyzed. We assumed the energy use intensity for space heating, the floor area per capita and the proportions of the urban population of different climate zones remain at the current level. Therefore, the urban population of the HSCW zone will be 274.67 million in 2020 and 327.22 million in 2030, that of the SC&C zone will be 354.75 million in 2020 and 422.62 million in 2030. Heating energy use of the SC&C zone would be 249.35 Mtce in 2020 and 389.78 Mtce in 2030. That of the HSCW zone would be 34.38 Mtce in 2020 and 117.73 Mtce in 2030. The results indicate that increasing heating floor area caused by urbanization could contribute to an increase in heating energy use by about 29.6% and 22.2% in 2020 and 2030 compared with the current level, respectively.
Based on this, Scenario 2 analyzed the impact of residents' usage pattern on heating energy use. Assuming residents in the HSCW zone change their usage mode from only using the heating devices when needed to use the devices for the entire heating season and improve the heating temperature to 16℃, then the energy consumption in the HSCW zone will be 58.46 Mtce in 2020 and 143.05 Mtce in 2030, indicating the usage mode contributes 46.32 Mtce in 2020 and 91.72 Mtce in 2030 to heating energy use.
Moreover, with the continuous urbanization and accompanying increase of household income and living standard, there is a hot debate on whether district heating should be extended to the HSCW zone.

In Scenario 3, we assume district heating is used in the HSCW zone and the energy use intensity of the district heating in the HSCW zone keeps the same path as that in the SC&C zone. Thus the heating energy use of the two regions in 2020 and 2030 would be 415.15 Mtce and 660.73 Mtce. The results indicate that the heating mode contributes 153.67 Mtce in 2020 and 219.61 Mtce in 2030 to heating energy consumption.

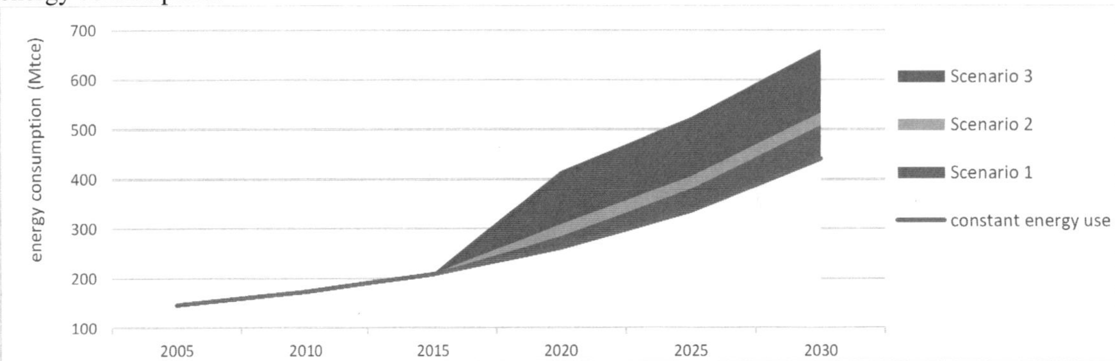

Figure 3. Scenario analysis of China's heating energy consumption

With the forecast of Scenario 3 which shown in Figure 3, the total energy use for heating in 2030 would be more than 2 times of that in 2015. As outlined in Revolution Strategy for Energy Production and Consumption (2016-2030) released by National Development and Reform Commission and National Energy Administration of China[21], China aims to limit the total amount of energy consumption to a target of 5 billion tce in 2020 and 6 billion tce in 2030. According to the NBS, the annual increase rate of residential heating energy use for 2014-2015 was 8.5%. If it keeps increasing at this rate, the residential heating energy use in 2030 will be more than 3 times that of the current level, namely, 708.85 million tce, which will amount to 11.8% of the total energy consumption of the country. The development target for China's energy system cannot accommodate this demand, therefore, the biggest challenge of China's heating sector is to find a suitable approach to limit the total energy use as well as provide an adequate living standard.

4. Possibilities of energy conservation in China's heating

As concluded above, the heating demand will increase as heating floor area increases and living standard improves. From 2001 to 2015, heating energy intensity increased rapidly in the HSCW zone, current heating energy policies have not paid enough attention to this region and the heating energy efficiency of this region is relatively poor[22]. In contrast, the energy use intensity of heating in the SC&C zone has shown a great decrease, indicating the great potential of energy conservation in the heating sector. The possibilities of energy conservation in the heating sector in China lie in these aspects.

(1) Improvement of building performance. In the SC&C zone, several design standards of energy efficiency were implemented to improve building envelope performance. For example, a standard established by MOHURD in 2010 set the target to reduce heat demand in new buildings by 65% compared to previous buildings (JGJ 26-2010)[23]. A standard of retrofitting existing residential buildings in the SC&C zone during 12th Five-year Plan period was also established by MOHURD. Besides, financial support has been provided by the Chinese government to improve the envelope insulation of existing buildings. These efforts greatly improved the insulation level of building envelopes and reduced heating energy intensity in the SC&C zone. While in the HSCW zone, the building performance is still very poor. Appropriate building design standards of energy efficiency need to be implemented in this region. Specific building performance efficiency policies applying to the HSCW zone should be improved to realize the potential for energy conservation in this area.

(2) Optimization of heating sources. Recently, in the SC&C zone, efficient heating sources such as large-scale combined heat and power (CHP) systems and industrial waste heat has expanded rapidly. High-efficiency CHP systems and boilers have replaced small coal boilers and household coal stoves at the regional level, while stoves, heat pump technology has developed rapidly and gas-source heating

systems are now used more widely. However, the efficiency of the heating devices in the HSCW zone is comparatively low. There are still no specific industry standards or regulations in regards to the energy efficiency of heating devices in the HSCW zone like electrical heaters[22]. Appropriate regulations and standards are needed to improve the efficiency of heating devices in this region.
(3) Adjustment in heating usage mode. The heating system in the SC&C zone is kept on for the entire heating season, without adjustment of the temperature for individual residence, resulting in serious overheating and network energy consumption, which consumes more than 30% of the total supplied space heating[1]. Therefore, flexible heating technology systems are inevitably called for. In the HSCW zone, residents' heating behaviors are much more diverse, thus the most important aspect to address in this region is the heating behaviors. In view of the top-level design, it is important to raise awareness of energy saving and encourage a "green" lifestyle. As for the detailed policies, some economic incentives like tiered energy pricing schemes should be promoted to encourage energy saving behaviors.

5. Conclusion

The main purpose of this research was to gain a panoptic description of heating energy use in China and to discuss the challenges caused by urbanization and possibilities of heating energy saving. An overview of China's heating sector was given and key characteristics of heating in different zones of China have been drawn in this study. Based on this, challenges of heating energy use caused by urbanization were analyzed, then the recommendations for future policy and technology development were given.
One of the limitations of this study concerns the quantitative analysis of the relationship between urbanization and energy usage mode as well as heating energy consumption. Another limitation of this study is the heating energy-saving technology was not discussed in detail. However, constant data for heat demand, heat resources, and heat grid are lacking to achieve these targets. Available data are strongly needed for further analysis. Further study would target these directions to study in a deep-going way if more heat metering devices were used and more specific data on heating were accessible to the public.

6. References

[1] Peng, Chen, Da Yan, Siyue Guo, Shan Hu, and Yi Jiang. 2015. "Building energy use in China: ceiling and scenario." *Energy and Buildings* no. 102:307-316.
[2] Xiong, Weiming, Yu Wang, Brian Vad Mathiesen, Henrik Lund, and Xiliang Zhang. 2015. "Heat roadmap China: new heat strategy to reduce energy consumption towards 2030." *Energy* no. 81:274-285.
[3] Building Energy Research Center of Tsinghua University. 2017. *China Building Energy Use 2017*. Beijing: China Architecture and Building Press.
[4] Chen, Xia, Li Wang, Lige Tong, Shufeng Sun, Xianfang Yue, Shaowu Yin, and Lifang Zheng. 2013. "Energy saving and emission reduction of China's urban district heating." *Energy Policy* no. 55:677-682.
[5] Wei, Bing, Song-Ling Wang, and Li Li. 2010. "Fuzzy comprehensive evaluation of district heating systems." *Energy Policy* no. 38 (10):5947-5955.
[6] Zhang, Qingyuan. 2004. "Residential energy consumption in China and its comparison with Japan, Canada, and USA." *Energy and buildings* no. 36 (12):1217-1225.
[7] Guo, Jin, Ying Huang, and Chu Wei. 2015. "North–South debate on district heating: Evidence from a household survey." *Energy Policy* no. 86:295-302.
[8] Hu, Shan, Da Yan, Ying Cui, and Siyue Guo. 2016. "Urban residential heating in hot summer and cold winter zones of China—Status, modeling, and scenarios to 2030." *Energy Policy* no. 92:158-170.
[9] Zhang, N. 2013. "Debate on heating in Southern China." *China Energy*:84-89.
[10] Ministry of Housing and Urban-Rural Development of the People's Republic of China. 1993. Thermal Design Code for Civil Buildings. Beijing: China Architecture & Building Press.

[11] Chen, Shuqin, Nianping Li, Hiroshi Yoshino, Jun Guan, and Mark D Levine. 2011. "Statistical analyses on winter energy consumption characteristics of residential buildings in some cities of China." *Energy and Buildings* no. 43 (5):1063-1070.

[12] Chen, Shuqin, Hiroshi Yoshino, Mark D Levine, and Zhenhai Li. 2009. "Contrastive analyses on annual energy consumption characteristics and the influence mechanism between new and old residential buildings in Shanghai, China, by the statistical methods." *Energy and Buildings* no. 41 (12):1347-1359.

[13] Lin, Borong, Zhe Wang, Yanchen Liu, Yingxin Zhu, and Qin Ouyang. 2016. "Investigation of winter indoor thermal environment and heating demand of urban residential buildings in China's hot summer–Cold winter climate region." *Building and Environment* no. 101:9-18.

[14] Hu, Shan, Da Yan, Siyue Guo, Ying Cui, and Bing Dong. 2017. "A survey on energy consumption and energy usage behavior of households and residential building in urban China." *Energy and Buildings* no. 148:366-378.

[15] Zhang, Shengyuan, Xiu Yang, Yi Jiang, and Qingpeng Wei. 2010. "Comparative analysis of energy use in China building sector: current status, existing problems and solutions." *Frontiers of Energy and Power Engineering in China* no. 4 (1):2-21.

[16] Guo, Siyue, Da Yan, Chen Peng, Ying Cui, Xin Zhou, and Shan Hu. 2015. "Investigation and analyses of residential heating in the HSCW climate zone of China: status quo and key features." *Building and Environment* no. 94:532-542.

[17] American Society of Heating, Refrigerating, and Air Conditioning Engineers 2015. Standard 55-2004, Thermal environmental conditions for human occupancy, atlanta: American society of heating, refrigerating, and air-conditioning engineers.

[18] Guo, Jin, Ying Huang, and Chu Wei. 2015. "North–South debate on district heating: Evidence from a household survey." *Energy Policy* no. 86:295-302.

[19] Yoshino, Hiroshi, Yasuko Yoshino, Qingyuan Zhang, Akashi Mochida, Nianping Li, Zhenhai Li, and Hiroyuki Miyasaka. 2006. "Indoor thermal environment and energy saving for urban residential buildings in China." *Energy and buildings* no. 38 (11):1308-1319.

[20] China, State Council of. 2016b. National Population Development Strategy (2016-2030).

[21] National Development and Reform Commission and National Energy Administration of China. 2016. Revolution Strategy for Energy Production and Consumption (2016-2030).

[22] Hu, Shan, Da Yan, Ying Cui, and Siyue Guo. 2016. "Urban residential heating in hot summer and cold winter zones of China—Status, modeling, and scenarios to 2030." *Energy Policy* no. 92:158-170.

[23] Ministry of Housing and Urban-Rural Development of the People's Republic of China. 2010. Design Standard for Energy Efficiency of Residential Buildings in Hot Summer and Cold Winter Zone.

Acknowledgments

This work was supported by the National Natural Science Foundation of China [grant number 71362026] and the Ministry of Education of Yunnan [grant number 2017YJS098]. The author would like to thank the China Scholarship Council for providing the fellowship [student number 201707030012].

Energy management methodology for energy sustainable actions in University of Campinas - Brazil

J G I Cypriano[1], L F Pinto[1], L C Machado[2], L C P da Silva[1] and L S Ferreira[3]

1 Energy Systems Department, School of Electrical and Computer Engineering, University of Campinas, Campinas, 400 Albert Einstein Avenue, SP, Brazil
2 Sustainable Campus Office, University of Campinas, Campinas-SP, Brazil
3 Innovation Department, CPFL Brazil Company, Campinas, 1755 Eng. Miguel Noel Nascentes Burnier Highway, SP, Brazil
Email:joao.ito@dsee.fee.unicamp.br,liafaria@dsee.fee.unicamp.br,
lui@dsee.fee.unicamp.br, lindi@unicamp.br, luanaferreira@cpfl.com.br

Abstract. Universities need to be open to changes in socioeconomic and environmental that are occurring in the world, to transform society around them. For this, one of the natural movements has been in sustainability, focusing on 2030 Agenda. In this context, this article shows the path that University of Campinas has been adopting in order to become a sustainable benchmark. In the past, there was an environmental movement that culminated in the sustainable policies creation and, earlier, an international commitment to other institutions. Currently, the Sustainable Campus Project is in implementation to develop energy sustainability and change the organizational culture. For the future, the transformations needed to promote a sustainable energy management system and continuous improvement of its actions are indicated. In the end, all those processes will allow a living laboratory in energy sustainability and a regional benchmark.

1. Introduction

In 2015, it was adopted by UN the 2030 Agenda for Sustainable Development with 17 goals, called Sustainable Development Goals (SDGs). The 7th SDG, "Ensure access to affordable, reliable, sustainable and modern energy for all", has three main targets for 2030, that are: universal access to affordable, reliable and modern energy services; increase the renewable energy generation in global energy; and, double global rate of improvement in energy efficiency. [1]

The Energy SDG is one of the most important goal, because it is interconnected and contributes to achieving 125 targets of the other SDGs (74 percent of all targets), supporting in poverty alleviation, economic growth, expand employment and curb carbon emissions, as shown in Figure 1. [2]

In this context, the University of Campinas is building a sustainable energy living lab in your main campus, Zeferino Vaz's University City, called: "Sustainable Campus Model at the University of Campinas—Brazil: A Living Lab integrating Renewable Generation, Electric Mobility, Energy Efficiency, Monitoring and Energy Demand Management"; or just Sustainable Campus Project. [3]

This Project was designed to support and promote new solutions to Universities and smart cities, being a regional benchmark case in Brazil and Latin America. Since August 2017, the Sustainable Campus project (SCP) was complemented with three new projects: one in energy efficiency labeling for buildings; another in electricity procurement for free market; and, the last one, in smart meter communication using RF Mesh and LoRa (long range communication) technologies. All these

Content from this work may be used under the terms of the Creative Commons Attribution 3.0 licence. Any further distribution of this work must maintain attribution to the author(s) and the title of the work, journal citation and DOI.
Published under licence by IOP Publishing Ltd

initiatives and knowledge need a full policy structure to support and maintain those actions in long term.

This article presents the sustainable campus model, which base energy sustainability initiatives and the next steps to continue the SCP's results, becoming in an energetically sustainable campus benchmark. The section 2, starts with the Unicamp's history that allowed a sustainable environment growing before Sustainable Campus Project. In section 3, the new infrastructure for measurement and data collection, that will provide a new vision in sustainable energy management, is detailed. The fourth section is about "Energy Management Structure", modeling the new infrastructure in continuous improvement for energy efficiency actions, including new actors and efficiency indicators. And last, the conclusion and expected results of this review.

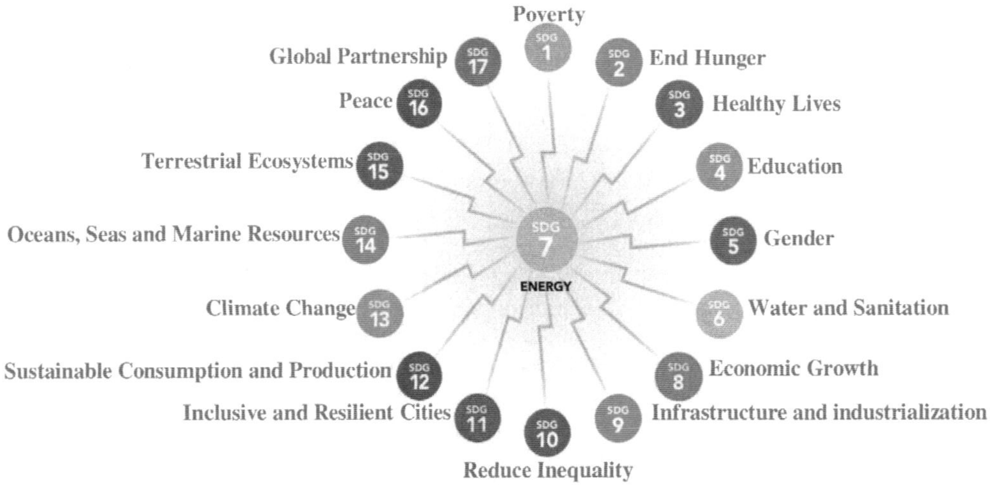

Figure 1. Relationship between Energy SDG and all remaining SDG. [2]

2. Movement towards sustainability

The University of Campinas (UNICAMP - Brazil), with only 52 years old, is known as "research factory" and training center for highly qualified professionals. It currently has 24 teaching and research units, accounting for 8% of all Brazilian academic research and 12% of national graduate studies.

With great representativeness and visibility in teaching and research, the University has been involved in environmental management as own policy since 2003, with the approval of Waste Management Group and Institutional Program for the Management of Biological, Chemical and Radioactive Waste.

After waste management initial actions, UNICAMP expanded its sustainability program, creating the Environmental Management Group (GGA in Portuguese) in 2006. Subsequently, an Environmental Management Program was implemented and finally, the Environmental Policy was launched in 2010.

In 2012, a technical group in energy efficiency proposed an Energy Management Program to GGA, becoming a future reference to SCP. And, in 2015, UNICAMP signed the ISCN-GULF Sustainable Campus Charter, from International Sustainable Campus Network (ISCN), that organizes a global support forum for key colleges, universities and corporate campuses, exchanging information, ideas and best practices to achieve sustainable operations in those institutions and integrating sustainability into research and teaching. With the signing of this Charter, UNICAMP had publicly committed itself to the sustainability and alignment of its research, teaching and administrative operations with the three principles established by the charter [4], starting the University movement towards sustainability.

Due to the results obtained by GGA, the increasing environmental awareness within the University and the Charter commitment, a broader proposal was formulated leading to the creation of a Sustainable University Management System (SUMS) and a Sustainable University Policy (SUP), with the Sustainable University Manager Group creation. [5]

2.1. Sustainable University Management Group (GGUS in Portuguese)

The SUMS is operated by GGUS and other complementary entities such as the Sustainable University Guidance Council (COUS in Portuguese) and the Technical Chambers (TCs), which should assist in the planning, implementation and management of SUMS, as shown in Figure 2.

The functions of GGUS are: set up, develop and implement policies, guidelines and standards for a sustainable university, based on continuous improvement and environmental, economic and social performance. COUS guarantee and approve the GGUS and TCs' actions.

The Technical Chambers' activities are: sustainable management programs and technical procedures elaboration; sustainability performance indicators and institutional goals definition; units' sustainability reports analyzation; and technical reports production.

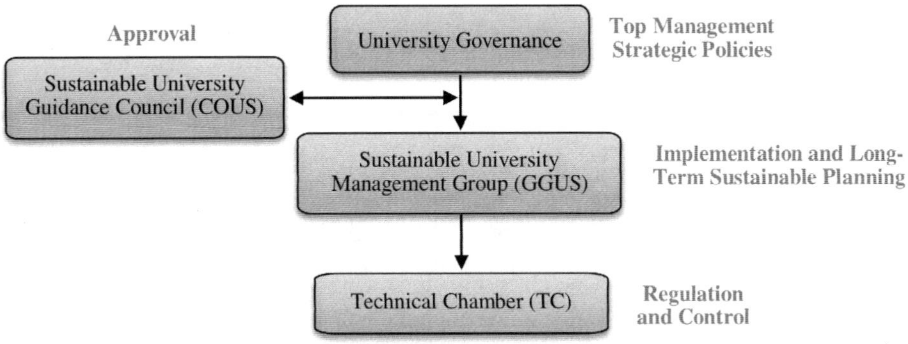

Figure 2. SUMS's Organizational Chart.

There are six technical chambers focused on: Energy; Water; Fauna and Flora; Urban Environment; Waste; and Environmental Education.

2.2. Energy Management Technical Chamber (CTGE in Portuguese)

CTGE is a TC created to improve and implement an Energy Management Program in the Campuses of Unicamp. This program aims to reduce energy consumption through the elaboration of management programs and technical procedures with the participation of all UNICAMP's entities. [5]

In 2016, CTGE conducted studies in energy management system implementation. Priority projects were modelled to medium-term improvements support in energy use and efficiency. These steps were adopted in the sustainable campus project a year later.

2.3. The Sustainable Campus Project

The sustainable campus project (SCP) is a partnership between UNICAMP and CPFL (local electricity distributor) placed in ANEEL's (Brazilian National Electricity Agency) R&D and Energy Efficiency program, signed in 2017 [3]. Initially the project started with 6 subprojects, however, three more subprojects have now been added [6], building a sustainable energy cycle around the university (Figure 3).

The subprojects (SPs) are:

- SP1 – Mini Operation Center (MOC): an intelligent data center for energy and environmental measurement, supporting all SCP subprojects;
- SP2 – Photovoltaic Generation: 534 kWp will be installed in four different university's locations;
- SP3 – Electric Vehicle: implementing a smart electric bus with sustainable recharge station;
- SP4 – Retrofit (EER): changing old equipment for labeled efficiency equipment;
- SP5 – Smart Energy Use (SEU): based on IoT sensors to measure internal environmental variables (as rooms and laboratories), in order to promote active energy efficiency actions by users;

- SP6 – Education: trainings, reports and a final book with the compendium of knowledge acquired in the SCP;
- SP7 – Energy Efficiency in Buildings (EEB): energy audit based on the Brazilian building labeling program;
- SP8 – Energy Procurement: electricity purchase in free market;
- SP9 – RF Mesh (RFM): network for transmitting data measured by the sensors.

It is expected that those subprojects will transform UNICAMP in a real Living Laboratory, approaching the academic research and new products development to market needs in sustainable energy solutions.

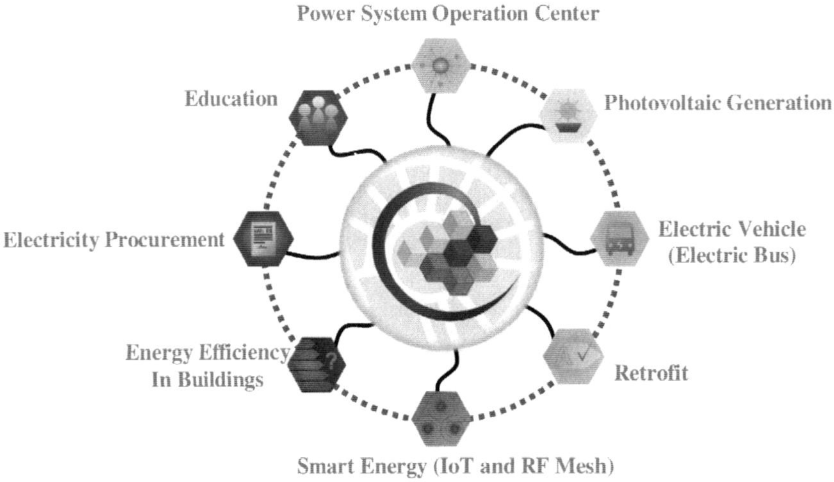

Figure 3. Sustainable Campus Model.

3. Improving the path to sustainable energy management

Despite the creation of GGUS and Sustainable Policies, it is necessary that University's management chain be integrated with the sustainable and priority projects applied (such as the sustainable campus project), in order to guarantee the benefits permanence implemented. The SCP is creating a new infrastructure to assure sustainable policies continuity for future projects and daily Campuses routine.

3.1. Infrastructure Construction for Sustainable Energy Management System - SEnMS

Five SCP subprojects will strengthen the SEnMS's infrastructure, with a reliable measurement, data flow and communication platforms. this smart chain will support the university's current demands, future and in progress projects. The benefits are:

- SP1 (Mini Operation Center) is providing real-time measurement per minute, through intelligent meters. The collected data will support decision making in energy efficiency, University's infrastructure and electrical system expansion planning, transformers predictive evaluation, as well as complementary studies developed by research groups in sustainable energy systems.
- SP4 (Retrofit) generated knowledge and expertise through the analysis of replacing inefficient equipment and building verification after and before the retrofit execution. This ensure new energy efficiency projects and a retrofit replication method.
- SP7 (Energy Efficiency in Buildings) is creating a new energy audit methodology for buildings, using: structural analysis, users' behavior and perception, air conditioning and lighting efficiency. This procedure will contribute to label buildings in energy efficiency and make units more sustainable and environmentally friendly.
- SP5 (Smart Energy Use) will address new concepts in smart buildings and user behavior for energy efficiency daily actions through IoT sensors. This system will be a model of electrical equipment user's awareness and their comfort and work environment improvement.

- SP9 (RF Mesh) will ensure continuous data flow infrastructure through a reliable communication platform.

3.2. ISO 50001: Energy Management System – Requirements with Guidance for Use

In order to improve energy performance by energy efficiency, use and consumption, ISO 50.001 establishes systems and processes methodologies to be incorporate in organizational infrastructure. The main Standard's structure is the continuous improvement based on PDCA (Plan-Do-Check-Act), that supports energy policy and energy management [7].

The PDCA must be part of organization context and practiced by all member, following the steps [8]:

- Plan: To achieve results that will lead to energy performance continuous improvement, is important the implementation of energy review, baseline establishment, energy performance indicators (EPIs), goals, targets and action plan.
- Do: To ensure competency and consider energy performance in the project and acquisition, based on Plan step, here is important realize the actions, operation controls, maintenance and communication expected.
- Check: Looking on energy performance direction, this step monitors, measures, analyzes, evaluates, audits and perform critical analysis of previous stages executions.
- Act: To solve nonconformity and continuously improve energy performance it is necessary to adopt corrective and definitive actions, besides documenting the acquired knowledge.

The four PDCA steps enhance the Energy Management System, providing routine and knowledge management and ensures sustainable actions continuity in long term. Figure 4 presents an example of PDCA with the implementation of subprojects (item 3.1), in order to incorporate its benefits from the University structure.

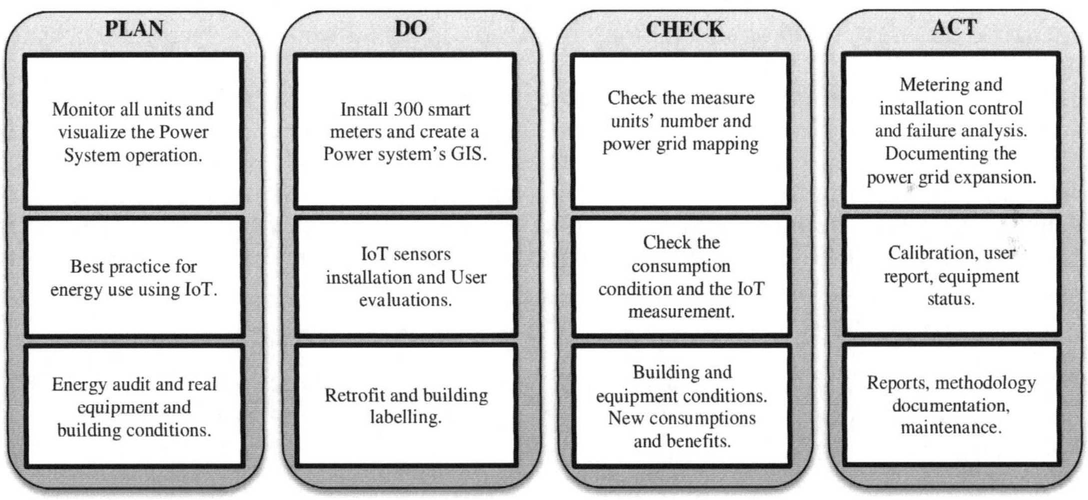

Figure 4. PDCA with Sustainable Campus Subprojects.

4. Benefits and improvement management structure

In this section, is presented an organizational model for energy management. This new architecture, based on ISO 50001 and supported by Sustainable Campus Project, will benefit the EnMS and strengthen continuous improvement throughout the University structure.

To achieve the sustainable policies intended outcomes, it is necessary include new agents and restructure the academic entities functions, aiming the energy sustainability. The suggested structure model is detail according to action's priority, on next items.

4.1. Strategic Sustainable Goals
The University's top management in each term, and/or annual reviews, must establish the University's sustainability guidelines, indicating the institution future path and clearly communicating them to the entire external and internal community.

4.2. Sustainable Policies
The GGUS, in its function, should define long-term planning to meet the guidelines established by University's top management, analyzing and suiting them according to Sustainable Policy and, finally, determine objectives and goals for each technical chamber.
Policy modifications, strategic reports to top management and tactical guidelines must be approved by COUS and disseminated to entire external and internal community.

4.3. Implementation
Each technical chamber (TC) will be responsible to regulate goals and targets for all entities, according GGUS's definition. All units must frequently report your results and actions to TCs. In case of CTGE, daily and monthly control metric should be adopted for managerial and operational control, using automatic reports and dashboards for quick actions.
CTGE must be closer to campus energy consuming units, following the continuous improvement actions and results adopted by each entity. Supporting them in training and tools for continuous improvement, daily management, energy efficiency and internal process checking.

4.4. Verification
The verification is necessary to ensure and review action's result from continuous improvement or upgrading investment. To this, each consumer needs to monitor their process by measuring or regularly variable checking that influence their results. The Mini Operation Center (MOC) will be responsible for energy verification infrastructure and to support all entities in energy consumption M&V.
Studies and energy review of main processes (those that directly influence unit's energy performance) must be controlled and measured by entity itself. This means that, consumers must deliver a regularly report to CTGE indicating your status, according to MOC metrics and CTGE goals.
Each year, CTGE shall perform an audit in each entity, to verify internal controls (maintenance, assets, equipment calibration, etc.) and nonconformity records, with corrective, prevention and improvements actions. The units themselves can create an internal committee to evaluate the audit items in order to guarantee their own processes.

4.5. Management Review
The COUS has the obligation to: monitor results generated by all technical chambers and GGUS, ensuring guidelines compliance; verify the strategic actions implemented; and sustainable policy review approval.

4.6. Organizational Continuous Improvement
Figure 5 presents the organizational continuous improvement structure, as a chapter synthesis.

5. Expected results
The Sustainable Campus project implementation will leverage the University's energy performance improvement in short-term. To guarantee those benefits in long-term it is necessary brings the University's organizational structure closer together, integrating top management, GGUS, Units and Professionals for a single objective, the Energy Sustainability
With SCP modernization and the initial structure of sustainable management system (SMS) in a strong structure will allow the ISO 50001 Standard application and an EnMS development, in accordance with University's sustainability policy.
Initially, the SCP works as energy measurement and verification infrastructure support (Do and Check methodology) from SMS. Your first data analysis is supporting the initial Sustainable Energy Policy

(SEP) and Sustainable Energy Management System (SEnMS) design (Plan methodology), by Energy Technical Chamber (CTGE). These Sustainable Energy policy and system will provide the planning activities background (Plan methodology). Finally, to complete PDCA cycle, the data, measurement, policy and a real time support system will able continuous improvement practice in daily Unicamp's processes (Action methodology).
As previously seen, subprojects (item 3.1) are extremely important for a new model development of University's energy management and for continuous improvement implementation inside the ISO Standard-based organizational culture. In the end, it will be possible to evaluate energy efficiency actions, use and consumption of energy, as well as to monitor energy performance.

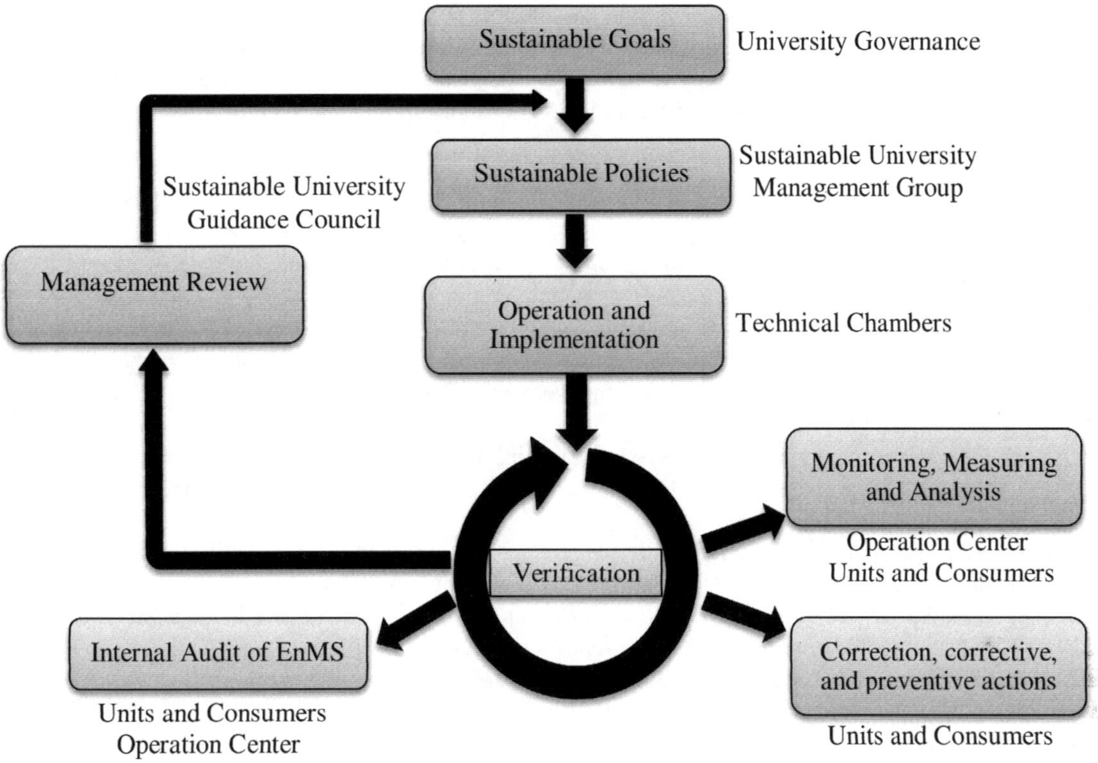

Figure 5. Organizational Continuous Improvement.

5.1. ISO Standard Benefits
According to [9], around 40% of future energy is expected to be from energy efficiency gains. By that way, a support to allow energy efficiency application and your continuous benefits is necessary.
The Standard implementation in the University's processes will encourage and strengthen the continuous benefits in energy efficiency actions, energy use and consumption, through: continuous improvement methodology, processes mapping and traceability of actions carried out. That is, the EnMS ISO-based ensures knowledge maintenance for energy performance into daily operation, changing individual traditional management to cultural business processes.

5.2. Living Lab in Sustainable Energy
Through Sustainable Campus Project and EnMS presented in this article, Unicamp will strengthen its living lab in sustainable energy, disseminating and integrating future projects and actions that may be replicated to other universities, cities and isolated communities.

In underway with new R&D and Continuous Improvement projects development, this process will add more structure and knowledge to our living laboratory. This open innovation ecosystem model will integrate users, academia and the market (public-private companies and startups).

5.3. Summary

As mentioned before, this article establishes a linkage between Sustainable Campus Project to ISO 50001, using one to support another, as example: measurement SCP structure with ISO performance baseline; Standard protocols to sustain energy efficiency actions; SCP organizational architecture to EnMS implementation; and continuous improvement to Sustainable Energy Efficiency in long-term.
Initially, SCP brought low-cost results that are launching the EnMS implementation in UNICAMP. In other words, before all SCP's infrastructure installation and only with energy efficiency idealized movement, some observation-based actions improved equipment retrofit benefits[10] and increased the top management interest in energy sustainability. In two years, we will have an integrated EnMS that can be used by another institutions, with low-cost execution.

6. References

[1] PNDU and IPEA. "2030 Agenda Platform". [Internet]. [Accessed September 2018]. Available from: http://www.agenda2030.com.br/.

[2] World Bank. 2017. *State of Electricity Access Report.* (Washington DC: World Bank Publications).

[3] L. C. P. da Silva, et al. 2018. Sustainable Campus Model at the University of Campinas—Brazil: An Integrated Living Lab for Renewable Generation, Electric Mobility, Energy Efficiency, Monitoring and Energy Demand Management. *Towards Green Campus Operations.* ed Springer. pp 457-472.

[4] University of Campinas. ISCN/GULF Sustainable Campus Charter. [Internet]. April 2015 [Accessed September 2018]. Available from: http://www.ggus.depi.unicamp.br/wp-content/uploads/2017/01/ISCN.pdf

[5] GGUS. Sustainable University Management Group. [Internet]. UNICAMP. [Accessed September 2018]. Available from: http://www.ggus.depi.unicamp.br

[6] GGUS. Sustainable Campus Projetct. [Internet]. UNICAMP. June 2018 [Accessed September 2018]. Available from: https://www.campus-sustentavel.unicamp.br

[7] ABNT. 2011. *NBR ISO 50001: Sistemas de gestão da energia – Requisitos com orientações para uso.* (Rio de Janeiro: ABNT).

[8] ABNT. 2018. *NBR ISO 50001: Sistemas de gestão da energia – Requisitos com orientações* para *uso.* (Rio de Janeiro: ABNT).

[9] International Energy Agency. 2014. *Energy Efficiency Indicators: Fundamentals on Statistics.* (Paris: OECD/IEA).

[10] Procel. Results from: Chamada Pública Procel Edifica 2018. [Internet]. Eletrobrás. February 2018 [Updated: August 2018, Accessed: September 2018]. Available from: http://eletrobras.com/pt/AreasdeAtuacao/Chamada_Publica_Procel_Edifica/Lista_de_Classificados_Final_Chamada_Publica_Procel_Edifica_2018.pdf

Acknowledgments

The National Council for Scientific and Technological Development (CNPq in Portuguese) has supported this work financially. Opinions, premises, conclusions and recommendations expressed in this document are those of the authors and do not necessarily reflect the opinions of the CNPq.
This work was developed under the Electricity Sector Research and Development Program PD-00063-3032 / 2017 - PA3032: "Sustainable campus model at the University of Campinas - Brazil: An integrated living lab for renewable energy, electric mobility, energy efficiency, monitoring and energy demand management", regulated by the National Electricity Agency (ANEEL in Portuguese), in partnership with CPFL Brazil (Local Electricity Distributor).

The impact of rural house's orientation on heating energy consumption in severe cold regions of China

T Shao[1] and H Jin[2]

1Assistant Professor, School of Mechanics, Civil Engineering and Architecture, Northwestern Polytechnical University, Xi'an, China
2 Professor, School of Architecture, Harbin Institute of Technology, Harbin, China
E-mail: shaoteng1989@qq.com

Abstract. The severe cold regions of China with characteristics of winter long, cold and snowy, summer short-lived, the heating period lasts for half a year. The heating energy consumption accounts for more than eighty percent of total energy consumption of rural residence. Because the rural houses are single-storey detached house, the heat of solar radiation has greater impact on energy consumption than urban buildings. Firstly, the temperature sensors were used to test the indoor thermal environment of different orientation rooms to clarify the effect of building orientation. Secondly, taking building orientation as main variable (set up twenty-four values, every 15°), building envelope's thermal performance as auxiliary variables (set four groups), the simulation software DesignBuilder was adopted to analyse the influence rule of building orientation on heating energy consumption under different conditions. It concluded that with the enhance of building envelope's thermal performance, the impact of building orientation on heating energy consumption and indoor temperature gradually increase, thus focus on the building orientation become more meaningful.

1. Introduction

Solar energy is one of the important renewable energy sources. The influence of orientation on building energy consumption mainly relies on the different utilization degree of solar radiant heat. Due to the uneven solar radiation of each orientation, the heat gain of rural house with different orientation will surely be different, so that the heating energy consumption required maintaining indoor comfortable temperature in winter is also different. Several studies have been carried out to show the correlation between orientation and energy consumption. Hussain et al (2011) [1] taking educational building for example, applied DEROB-LTH software to analyse the energy consumption and PMV values of rooms with different orientations. Abanda et al (2016) [2] used BIM technology to analyse the effect of building orientation on whole energy consumption, concluding that suitable orientation has great energy saving potential in the whole life cycle, and with 30 years as the operating cycle, the best oriented architecture can save £878 compared to the worst oriented building. Tokbolat et al (2013) [3] selecting high-rise residence of Kazakhstan as research object, analysed the effect of orientation on building energy consumption. Dong Tan et al (2011) [4] used PKPM software to conclude the building energy consumption of 13 different orientations, and the energy saving contribution rate of each orientation for nine-story and eighteen-story residential buildings. Yu Wei (2011) [5] adopted EnergyPlus software analysing residential building energy consumption and indoor thermal comfort of eight orientations. Building orientation is referred to as one of the most important factors among parameters for passive design of buildings (Morrissey et al, 2011 [6] and Florides et al, 2002 [7]). However, existing research mainly focus on the urban public and residential buildings, few on the

Content from this work may be used under the terms of the Creative Commons Attribution 3.0 licence. Any further distribution of this work must maintain attribution to the author(s) and the title of the work, journal citation and DOI.
Published under licence by IOP Publishing Ltd

rural houses, appeared regional characteristics, as well as usually aims at single condition. If the design criteria for other factors are changed, the impact of orientation on building energy consumption how to change is worthy of further study.

This paper, taking rural houses of severe cold regions as research object, sets building orientation as main variable, building envelope's thermal performance as auxiliary variables, and adopts DesignBuilder software to analyse the influence rule of building orientation on energy consumption and indoor temperature under different conditions.

2. Methods

2.1. Field testing

A two-story detached rural house which is uninhabited and unheated was selected to carry out field test between February 25 and 27, 2015. Solar radiation is the main factor affecting indoor thermal environment, avoiding other factors interfere. The building appearance and layout of testing points are as shown in Fig.1 and Fig.2. BES-02 temperature sensors (temperature precision: ≤0.5℃) are applied to continuous monitoring indoor thermal environment. The indoor testing points are arranged in the plane geometry centre of each room, with 1.6m height, and the data is automatically recorded every 5min. The instruments are wrapped in aluminium foil to prevent the effects of solar radiation.

Figure 1. The appearance of testing house

Figure 2. The layout of testing points

2.2. Numerical simulation

DesignBuilder software is used to simulate the heating energy consumption of rural houses in severe cold regions. The software taking EnergyPlus as energy consumption simulation engine has passed the building envelope thermal performance and energy consumption of ASHRAE Standard (ASHRAE, 2006 [8]). According to the survey results of rural houses in severe cold regions (Teng Shao et al, 2017 [9]), the form and envelop structure of benchmark building are as shown in Fig.3 and Table 1.

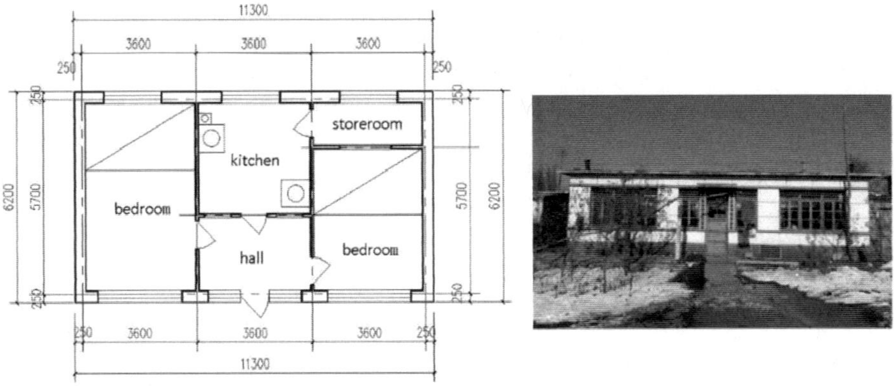

Figure 3. The plane and appearance of benchmark building

Table 1. The building envelope parameters of benchmark building

Building envelope	Construction	U-value ($W/m^2 \cdot K$)
Exterior wall	20mm cement mortar + 370 mm brick + 20mm composite mortar	1.58
Roof	suspended ceiling + wood joist + 100mm plant ash + wood roof truss + plank + waterproof layer + tiled roof	0.93
Exterior window	single wooded window	4.70
Exterior door	single wood door	3.50
Ground	20mm cement mortar + concrete cushion + rammed earth	3.00

It takes building orientation as main variable, building envelope's thermal performance and south window-wall area ratio as auxiliary variables, and explores the influence of building orientation on heating energy consumption under multifactorial interaction. Setting 0° as the south orientation, every 15° set a model, a total of twenty-four cases. The performance parameters of building envelope are selected according to the survey data and relevant energy saving design standards, including four groups parameter. The values of variables are as show in Table 2.

Table 2. The values of variables

Building orientation	Building envelope thermal performance （$W/m^2 \cdot K$）					
	Wall	Roof	Window	Ground	Door	Reference standard
0° (south)	1.58	0.93	4.70	3.00	3.50	A. Traditional rural houses (uninsulated measures)
15°	0.50	0.45	2.00	0.30	2.00	B. <Design standard for energy efficiency of rural residential buildings (GB/T50824-2013)>
30°	0.30	0.25	1.80	0.14	1.50	C.<Design standard for energy efficiency of residential buildings in severe cold and cold zones (JGJ26-2010)>
45°-345°	0.10	0.10	0.80	0.10	1.00	D.<Criteria for the Passive House, German>

The rural house is considered as a single zone in the simulations carried out by the DesignBuilder program for the calculation of heating energy consumption during winter heating season (October 20 to the next year April 27, a total of 175 days). This reduces the required simulation time with negligible effect on the accuracy of the results. According to the *Design standard for energy efficiency of rural residential buildings* (GB/T50824-2013), the indoor calculation temperature is set to 14 ℃, and the ventilation rate is $0.5h^{-1}$. Using the radiator and heated kang as internal heat sources. The heat dissipation of equipment, human clothing and metabolism, illumination and so on are set according to the actual situation of rural houses.

3. Results and discussions

3.1. Field testing result

Taking air temperature as the evaluation index of indoor thermal environment, the test data of February 26 are selected to compare and analyse. To compare the indoor temperature of different orientation rooms, two groups are selected for comparative analysis. Group1: bedroom and kitchen of first floor, the bedroom on the south side and the kitchen on the north side. Group2: secondary bedroom and study of second floor, the secondary bedroom with south window and the study with north window. The temperature changing curves of the two groups are shown in Fig. 4. The average temperature of bedroom (south room) is -1.07℃, and the maximum and minimum temperature respectively are 0.4℃ and -1.7℃. The average temperature of kitchen (north room) is -1.82℃, and the maximum and minimum temperature respectively are-0.88℃ and -2.32℃. The results show that the average, maximum and minimum temperature of south room are respectively higher 0.75℃, 1.28℃ and 0.62℃ than north room.

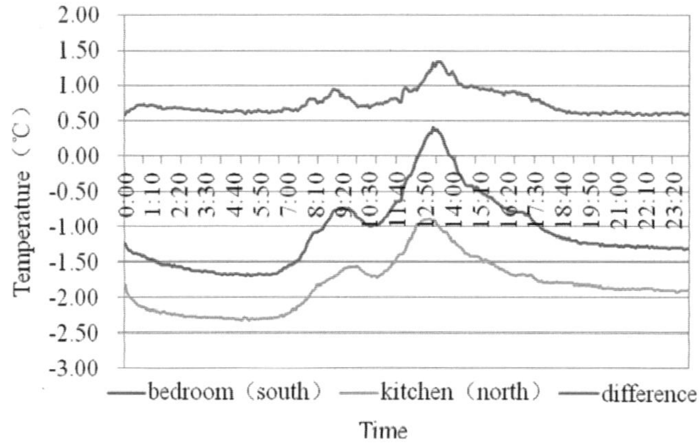

a)

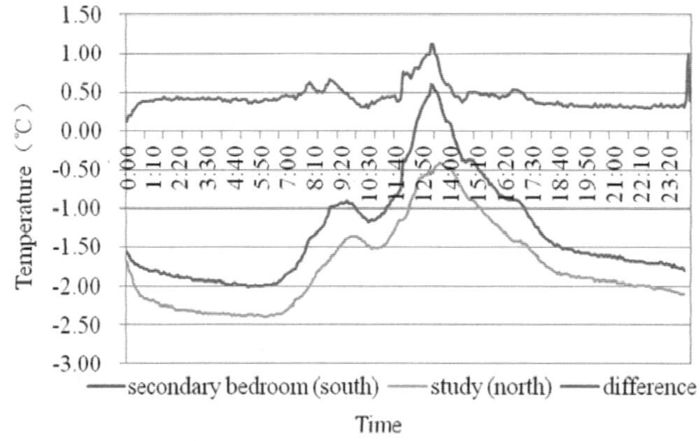

b)

Figure 4. The changing curve of indoor temperature. a) Group 1, b) Group 2.

As shown in Fig.4 b), the average temperature of secondary bedroom (with south window) is -1.31℃, and the maximum and minimum temperature respectively are 0.6℃ and -2.01℃. The average temperature of study (with north window) is -1.74℃, and the maximum and minimum temperature respectively are-0.40℃ and -2.40℃. The results show that the average, maximum and minimum temperature of south room are respectively higher 0.43℃, 1.0℃ and 0.39℃ than north room. Among the time 12:00 to 14:00, the sunshine directly on the south room, thus the temperature difference of two groups reached the maximum, other times the temperature difference remaining stable. To sum up, the solar radiation has a certain heating effect on the interior. The temperature of south room is higher than the north room. Building orientation has effect on the indoor thermal environment, also play a role in reducing heating energy consumption in winter.

3.2. Simulation result

The changing curve of energy consumption absolute value with building orientation under four different type envelopes are as shown in Fig.5.

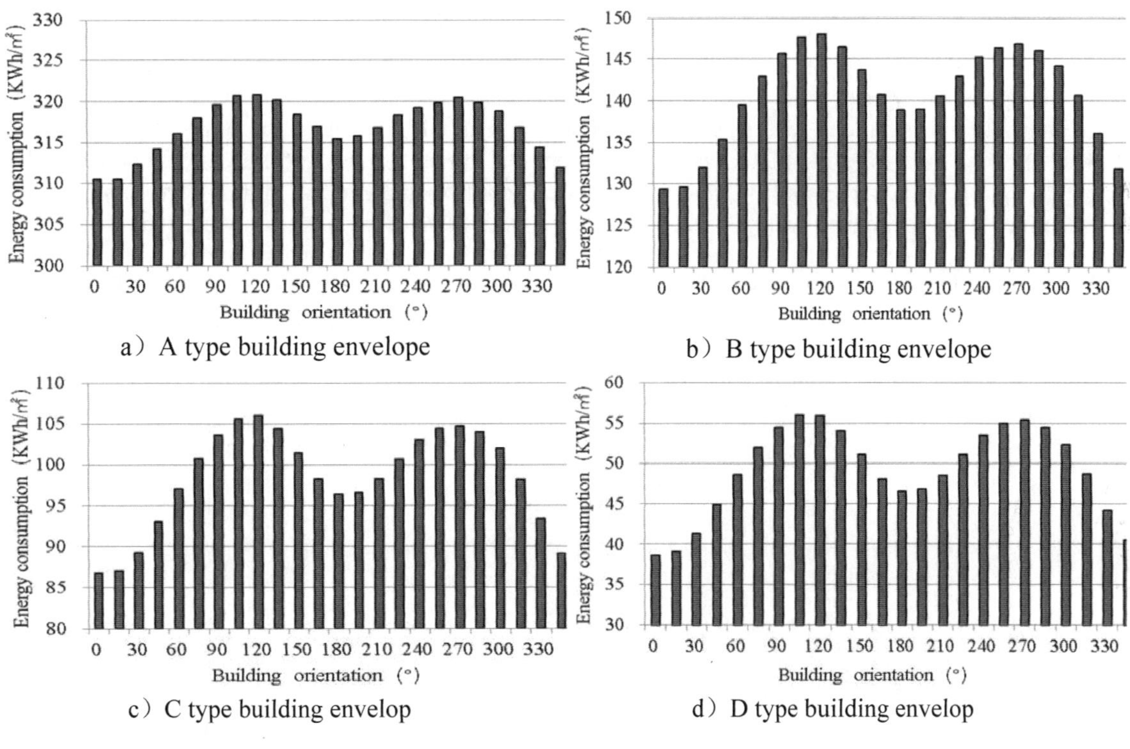

Figure 5. The changing curve of energy consumption absolute value with building orientation under four different types building envelope

As shown in Fig.5, when the building orientation is 0° (south), the energy consumption reaches to the minimum value, with the orientation rotating from 0° to 360°, the energy consumption presenting the changing rule "low-high-low-high-low". As shown in Fig.6, when the building orientation is 90° and 270° (west and east), the energy consumption is higher because of the less gain of solar radiation heat. It's important to note that the benchmark building with window-wall ratio of frontage is 0.4 and of rear face is 0.3, when the building orientation is 180°, that is the frontage faces north and the rear face towards south, a certain amount of solar radiation heat also can be obtained, but the window-wall ratio less than frontage, the heat gain of solar radiation less than the condition of 0° orientation. Thus, the energy consumption is in the middle level, not reach to the minimum value.

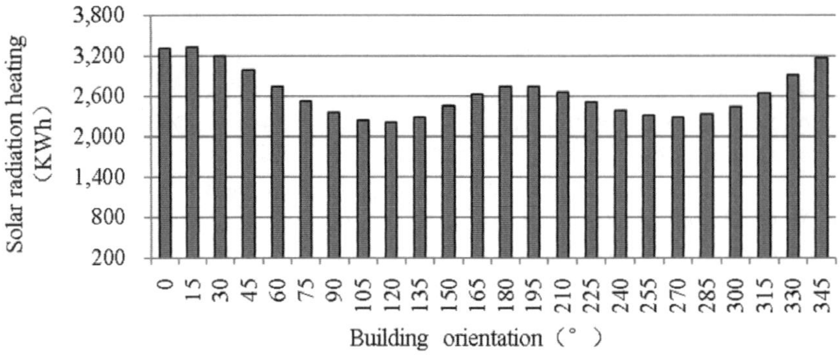

Figure 6. Solar radiation heating of different building orientation

In general, the changing rule of energy consumption with building orientation under four types building envelope is consistent. However, with the change of building envelope's thermal performance, the change rate of every orientation's energy consumption is different. Using relative value to show the change of every orientation's energy consumption, the energy consumption value of south is set as one and other orientations' energy consumption are expressed by relative value. As shown in Fig.7, with the enhance of building envelope's thermal performance, the difference of relative value gradually increases, indicating that the change of building orientation has more effect. For example, when the A type building envelope is adopted, the relative value of north's energy consumption is 1.02, while applying the D type building envelope, this vale increases to 1.20. Therefore, when the building envelope's thermal performance is improved, the rural houses' orientation should be as south as possible, the energy saving effect is more significant, thus focusing on the building orientation become more meaningful.

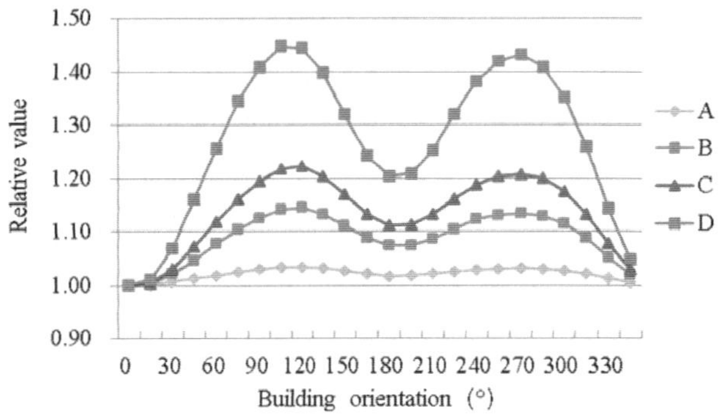

Figure 7. The changing curve of energy consumption relative value with building orientation under four types building envelope

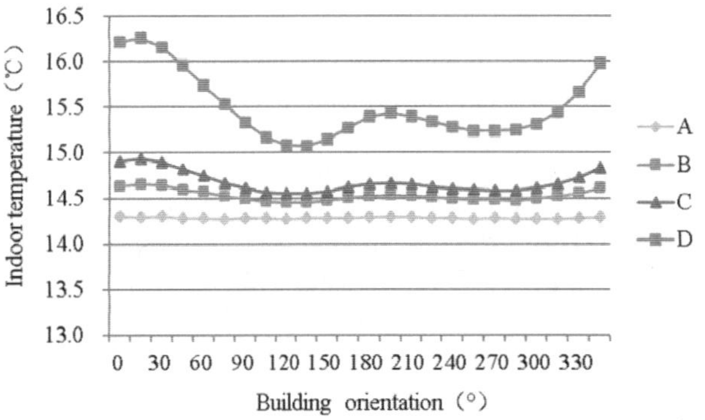

Figure 8. The changing curve of indoor temperature with building orientation under four types building envelope

The indoor temperature changing rule with building orientation is shown in Fig.8. With the improvement of building envelope's thermal performance, the air temperature gradually increases, but the fluctuation rules are different. For example, when the D type building envelope is adopted, the air temperature is highest, but the temperature of different orientations has obvious difference, the standard deviation of temperature data is 0.36. The building orientation has a significant effect on indoor temperature. While applying the A type building envelope, the air temperature is lowest, but the temperature of different orientations is basically the same, the standard deviation of temperature data is only 0.009, indicating that the building orientation has a minimal effect on indoor temperature. It also proves that the higher thermal performance the building envelope, the more attention should be paid.

4. Conclusions

Through the field testing and numerical simulation, it can be concluded that building orientation has an impact on energy consumption and indoor thermal environment in winter, while the design criteria for other factors are changed such as building envelope's thermal performance, the effect extent of building orientation on energy consumption is also changing.

According to the testing results, the solar radiation has a certain heating effect on the interior, and the temperature of south room is higher than the north room. On the basis of simulation results, the changing rule of energy consumption with building orientation under four types building envelop is consistent, but with the improvement of building envelope's thermal performance, the change rate of every orientation's energy consumption and the indoor temperature fluctuation increase, that is the building orientation has greater impact on heating energy consumption and indoor temperature. It's more meaningful to focus on the building orientation.

5. References

[1] Hussain H. Alzoubi and Amal T. Malkawi. 2011. Investigating the Effect of Building Orientation on Thermal Comfort and Energy consumption in Educational Buildings. *Global Conference on Renewable energy and Energy Efficiency for Desert Regions,* 1-10

[2] Abanda F H and Byers L. 2016.An investigation of the impact of building orientation on energy consumption in a domestic building using emerging BIM (Building Information Modeling). *Energy*, 97, 517-527

[3] Tokbolat S, Tokpatayeva R et al. 2013. The Effects of Orientation on Energy Consumption in Buildings in Kazakhstan. *Journal of Solar Energy Engineering*, 135(4), 1-8.

[4] Dong Tan, Li Nianping et al. 2011. The influence and analysis of residential building orientation on energy consumption in Changsha. *Proceedings of the conference on building environment and equipment technology*, 403-405

[5] Yu Wei. 2011. The low Energy-Consumption Strategy for Improving Indoor Thermal Environment Quality in Residential Building. *Ph.D. Thesis*, Chongqing University.

[6] Morrissey J, Moore T et al. 2011. Affordable passive solar design in a temperate climate: An experiment in residential building orientation. *Renewable Energy*, 36(2), 568-577.

[7] Florides G A, Tassou S A et al.2002. Measures used to lower building energy consumption and their cost effectiveness. *Applied Energy*, 73(3), 299-328.

[8] ASHRAE. 2006. ANSI/ASHRAE Standard 140-2004, Building Thermal Envelope and Fabric Load Tests-DesignBuilder Version 1.2.0. Atlanta: American Society of Heating, Refrigerating, and Air-Conditioning Engineers, Inc

[9] Teng Shao, Hong Jin et al. 2017. Current Situation and Improving Strategies for Northeast China's Rural Housings. *Open House International*, 42(4): 70-77.

Acknowledgement

This paper is support by the Fundamental Research Funds for the Central Universities (Grant No. G2018KY0308) and the National Natural Science Foundation of China (Grant No. 51378136).

IOP Conf. Series: Earth and Environmental Science **257** (2019) 012036 doi:10.1088/1755-1315/257/1/012036

Study on the effect of ethanol content on the fire risk of nitrocellulose

R C Wei[1,2], S S Huang[1,2], Z Wang[1], R Yuen[2], and J Wang[1,*]

1 State Key Laboratory of Fire Science, University of Science and Technology of China, Hefei, Anhui 230026, PR China
2 Department of Civil and Architectural Engineering, City University of Hong Kong, Hong Kong 999077, PR China
* Corresponding author
E-mail: wangj@ustc.edu.cn

Abstract. Obtaining the burning properties of nitrocellulose with different ethanol contents is necessary to reduce the possibility of nitrocellulose accidents in the process of handing, storage and usage. In this study, the effect of humectant content on the burning properties of NC was tested by the ISO 5660 Cone Calorimeter Test. It was noted that the amount of the ethanol content had little effect on the ignition time of nitrocellulose mixture. Both the maximum mass loss rate and the average mass loss rate of the sample decreased with an increasing ethanol content and gradually closed to fixed values. Also, it was concluded that the peak heat release rate of the sample was mainly contributed by the nitrocellulose component.

1. Introduction

Nitrocellulose (NC) is a typical industrial ingredient and is widely used in both military and civilian industries [1, 2]. NC is usually kept wet in water or an alcohol humectant. Once exposed to dry conditions, NC is shock sensitive and can ignite spontaneously and explode when exposed to heat, flames, ignition sources, sunlight or oxidizing agents (such as perchlorates, peroxides, permanganates, chlorates, nitrates, chlorine, bromine and fluorine) [3, 4]. Defining the effect of the ethanol content on the burning behaviors of NC can improve our understanding of the potential dangers of NC products in the storage and transportation process, and optimize the content of product components.

The burning behaviors of NC-humectant mixtures have been investigated by a few researchers. Based on the measurement of ISO 5660 Cone Calorimeter, recent studies [5, 6] found that the fire hazard of NC-humectant mixtures satisfies NC-isopropanol > NC-ethanol > NC-water. Liu et al. [7] experimentally examined the heat release rate (HRR) of NC-isopropanol and NC-ethanol and proposed a simplified method to predict the HRR of industrial NC according to flame height. Furthermore, a sequence of experiments on NC-isopropanol mixture with eight consecutive bulk densities was performed by Liu et al. [8], and it was concluded that, with increasing bulk densities, the fuel regression rate in the quasi-steady burning stage exhibit two distinctive stages: constant stage and decreasing stage. However, the above studies focused only on the burning behaviors of NC with a fixed humectant content and scarce work concerning the effect of different humectant contents on the behaviors of NC-humectant fire. In order to fill the gap in this field and raise awareness of NC products, it is necessary to carry out the present study.

NC can be preserved with the addition of ethanol, while with the loss of ethanol, the spontaneous ignition of NC will be inevitably triggered to cause uncontrollable fire or explosion, which has been asserted to be primary causes for the disastrous explosion in Tianjin Port in China on August 12, 2015.

Content from this work may be used under the terms of the Creative Commons Attribution 3.0 licence. Any further distribution of this work must maintain attribution to the author(s) and the title of the work, journal citation and DOI.
Published under licence by IOP Publishing Ltd

IOP Conf. Series: Earth and Environmental Science **257** (2019) 012036 doi:10.1088/1755-1315/257/1/012036

This disaster resulted in 165 deaths, 8 missing persons, 798 injuries and CNY 6.866 billion direct economic loss [9, 10]. To provide basic knowledge for fire behaviors of NC-ethanol mixtures with different ethanol contents, an attempt was made in the present study to perform fire tests on NC-ethanol mixtures by means of ISO 5660 Cone Calorimeter.

2. Experiments

In this work, the samples were prepared by mixing different mass of ethanol into pure NC (11.96 % N). Table 1 shows the weight of each element in the samples. The total mass of each sample is 10.0 ± 0.2 g.

Table 1. The experimental configurations.

Samples	No.	Ethanol content/wt.%	NC (wt.)/g	Ethanol (wt.)/g
NC-10	1	10	9.0	1.0
NC-20	2	20	8.0	2.0
NC-30	3	30	7.0	3.0
NC-40	4	40	6.0	4.0
NC-50	5	50	5.0	5.0

The ISO 5660 Cone Calorimeter Test [11] has proven to be an effective measurement method to assess the fire risk of NC mixtures. [12] A schematic view of the ISO 5660 Cone Calorimeter is shown in Figure 1. In the present study, a series of experiments was performed using the Cone Calorimeter provided by City university of Hong Kong. Before the start of the experiment, different masses of ethanol and pure NC were thoroughly mixed according to the ratios shown in Table 1. As presented in Figure 2 (a), the sample was wrapped in aluminum foil and placed in a steel square pan that was 100 mm wide and 20 mm high. The experiments were performed in a uniform radiant heat flux of 5 kW/m^2 over the sample surface. The CO and CO_2 produced after the burning of the sample is recorded by the data acquisition system after passing through soot filter and water filter. As shown in Figure 2 (b), there is basically no combustion residue for each sample. Fire risk parameters, such as ignition time t_{ig} and HRR of the NC with different ethanol content were recorded.

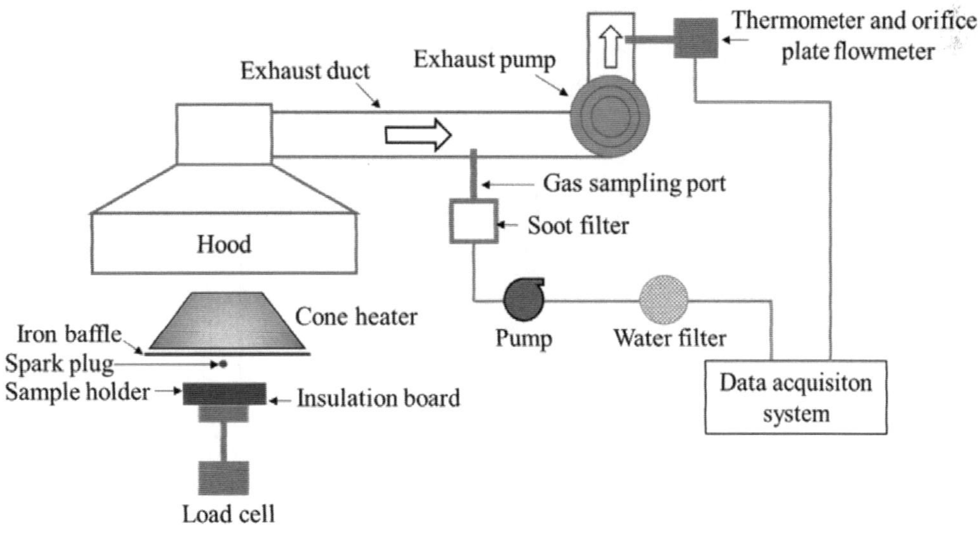

Figure 1. The configuration of the cone calorimeter.

(a) (b)

Figure 2. Comparison of 20 % ethanol sample. (a) Before burning; (b) After burning

3. Results and Analysis

3.1. Ignition time

The t_{ig} of the sample as a function of ethanol content is plotted in Figure 3. It can be seen that the amount of the ethanol content has little effect on the ignition time t_{ig} of the sample and the t_{ig} fluctuates around 21.7 s. This phenomenon is most likely attributed to that t_{ig} mainly depends on the volatile properties of treated materials and ignition would occur when the fuel vapor concentration around the electrical spark reaches the combustion limit. [13] As indicated in previous researches [6, 14] and Figure 4 [13, 15], the amount of ethanol volatilized is greater than the amount of decomposition products of NC under 5 kW/m^2 external radiant heat flux, thus the t_{ig} of NC-ethanol mixture is dominated by ethanol.

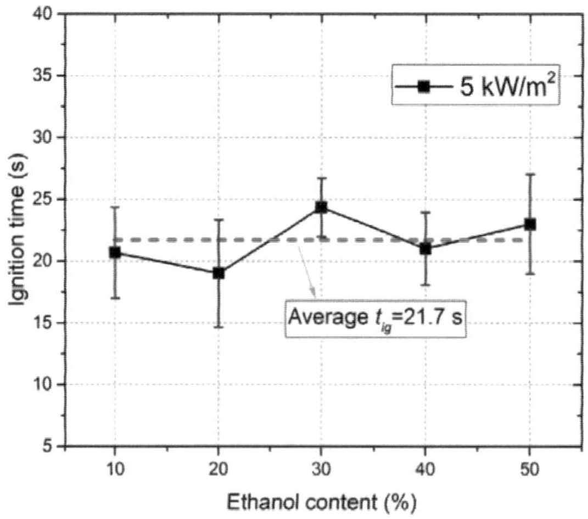

Figure 3. Ignition time of samples as a function of ethanol content under external radiant heat flux of 5 kW/m^2.

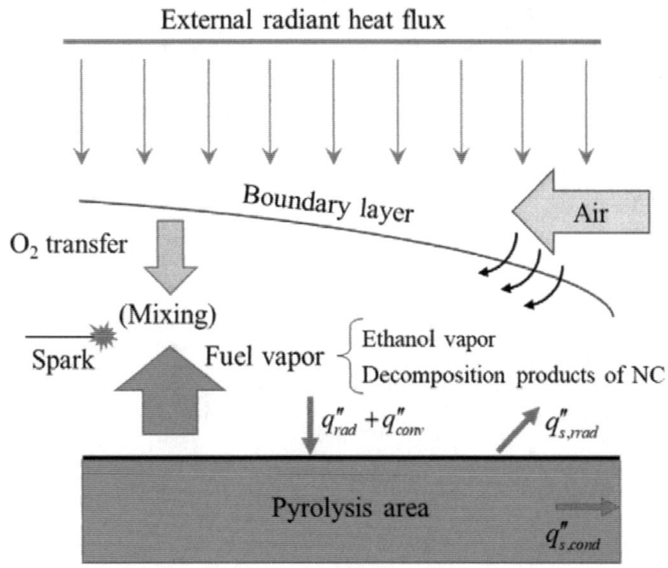

Figure 4. The heat transfer and ignition of NC-ethanol under external radiant heat flux.

3.2. Burning process

The mass loss rate (MLR) curves of NC-ethanol mixtures are plotted in Figure 5. The burning period increased, while the peak value of the MLR decreased with an increasing ethanol content. The maximum MLR $\dot{m}_{max}$ and average MLR $\dot{m}_{ave}$ of samples are analyzed and plotted in Figure 6. It can be seen from Figure 6 that both $\dot{m}_{max}$ and $\dot{m}_{ave}$ decreased with the increase of ethanol content and approached fixed values.

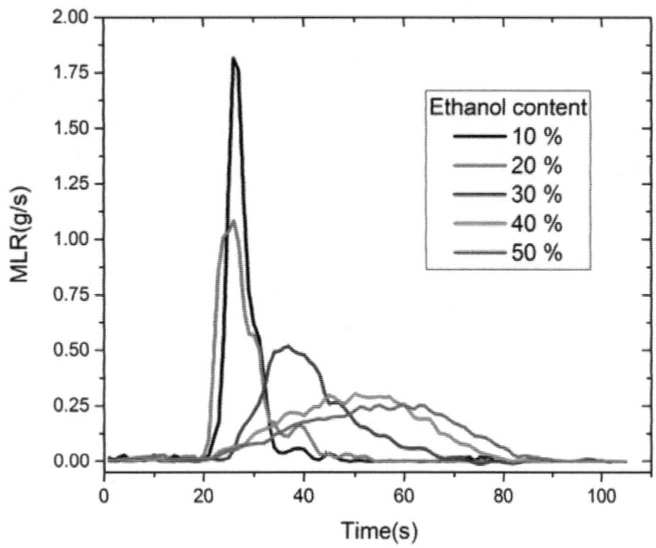

Figure 5. Mass loss rate of NC-ethanol mixture with different ethanol contents.

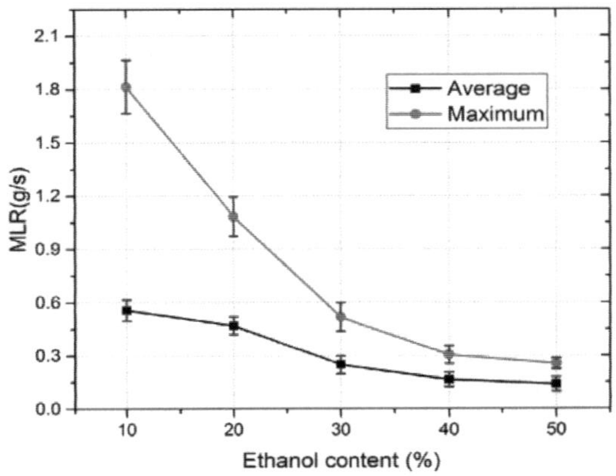

Figure 6. The $\dot{m}_{ave}$ and the $\dot{m}_{max}$ of sample with the increase in ethanol content.

The HRR histories of samples with varying ethanol contents, as measured by the cone calorimeter, are shown in Figure 7. The HRR histories had the same trend as the burning rate. For the NC mixture fire, the total MLR of the mixture can be expressed as:

$$\dot{m} = \dot{m}_1 + \dot{m}_2 \tag{1}$$

where Δh_1 and Δh_2 are the effective combustion heats of ethanol component and NC component, respectively. The values of Δh_1 and Δh_2 are 29.68 kJ/g and 4.17 kJ/g, respectively. [14, 16, 7]

Then, the HRR of NC-ethanol mixtures can be expressed as:

$$\begin{aligned} \dot{m}_1 &= \left(\dot{q} - \Delta h_2 \cdot \dot{m}\right)/\left(\Delta h_1 - \Delta h_2\right) \\ \dot{m}_2 &= \left(\dot{q} - \Delta h_1 \cdot \dot{m}\right)/\left(\Delta h_2 - \Delta h_1\right) \end{aligned} \tag{2}$$

Based on the measured MLR and HRR data, MLR corresponding to each component of the mixtures can be calculated. The above method was applied to the NC-ethanol mixture with 30 % ethanol content. The result is shown in Figure 8. For NC-ethanol mixture with 30 % ethanol content, the MLR of NC agreed well with the total MLR mixture. Therefore, it can be inferred that for mixtures with an ethanol content of less than 30 %, the MLR was mainly contributed by NC.

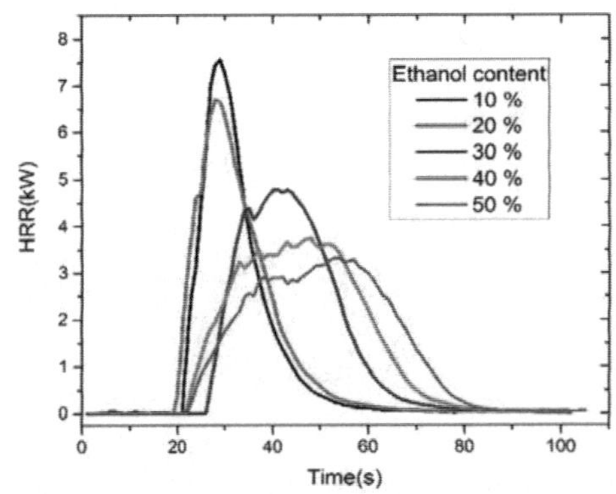

Figure 7. The HRRs of NC-ethanol mixtures with different ethanol

content under an external radiant heat flux of 5 kW/m^2.

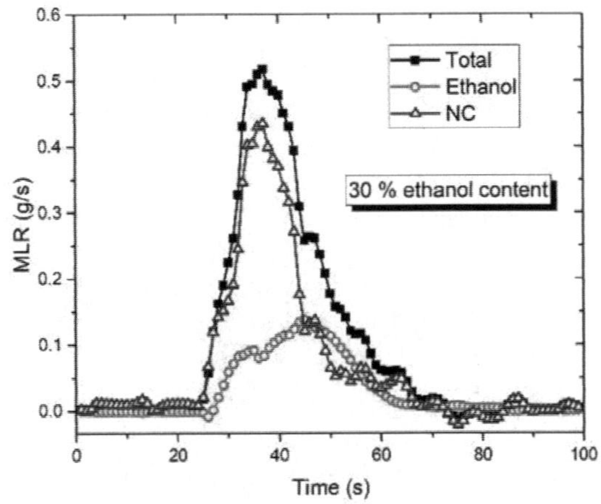

Figure 8. The calculated MLRs of ethanol component and NC component in the NC-ethanol mixture.

4. Conclusions

In order to investigate the effect of ethanol content of NC-ethanol mixture on its burning behaviors, a series of experiments on NC-ethanol mixture with five consecutive ethanol content were carried out in the present study. Primary measurements of ignition time, MLR and HRR were analyzed and discussed in detail. The major conclusions are drawn as follows.

The ignition time of NC-ethanol mixture containing different ethanol contents (10 % - 50 %) fluctuated around 21.7 s, and the ignition time was mainly dominated by ethanol which is volatile and has a lower ignition point. Both $\dot{m}_{max}$ and $\dot{m}_{ave}$ of samples decreased with the increasing ethanol content. For samples with an ethanol content of less than 30%, the peak of MLR was mainly contributed by NC.

This study only focused on the effect of ethanol content on the burning characteristics of NC. It is well known that water and isopropanol are also commonly used as humectants for NC. It is also necessary to study the effects of different isopropanol content and different water content on the fire risk of NC in future work. In addition, only the fire risk of small-scale samples is concerned in the present study. NC products are generally stored in large-scale barrels and are more dangerous in the event of a burning accident. Therefore, large-scale NC burning is worthy of study in the future.

5. References

[1] Wei R, Huang S, Wang Z, He Y, Yuen R and Wang J 2018 *Propellants, Explos. Pyrotech.* **43** 1-8 https://doi.org/10.1002/prep.201800149
[2] Wei R, Huang S, Wang Z, Yuen R and Wang J 2018 *J. Loss Prev. Process. Ind.* **56** 289-299
[3] Wei R, Huang S, Huang Q, Ouyang D, Chen Q, Yuen R and Wang J 2018 *J. Therm. Anal. Calorim.* 1–10 https://doi.org/10.1007/s10973-018-7410-9
[4] Wei R, Huang S, Wang Z, Wang X, Ding C, Yuen R and Wang J 2018 *J. Therm. Anal. Calorim.* 1–10 https://doi.org/10.1007/s10973-018-7653-5
[5] He Y, He Y, Liu J, Li P, Chen M, Wei R and Wang J 2018 *J Hazard Mater.* **340** 202-212
[6] Wei R, He Y, Zhang Z, He J, Yuen R and Wang J 2018 *J. Therm. Anal. Calorim.* **133(3)** 1291-1307
[7] Liu J and Chen M 2018 *Appl. Sci.* **8** 910

[8] Liu J, Wang J, Xie Q, Wang J, Tao C and Yuen R 2018 *Propellants, Explos. Pyrotech.* **43(5)** 445-452
[9] Fu G, Wang J and Yan M 2016 *Process Saf. Prog.* **35** 216-220
[10] Zhao B 2016 *Nat. Hazards* **84** 707-713
[11] ISO 5660-1: Reaction-to-fire tests - Heat release, smoke production and mass loss rate - Part 1:Heat release rate(cone calorimeter method) 2002
[12] Wei R, He Y, Liu J, He Y, Mi W, Yuen R and Wang J 2017 *Materials (Basel)* **10** 316
[13] Wang X, He Y, Zhou T, Chen Q, Ding C and Wang J 2018 *Fire Technol* 1-29 https://doi.org/10.1007/s10694-018-0776-1
[14] Wei R, Huang S, Wang Z, Wang C, Zhou T, He J Yuen R and Wang J 2018 *J. Therm. Anal. Calorim.* 1–17 https://doi.org/10.1007/s10973-018-7521-3
[15] Jiang L, Miller C, Gollner M, Sun J. 2017 *P. Combust. Inst.* **36(2)** 2987-2994
[16] Jessup R and Prosen E 1950 *J Res Natl Bur Std.* **44** 387

Acknowledgments
This research was supported by the National Key R&D program of China (No. 2016YFC0802500) and a grant from the Research Grant Council of the Hong Kong Special Administrative Region, China (contract grant number CityU 11301015).

IOP Conf. Series: Earth and Environmental Science **257** (2019) 012037 doi:10.1088/1755-1315/257/1/012037

Investigation of intake pressure and fuel injection timing effect on performance characteristics of diesel engine

Willyanto Anggono[1,2,3], Wataru Ikoma[1], Haoyu Chen[1], Zhiyuan Liu[1], Mitsuhisa Ichiyanagi[1], Takashi Suzuki[1] and Gabriel Jeremy Gotama[2,3]

1 Department of Engineering and Applied Sciences, Sophia University, Tokyo, Japan
2 Mechanical Engineering Department, Petra Christian University, Surabaya, Indonesia
3 Centre for Sustainable Energy Studies, Petra Christian University, Surabaya, Indonesia
E-mail: willy@petra.ac.id

Abstract. The low fuel consumption and efficient power usage in diesel engine has made diesel engine as the more appealing option in comparison with other type of engines. To raise the performance of DI (direct injection) diesel engine further, injection timing and intake pressure variation were studied under 2000 RPM (rotation per minute) engine rotation. This experiment used DI diesel engine with single cylinder. The length of stroke was set to 96.9 mm, the diameter of bore was set to 85 mm and the compression ratio of the engine was 16.3. The variations of main injection timing were set for 1° after TDC (top dead centre) as advanced injection timing, and 3° after TDC as retarded injection timing. Boost pressures for intake pressure were varied with 20 in increments and started from 0 KPa to 60 KPa. In-cylinder pressure characteristics and heat release rate were used to evaluate the engine performance. The experiment indicated, as the boost pressure raises, the heat release rate and in-cylinder pressure are increased. The main injection timing advancement from 3° after TDC to 1° after TDC causes increase to the peak of in-cylinder pressure after TDC in DI diesel engine. This phenomenon is due to the slower combustion in retarded injection timing. For heat release rate, the advancement of injection timing causes the differences between various intake pressures to be more apparent.

1. Introduction

Compression ignition (CI) engine, due to their excellent fuel efficiency and durability, has become the popular power plant for automotive application. This is globally the most accepted type of internal combustion engine used for powering agricultural implements, industrial applications, and construction equipment along with marine propulsion. However, emissions from diesel engines have been focused in increasingly stringent emission regimes because of their adverse health impact on humans. Diesel particulates are classified as 'probable carcinogen'. Under tremendous pressure to comply with increasingly stringent emission norms adopted worldwide, mass emissions of particulate matter (PM) from diesel engines have been significantly reduced by automotive OEMs (original equipment manufacturers) by employing improved exhaust gas after-treatment technologies [1]. In diesel engines, it is rather difficult to lower NOx and PM emissions simultaneously due to soot-NOx trade-off. High NOx and PM emissions are still the main obstacle in the development of next generation conventional diesel engines.

Content from this work may be used under the terms of the Creative Commons Attribution 3.0 licence. Any further distribution of this work must maintain attribution to the author(s) and the title of the work, journal citation and DOI.
Published under licence by IOP Publishing Ltd

Combustion, performance and emission characteristics of diesel engines depend on several factors including fuel injection timing (FIP), start of injection (SOI), fuel quantity injected, number of injections (post- and pilot-), design of combustion chamber and nozzle spray patterns. High-pressure direct injection (HPDI) seems to be one of the most efficient ways to comply with the stringent global emission norms. FIP for different generation of diesel engines varies from 200 to 2000 bars. Kato et al. [2] demonstrated using high fuel injection pressures as a means to reduce PM emissions without increasing NOx emissions. High FIP seem to induce a very different spray structure than low FIP sprays used earlier [3]. This is mainly due to cavitation created in the nozzles at high FIP, which results in significantly faster atomization [4].

Other studies [5,6] showed that higher FIP improve fuel–air mixing, followed by faster combustion, which directly influences pollutant formation. Diesel spray characterization is usually done for parameters such as spray tip penetration, spray angle, droplet velocities, droplet sizes and distributions, and global spray structure. A good understanding of these characteristics is essential for increasing the combustion efficiency and reducing the environmental impact. High pressure difference across the injector nozzle is necessary to atomize the liquid fuel into small droplets in order to enable rapid vaporization as well for high jet penetration in the combustion chamber [7,8]. Droplets size distribution of a spray fundamentally affects CI engine combustion. Smaller fuel droplets vaporize rather quickly compared to larger droplets however their penetration is shorter therefore the size distribution needs to be optimized. Chen et al. reported that small droplets and high penetration depth of fuel jet enhances the fuel–air mixture quality, which provides shorter ignition delays and more complete combustion [8,9]. Lower FIP gives larger droplet diameters, and thus increasing ignition delay during combustion [9]. This also leads to higher cylinder pressures, which ultimately results in higher NOx emissions. When FIP increases, spray droplet diameter distribution reduces. This leads to improved fuel–air mixture formation because of superior mixing during ignition delay, therefore smoke and CO emission are reduced [10]. However, if FIP is too high, ignition delay period becomes too short. Hence, possibility of homogeneous mixing decreases and as a result, combustion efficiency is reduced [11]. Bruneaux [12] investigated spray characteristics of common rail direct injection (CRDI) system in a high pressure, high temperature cell, which created conditions existing in a typical diesel engine. An increase of FIP was found to enhance the fuel atomization at the nozzle outlet, resulting in more distributed vapor phase, which improves mixing. Hence the fuel injection strategy is an important parameter in diesel engines to optimize the combustion, performance and tailpipe emissions.

These injection parameters also affect the particulate emission from diesel engines. High compression ratios, along with relatively high oxygen concentration in diesel combustion chamber deliver excellent thermal efficiency, and low CO and HC emissions in contrast with a comparable gasoline engine [13]. However, mass of particulates emitted from diesel engines are generally 10–100 times higher than SI engines [14–16]. Particulates are of concern from engine performance, durability and harmful environmental impact. Higher particulate emissions result in reduced fuel economy because of fuel loss due to the incomplete combustion. Interaction of these particulates results in increased wear of the engine components. Agarwal et al. [16] carried out experiments to investigate the characteristics of particulates and concluded that lubricating oil contaminated by diesel soot is a key factor responsible for higher engine wear [17].

Particulates have adverse environment impacts such as they affect human and live-stock health, lead to poor visibility, and soil the buildings. While attempting engine optimization, it is required to consider particulate numbers as well with the particulate mass. Methods used for reducing the particulate mass such as increasing FIP, use of variable geometry turbochargers (VGT) and diesel particulate filters (DPF) tend to increase particulate numbers by reducing their size, which is likely to be more harmful for human health [18,19]. A serious study on the diesel particulate characterization is important because a significant proportion of diesel particulates have aerodynamic diameters less than 1 μm. Diesel particulates in this size range have a high probability of being inhaled and deposited in the respiratory tract, and potentially cause respiratory diseases and consequently damage the lungs [20,21]. Particles emitted from diesel engines can be completely characterized by gravimetric measurements, particle number size distribution, particle surface area-size distribution, particle volume-size distribution, soluble organic fraction (SOF), elemental trace metals, elemental carbon (EC), organic carbon (OC), total carbon, and polycyclic aromatic hydrocarbons (PAHs) [22–26]. Agarwal et al. [16] reported an increase in particle number concentration at lower engine loads and particle number concentration reduction at higher engine loads with addition of 20% biodiesel to diesel.

In the present investigation, a flexible single cylinder research engine was used to experimentally evaluate the effect of fuel injection timings and FIP on combustion, emissions and performance. Mineral diesel was used as test fuel. This engine is capable of precisely controlling fuel injection parameters such as FIP, SOI and injected fuel quantity. The effect of variations in these parameters on engine combustion, performance and emission characteristics is evaluated. For particulate size and number distribution, engine exhaust particle sizer (EEPS) was used.

2. Experimental setup and procedure

In this experiment, diameter of the engine bore was set to 85 mm, the length of the stroke was set to 96.9 mm and the compression ratio of the engine was set to 16.3. The piston and head of the cylinder were manufactured using aluminum alloy while the dry liner in the cylinder is manufactured with cast iron as material. Diagram showing the schematic of the experiment is displayed in figure 1. The complete specification of the DI diesel engine with single cylinder utilized in this experiment is displayed in table 1. Kistler Japan Type 6052 pressure sensor was used to measure in-cylinder pressure in this experiment.

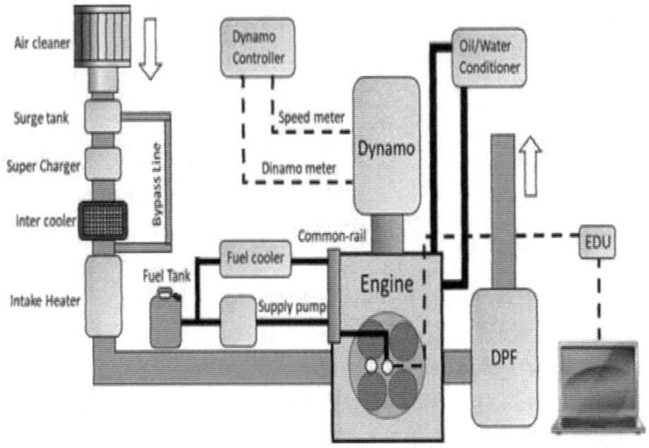

(a)

(b)

Figure 1. Schematic of (a) the experimental system and (b) experimental system construction.

Signals that originate from coaxial heat flux meter with high-speed response and coaxial thermocouple were amplified and documented inside data logger simultaneously to obtain the heat release rate (HRR) data. The heat release rate data were obtained along with the signal from in-cylinder pressure sensor in every crank angle (CA) position.

Table 1. Specifications of the engine used in the experiment.

Parameter	Value / Description
Type of engine	Single cylinder engine
Length of connecting rod [mm]	150.46
Stroke [mm]	96.9
Bore [mm]	85
Capacity of the cylinder [cc]	550
Ratio of compression [-]	16.3
Duration of Intake valve opening [deg.]	From 347 to −120 (Comp. TDC is 0 deg.)
Duration of Exhaust valve opening [deg.]	From 122 to −330 (Comp. TDC is 0 deg.)
Offset of the cylinder [mm]	6.5
Offset of the piston-pin [mm]	0.8

Table 2. Experiment condition at 2000 rotation speed with boost pressure of 0 KPa, 20 KPa, 40 KPa, and 60 KPa.

Rotation speed [RPM]	Injection step	Injection timing [deg.]	Injection time[ms]	Injection quantity [g/st]	Total injection quantity[g/st]
2000	pilot	-25	0.16	0.0014	0.0398
	pre	-15	0.18	0.0024	
	main	3	0.83	0.0360	
	pilot	-25	0.16	0.0014	0.0398
	pre	-15	0.18	0.0024	
	main	1	0.83	0.0360	

The intake pressures in this experiment were varied from naturally aspirated (0 KPa) to 60 KPa with 20 KPa increment. The boost pressures originated from utilizing supercharger. Naturally aspirated condition was examined to understand further the outcome of supercharging in DI diesel engine. As for the main injection timing, the timings were varied into two, advanced injection timing located 1° after TDC and retarded injection timing located 3° after TDC. Details regarding injection timing, quantity, and time is provided in table 2. The evaluation of engine performance was conducted through examining both heat release rate and in-cylinder pressure of the engine.

3. Results and discussion

The results of injection timing and intake pressure effect on in-cylinder pressure are shown in figure 2. Figure 2 shows main injection step of 3° and 1° after TDC, with 2000 RPM engine speed and various intake pressures. This figure indicates the rise of in-cylinder pressure with the rise of intake pressure. The results are similar with investigation done by Lee et al. that propose the rise of maximum in-cylinder pressure under supercharged condition with engine speed below 3000 RPM [27].

Noticeable difference between main injection of 3° and 1° after TDC, as seen in figure 3, is found in the timing of maximum in-cylinder pressure. The maximum in-cylinder pressure during main injection step of 3° after TDC is located during TDC while the maximum in-cylinder pressure during main injection of 1° after TDC is located few degrees after TDC. This phenomenon occurs due to the slower combustion reaction with main injection timing of 3°. The slower combustion reaction reduces the rise of in-cylinder pressure value because of the fuel mainly burns after TDC. The lower value of maximum in-cylinder pressure in main injection timing of 3° after TDC causes the in-cylinder pressure during TDC to be higher than the aftermath in-cylinder pressure rises [28-30].

The results of injection timing and intake pressure effect on heat release rate are shown in figure 4. Figure 4 shows the main injection step of 1° and 3° after TDC with 2000 RPM engine speed and various intake pressures. This figure indicates the rise of heat release rate is proportional to the rise of intake pressure. The results are similar with investigation done by Lee et al. that propose the rise of maximum in-cylinder pressure under supercharged condition with engine speed below 3000 rpm [27].

Noticeable difference between main injection of 1° and 3° after TDC is found in the rise of heat release rate before TDC. Heat release rate raises before TDC in all level of intake pressures. In main injection timing of 3° after TDC, the differences of heat release rate value between various intake pressures are miniscule. Contrary to this, with main injection timing of 1° after TDC, the heat release rate value and timing differences are more apparent. Both injection timings have similar pattern with higher intake pressure having lower heat release rate before TDC.

In figure 5, the effect of injection step advancement from 3° and 1° to the heat release rate is displayed with respond to certain level of intake pressure. The maximum heat release rate between 3° and 1° are quite similar with small degree of differences. It is also found that the maximum heat release rate between 3° and 1° occurred in different time. Maximum heat release rate occurs earlier in 1° injection timing. Another point of interest is seen in the trend of heat release rate of both 3° and 1° injection timing. It is observed that the heat release rate of 3° overcomes the heat release rate of 1° after certain degree from TDC. These phenomena are caused by the slower combustion rate of 3° which delays the enormous increase of heat release rate to higher crank angle timing.

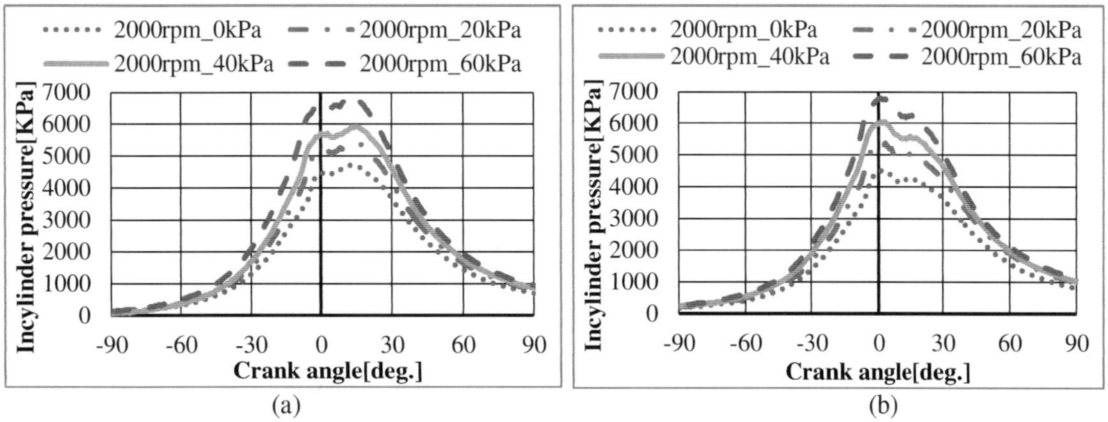

Figure 2. The in-cylinder pressure of DI diesel engine with injection timing of (a) 1° and (b) 3° after TDC under 2000 RPM and various intake pressures.

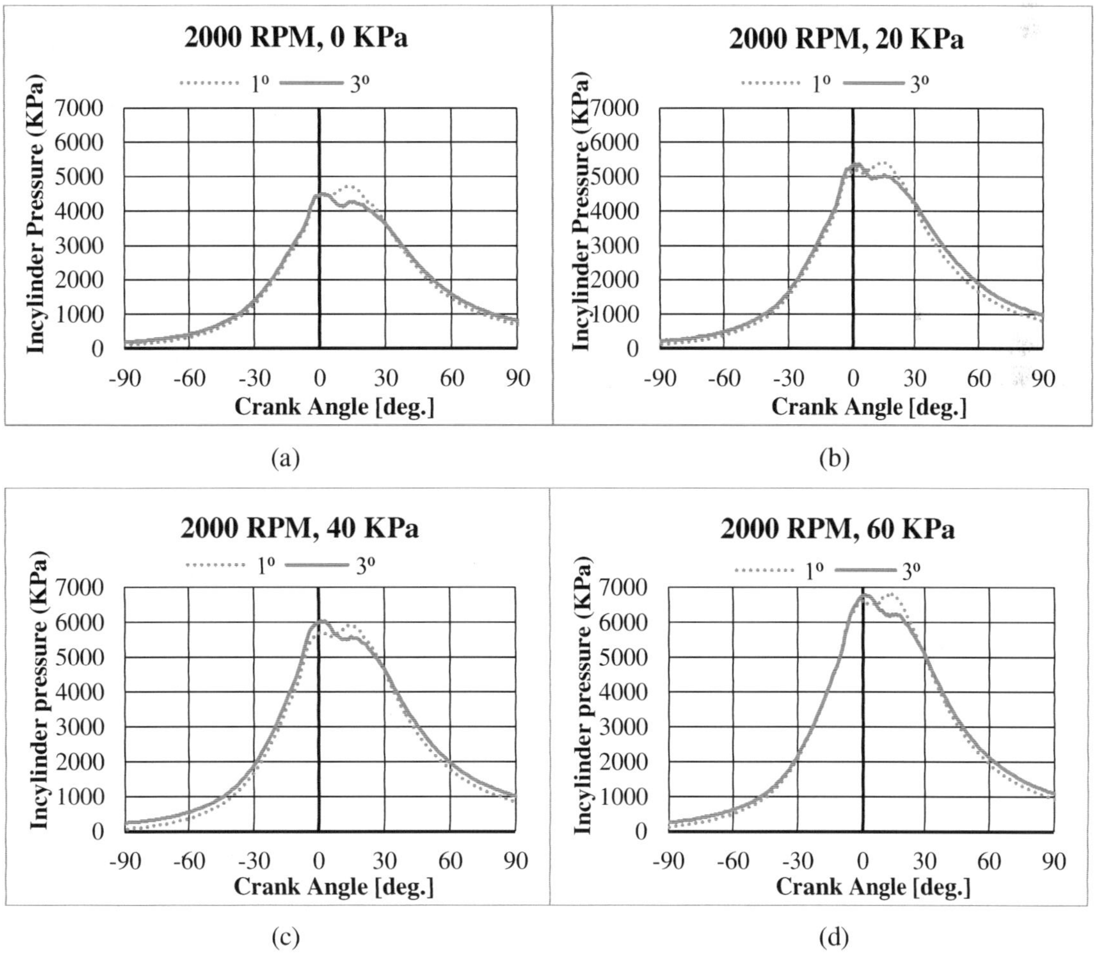

Figure 3. The in-cylinder pressure comparison of DI diesel engine with injection timing of 1° and 3° after TDC under 2000 RPM and various intake pressures (a) 0 KPa (b) 20 KPa (c) 40 KPa (d) 60 KPa.

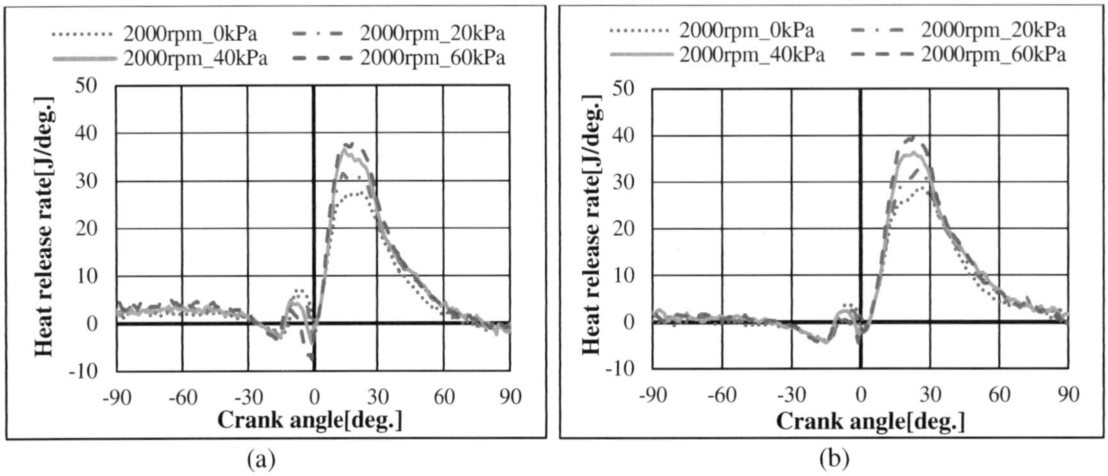

Figure 4. The heat release rate of DI diesel engine with injection timing of (a) 1° and (b) 3° after TDC under 2000 RPM and various intake pressures.

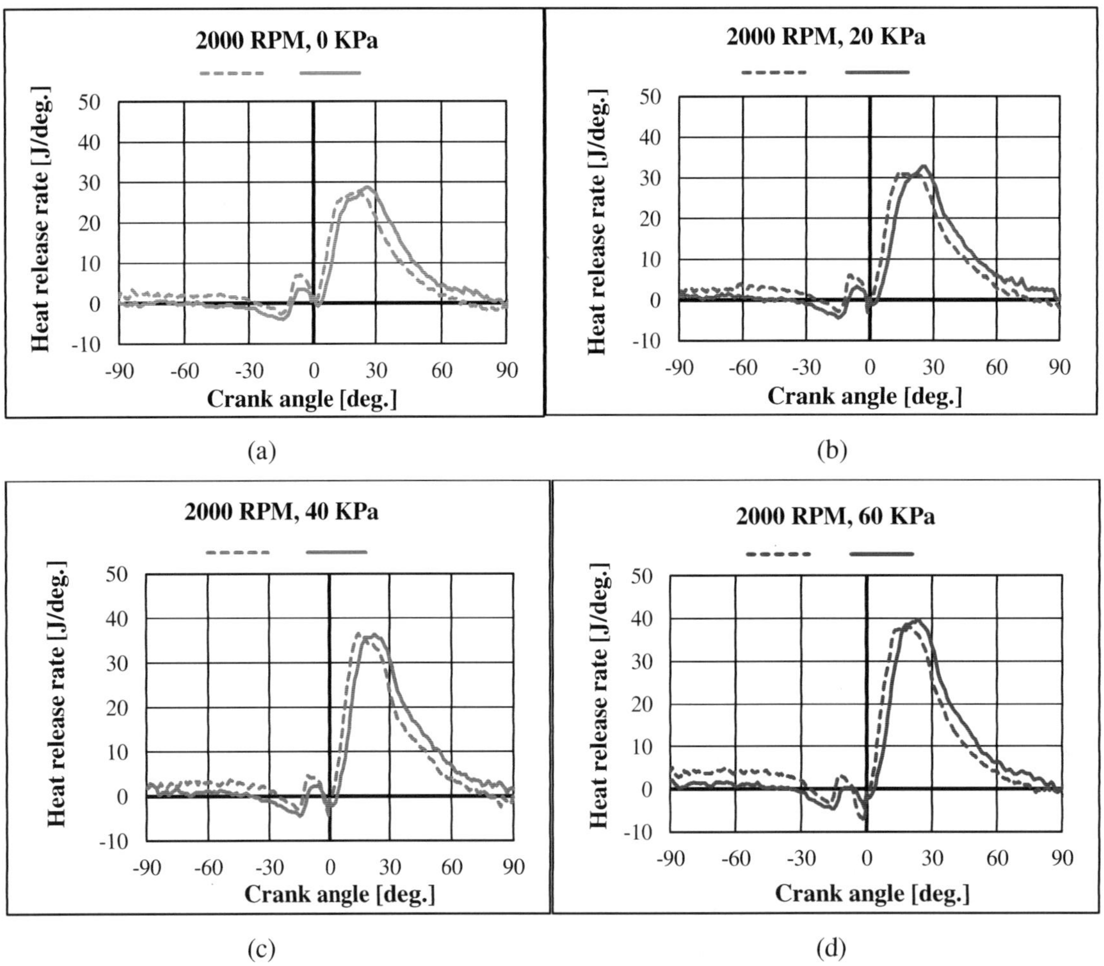

Figure 5. The heat release rate of DI diesel engine with injection timing of 1° and 3° after TDC under 2000 RPM and various intake pressures (a) 0 KPa (b) 20 KPa (c) 40 KPa (d) 60 KPa.

4. Conclusion

From these investigations, it can be concluded that the injection timing and intake pressure highly affect the in-cylinder pressure and heat release rate performance characteristics in DI diesel engine. The rise of intake pressure from naturally aspirated to 60 KPa boosts the in-cylinder pressure and heat release rate of DI diesel engine. The main injection timing advancement from 3° to 1° after TDC causes increase to the rise of in-cylinder pressure after TDC in DI diesel engine. In retarded injection timing, the rise of in-cylinder pressure during TDC is higher compared to the rise of in-cylinder pressure after TDC. In the advanced injection timing, the rise of in-cylinder pressure during TDC is lower compared to the rise of in-cylinder pressure after TDC. For heat release rate, the occurrence of maximum heat release rate appears earlier in advanced injection timing compared to retarded injection timing. Both phenomena are due to the slower combustion in the retarded injection timing condition.

5. References

[1] Kirchstetter T W, Harley R A, Kreisberg N M, Stolzenburgh M R and Hering S V 1999 On road measurement of fine particulate and nitrogen oxide emissions from light and heavy-duty motor vehicles *Atmos. Environ.* **33** 2955–68
[2] Kato T, Tsujimara K, Shintani M, Minami T and Yamaguchi L 1989 Spray characteristics and combustion improvement of a DI diesel engine with high pressure fuel injection *SAE* **890265**
[3] Baritaud T, Heinze T and Coz J 1994 Spray and self-ignition visualization in a DI diesel engine *SAE* **940681**
[4] Bruneaux G, Verhoeven D and Baritaud T 1999 High pressure diesel spray and combustion visualization in a transparent model diesel engine *SAE* **1999-01-3648**
[5] Dec J 1997 A conceptual model of DI diesel engine *SAE* **970873**
[6] Agarwal A K, Chaudhary V, Agarwal A and Shukla P C 2013 Comparative study of macroscopic spray parameters and fuel atomization behaviour of SVO (Jatropha), its biodiesel and blends *Therm. Sci.* **17**(1) 217–32
[7] Birch S 2000 Fuel injection: the inside story *Tech Briefs SAE*
[8] Chen P, Wang W, Roberts W L and Fang T 2013 Spray and atomization of diesel fuel and its alternatives from a single-hole injector using a common rail fuel injection system *Fuel* **103** 850–61
[9] Wang X, Huang Z, Kuti O A, Zhang W and Nishid K 2010 Experimental and analytical study on biodiesel and diesel spray characteristics under ultra-high injection pressure. *Int. J. Heat Fluid Flow* **31** 659–66
[10] Zhang G, Qiao X, Miao X, Hong J and Zheng J 2012 Effects of highly dispersed spray nozzle on fuel injection characteristics and emissions of heavy-duty diesel engine *Fuel* 2012 **102** 666–73
[11] Celikten I 2003 An experimental investigation of the effect of the injection pressure on engine performance and exhaust emission in indirect injection diesel engines *Appl. Therm. Eng.* **23** 2051–60
[12] Bruneaux G 2001 Liquid and vapour spray structure in high-pressure common rail diesel injection *Atomization Spray* **11** 533–56
[13] McClellan R O 1986 Health effects of diesel exhaust: a case study in risk assessment *Am. Ind. Hygiene. Assoc. J.* 1986 **47**(I) 1–13
[14] Kittelson D B 1998 Engines and nanoparticles: a review *J. Aer. Sci.* 1998 **29**(5/6) 575–88
[15] Gupta T, Kothari A, Srivastava D K and Agarwal A K 2010 Measurement of number and size distribution of particles emitted from a mid-sized transportation multipoint port fuel injection gasoline engine *Fuel* **89** 2230–3
[16] Agarwal A K, Gupta T and Kothari A 2011 Particulate emissions from biodiesel vs diesel fuelled compression ignition engine *Renew. Sustain. Energy Rev.* **15** 3278–300
[17] George S, Balla S and Gautam M 2007 Effect of diesel soot contaminated oil on engine wear *Wear* **262** 1113–22
[18] Donaldson K, Li X Y and MacNee W 1998 Ultrafine (nanometre) particle mediated *J. Aer. Sci.* **29**(5/6) 553–60.

[19] Kim H, Sung Y, Jung K, Choi B and Lim M T 2008 Size distributions and number concentrations of particles from the DOC and CDPF *J Mech Sci Tech* **22** 1793–9
[20] Dockery D W et al. 1993 An association between air pollution and mortality in six US cities *The New England J. of Med.* **329** 1753–9.
[21] Pope C A et al. 1995 Particulate air pollution as a predictor of mortality in a prospective study of US adults *Am. J. Resp. Crit. Care. Med.* **151** 669–74
[22] Burtscher H 2005 Physical characterization of particulate emissions from diesel engines: a review *Aer. Sci.* **36** 896–932
[23] Neeft J P A, Makkee M and Moulijn J A 1996 Diesel particulate emission control *Fuel Process Tech.* **47** 1–69
[24] Maricq M M 2007 Chemical characterization of particulate emissions from diesel engines: a review *Aer. Sci.* **38** 1079–118
[25] Dwivedi D, Agarwal A and Sharma M 2006 Particulate emission characterization of a biodiesel vs. diesel-fuelled compression ignition transport engine: a comparative study *Atmos. Environ.* **40** 5586–95
[26] Kittelson D et al 2000 Diesel Aer Sampling Atm. *SAE* **2000-01-2212**
[27] Lee C S, Lee K H, Whang D H, Choi S W and Cho H M 1997 Supercharging Performance of a Gasoline Engine with a Supercharger *KSME Int. J.* **11**(5) 556-564
[28] Raeie N, Emami S and Sadaghiyani O K 2014 Effects of injection timing, before and after top dead center on the propulsion and power in a diesel engine *Prop. and Power Res.* **3**(2) 59-67
[29] Ganapathy T, Gakkhar R P and Murugesan K 2011 Influence of Injection Timing on Performance, Combustion and Emission Characteristics of Jatropha Biodiesel Engine *App. Ene.* **88** 4376-4386
[30] Park S H, Youn I M and Lee C S 2011 Influence of Ethanol Blends on the Combustion Performance and Exhaust Emission Characteristics of a Four-Cylinder Diesel Engine at Various Engine Loads and Injection Timings *Fuel* **90** 748-755

Acknowledgments
This research was supported by Sophia University, Japan under STEC2018 Research Grant program and Petra Christian University, Indonesia. Authors would like to express their gratitude toward these institutions.

IOP Conf. Series: Earth and Environmental Science **257** (2019) 012038 doi:10.1088/1755-1315/257/1/012038

Efficiency improvement and economic analysis of micro hydro power plant by using twin vertical hydro turbine

K Tanutwutthigorn and R Suntivarakorn

Department of Mechanical Engineering, Faculty of Engineering, Khon Kaen University, Khon Kaen, 40002, Thailand
E-mail: ratchaphon@kku.ac.th

Abstract. This research aimed at improving the efficiency of and comparing the production costs of micro hydropower plants by using twin vertical hydro turbines as low headwater sources. This micro hydropower plant was designed with the highest capacity of 3 kW and utilized a directly connected transmission system and induction generator by modifying two 3 HP induction motors. At the inlet area of the turbines, guide vanes were installed on the 25 turbines at 24°. The ratio of turbine diameter was equal to 0.68. After the experiment was completed, it was found that the highest electrical power generated from the plant had been 1.38 kW with an efficiency of 49% at head 0.7 m. When compared to the plant without any installed guide vanes, the increase was 26.94%. Regarding the electrical production cost, the total cost was 5,844 USD or 1,948 USD/kW.

1. Introduction

Micro hydropower plants, which are used for low headwater sources, are lower-cost operating systems at less than 2,500 USD/kW [1]. These are suitable for residential areas that are far away from the transmission system, but which have a water source nearby. European countries, currently, have been widely showing interest in and developing micro hydropower plant technology. Micro hydropower plants have power generation capacities that do not exceed 100 kW with water heads less than 10 m [2]. Therefore, the selected turbines, which are appropriate for the head level and water flow rate, play an important role in highly efficient power generation and can consequently lower the production costs. A suitable type of turbine and proper installation are essential factors that can yield a highly efficient power plant. A micro hydropower plant is a system that operates with headwaters lower than 10 m, where such a head is found to work appropriately with a cross flow turbine. Cross flow turbines are lower in cost due to their simple designs and compared to the ones by Kaplan and Francis, they neither require the latest production technology nor are modern machines needed [3]. The results showed that cross flow turbines have an average efficiency of 78% between 3-20 m at the head. However, the design of the cross flow turbines is an impulse type, which results in poorer efficiency when utilizing a head lower than 3 m. Hence, in order to study the factors; which influence cross flow turbine efficiency, such as the numbers of blades, the inlet angle of attack, and the proper water turbine size, etc.; a water turbine, which is more appropriate for the water source, needs to be created in order to achieve a greater degree of efficiency from the power plant. Mockmore and Merryfield studied cross flow turbines with 20 blades, a 16° inlet angle of attack, and a diameter ratio of 0.66. Their findings suggested that an efficiency of 68% was the highest, which could be achieved from a water turbine. In contrast, V. R. Desai and Nadim M. Aziz conducted experiments to determine the influences of significant variables, which had been made to the cross flow turbines.

Content from this work may be used under the terms of the Creative Commons Attribution 3.0 licence. Any further distribution of this work must maintain attribution to the author(s) and the title of the work, journal citation and DOI.
Published under licence by IOP Publishing Ltd

Their study revealed that water turbines with a 24° inlet angle of attack, with 25 blades, and with a diameter ratio of 0.68 had given the highest efficiency of 88%. At present, there are hydropower plants, which are being created with twin vertical hydro turbines. Such types of plants are designed to be located near low headwater sources like irrigation canals, floodgates, and waste water treatment facilities in factories where the water head ranges from 0.7 - 2 m in order to generate 1.5 kW of power and to achieve a water turbine efficiency of 54.26% [4]. Obviously, from the present research, the results indicate a greatly lower degree of efficiency. Therefore, studying the variables that can influence the efficiency of water turbines, as well as examining the weaknesses of the cross flow turbine power plants, can assist in establishing some guidelines to better develop the plants' efficiency. In addition, they can lower the costs of power generation. Furthermore, these efforts can stimulate interest, increase motivation, and promote the future employment of micro hydropower plants in low headwater sources.

2. The Design of the Cross Flow Turbine

2.1. The design of the cross flow turbine

Mockmore and Merryfield suggested their cross flow turbine design with the inlet blade angle (β_1) as calculated below:

$$\beta_1 = \tan^{-1}(2\tan\alpha_1) \tag{1}$$

The curvature of the blades was determined by:

$$\rho = \left[\left(r_1^2\right) - \left(r_2\right)^2\right] / (2r_1 \cos\beta_1) \tag{2}$$

In which r_1 and r_2 are the radiuses of the inside and outside edges of the water turbine (m) and its central angle was derived from the following:

$$\delta = 2\left(\tan^{-1}\left(\frac{\cos\beta_1}{\left(\sin\beta_1 + \left(r_2/r_1\right)\right)}\right)\right) \tag{3}$$

The diameter of the blades' outside edges was calculated using the following equation:

$$D_1 = \left(\frac{30V_1\cos\alpha_1}{\pi N}\right) \tag{4}$$

In which V_1 is the inlet velocity (m/s), and N is the turbine's speed (rpm), then water turbine's height was derived from:

$$L = \frac{2Q_1}{D_1V_1\left(\theta\pi/180\right)} \tag{5}$$

Q_1 indicated the inlet water flow rate (m3/s) and θ represented the inlet angle to the runner (degrees). Then the water turbine was designed to generate 3 kW of power as illustrated in Table 1.

Table 1. The typical water turbine profile from the given parameters

Design Parameters	Values	Descriptions of the Design Parameters
α_1	24	Inlet angle of attack (degrees) [5]
N_b	25	Number of blades [5]
D_2/D_1	0.68	Diameter ratio [5]
β_1	41.68	Inlet Blade angle (degrees)
δ	58.08	Central Angle (degrees)
ρ	0.0899	Curvature of the blade (m)

The results, obtained from the cross flow turbine design, indicated that the blade inlet angle was 41.68°, the center of the angle was 58.08°, the curvature of the blades was 0.0899 m, and that water turbine's diameter and height were 0.5 m. as illustrated in Fig. 1.

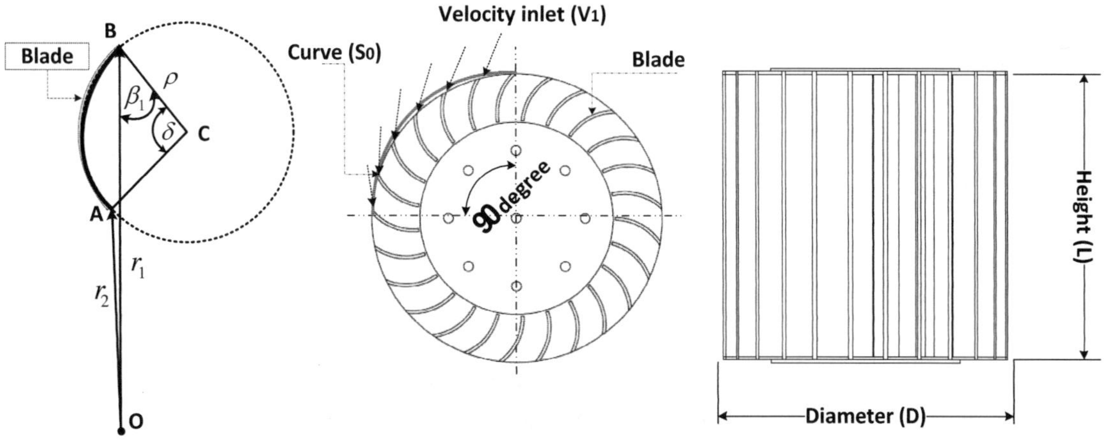

Figure 1. The typical water turbine profile design from the given parameters.

2.2 Guide vane design

The purpose of the guide vane is to assist in more appropriately controlling the inlet angle of attack. Vincenzo Sammartano, et al. [5] suggested their equations and design for guide vanes. They found that the radius of the guide vanes would change by degrees (BC), as shown in Fig. 2 (a), and that it could be calculated using the correlated equations (6)-(9):

$$r(\theta) = K.\theta + \frac{D_1}{2} \tag{6}$$

When K is a constant, the following could be determined from equation (7):

$$K = \frac{1}{\lambda - \gamma}\left[\frac{S_0 \cos\alpha_1 + r_1}{\cos\gamma} - r_1\right] \tag{7}$$

In which λ is the inlet angle of turbine, when considering Fig. 2 is (a) and (b), they referred to rectangular AOD and γ was determined using equation (8):

$$\gamma = \tan^{-1}\frac{S_0 \sin\alpha_1}{S_0 \cos\alpha_1 + r_1} \tag{8}$$

The nozzle's height (S_0) was derived from equation (9):

$$S_0 = \lambda(r_1 \sin\alpha_1) \tag{9}$$

Fig. 2 (a) and (b) represent the elements and installation positions of the guide vanes from equations (6) – (9). These guide vanes more appropriately offer the inlet angle of the turbine and offer it equally to all vanes.

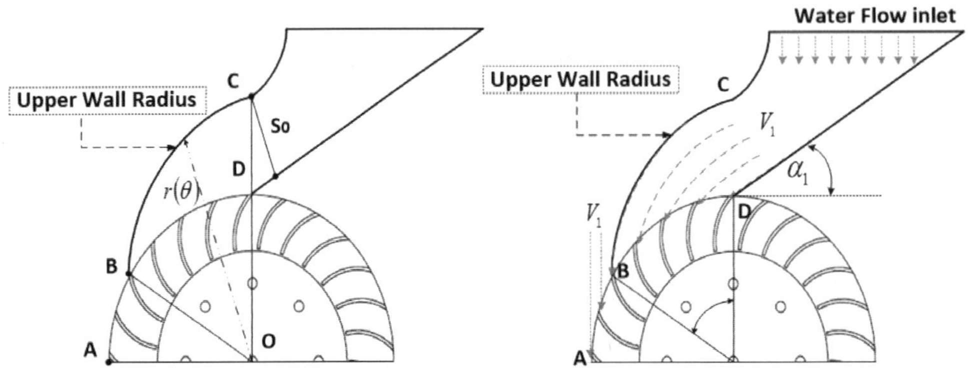

Figure 2. (a) shows the turbine's curvature to obtain hydropower and (b) shows the inlet profile into blades after the nozzle curvature installation.

2.3 The construction of the hydro power plant with twin vertical turbines

Then, the hydro power plant was constructed with twin vertical turbines (see Fig. 3). The major portions of the configured hydro power plant consisted of the following: the runner, chambers (No. 1-3), and the guide vanes, made from 304 stainless steel and 1.5 mm thick. The transmission system was supplied with power from turbine's shaft axis to the direct-drive electric generator, which was generated by the direct injection system from the 2 induction motors 3HP (3 phase).

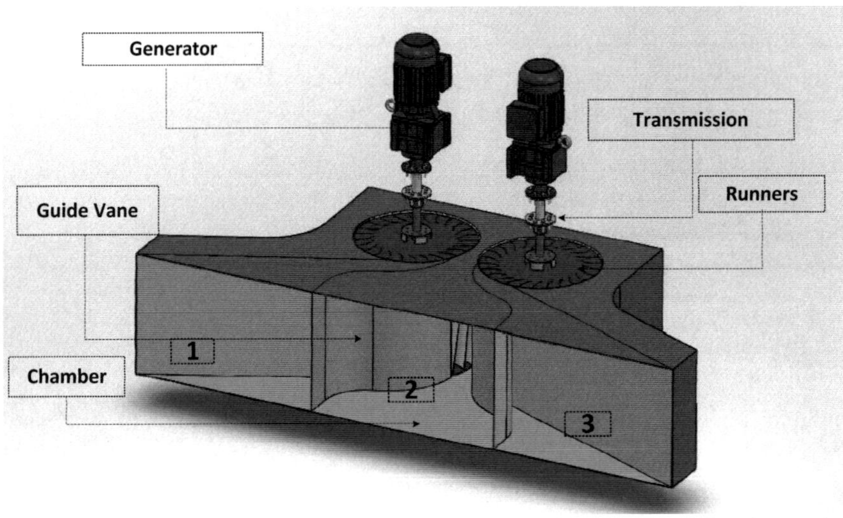

Figure 3. An illustration of the hydro power plant at low head level built with twin vertical turbines.

2.4 Electrical efficiency calculation

After the hydro power plant had been installed and the irrigation canal blocked, there was headwater found in system (*H*) and a flow rate (Q_1) flowing through turbine. The power produced from the generator demonstrated the results of an electric current (*V*) and a voltage (*I*), which were displayed on the digital screen. Hence, the electrical efficiency, generated from this hydropower plant, could be calculated as follows:

$$\eta_{system} = \frac{VI}{\rho g Q_1 H} \tag{10}$$

2.5 The installation of and experiments on the micro hydropower plant

The power plant was designed to be located in the actual setting of an irrigation canal, which was blocked so that headwater could accumulate in the plant (see Fig. 4). The irrigation canal was 2.7 m in width, 1.8 m in depth, and the depth of water before the plant was installed was 1 m. Moreover, the flow rate within the canal was equal to 2.1 m3/sec. The power plant consisted of a set of turbines, a generator, and an electricity meter, which displayed the amount of power and the electrical load by using two 1,500 W fluorescent bulbs. For the convenience of the experiment, the power plant was designed so that it could be moved in and out of irrigation canal. Once installation was completed, the water head would continue to rise. Water velocity (V_1) would be due to head accumulation and inletting with the angle of the blade (β_1). Inlet angle of attack was equal to α_1 and hydropower, which was transmitted to blade, resulted in turbine rotation with the speed of N. Then further forward power was generated by the direct connection transmission system via a gearbox with a gear ratio 1:15, which led to higher speeds at the shaft axis. This caused the generator axis to rotate with speed, making it able to generate power. The generated power was represented as voltage and electric current on the digital screen (Fig. 4). The process of data collection was divided into 7 trials, which were distinguished by the head levels of 0.25, 0.3, 0.4, 0.5, 0.55, 0.6, and eventually 0.7 m. The highest head that the irrigation canal was able to support was 0.7 m. From each trial, the inlet flow rate was recorded by employing a digital water velocity meter, Model BA1100, which displayed the flow rate in a digital format. The water velocity was measured at between 0.1-6.1 m/sec and afterwards, water head was brought to be calculated as the energy input into the plant. The power, obtained from the power plant, was recorded as the electric current and voltage. Then, this data was used to calculate the power efficiency.

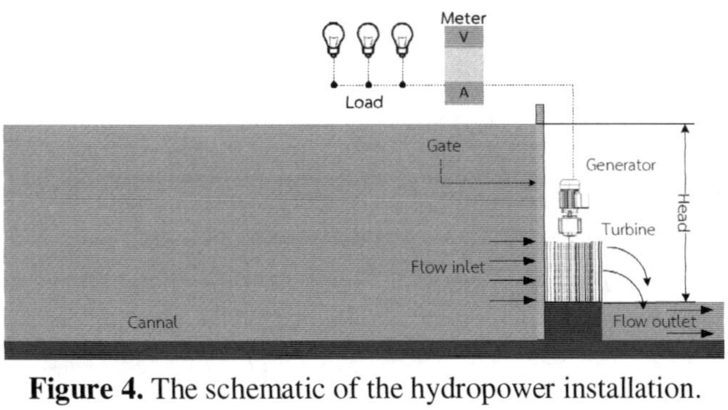

Figure 4. The schematic of the hydropower installation.

3. Results

Fig. 5 (a) indicates the correlation between the generated power and water head level within power plant. After the plant was installed by blocking the irrigation canal, the initial head in plant was equal to 0.25 m, which indicated that this head level was not sufficient to generate any power. Consequently, when the water head level was increased to 0.3, 0.4, 0.5 0.55 0.6, and 0.7 m, the power plant was considerably able to utilize the increase in order to generate 36.40, 129.30, 457.80, 950.60, 1135.20, and 1379.90 W, respectively.

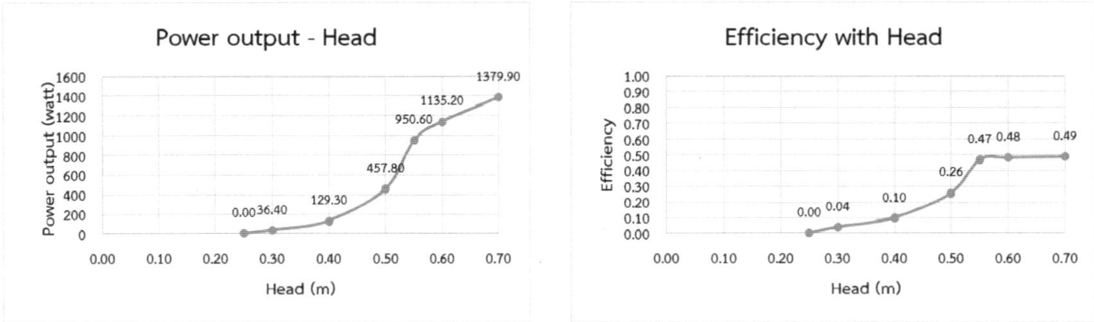

Figure 5. The correlation between the power generated and the water head level within the power plant.

After that, data, obtained from these experiments, was calculated as the efficiency of the power plant. The curve of the electric efficiency from the hydro power plant looked similar to a parabola from head between 0.25-0.55 m. Furthermore, its efficiency signified the constants at the head level of 0.55 onwards and the highest efficiency indicates was 49% as Fig. 5 (b).

The useful data, received from our experiment, allowed for the power curve of 1,500 W to be plotted, so that the power consumption to generate could be compared. Fig. 6 shows the flow rates with guide vane installation (solid line) and without guide vane installation (broken line).

The results indicated the following: with the guide vane installation, the efficiency was equal to 49.6% from a hydropower consumption of 2.952 kW (A), and without the guide vane installation, the efficiency was equal to 38.6% from a hydropower consumption of 3.796 kW (B). The increased efficiency was determined to be approximately 26.94%.

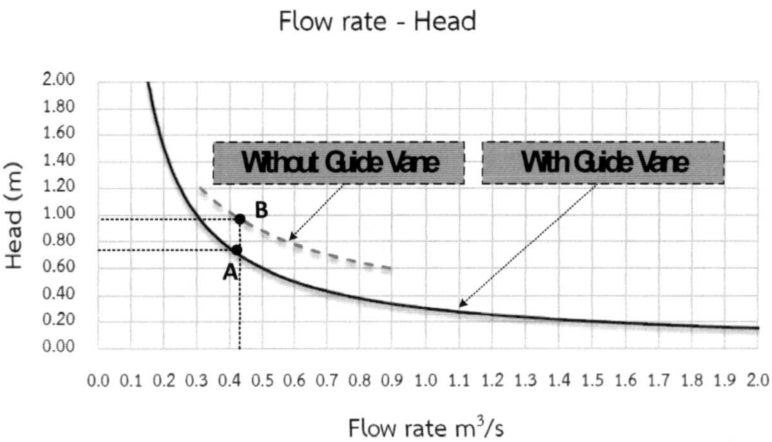

Figure 6. A comparison of the power consumption with and without the guide vane installation.

Table 2 represents the costs of building and installing the power plant, which was divided into four parts: (1) the cost of the turbine generator was at 41%, (2) the cost of electrical control system was at 45%, (3) the cost of the civil work and the installation as 6%, and (4) the cost for management was 5%. When these ratios were considered, it was decided that the cost of civil labor for installing the plant in the irrigation canal had been relatively low due to its convenience and the fact that this type of system is less time consuming than installing a water-diversion system. Moreover, the costs for the turbine generator and the electrical control equipment were close (only a 4% difference). The reason for the similar costs might stem from the fact that they are the essential parts of a micro hydro power plant.

After analysing these four costs, the total expenditures had been 5,844 USD or 1,948 USD/kW. These figures suggest that the cost from the present study was lower than the cost estimated by A.H. Elbatran, et al., [1] who determined the cost of building a micro hydro power plant to be 2,500 USD/kW.

Table 2. The ratio of cost for power generating from hydro power plants.

Types of Cost	A.H. Elbatran [1]	This study
Civil labor (%)	40	6
Turbine Generator set (%)	30	41
Control Equipment (%)	22	45
Management (%)	8	5
Total budget (USD/kW)	2,500	1,951

4. Conclusion

The power plant design for low water head utilizes twin vertical turbines with a maximum power generation 3 kW. After that, the plant was actually installed by blocking a concrete wall irrigation canal with a size of 2.7 m and a water depth of 1.8 m. The flow rate in the canal before installation was 2.1 m3/sec measured at 1 m of water depth. The experiments were carried out at separate head water levels of 0.2, 0.3, 0.4, 0.5, 0.6, and 0.7, respectively. This power plant was designed with and inlet angle at guide vanes of 24° and with 25 blades. The height and the outside diameter of water turbine was 0.5 m, and the ratio of turbine diameter was 0.68. The results found that with the guide vane installation, the efficiency had increased and constants were found at the water head level of 0.55 onwards. The plant system indicated that the highest efficiency was 49% at a water head level of 0.7 m. Finally, the cost for building and for installation was 1,948 USD/kW.

5. Acknowledgments

The authors would like to express their gratitude to National Research Council of Thailand, The Thailand Research Fund (TRF), Energy Conservation Promotion Fund, Energy Policy and Planning, Ministry of Energy for financial support, and Farm Engineering and Automation Technology Research Group (FEAT), Khon Kean University for tool and equipment supported in the study.

6. References

[1] Elbatran AH, Yaakob OB, Ahmed YM and Shabara HM. Operation, performance and economic analysis of low head micro-hydro power turbines for rural and remote areas: A review. Renewable and Sustainable Energy Reviews 2015; 43:40–50.

[2] Yaakob OB, Ahmed YM, Elbatran AH and Shabara HM. A review on micro hydro gravitational vortex power and turbine systems. Jurnal Teknologi 2014; 69: 1–7.

[3] Pimnapat Iemsomboona, Patib Trirath and Bhumkittipichc Krischonme. Performance study of micro hydro turbine and PV for electricity generator. Bunnasopit School, Nanprovince, Thailand, 10th (EMSES 2012). Energy Proc 2013; 34: 235– 42.

[4] Types and Power Curve [internet]. Japan: Ultra-low head micro hydro turbine system STREAM Vertical Dual Cross-Flow Turbine; [cited 2014 July 02]. Available from: http://www.jagseabell.jp/english/index.html.

[5] Vincenzo S, Costanza A, Armando C, Oreste F and Tullio T. Banki-Michell Optimal Design by Computational Fluid Dynamics Testing and Hydrodynamic Analysis. Energies 2013; 6: 2362-85.

Experimental investigation of the effect of *Nephelium Lappaceum* seed biodiesel to the automotive diesel engine performance

Willyanto Anggono[1,2], Sutrisno[1,2], Fandi D Suprianto[1,2], Muji Setiyo[3], Rendy Wibisono[1,2] and Gabriel Jeremy Gotama[1,2]

1 Mechanical Engineering Department, Petra Christian University, Surabaya, Indonesia
2 Centre for Sustainable Energy Studies, Petra Christian University, Surabaya, Indonesia
3 Department of Automotive Engineering, Muhammadiyah University of Magelang, Magelang, Indonesia
Email: willy@petra.ac.id

Abstract. To decrease the reliance of fossil fuel and at the same time utilizing organic waste, an investigation regarding the viability of *Nephelium lappaceum* (rambutan fruit) seed oil as biodiesel has been conducted. The investigation was carried out by comparing the fuel characteristics of *Nephelium lappaceum* seed biodiesel with pure petro-diesel produced in Indonesia. The outcome of this investigation suggested that *Nephelium lappaceum* seed biodiesel has comparable fuel characteristics as that of commercial diesel fuel. Further comparison was also conducted by testing these fuels in diesel engine. The series of tests compared these fuels in terms of fuel consumption, brake mean effective pressure, thermal efficiency, torque, and power. The result indicates that *Nephelium lappaceum* seed biodiesel has better engine performance than pure petro-diesel. Overall, the output of this study proves the feasibility of *Nephelium lappaceum* seed biodiesel as environmentally friendly substitute for pure petro-diesel fuel in Indonesia.

1. Introduction

The economic growth is strongly affected by many factors, including the availability of energy resources. The demand of fossil fuel raises while the supply for it are limited [1-3]. The most prominent hurdle faced by society in terms of energy are the scarcity of fossil fuel and its damage to environment from utilizing it [4,5]. One solution to resolve this crisis comes by developing alternative source of energy, mainly known as renewable energy [6].

One of many examples of renewable energy which may solve the energy problem is biodiesel [7,8]. Biodiesel is composed of long-chain of fatty acids with alcohol attached to them [9,10]. Biodiesel is a type of biofuel, and as a biofuel, it is an intriguing object of research around the world [11]. Biodiesel requires organic substances as raw material and they are easily obtained from vegetation crops and natural wastes found around society [12]. Because of the abundant resources of its materials (350 crops) [13-15], biodiesel production is highly supported [16]. Most of these materials are easily found in rural area and may benefit the community of those area [17]. To obtain biodiesel, transesterification technique is known to be the most common and most important technique to be utilized [18,19].

Content from this work may be used under the terms of the Creative Commons Attribution 3.0 licence. Any further distribution of this work must maintain attribution to the author(s) and the title of the work, journal citation and DOI.
Published under licence by IOP Publishing Ltd

Compared to fossil fuel, biodiesel prevails in term of renewability, ease of utilization, lower emission, and energy efficacy. The environmentally friendly and biodegradable characteristics of biodiesel support biodiesel viability as solution for energy crisis problem [3,20]. Another strong advantage offered by biodiesel is its capability to be blended with common fuels, regardless of the ratio between them. Utilization of biodiesel only demand small to no further upgrade and adjustments in the engine [21-23]. Combining biodiesel with diesel was discovered to add some numbers of advantages to diesel engine such as lower brake specific fuel consumption, higher power and torque, and higher brake thermal efficiency [7,24]. Renewable energy like biodiesel also helps in resolving the rise of energy consumption and hydrocarbon price in developing countries by offering a low-cost alternative [25].

While the advantages of biodiesel seem prominent, it has its drawback. Since biodiesel uses organic materials including food crops, it may diminish the stock for food supply and deemed as wasting food resources. Solution to this problem includes the use of non-edible raw material for biodiesel like animal tallow and shrubs [26,27]. Other non-edible raw material includes *Nicotiana tabacum, Cerbera manghas, Calophyllumin ophyllum, Jatropha curcas,* and *Hevea brasiliensis* which require additional process because of their elevated free unsaturated fat content [28-30].

The diversity of flora and abundant number of organic matters in Indonesia provide favorable condition for biodiesel development. Rambutan fruit with its scientific name *Nephelium lappaceum* is commonly found in Indonesia. Studies have suggested the viability of producing biodiesel from *Nephelium lappaceum* seed oil [31]. The extraction of oil from *Nephelium lappaceum* was found to be 40% by Wong et al, 41.3% by Wanayanuwattikun et al and 42 % by Nguyen et al [31-33]. Wanayanuwattikun et al discovered the cetane index of *Nephelium lappaceum* seed oil to be 61.12 [33]. Albeit all these studies, engine testing for biodiesel derived from *Nephelium lappaceum* seed oil has not been conducted before. By conducting an engine testing, the potential of *Nephelium lappaceum* seed oil as an alternative fuel for diesel engine can be determined further.

2. Experimental methods

Nephelium lappaceum seeds were dried and crushed. Afterward, the process of extracting the oil from these seeds were done by using soxhlet instrument. After obtaining the seed oil, the oil was purified by using rotary evaporator to separate the remnant of n-hexane in the oil. After purified seed oil was obtained, the transesterification process was commenced using magnetic hotplate stirrer.

After completing the transesterification process, the oil was idled for 24 hours to separate the glycerol and methyl esters. Subsequently, the methyl ester was heated to evaporate methanol and n-hexane. The obtained methyl ester was combined with petro-diesel to obtain biodiesel. Since the methyl ester is obtained from *Nephelium lappaceum* seed, the biodiesel is named *Nephelium lappaceum* seed biodiesel. The methyl ester and pure petro-diesel were mixed under 10:90 ratio.

Figure 1. Photographic view of combustion engine laboratory in Petra Christian University.

The obtained *Nephelium lappaceum* seed biodiesel was tested for its fuel characteristics and engine performances. The fuel characteristics tests were carried under ASTM (American Society for Testing and Materials) standards in UPPS (*Unit Produksi Pelumas Surabaya*, Surabaya Lubricant Production Unit) laboratory in Pertamina. The engine performance test of *Nephelium lappaceum* seed biodiesel was performed using diesel engine and water brake dynamometer. The engine performance test was carried in combustion engine laboratory under management of Petra Christian University. The condition of combustion engine laboratory is shown in figure 1. The specification of diesel engine and water brake dynamometer are provided in table 1 and table 2, respectively.

Table 1. Diesel engine specification.

Items	Description
Brand and type	Automotive diesel engine ISUZU (4JA - I. OHV)
Type of combustion	Direct injection
Ratio of compression	18.4
Volume of the cylinder	2499 cm^3
Pressure of compression	31 kg/cm^2
Bore	93 mm
Stroke	92 mm
Peak output	86 ps / 3900 RPM (Rotation per minute)
Number of cylinder and arrangement	4 - in line
Speed during idle	750 RPM
Number of holes in nozzle	4
Pressure during injection	182 kg/cm^2
Timing of injection	12° before TDC (Top dead center)
Type of pump injection	Bosch Distributor
Type of nozzle	Hole

Table 2. Water brake dynamometer specification.

Items	Description
Brand and type	Zollner (3n19A)
Balancer of weight	Sluice gate
Impeller number	1
Direction of rotation	1 direction
Peak speed	7500 RPM
Peak performance	120 kW

The engine performance and fuel characteristics tests were also carried on pure petro-diesel produced by local fuel company in Indonesia. The fuels were tested in order to compare their result with *Nephelium lappaceum* seed biodiesel. The comparison is used to determine the performance of *Nephelium lappaceum* seed biodiesel and its viability as a substitute for pure petro-diesel.

3. Results and discussion

3.1. Fuel characteristics

The characteristics of fuels were tested using ASTM standards. The results of the fuel characteristics tests are displayed in table 3. In this table, pure petro-diesel is abbreviated to PPD and *Nephelium lappaceum* seed biodiesel is abbreviated to NLSB. The test result of density testing under ASTM D-1298 standard suggested that *Nephelium lappaceum* seed biodiesel has higher density of 0.8331 Kg/L compared to pure petro-diesel with density of 0.8326 Kg/L. The test results of kinematic viscosity under ASTM D-445 standard suggested that *Nephelium lappaceum* seed biodiesel has higher kinematic viscosity level of 2.55 cSt compared to pure petro-diesel with kinematic viscosity level of 2.49 cSt.

The test result of flash point testing under ASTM D-93 standard suggested that pure petro-diesel has higher flash point with value of 73 °C compared to *Nephelium lappaceum* seed biodiesel with flash point of 58 °C. The test result of pour point testing under ASTM D-97 standard suggested that *Nephelium lappaceum* seed biodiesel has higher pour point of -4 °C compared to pure petro-diesel with pour point of -9 °C. The test result of Sulphur content testing under ASTM D-4294 standard suggested that pure petro-diesel has higher sulfur content of 0.047 %wt compared to *Nephelium lappaceum* seed biodiesel with sulfur content of 0.041 %wt. The test result of color testing under ASTM D-4294 standard suggested that pure petro-diesel and *Nephelium lappaceum* seed biodiesel has similar color point of 2.

Table 3. Summary of fuel characteristics for pure petro-diesel (PPD) and *Nephelium lappaceum* seed biodiesel (NLSB).

Parameters	PPD	NLSB
Density (kg/L)	0.8326	0.8331
Kinematic viscosity (cSt)	2.49	2.55
Flash point (°C)	73	58
Pour point (°C)	-9	-6
Sulfur content (%wt)	0.047	0.041
Color characteristic	2.5	2.5
Water content (ppm)	157	117
Cetane index	51	51
Distillation point (°C)	234	337

The test result of water content testing under ASTM D-6304 standard suggested that pure petro-diesel has higher water content with value of 157 ppm compared to *Nephelium lappaceum* seed biodiesel with water content of 117 ppm. The water content follows the standard of EN 14214 which is below 500 mg/kg or 500 ppm [31]. The test result of cetane index testing under ASTM D-4737 standard suggested that pure petro-diesel and *Nephelium lappaceum* seed biodiesel has similar cetane index of 51. This value of cetane number is acceptable under standard of ASTM D6751 which requires the cetane index of diesel fuel to be higher than 47 [31]. The test result of distillation point testing under ASTM D-86 standard suggested that *Nephelium lappaceum* seed biodiesel has higher distillation point of 337 °C compared to pure petro-diesel with distillation point of 234 °C.

3.2. Engine performances

The results of the engine performance tests are displayed from figure 2 to figure 6. Similar with fuel characteristics test results, pure petro-diesel is abbreviated to PPD and *Nephelium lappaceum* seed biodiesel is abbreviated to NLSB.

Power parameter shows the capability of engine in performing movement under certain load. Engine testing result for power value under various range of RPM is presented in figure 2. The result suggested that *Nephelium lappaceum* seed biodiesel has higher average power value of 11.98 HP (horse power) compared to pure petro-diesel with average power value of 11.25 HP. *Nephelium lappaceum* seed biodiesel has higher average power of 6.1 % than pure petro-diesel.

Torque level indicates the capability of an engine to produce movement. Engine testing result for torque value under various range of RPM is presented in figure 3. The result suggested that *Nephelium lappaceum* seed biodiesel has higher average torque value of 34.78 Nm compared to pure petro-diesel with average torque value of 32.65 Nm. *Nephelium lappaceum* seed biodiesel has higher average torque of 6.14 % than pure petro-diesel.

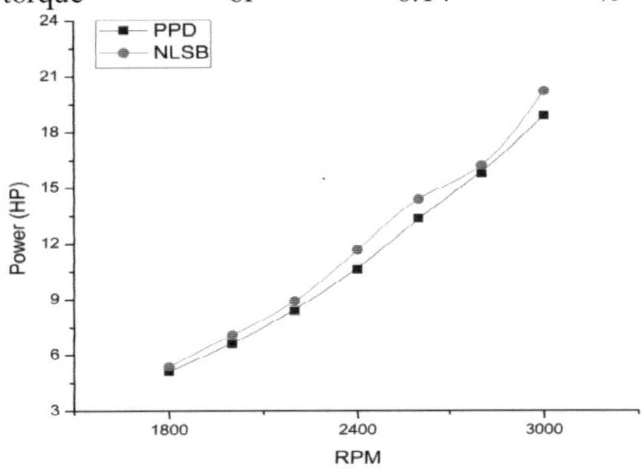

Figure 2. Power engine test results of various diesel fuels with respond to RPM changes.

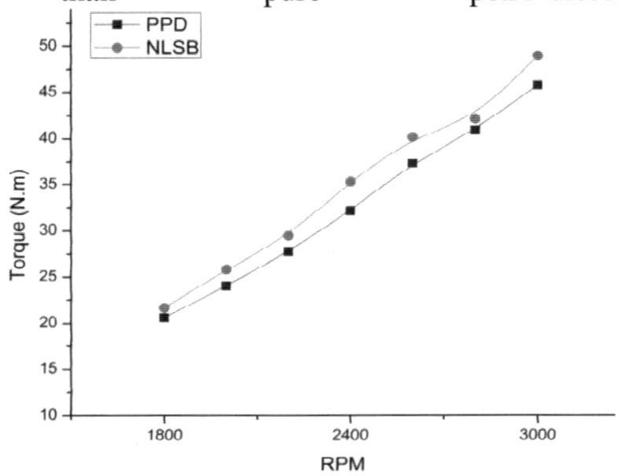

Figure 3. Torque engine test results of various diesel fuels with respond to RPM changes.

Break mean effective pressure (BMEP) is one of many parameters used to assess the engine performance of fuel. Engine testing result for BMEP value under various range of RPM is presented in figure 4. The result suggested that *Nephelium lappaceum* seed biodiesel has higher average BMEP value of 1.72×10^{-6} kg/cm^2 compared to pure petro-diesel with average BMEP value of 1.62×10^{-6} kg/cm^2. *Nephelium lappaceum* seed biodiesel has higher average BMEP of 6.13 % than pure petro-diesel.

Specific fuel consumption (SFC) indicates the number of fuel utilized in providing the engine with power in an hour. Engine testing result for SFC value under various range of RPM is presented in figure 5. The result suggested that *Nephelium lappaceum* seed biodiesel has higher average SFC value of 0.3757 kg/HP.hour compared to pure petro-diesel with average SFC value of 0.4014 kg/HP.hour. *Nephelium lappaceum* seed biodiesel has lower average SFC of 6.85 % than pure petro-diesel.

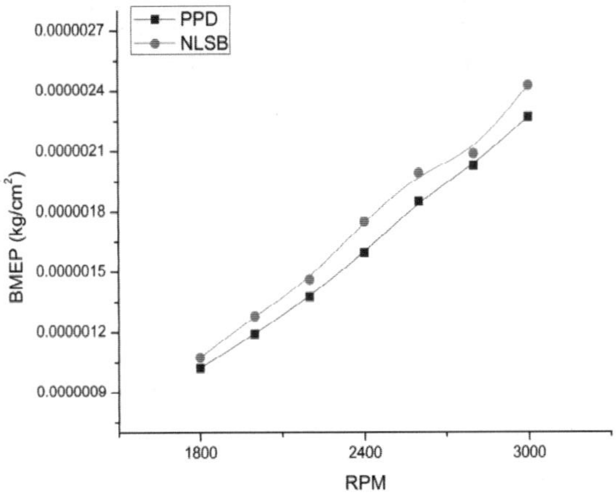

Figure 4. BMEP engine test results of various diesel fuels with respond to RPM changes.

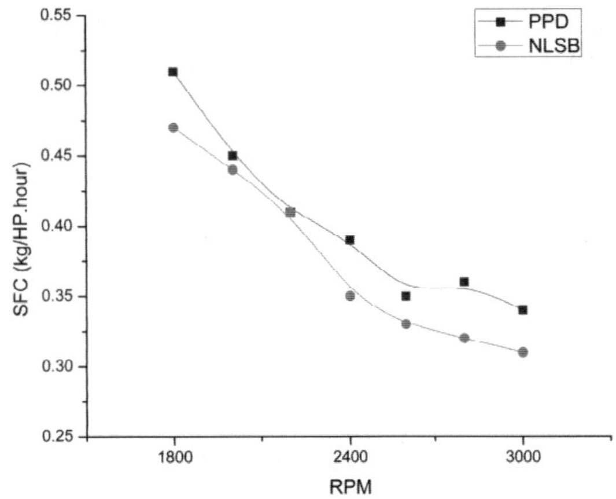

Figure 5. SFC engine test results of various diesel fuels with respond to RPM changes.

Thermal efficiency is parameter used to indicate the economical advantage of an engine when operating using certain fuel. Engine testing result for thermal efficiency value under various range of RPM is presented in figure 6. The result suggested that *Nephelium lappaceum* seed biodiesel has higher average thermal efficiency value of 17.38 % compared to pure petro-diesel with average thermal efficiency value of 16.12 %. *Nephelium lappaceum* seed biodiesel has higher average thermal efficiency of 7.26 % than pure petro-diesel.

The engine performance tests in *Nephelium lappaceum* seed biodiesel bolster the fuel characteristics investigation of *Nephelium lappaceum* seed biodiesel. Lower fuel flow caused by the low viscosity of *Nephelium lappaceum* seed biodiesel explains its higher power value compared to pure petro-diesel [34]. The higher power, thermal efficiency and lower specific fuel consumption in *Nephelium lappaceum* seed biodiesel compared to pure petro-diesel may be instigated from the lower water content of *Nephelium lappaceum* seed biodiesel [35].

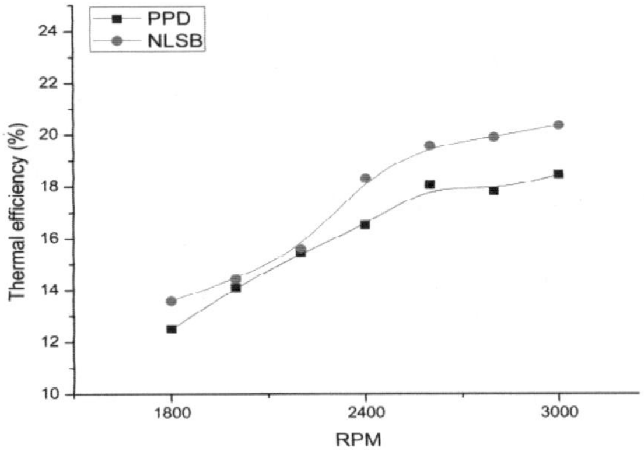

Figure 6. Thermal efficiency engine test results of various diesel fuels with respond to RPM changes.

4. Conclusion

Based on the investigations, it can be concluded that *Nephelium lappaceum* seed biodiesel is a viable substitute for pure petro-diesel. The fuel characteristics of *Nephelium lappaceum* seed biodiesel is comparable with pure petro-diesel which means it meets the standard of an engine fuel. From the engine performance test, the *Nephelium lappaceum* seed biodiesel displays better performance compared to pure petro-diesel. In term of power, torque, BMEP, SFC, and thermal efficiency, *Nephelium lappaceum* seed biodiesel is better compared to that of pure petro-diesel. Another advantage of *Nephelium lappaceum* seed biodiesel is found in its low sulfur content. The low sulfur content may reduce the SOx emission in the exhaust. The lower SFC and higher thermal efficiency in *Nephelium lappaceum* seed biodiesel may help the effort to solve the energy crisis faced by society today.

5. References

[1] Canakci M and Van Gerpen J 2001 *Trans. Ame. Soc. Agri. Eng.* **44** 6 1429–36
[2] Khalid, Jaat N, Sapit A, Razali A, Manshoor A, Zaman I and Abdullah A A 2015 *Int. J. Auto. and Mech. E. (IJAME)* **11** 2447–57
[3] Kumaran P, Gopinathan M and Kantharrajan S 2014 *Int. J. Auto. and Mech. E. (IJAME)* **10** 2112–21
[4] Rostami S, Ghobadian B and Kiani Deh Kiani M 2014 *Int. J. Auto. and Mech. E. (IJAME)* **10** 1945–58
[5] Anggono W, Wardana I N G, Lawes M, Hughes K J, Wahyudi S and Hamidi N 2012 *J. App. Sci. Res* **8**(8) 4126–32
[6] Anggono W, Suprianto F D, Sutrisno, Gotama G J, Evander J and Kasrun A W 2018 *ARPN J. Eng. and App. Sci.* **13**(3) 1080–84
[7] Azad A K, Rasul M G, Giannangelo B and Islam R 2015 *Int. J. Auto. and Mech. E. (IJAME)* **12** 2866–81
[8] Hariram V and Mohan Kumar G 2013 *J. Eng. Sci. and Tech* **8**(1) 77–92
[9] Demirbas A 2009 *Ene. Conv. and Man.* **50**(1) 14–34
[10] Azhari Mohd Ghazi T I, Gunam Resul M F M, Yunus R and Shean Yaw T C 2008 *J. Eng. Sci. and Tech.* **3**(2) 138–45
[11] Anggono W, Wardana I N G, Lawes M, Hughes K J, Wahyudi S, Hamidi N and Hayakawa A 2016 *AIP Conf. Proc.* **1717** 030001 DOI: 10.1063/1.4943425.
[12] Silitonga A S, Atabani A E, Mahlia T M I, Masjuki H H, Badruddin I A and Mekhilef S 2011 *Ren. and Sust. Ene. Rev.* **15** 3733–56
[13] Demirbas A 2008 *Fuel* **87** 1743–48
[14] Singh S P and Singh D 2010 *Ren.and Sust. Ene. Rev.* **14** 200–16
[15] Yaakob Z, Mohammad M, Alherbawi M, Alam Z and Sopian K 2013 *Ren. and Sust. Ene. Rev.* **18** 184–93
[16] Ikwuagwu O, Ononogbu I C and Njoku O U 2000 *Ind. Crops and Prod.* **2**(1) 57–62
[17] Bojan S G, Chelladurai S and Durairaj S K 2011 *ARPN J. Eng. and App. Sci.* **6**(8) 73–8
[18] Talha N S and Sulaiman S, *ARPN J. Eng. and App.Sci.* **11**(1) 439–48
[19] Ramaraju A and Ashok Kumar T V 2011 *ARPN J. of Eng. and App.Sci.* **6**(4) 1–6
[20] HO K C, Shahbaz K, Rashmi W, Mjalli F S, Hashim M A and Alnashef I M 2015 *J. Eng. Sci. and Tech.* **EURECA 2014 Special Issue January** 98–111
[21] Salvi B L and Panwar N 2012 *Ren. and Sus. Ene. Rev.* **16** 3680–89
[22] Sarin R, Sharma M, Sinharay S and Malhotra R K 2007 *Fuel* **86** 1365–71
[23] Azad A K, Ameer Uddin S M and Alam M M 2012 *Int. J. Auto. and Mech. E. (IJAME)* **5** 576–86
[24] Asokan M A, Senthu Prabu S, Kamesh S and Khan W 2018 *Energy* **145** 238–45
[25] Anggono W 2017 *Int. J. of Ren. Ene. Res.* **7**(1) 306–10
[26] Benjumea P, Agudelo J and Agudelo A 2008 *Fuel* **87** 2069–75
[27] Chen Y H, Chen J H, Chang C Y and Chang C C 2010 *Bio Res. Tech.* **101** 9521–26
[28] Said N H, Ani F N and Said M F M 2015 *J. of Mec. Eng. and Sci. (JMES)* **8** 1302–11
[29] Borugadda V B and Goud V V 2012 *Ren. and Sust. Ene. Rev.* **16** 4763–84
[30] Atabani A E, Silitonga A S, Ong H C, Mahlia T M I, Masjuki H H, Badruddin I A and Fayaz H 2013 *Ren. and Sust. Ene. Rev.* **18** 211–45
[31] Nguyen H, Nguyen M T, Nguyen T D and Nguyen P T 2016 *Env. Prog. & Sust. Ene.* **35**(2) 603–10
[32] Wong C S and Othman R Int. 2014 *J. of Eng. and Tech. (IJET)* **6**(6) 2773–77
[33] Pwinayanuwattikun P, Kaewpiboon C, Piriyakananon K, Tantong S, Thakerkankit W, Chulalaksananukul W and Yongvanich T 2008 *Biom. and Bioe.* **32** 1279–86

IOP Conf. Series: Earth and Environmental Science **257** (2019) 012039 doi:10.1088/1755-1315/257/1/012039

[34] Utlu Z and Kocak M S 2008 *Ren. Ene.* **33** 1936–41
[35] Awad O I, Mamat R, Noor M M, Yusop F and Yusri I M 2017 *WSEAS Trans. Env. and Dev.* **13** 120–8

Acknowledgments
This research was supported by Petra Christian University, Indonesia and *Direktorat Jendral Pendidikan Tinggi Kementerian Riset Teknologi dan Pendidikan Tinggi Republik Indonesia* (Directorate General Ministry of Research, Technology and Higher Education, Republic of Indonesia) as part of *Hibah Penelitian Produk Terapan 2015-2017* (Applied Product Research Grant 2015-2017) program. Authors would like to express their gratitude toward these institutions.

IOP Conf. Series: Earth and Environmental Science **257** (2019) 012040 doi:10.1088/1755-1315/257/1/012040

The effects of blade number and turbine baffle plates on the efficiency of free-vortex water turbines

P Sritram and R Suntivarakorn *

Department of Mechanical Engineering, Faculty of Engineering, Khon Kaen University, Khon Kaen, Thailand 40002
E-mail: Ratchaphon@kku.ac.th *

Abstract. This article presents the results of the study on the effect of blade number and turbine baffle plates on the efficiency of a free-vortex water turbine. The laboratory experimentation performed to determine the power generation efficiency. The 2 to 7 blade water turbines were built and tested to find the most appropriate number of blades, and the result showed the 5 blade turbine being appropriate because it yields the highest torque from receiving impact from water flow. Next, the baffle plates were designed and attached to the top and bottom of the turbine blades. Four different sizes of space from 25% to 100% of the curve area around the blades were used. Experiments were carried out at the water flow rates of 0.04 to 0.06m3/s. The finding showed the 50% proportion of the curve area being most appropriate, and the blades installed with top and bottom baffle plates had the highest efficiency of 43.83%, which was 6.59% higher than without baffle plates. It was also found that when the water flow rate increased, the system efficiency became higher.

1. Introduction

It is widely known that water is a clean energy source that does not create environmental pollution. However, there is a present constraint in hydrological power generation since it requires construction of a large dam, which has a great impact on the natural environment around the dam. Still, power generation from a river or a small water resource where water flows all year round is an alternative energy production that receives increased interest. There are now 2 methods of small-scale hydrological power generation. The first is the Bypass Method that diverts water from the main feed canal to the constructed power generation system. The second is the Open Canal Method, which involves installation of the generator across the canal without any other additional structure.

The advantage of the Bypass Method in electricity generation is that after the process is completed, the water can be discharged back to its original source and can still be used for other purposes. Nowadays, more and more researchers become interested in conducting a study of this method, for instance, power generation from free-vortex water flow or from a screw-type water turbine. The free-vortex water flow system is a system from which electricity is generated by the centrifugal flow of water through a cylindrical vortex pool that has a water outlet or hole at the bottom center. This system generates electricity at low head water, relying on the vortex flow of water, the velocity of which is accelerated by gravity at the outlet. This system was first studied by Franz Zotloterer, who built a low-head power plant based on the principle of the kinetic energy of vortex flow of water in the pool. Additional studies were later conducted on vortex water flow in the pool [1-2] and generation of power from the free vortex flow of water in a cylindrical pool [3-5]. The study showed that the outlet or discharge hole should be 0.2-0.3 times of the diameter of the pool, where free vortex occurs [6]. There are other research studies conducted on different variables affecting power generation from free-

Content from this work may be used under the terms of the Creative Commons Attribution 3.0 licence. Any further distribution of this work must maintain attribution to the author(s) and the title of the work, journal citation and DOI.
Published under licence by IOP Publishing Ltd

vortex water flow based on Computational Fluid Dynamics, which demonstrated that at the highest flow, the turbine efficiency is 40% at the speed between 28 and 38 rpm [7]. A laboratory experiment showed the highest overall efficiency of 30% at the turbine speed of 40-50 rpm [8].

Additionally, there were studies of different approaches to increase the efficiency of a free-vortex water turbine such as the change of blade material, modification of the vortex pond's shape or the change of the blade design. Dhakal et al. studied the structural analysis of the pool characteristics that will create vortex flow of water and increase the propelling force of the water turbine. This study found that a conical pool was able to increase the work efficiency. The highest percentage of efficiency, when compared to a cylindrical pool, was 9.09% at the flow rate of 0.01 m^3/s and the height of 0.85 m [9-10]. Other research studies were conducted on the change of material used to construct the water turbine, and new materials used resulted in increase of power generation from the free-vortex water turbine. Sritram et al. found that after changing the material of the turbine from iron to aluminium, the power generation efficiency of an aluminium turbine was on average 8.14% higher than the former turbine [11]. Suntivarakorn et al. designed a turbine with a twisted angle and the turbine size that matches the vortex flow of water. This turbine could increase the efficiency of over 15% when compared to the former turbine [12]. In addition, it has been shown that the number of blades and adjustment or installation of supplementary equipment on the blades could increase the efficiency. The study of Savonius wind turbine installed with baffle plates could increase 36% higher turbine efficiency than the old turbine [13].

The assumption in this research was that if the number of water turbine blade is higher, the system efficiency should be higher. However, the past studies have not clarified the number of blades that affect the efficiency, nor if installation of baffle plates at the water turbine blades allows the turbine to receive the impact from water and increase the turbine efficiency. Thus, this study was interested in increasing the efficiency of small-scale power generation from a free-vortex flow of water through the study of impact from the number of blades and installation of baffle plates. The study was based on laboratory experiment. The objectives were to know the effect of the number of blades and the characteristics of the baffle plates installed on the efficiency of small-scale power generation from a free-vortex flow of water.

2. Equation of Calculation

2.1. Power and torque obtained from the turbine [14-17]

Real power obtained from water current (P_m) could be calculated by:

$$P_m = \rho g Q H \tag{1}$$

Torque (T) from the turbine could be found by:

$$T = \frac{60 P_m}{2\pi N} \tag{2}$$

When ρ = water density, 1,000 kg/m^3, g= acceleration by gravity, 9.81 m^2/s, Q= rate of flow (m^3/min), N = revolutions of the shaft (rpm), and H = water head (m).

Calculation for efficiency could be done from the proportion between the real electrical energy obtained and dynamic energy from the equations:

$$\eta = \frac{P_e}{P_m} \tag{3}$$

$$P_e = EI \tag{4}$$

When E = voltage (V), I = current (A) and P_e = electrical power (W).

IOP Conf. Series: Earth and Environmental Science **257** (2019) 012040 doi:10.1088/1755-1315/257/1/012040

3. Equipment and Experimental Method

3.1. Equipment for testing the free-vortex water system

Figure 1. Power generation from free-vortex water flow.

This study was performed on a small-scale power generation from a free-vortex water flow in the laboratory of Department of Mechanical Engineering, Khon Kaen University. The system is composed of an upper storage tank of 1.5×1.5×1.2 m connected with a gutter and a cylindrical pool 1 m in diameter and 1 m in height. There is a water outlet hole of 0.2 m diameter at the bottom of the pool and a water pump adjustable within 0.04-0.06 m^3/s. In Figure 1, the flow of water is pumped from a storage tank below through a flow rate control valve. The flow rate sonic gauge was used in order to control the rate of flow of water fed into the system. The water flows to the upper tank and then to the gutter before reaching the vortex pool. At the vortex pool, a turbine was installed with its shaft connected to the generator's shaft to generate electricity towards the load by means of the centrifugal force of the vortex in the pond. This load is exerted on the turbine. Then the water flows centrifugally to the lower water storage tank to be pumped back continuously for use in the experiment.

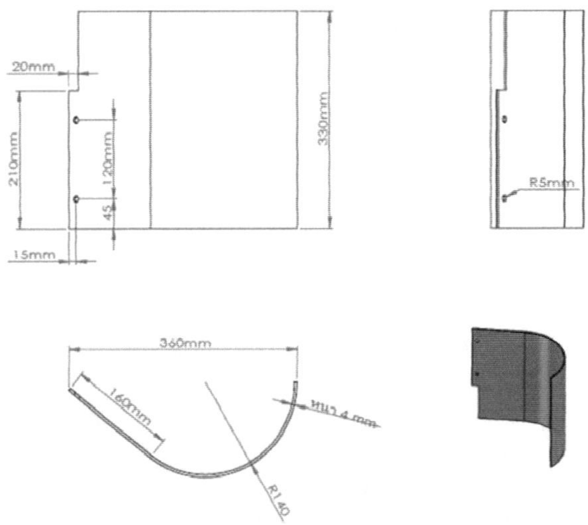

Figure 2. The blades of the water turbine in the test.

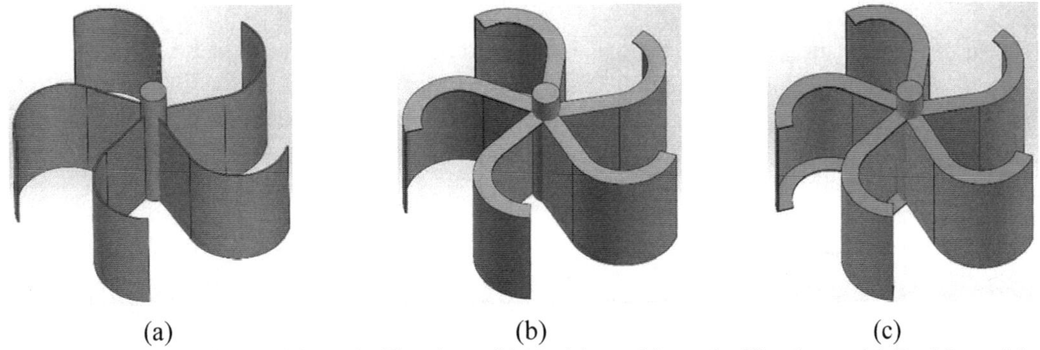

Figure 3. (a) Turbine with no baffle plate, (b) Turbine with top baffle plates, (c) Turbine with top and bottom baffle plates

3.2. Blades

The turbine in the test has a diameter of 45 cm and a height of 32 cm. The blades were made of iron and are curved in the shape as shown in Figure 2. The blade numbers tested to find the turbine efficiency were 2 to 7 blades. After the suitable number was found, the baffle plates were installed at the blades to find the highest efficiency among the 3 cases tested: 1) blades with no baffle plate, 2) blades with top baffle plates, and 3) blades with both top and bottom baffle plates, as shown in Figure 3.

3.3. Experimental method

In the finding of the suitable number of blades for the water turbine, the efficiency of the turbine operation was tested with the turbine having from 2 to 7 blades. This was a laboratory test of power generation from a free-vortex water flow. Electrical load used consisted of four 25 W bulbs. The rate of water flow was set at 0.06 m^3/s. The load was connected with the power generation system, with the initial load of 25 W. The water head, revolution of the turbine, voltage and electric current from the generator were measured. Next, the load was increased to 50, 75, and 100 W. The experiments began on the performance of the turbine with 2 blades through to 7 blades. The results were calculated to find the torque and efficiency values, including water flow patterns in the turbine before the highest efficiency level and the most suitable number of blades determined.

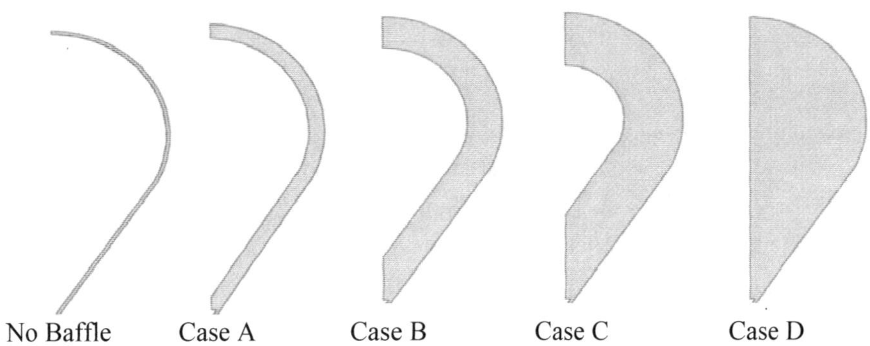

Figure 4. Proportion of baffles fit on 5 blades.

After knowing the suitable number of blades, the baffle plates were designed in 5 shapes. The proportions of the baffles to be attached to the blades were from 0% to 100% as shown in Figure 4.

IOP Conf. Series: Earth and Environmental Science **257** (2019) 012040 doi:10.1088/1755-1315/257/1/012040

The proportion of 100% means the surface area between both ends of the turbine was as stated in Table 1.
Wichian et al. studied the CFD program was used for the analysis of the right proportion of baffle plates on turbine blades. This study found that a water turbine Case B, with 50% proportion of baffle plate contact on the blades received impact that matched the mass distributed on the turbine [17].

Table 1. Showing correlation of space with baffle plates.

	No Baffle	Case A	Case B	Case C	Case D
Proportion of baffle fitting on blade (%)	0	25	50	75	100
Area of baffle plate (cm^2)	0	100.8	210.05	322.32	438.38

Next, the turbines were built for the experimental purpose with the number of baffle plates and proportion of fitting areas. Thus, there were 3 water turbines built for the testing: Case 1 turbine blades with no baffle plates; Case 2 turbine blades with top baffle plates; and Case 3 turbine blades with top and bottom baffle plates, as illustrated in Figure 3. In the experiment, the turbine blades with baffles were installed in the small power generation system using a small free-vortex flow of water. The efficiency of the turbine was found in the same way as in the finding of the suitable number of blades. The water flow rates in the experiments were set at 0.04, 0.05, and 0.06 m^3/s, and all of the three types of turbines were tested to find the flow force, turbine's torque, electricity and efficiency of electricity generation.

4. Results and Discussion

4.1. Effects of the number of blades on turbine's torque
The graph in Figure 5 shows the comparison of the torques obtained from the turbines with different numbers of blades (2-7 blades). The turbine with 5 blades produced the highest torque. This was consistent with the finding of Wanchat et al. who performed the test in a laboratory. In Figure 6, when the test was performed in the loads of 0-100 W, the turbine with 5 blades was still found to yield the highest torque. It can be concluded that the flow of water exerted on the turbine with different numbers of blades yields different torques. The turbine with 5 blades is the best to receive impact from water flow.

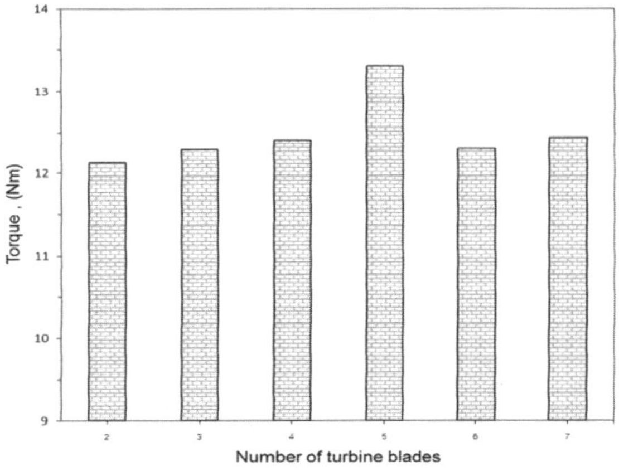

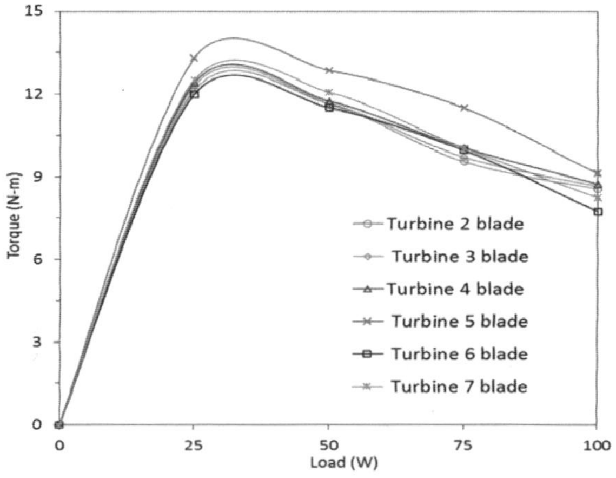

Figure 5. Torques of turbines with 2-7 blades.

Figure 6. Torques from turbines with 2-7 blades at the load of 0-100 W.

Observation of the channel of water flow that impacted on the blades showed that water eddy circulated around the pool and centrifugally through the hole at the bottom. The blades of the turbine then received the impact from the centrifugal flow of water. It was found that when the number of blades increased until 5 blades, the surface area contacting with water was greater, resulting in higher torque. But when there were 6 or 7 blades, the distance between blades was less, making the impact of water current on the blades decrease. Moreover, water flow through the hole developed resistance against the movement of the next blade in series, resulting in resistance of the water eddy and hence reducing the torque produced by the turbine, as illustrated in Figure 7.

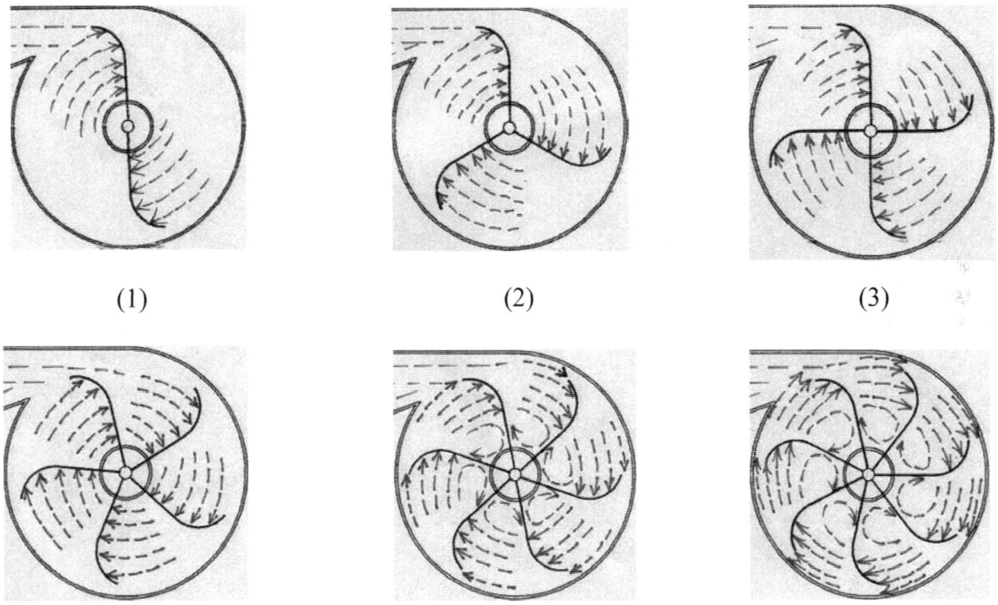

(1) (2) (3)

(4) (5) (6)

Figure 7. Sketches of water direction impacting the blades of the turbines with: (1) 2 blades, (2) 3 blades, (3) 4 blades, (4) 5 blades, (5) 6 blades, (6) 7 blades

The results of the tests were extended in construction of a power generation system using a free-vortex flow of water, including a concrete pool having a diameter of 4 m and a height of 1 m. The 5 blade turbine was installed and connected to an electricity generator. It was found that at the flow rate of 0.6 m^3/s, the highest capacity of electricity generation was 1.7 kW, and the efficiency was 31.65% [18].

4.2. Effect of baffle plates on turbine efficiency

The analysis of the water turbine efficiency to compare the 3 types of turbines, namely, Case 1 no baffle plates, Case 2 with top baffle plates, and Case 3 with top and bottom baffle plates showed that at the highest flow rate of 0.6 m^3/s, the turbine Case 3 yielded higher torque than the other turbines as shown in Figure 8. In Figure 9, the turbine Case 3 yielded the greatest electricity power because the baffle plates retained more water mass impacting on the surface area of each blade, and therefore the torque was higher than other cases at all of the loads fed. Considering electricity power generated in Figure 8, it can be seen that Case 2 and Case 3 of turbines did not differ much in terms of yields. However, the blades with baffle plates were more effective than the blades without baffle plates.

From the experiment, the highest efficiency was found at the flow rate of 0.06 m^3/s and power load of L = 25 W. Case 1, Case 2 and Case 3 of turbines tested showed the efficiency of 37.24%, 42.13%, and 43.83%, respectively. It can be seen that the efficiency of turbines with baffle plates on the blades was higher than the turbine with no baffle plates. The two cases with baffle plates also showed quite a similar efficiency. The tendency of the system efficiency was increased with the increase of water flow rate.

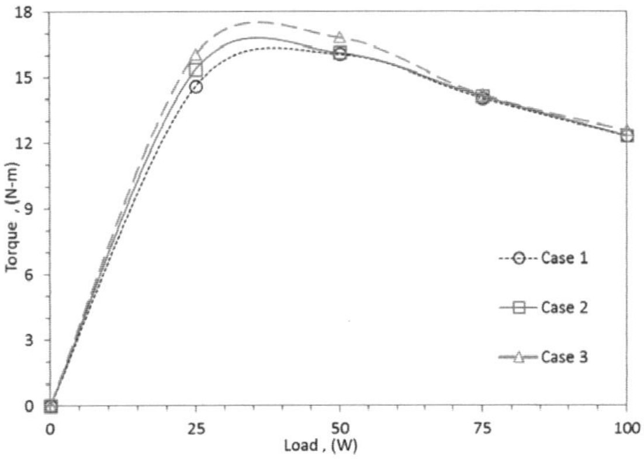

Figure 8. Turbine torques at Q = 0.06 m^3/s.

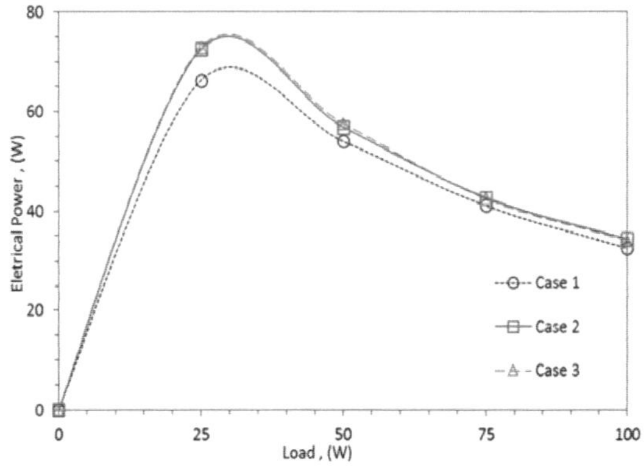

Figure 9. Electricity power generated at Q = 0.06 m^3/s.

5. Conclusions

The study showed that the number of blades and baffle plates attached to the blades of the water turbine affect the efficiency of the turbine. The water turbine with 5 blades was found the most appropriate for use and it also yielded the highest torque as the distance between blades was effective for receiving the exertion of water flow on the blades. When baffle plates were attached to the blades, the efficiency of turbine increased because the baffle plates assisted in retaining the right amount of water and yielding higher torque. It was fond that the 50% proportion of the curve area being most appropriate and with the flow rate of 0.06 m^3/s and the load of 25 W, the efficiency of turbine with top and bottom baffle plates was 43.83%, and it was 6.59% higher than that without baffle plates.

6. Acknowledgements

Researchers are grateful to the Farm Engineering and Automation Technology Research Group, Department of Mechanical Engineering Khon kaen University for the support of tools and equipment used in this research.

7. References

[1] Shi X M, Yang F, Dai R, Chen T J and Wu YL 2012 Simulation of free-surface vortex produced by a rotating cylindrical wall below a static barrel *IOP Conference Series: Earth and Environmental Science* vol. 15 052034

[2] Shabara H M, Yaakob O B, Yasser M Ahmed, Elbatran A H and Muhammad S M Faddir 2015 CFD Simulation of Water Gravitation Vortex Pool Flow for Mini Hydropower Plants *Jurnal Teknologi (Sciences & Engineering)* Vol. 5 p 77–81

[3] Kueh T K, Beh S L, Rilling D and Ooi Y 2014 Numerical Analysis of Water Vortex Formation for the Water Vortex Power Plant *International Journal of Innovation: Management and Technology* Vol. 5 p 111-115

[4] Li H F, Chen H X, Ma Z and Zhou Y 2009 Formation and influencing factors of free surface vortex in barrel with a central orifice at bottom *Journal of Hydrodynamics* Vol. 21(2) p 238-244

[5] Chen Y L, Wu C, Ye M and Ju X M 2007 Hydraulic characteristics of vertical vortex at hydraulic intakes *Journal of Hydrodynamics Ser. B* Vol. 19(2) p 143-149.

[6] Wanchat S and Suntivarakorn R 2012 Preliminary design of a vortex pool for electrical generation *Advanced Science Letters* Vol.13 p173–177.

[7] Shabara H M, Yaakob O B, Yasser M Ahmed, Elbatran A H and Muhammad S M Faddir 2015 CFD Validation for Efficient Gravitational Vortex Pool. System *Jurnal Teknologi (Sciences & Engineering)* Vol. 5 p 97–100

[8] Wanchat S, Suntivarakorn R, Wanchat S, Tonmit K and Kayanyiem P 2013 A Parametric Study of a Gravitation Vortex Power Plant *Advanced Materials Research* p 811-817

[9] Dhakal S, Timmilsina A B, Dhakal R, Fuyal D, Bajracharya T R, Pandit H A, Amatya N and Nakarmi A M 2015 Comparison of Cylindrical and Conical Basins with Optimum Position of Runner: Gravitational Water Vortex Power Plant *Renewable and Sustainable Energy Reviews* Vol. 48 p 662-669

[10] Dhakal S, Timmilsina A B, Dhakal R, Fuyal D, Bajracharya T R and Pandit H A 2014 Effect of Dominant Parameters for Conical Basin: Gravitational Water Vortex Power Plant *Proceedings of IOE Graduate Conference* p 380-386

[11] Sritram P, Treedet W and Suntivarakorn R 2015 Effect of Turbine Materials on Power Generation Efficiency from Free Water Vortex Hydro Power Plant *IOP Conference Series: Material Science and Engineering* Vol. 103 p 012018

[12] Suntivarakorn R, Wanchat S and Monnatrakul W 2016 An Experimental Study of Electricity Generation Using a Horizontal Spiral Turbine. *Energy Procedia* Vol. 100 p 532-536

[13] Jeon K S, Jeong J I, Pan J K and Ryu K W 2014 Effects of End Plates with Various Shapes and Sizes on Helical Savonius Wind Turbines. *Renewable Energy* Vol. 79 p 167-176

[14] Chinnajai S and Chaitep S 2009 Performance Enhancement of Vertical-Axis Wind Turbine by Design of Blade and Inlet Air Scoop. *KKU Engineering Journal* Vol. 36 p 97-106

[15] Sritram P and Suntivarakorn R 2017 Comparative Study of Small Hydropower Turbine Efficiency at low head water *Energy Procedia* Vol.138 p 646–650

[16] Wichian P and Suntivarakorn R 2016 The Effects of Turbine Baffle Plates on the Efficiency of Water Free Vortex Turbines *Energy Procedia* Vol. 100 p 198–202

[17] Suntivarakorn R, Wanchat S and Suwanasri S 2016 An Experimental Study of Electricity Production Using a Vortex Turbine in a Gravitational Water Vortex Power Plant. *GMSARN Int. Conf. on Innovative Energy, Environment, and Development in GMS* p 1-4

IOP Conf. Series: Earth and Environmental Science **257** (2019) 012041 doi:10.1088/1755-1315/257/1/012041

Scaling of heights and widths of laminar jet diffusion flames under subatmospheric pressures

Pingchuan Ma[1], Haihang Li[2], jian Wang[1]

1 State Key Laboratory of Fire Science, University of Science and Technology of China, Hefei, China.
2 China Jiliang University

Abstract. The luminous shapes were presented to examine the scaling of flame heights and widths of laminar jet diffusion flames of burning methane, ethylene, and propane in a subatmospheric pressure chamber. All the flames examined were stable with no soot emitting. The Reynolds scaling mentioned below was generally suitable for the three hydrocarbon fuels although slopes were becoming steeper as soot formation elongated the flame height at high *Re* under high pressure. The regions of Froude number where buoyancy is dominant or buoyancy is unimportant were demonstrated by following Reynolds–Froude height scaling of Altenkirch's model. In Froude scaling of width, the unchanged slopes between dimensionless width and *Fr* with pressure increase indicated the good linearity of flame width changing on pressure. Finally, regression of the form for flame width was also considered.

1. Introduction

Flame height and width are essential properties of diffusion flames to be modeled under different pressures [1-6] and gravity levels [2,7]. Linear scaling between normalized height and dimensionless numbers was commonly used in those studies [8-10]. $L/d \propto Re$ is the classical expression to expound the proportionality between flame height and fuel mass flow rate [9-11] which derived by Jost [12] through equating diffusion time with convection time according to Einstein's diffusion equation. But the expression cannot reflect the effect of pressure on flame height since Reynoldes number is independent with pressure.

By adding Froude number, $Fr = u^2/gd \propto P^{-2}$, Altenkirch et al. [5] introduced both Reynolds and Froude numbers to discuss the role of buoyancy at elevated gravity. After obtaining the slopes of $(L/d)Re^{-2/3}Fr^{-1/3}$ versus Fr, Altenkirch et al. [5] found regions where buoyancy was dominant and where buoyancy was not important. But Sunderland et al. [10] concluded that the Reynolds-Froude height scaling of Altenkirch et al. [5] is more appropriate for slot burner flames because it neglects transverse curvature effect.

Subatmospheric and superatmospheric conditions also provide various buoyancy levels for testing the above scaling laws. Recently, Hu et al. [13] scaled normalized flame height with heat release rate at two different atmospheric pressures in a cubic fire enclosure. At reduced pressures luminous flame heights and widths were considered to examine the relative merits of Reynolds, Froude and Reynolds-Froude scaling laws in this paper.

2. Experimental setup

2.1. Subatmospheric pressure chamber and burner

Content from this work may be used under the terms of the Creative Commons Attribution 3.0 licence. Any further distribution of this work must maintain attribution to the author(s) and the title of the work, journal citation and DOI.
Published under licence by IOP Publishing Ltd

The experiments were conducted in a subatmospheric pressure chamber of 3 × 2 × 2 m in the internal size. Low pressures can be obtained by depressurizing the sealed chamber space. An observation window was embedded on the chamber wall to facilitate observation on ignition and flame stability. The burner configuration is Burke-Schumann type and made of stainless steel, but no air co-flow was used during the experiments. The fuel tube of the burner has an inner diameter of 5.0 mm with a tapered exit to reduce the formation of turbulent eddies. Sintered metal foam elements placed in the fuel nozzle minimize the instabilities in the initial fuel flow and create a top-hat velocity profile on the exit.

2.2. Mass flow controller

Methane, propane, and ethylene were selected (all with purities ≥ 99.9 %) as studied fuels with increasing carbon ratio. Fuel was supplied from compressed cylinder into the burner and its flow rate was metered by a thermal based mass flow controller (Sevenstar D07-19B). The range of the controller is 0-1000 sccm with 0.1 sccm resolution. The accuracy of the mass flow rate measured is within ±1 % and the repeatability is ±0.2 %.

2.3. Fuel flow rates

The volumetric flow rate (sccm) of the fuel is defined as the gas jet volume under the standard condition (1.0 atm, 0 ℃). And the mass flow rates of the fuels maintain constant at all pressures, Table. 1. The fuel mass flow rates were designed after trial tests to insure soot was completely oxidized within the visible flame envelope and the flame was relatively stable at all pressures considered. The estimated Reynolds numbers (Re) indicate that all the examined flames were in laminar status. The flames remained stable for a significant amount of time, but spontaneous fluctuations in the form of tip flicking occurred at a regular interval throughout the experiment. The flame images of stable periods were analyzed to determine the flame shape (height and width) for every test.

Table 1. Summary of test parameters.

Parameter	Fuel flow rate [1], sccm	Exit velocity, m/s	Re	Fr
Methane	25.0-150.0	0.021-0.425	7.4-44.4	0.009-3.7
Ethylene	12.5-150.0	0.011-0.425	7.1-84.7	0.002-3.7
Propane	6.25-100.0	0.005-0.283	7.0-111.4	0.0006-1.6

1 Volumetric fuel flow rate under 1.0 atm and 0 ℃ (same in this paper without special note).

3.Results and discussion

3.1. Flame height

As mentioned in the Introduction, many theoretical analyses [12] and experimental results [9-11] have reached a consensus about the Reynolds scaling of

$$L/d \propto Re \,. \tag{1}$$

For a given fuel mass flow rate, Eq. (1) indicates that flame height is independent of pressure on account of $Re = \rho u d/\mu = 4\dot{m}/\pi\mu d \propto P^0$. However, many experiment data have reported that flame height changed with pressure [7,14 , 15]. So this prediction of independence from pressure is of particular interest here.

Figure 1 shows the relationships between normalized flame height L/d and Reynolds number Re. This type of coordinate axis followed a long tradition [9-11]. We can imagine that the decreasing air entrainment by buoyancy under reduced pressure should make the flame become longer. This speculation can be confirmed by data of relatively low Reynolds (i.e. low flow rate). As Reynolds increased, this shrank difference giving a concentrating appearance of scatters. This happens because Froude number (the ratio of momentum to buoyancy forces) generally increases with Re [9], which weakening the effect of buoyancy at higher *Re*. This can be seen in Figure 1 of Reynolds number less than or equal to 33, 42, and 42 for methane, ethylene, and propane flames respectively.

In fact, the increasing soot formed at higher *Re* contributes a great deal to the narrowing of the flame height differences. As shown in Figure 1, when *Re* exceeded the above values, the relative size of flame height between lower pressure and higher pressure changed. This is mainly because at higher fuel flow rate, the excessive soot generated under higher pressure greatly elongated the flame height [16]. This phenomenon was also observed under elevated pressures in [17], where soot volume fraction is proportional to pressure of 1.2 power. And under some moderate fuel flow rates, flame heights increased with pressure below their smoke points [18].

The combined effect of increased buoyancy and soot formation made the slopes of all the hydrocarbon fuels bigger as pressure increase, Figure 1 When the chamber pressure increased from 0.3 to 1.0 atm, the average slope of the three fuels changed from about 0.71 to 1.09.

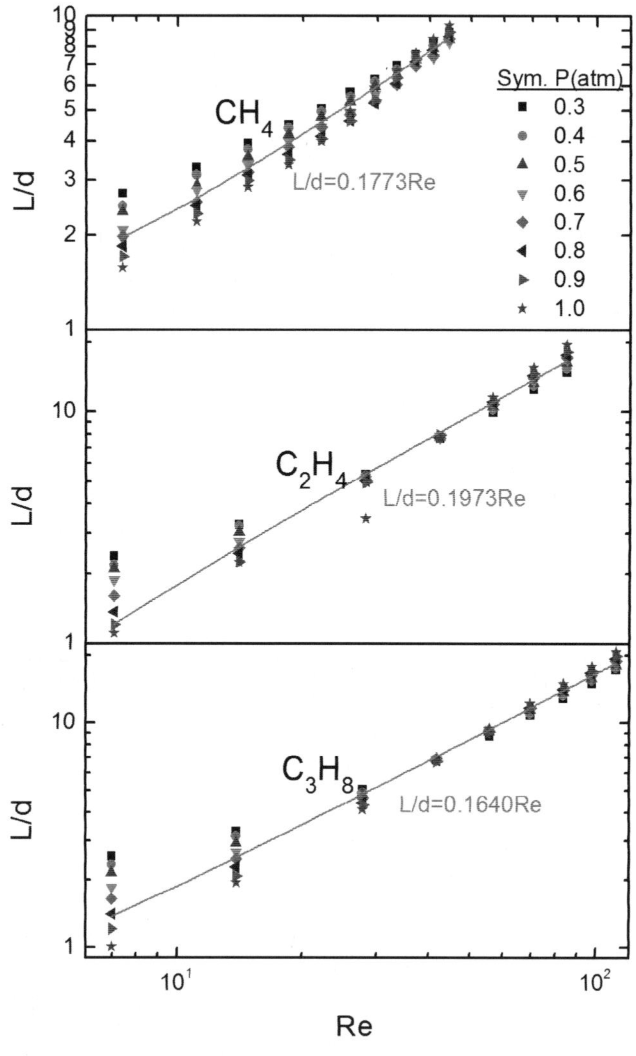

Figure 1. Normalized flame height as a function of Reynolds number for CH_4, C_2H_4, C_3H_8 flames under subatmospheric pressures. The lines and equations shown are best fits predicted on unity slope.

Altenkirch et al. [5] simplified the laminar gas jet diffusion flame as a cylinder and used a free convection boundary layer method. By establishing the balance of the oxidizer and fuel, it finally led to

$$L/d \sim Re^{2/3} Fr^{1/3} (W/d)^{-4/3} . \quad (2)$$

Suggesting W/d is constant, Altenkirch et al. [5] arrived two limits [10] using their results at elevated gravity:

$$\text{Nonbuoyant limit (large } Fr\text{)}, \; L/d \sim Re^{2/3} ; \quad (3a)$$

$$\text{Buoyant limit (small } Fr\text{)}, \; L/d \sim Re^{2/3} Fr^{1/3} . \quad (3b)$$

Figure 2 shows the relationship between $(L/d)Re^{-2/3}Fr^{-1/3}$ and Fr, this type of coordinate axis also followed a long tradition as [5, 10]. It can be seen that the linearity of methane is the best among the three fuels, and the divergence seemed to increase with increasing sooting tendency of hydrocarbons. The fitting slopes show $(L/d)Re^{-2/3}Fr^{-1/3}$ decreasing with Fr to -0.25, -0.17, -0.19 for methane , ethylene, and propane respectively. These slope differences have also been found by [10], whose values were -0.26 and -0.23 for methane and ethane, respectively. And the average slope for methane, ethane, and propane was -0.25 in [5] and -0.23 in [9].

As for methane flames, the slopes changed from -0.35 to -0.18 as pressure increased from 0.3 to 1.0 atm. This can be illustrated by Froude number changing in Figure 2: The slope at the largest Fr was about -1/3 showing $(L/d)Re^{-2/3}Fr^{-1/3}$ is changing with $Fr^{-1/3}$ .The largest contribution of Fr at 0.3 atm indicates buoyancy is unimportant in determining flame height. The phenomenon observed in [18-21]in which flame heights is constant in different pressure is this case. In stark contrast to the largest Fr, $(L/d)Re^{-2/3}Fr^{-1/3}$ remained nearly constant when $Fr<0.02$ indicating buoyancy dominant tendency was reached [5].

With regard to ethylene and propane flames, the absolute slopes between $(L/d)Re^{-2/3}Fr^{-1/3}$ and Fr were becoming smaller at higher pressures resulting from the increasing buoyancy effects. Similarly to methane flames, $(L/d)Re^{-2/3}Fr^{-1/3}$ kept nearly constant when $Fr < 0.01$ reflecting buoyancy became dominant in this region for propane flames. But for propane flames, the data of the smallest fuel flow rate above the blue dotted line did not meet the level off trend. The flame images of this flow rate were found to be too crescent that the assumption of cylindrical shape of flame is no longer reasonable. This disarray distribution phenomenon was also observed for ethylene flames. And for ethylene we might infer that $(L/d)Re^{-2/3}Fr^{-1/3}$ would be level off below a certain Fr if the burner diameter was suitable.

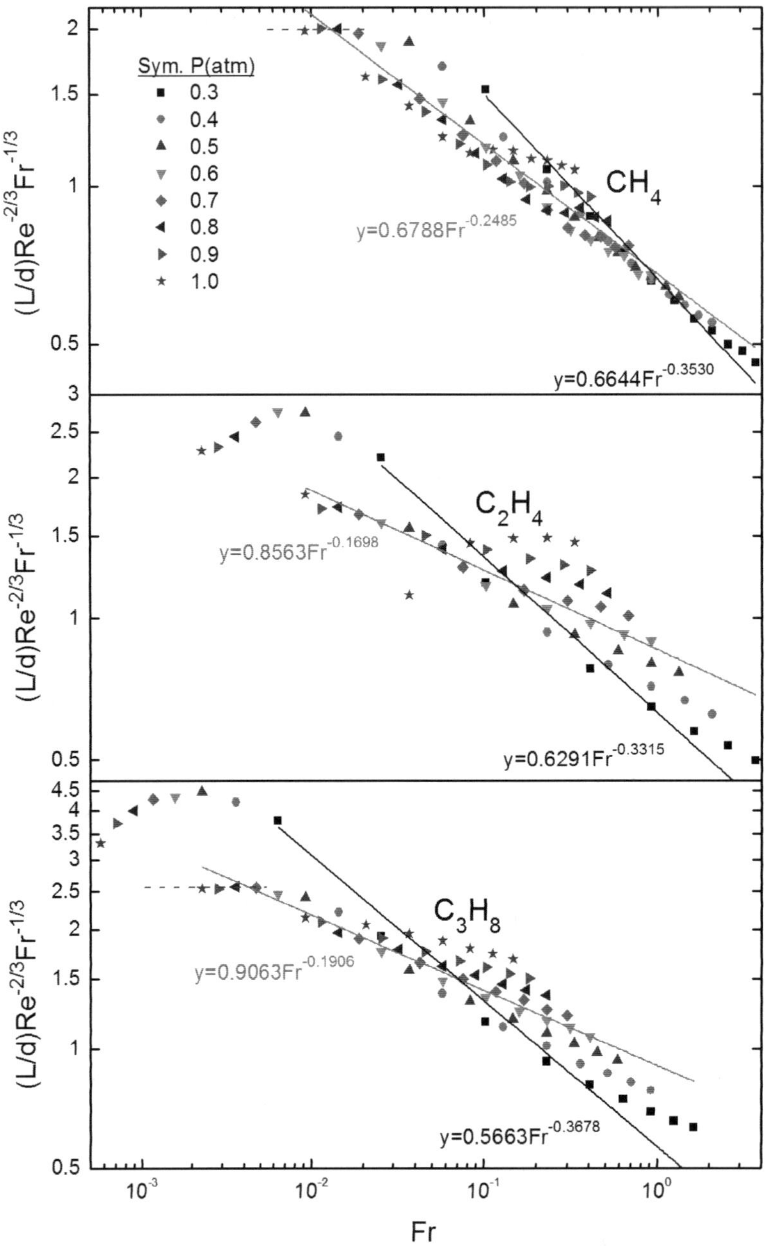

Figure 2. Flame height relationship for CH4, C2H4, C3H8 flames under subatmospheric pressures. The lines and equations shown are best fits, where y is defined as $(L/d)Re^{-2/3}Fr^{-1/3}$.

3.2. Flame width

Davis et al. [22] reported that Froude number is the appropriate scaling parameter for gravitational variations in laminar jet diffusion flames. We will verify whether this scaling relationship is suitable for pressure change.

Figure 3 shows the relationship between *W/d* and *Fr* for the present flames. The sooting tendency of methane, propane, and ethylene can be also seen in normalized width verse *Fr*, which can be demonstrated by increased scattering level in data. The power of *W/d* dependence on *Fr* were 0.16, 0.13, and 0.13 for methane , ethylene, and propane respectively in Figure 3, approximating to 0.16 in [9] averaged from methane, ethane, and propane. Besides, the slopes between *W/d* and *Fr* of all the fuels were not changing with pressure and the level off phenomenon didn't happen as in Froude scaling of height, both of which were indicating the good linearity of flame width changing on pressure.

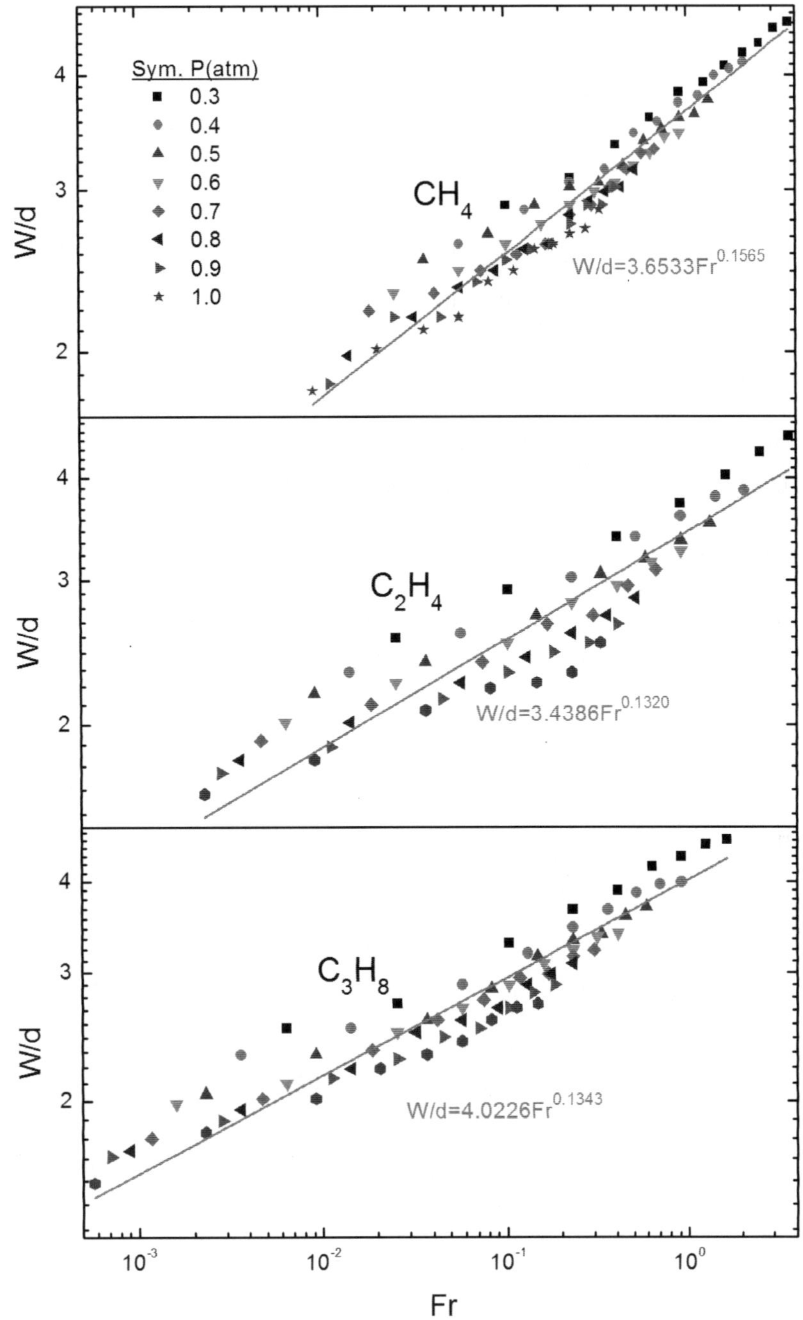

Figure 3. Normalized flame width as a function of Froude number for CH4, C2H4, C3H8 flames under subatmospheric pressures. The lines and equations shown are best fits.

Following the correlation of flame widths in [9] and references therein, regressions of the form $W/d \propto Fr^a Re^b$ were also considered in this paper. As mentioned in [9], several theories predicted $a = -b = 0.25$, or $W/d \propto St^{0.25}$. According to the multiple linear regression of methane, ethylene, and propane flames, the values of a were 0.1964, 0.2175, and 0.2047, b were -0.1434, -0.2420, -0.1973 here.

4.Conclusions

The luminous shapes of laminar jet diffusion flame were measured to examine the scaling of flame height and width in a subatmospheric pressure chamber. The mass flow rates of methane, ethylene, and propane were kept constant at all pressures and the corresponding flames were no soot emitting. The major conclusions were as follows:

- The Reynolds scaling of $L/d \propto Re$ was generally suitable for the three hydrocarbon fuels. Under reduced pressure, the air density will become lower, which contributes to a weeker buoyancy. And the decreasing air entrainment by a weaker buoyancy made the flame longer at relatively low Re. As Re increased, the weakening effect of buoyancy showed a concentrating height difference. And excessive soot generated under higher pressure greatly elongated the flame height at the higher Reynolds number.
- The -1/3 power of Fr contribution to $(L/d)Re^{-2/3}Fr^{-1/3}$ indicates that buoyancy is unimportant in determining flame height at the largest Fr and at 0.3 atm. Thus, for a relative larger *Fr*, flame heights of a laminar diffusion flame will not change with pressure. Oppositely, at the smallest *Fr*, $(L/d)Re^{-2/3}Fr^{-1/3}$ remained nearly constant where buoyancy dominant tendency was reached.
- In Froude scaling of width, the slopes between *W/d* and *Fr* of all the fuels were not changing with pressure and the level off phenomenon didn't happen, both of which were indicating the good linearity of flame width changing on pressure. Finally, regression of the form $W/d \propto Fr^a Re^b$ for flame width was also found rational here.

5.Reference

[1] S.P. Burke, Schumann, T.E.W. 1928 Diffusion Flames Ind. Eng. Chem. 20(10) 998-1004.
[2] J.-M. Most, P. Mandin, J. Chen, P. Joulain 1996 Influence of gravity and pressure on pool fire-type diffusion flames Proc. Combust. Inst. 26(1) 1311-1317.
[3] T. YUAN, DUROX, D., VILLERMAUX, E. 1993 The Effects of Ambient Pressure Upon Global Shape and Hydrodynamic Behavior of Buoyant Laminar Jet Diffusion Flames Combust. Sci. Technol. 92 69-86.
[4] J. Gong, X. Zhou, Z. Deng, L. Yang 2013 Influences of low atmospheric pressure on downward flame spread over thick PMMA slabs at different altitudes Int J Heat Mass Tran 61 191-200.
[5] R.A. Altenkirch, Eichhorn, R., Hsu, N.N., Brancic, A.B., Cevallos, N.E. 1977 Characteristics of laminar gas jet diffusion flames under the influence of elevated gravity, Proc. Combust. Inst. 16(1) 1165-1174.
[6] Thomson, K. A., Gülder, Ö. L., Weckman, E. J., Fraser, R. A., Smallwood, G. J., & Snelling, D. R. 2005 Soot concentration and temperature measurements in co-annular, nonpremixed CH4/air laminar flames at pressures up to 4 MPa. Combustion and Flame 140(3) 222-232.
[7] J. Baker, Srireddy, K., Varagani, R. 2003 Buoyancy-Controlled Laminar Diffusuion Slot Flame Heights: A Comparison of Theoretical Predictions and Microgravity Results Microgravity Sci. Tec. 14(4) 27-35.
[8] F. Khaldi, Gillon, P. 2003 Laminar jet diffusion flame behavior under a strong magnetic field gradient Proceedings of the European Combustion Meeting (Orléans, Frankreich) pp. 1-6.
[9] P.B. SUNDERLAND, MENDELSON, B. J., YUAN, Z.-G., URBAN, D. L. 1999 Shapes of Buoyant and Nonbuoyant Laminar Jet Diffusion Flames Combust. Flame 116 376-386.
[10] P.B. Sunderland, J.E. Haylett, D.L. Urban, V. Nayagam 2008 Lengths of laminar jet diffusion flames under elevated gravity Combust. Flame 152 60-68.

[11] J.R. Camacho, Choudhuri, Ahsan R. 2006 Shapes of elliptic methane laminar jet diffusion flames J. Eng. Gas. Turb. Power 128 1-7.
[12] D. Drysdale 1999 An Introduction to Fire Dynamics ed John Wiley & Sons Ltd (Chichester, England) chapter2 pp65-71.
[13] L.H. Hu, F. Tang, M.A. Delichatsios, Q. Wang, K.H. Lu, X.C. Zhang 2013 Global behaviors of enclosure fire and façade flame heights in normal and reduced atmospheric pressures at two altitudes Int J Heat Mass Tran 56(1–2) 119-126.
[14] P.M. Mandatori, O.L. Gülder 2011 Soot formation in laminar ethane diffusion flames at pressures from 0.2 to 3.3 MPa Proc. Combust. Inst. 33 577-584.
[15] G. Intasopa 2011 Soot Measurements in High-Pressure Diffusion Flames of Gaseous and Liquid Fuels (Toronto ,University of Toronto) chapter4 pp36-37.
[16] I.M. Miller, H.G. Maahs 1997 High-pressure flame system for pollution studies with results for methane-air diffusion flames (Washington National Aeronautics and Space Administration.) pp 121-124
[17] [L.L. McCrain, Roberts, W.L. 2005 Measurements of the soot volume field in laminar diffusion flames at elevated pressures, Combust. Flame, 140(1-2) 60-69.
[18] A.E. Karatas, Gülder, Omer L. 2012 Soot formation in high pressure laminar diffusion flames, Prog. Energy Combust., 38 818-845.
[19] Daca, A. E., & Gülder, Ö. L. 2017 Soot formation characteristics of diffusion flames of methane doped with toluene and n-heptane at elevated pressures. Proceedings of the Combustion Institute 36(1) 737-744.
[20] Bento, D. S., Thomson, K. A., & Gülder, Ö. L. 2006 Soot formation and temperature field structure in laminar propane–air diffusion flames at elevated pressures. Combustion and Flame 145(4) 765-778.
[21] Karataş, A. E., Intasopa, G., & Gülder, Ö. L. 2013 Sooting behaviour of n-heptane laminar diffusion flames at high pressures. Combustion and Flame 160(9) 1650-1656.
[22] R.W. DAVIS, MOORE, E.F., SANTORO, R.J., NESS, J.R.,1990 Isolation of Buoyancy Effects in Jet Diffusion Flame Experiments, Combust. Sci. Technol., 73 625-635.

Acknowledgments
This research was supported by the Open Foundation of State Key Laboratory of Fire Science (No. HZ2016-KF13).

IOP Conf. Series: Earth and Environmental Science **257** (2019) 012042 doi:10.1088/1755-1315/257/1/012042

Multi-objective optimization of SOFC systems

Xiaojuan Wu[1], Ling He[2], Danhui Gao[2] and Yuanyuan Zhu[3]

1 Associate professor, School of Automation Engineering, University of Electronic Science and Technology of China, Chengdu, China
2 Student, School of Automation Engineering, University of Electronic Science and Technology of China, Chengdu, China
3 Lecturer, Department of biology and chemical engineering, ZhiXing College of HuBei University, Wuhan, China

Abstract. For solid oxide fuel cell (SOFC) development, maximizing its electrical efficiency and minimizing its cost are two important optimization objects. A new optimization strategy is proposed in this work, which can maximize the SOFC electrical efficiency and minimize the cost in the case of an air leakage fault. The proposed optimization method involves a fault diagnosis module, a switching module and two backup optimizers. The fault diagnosis part is employed to identify the SOFC current fault type, and the switching module is used to select the appropriate backup optimizer. For the efficiency and cost are two conflicting objectives, the multi-objective optimization strategy based on a non-dominated sorting particle swarm optimization algorithm is applied to determine the trade-off solutions. The optimization results show the proposed method can achieve the maximum efficiency and the minimum cost in the case of SOFC normal, and even in the air leakage fault.

1. Introduction

Solid Oxide Fuel Cell (SOFC) is one of the most promising technologies among various new energy resources. However, efficiency and cost are two major hurdlers to make SOFC unattractive as an alternative option for electricity users. For the users, they may wish to maximize the SOFC electrical efficiency, but meanwhile to minimize its costs.

Efficiency and cost are conflicting to each other, because the higher efficiency would increase the cost. Therefore, to obtain the trade-off solutions, Pareto front has been introduced to the SOFC system, on which all points are potentially an optimum solution for the efficiency and cost optimization. In Ref.[1], by combing the efficiency and cost with different weights, a plurality of Pareto solution was obtained for the SOFC system. In Refs.[2-4], based on genetic algorithm, a series of Pareto solution was acquired for the SOFC efficiency and cost optimization. However, the existing optimization strategies are designed in the normal operation, which can not consider the fault effect on the instantaneous performance of the SOFC system.

For the SOFC system, faults may happen anytime and anywhere, which may further result in the performance losses and even system failure [5]. In the last decades, plenty of fault diagnosis approaches have been advocated to improve the SOFC system stability [6-8]. Even if many optimization techniques and fault diagnosis methods have been developed and implemented on the SOFC system, combining optimization techniques with fault diagnosis technique is still lacking. As far as we know, no literature is to study the maximum efficiency and the minimum cost for the SOFC system under fault state.

Content from this work may be used under the terms of the Creative Commons Attribution 3.0 licence. Any further distribution of this work must maintain attribution to the author(s) and the title of the work, journal citation and DOI.
Published under licence by IOP Publishing Ltd

Motivated by the above requirement, simultaneous fault diagnosis and online optimization strategy is proposed in this study, which can maximize the efficiency and minimize the cost in the case of SOFC normal operation and the air leakage fault. The proposed tactics contains a fault diagnosis module, a switching part and two backup optimizers. The diagnosis module is used to detect the SOFC system current fault(air leakage faults or normal).The switching module is employed to choose the suitable backup optimizer. A non-dominated sorting particle swarm optimization (NSPSO) algorithm is presented to design the two backup optimizers in the case of air leakage fault and normal operation.

The paper hereafter is formed as below. The SOFC models in the case of normal operation and air leakage fault are presented in Section 2. The proposed multi-objective optimization strategy for the SOFC system is explained in Section 3. In Section 4, the optimization results from the concerned SOFC system are given. Section 5 draws the conclusion.

2. SOFC model

A typical SOFC system is considered in this study, which consists of an air compressor, an air heat exchanger, a fuel heat exchanger, a stack and a burner [9]. In the following, the main physical equations have been given above, and further details about the SOFC model can be retrieved from the literature [10-11].

The SOFC electrical efficiency is evaluated by [10]:

$$\eta = \frac{VI - P_{cp}}{W_{fuel,in} \cdot LHV_f} \tag{1}$$

Where, V is the stack voltage, which is obtained by [12-15]:

$$V = E - \eta_{ohm} - \eta_{act} - \eta_{con} \tag{2}$$

In the normal operation, the cold gas inlet flow rate in the air heat exchanger is given by:

$$W_c = W_{air,in} \tag{3}$$

Under the air leakage fault, the cold inlet flow rate of the air heat exchanger is substituted by the following equation:

$$W_c = W_{air,in} - W_H \tag{4}$$

Where, W_H is the air leakage flow rate between the air compressor and the air heat exchanger, which is expressed by [16]:

$$W_H = \begin{cases} \dfrac{\pi C_D D_H^2 P_{a,out}}{4\sqrt{T_8 R / M_a}} \left(\dfrac{P_0}{P_{a,out}}\right)^{1/\gamma} \sqrt{\dfrac{2\gamma}{\gamma - 1}\left[1 - \left(\dfrac{P_0}{P_{a,out}}\right)^{(\gamma-1)/\gamma}\right]} & if\ \dfrac{P_0}{P_{a,out}} > \left(\dfrac{2}{\gamma+1}\right)^{\gamma/(\gamma-1)} \\ \dfrac{\pi C_D D_H^2 \sqrt{\gamma} P_{a,out}}{4\sqrt{T_8 R / M_a}} \left(\dfrac{2}{\gamma+1}\right)^{(\gamma+1)/[2(\gamma-1)]} & if\ \dfrac{P_0}{P_{a,out}} \le \left(\dfrac{2}{\gamma+1}\right)^{\gamma/(\gamma-1)} \end{cases} \tag{5}$$

For the SOFC system, the cost balance model can be expressed as follows [11]:

$$C_2 \cdot Ex_2 + C_5 \cdot Ex_5 = C_1 \cdot Ex_1 + C_4 \cdot Ex_4 + Z_{h_1} \tag{6}$$

$$C_6 \cdot Ex_6 + C_{10} \cdot Ex_{10} = C_5 \cdot Ex_5 + C_9 \cdot Ex_9 + Z_{h_2} \tag{7}$$

$$C_8 \cdot Ex_8 = C_7 \cdot Ex_7 + C_{12} \cdot Ex_{12} + Z_c \tag{8}$$

$$C_3 \cdot Ex_3 + C_{11} \cdot Ex_{11} + C_{13} \cdot Ex_{13} = C_2 \cdot Ex_2 + C_{10} \cdot Ex_{10} + Z_s \tag{9}$$

$$C_4 \cdot Ex_4 = C_3 \cdot Ex_3 + C_{11} \cdot Ex_{11} + Z_b \tag{10}$$

Where, Z_{h_1}, Z_{h_2}, Z_c, Z_s and Z_b are the cost rates for the fuel heat exchanger, air heat exchanger, compressor, stack and burner respectively, which are calculated by Ref. [10]. Ex_k is the total exergy rate for the $k-th$ stream, which is estimated from Ref.[11]. C_k is the cost per exergy unit for the stream of the SOFC system.

3. Multi-objective optimization strategy for the SOFC system

The proposed optimization strategy for the SOFC system consists of three parts: fault diagnosis module, switching module and optimization part. In the following subsections, each module is described in details.

3.1. Fault diagnosis module

The diagnosis module is presented to classify the current fault type for the SOFC system. For the air in the pipeline linking the air compressor to the air heat exchanger occurs leakage, the inlet gas flow rate of the air heat exchanger will be reduced compared to the normal operation. Thus the air heat exchanger inlet flow rate residual δ_W is selected as an indicator to diagnose the fault, which is given as below:

$$S = \begin{cases} 1 & \delta_W = W_{c(N)} - W_c = 0 \\ 2 & \delta_W = W_{c(N)} - W_c \neq 0 \end{cases} \tag{11}$$

Where, W_c is the real inlet flow rate of the air heat exchanger, and $W_{c(N)}$ is the cold inlet flow rate of the air heat exchanger in the case of normal operation. When the residual δ_W is zero, the symptom S is 1, which indicates that the SOFC system is in normal running. Otherwise, the symptom S is 2, which shows that the air leakage fault occurs.

3.2. Switching module

Based on the fault diagnosis results, the switching module selects the appropriate backup optimizer to optimize the electrical efficiency and cost for the SOFC system, which is designed as follows:

$$u = \begin{cases} u_N & (S=1) \\ u_F & (S=2) \end{cases} \tag{12}$$

When the Boolean variable S is equal to 1, the optimization law u_N in the case of SOFC normal operation is chosen to optimize the efficiency and cost. While if the Boolean variable S is changed to 2, the optimization law u_F in the case of air leakage fault is designed to optimize the efficiency and cost. How to obtain the optimization law u_N and u_F will be discussed in the following subsection.

3.3. Optimization module based on non-dominated sorting PSO algorithm

The optimization goal is to maximize the electrical efficiency and minimize the cost. Therefore, the following fitness function is formulated:

$$\max \eta \tag{13}$$

$$\min C_{13} \tag{14}$$

To limit the results to a range of feasible points, the following inequality constraints are considered in this study: 1) the burner outlet temperature $T_4 \leq 1400K$; 2) the stack inlet temperature difference $\Delta T \leq 200K$.

Particle swarm optimization(PSO) method is a stochastic optimization technique, which is proposed by Eberhart and Kennedy in 1995. In the PSO algorithm, each particle has a leader which is used to update its position and velocity. However, each particle has a series of various leaders in the NSPSO algorithm.The steps of the non-dominated sorting PSO algorithm is given as below[17]: 1) Generate an initial population;2) *Generate* an offspring population; 3) Update the personal best position; 4) Update the global best position.

4. Optimization results

4.1. Case 1: SOFC normal operation

In the SOFC normal operation, the switching signal S is kept 1, and the optimization law in the normal operation is executed on the SOFC system. To evaluate the proposed method, the non-dominated sorting genetic algorithm(NSGA) is employed in this study to compare with the proposed NSPSO. The parameters of the NSGA are set as follows: population:150, iteration steps:200, crossover probability:0.9, and mutation probability:0.1. Using the two optimization method, a series of Pareto optimal solutions is achieved, which is shown in figure 1(a). The blue line indicates the optimization results with the NSPSO algorithm, and the red line depicts the results with the NSGA method. Minimizing the cost and maximizing the electrical efficiency are obviously in conflict with each other. Increasing the efficiency up to about 34% does not significantly increase the unit cost of the SOFC electrical power, while increasing it from 34% to 37% results in a moderate increase in the unit cost of the SOFC electrical power. Increasing the efficiency beyond 37%, leads to a steep rise in the cost. In order to compare the two algorithm, the following four generic metrics are calculated: 1) Number of non-dominated solution(NDS); 2) Spread ; 3) Spacing; 4) Maximum spread.

From figure 1(a), the optimization result with the NSPSO method is obviously better than the NSGA algorithm. Further, the metrics previously described are calculated. The number of the NDS in the two algorithm are both 150. The spread metric with the proposed algorithm is 0.4858, however this is 0.7205 with the NSGA method, which implies the uniformity of the solution is better. The maximum spreads are respectively 0.0223 and 0.0211 with the proposed algorithm and the NSGA. The spacing with the proposed algorithm is 7.5388×10^{-5}, however the spacing with the NSGA method is 8.3082×10^{-5}. All these metrics indicate that the proposed NSPSO algorithm is superior to the NSGA method.

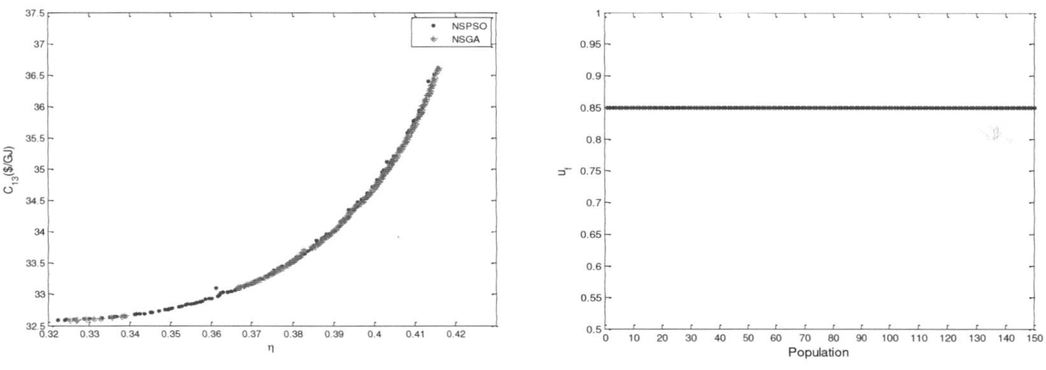

(a)The Pareto solution under SOFC normal operation (b) The optimal fuel utilization

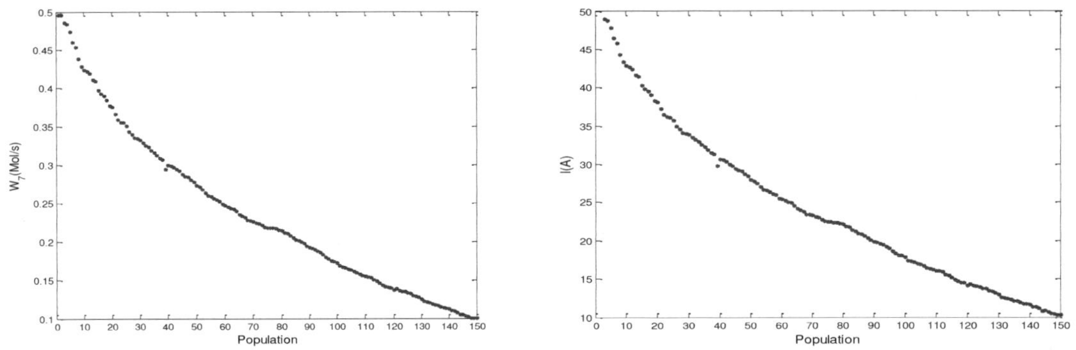

(c) The optimal air inlet flow rate (d) The optimal current

Figure 1. Optimization results under SOFC normal operation

In the optimization process, the variations of the decision variables(u_f , $W_{fuel,in}$ and I) are displayed in figure 1(b,c,d). Each point on the horizontal axis directly corresponds to an optimal solution in figure 1(a). For increasing the fuel utilization results in the improvement of efficiency and cost objective functions, the maximum value of the fuel utilization is selected, which is given in figure 1(b). The air inlet flow rate and the current cause a conflict between the defined efficiency and cost objective functions. An increase in the air inlet flow rate shifts the SOFC efficiency to the better optimization result, however, leads to the cost increase. Therefore, the current and the inlet air flow rate are the most widely distributed parameters over their allowable domain values. The current is varied from 10A to 50A, and the inlet air flow rate is changed from 0.1mol/s to 0.55mol/s.

In figure 1(a), each point in the Pareto front is an optimum non-dominated solution. The manufacturer needs to consider many possible "trade-off" solutions and select the optimum one based on their actual needs and interests. In this study, the TOPSIS approach is employed to choose the best optimum solution based on the following steps[18]:

1) The cost and efficiency data in figure 1(a) are presented in a non-dimensionalized form, which is given in figure 2.

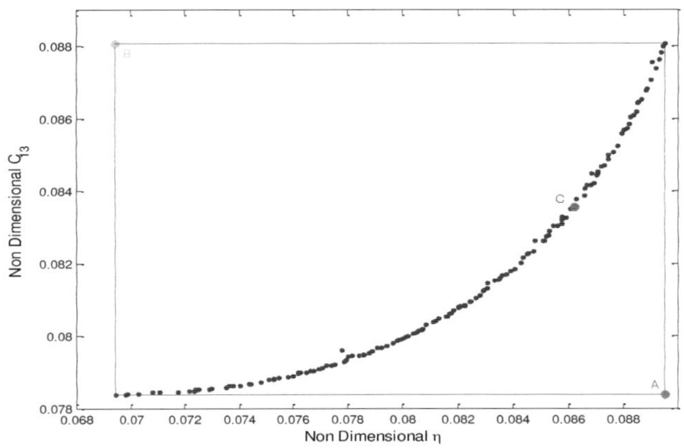

Figure 2. The optimal point under SOFC normal operation

2) Find an ideal solution and a non-ideal solution. The ideal solution is an imaginary point where each objective can obtain the optimum value, which is labeled point A in figure 2. Point A represents the lowest cost with 32.59$/GJ as well as the highest efficiency with 41.57%. The non-ideal solution is

also an imaginary point where each objective has its worst value, which is marked point B in figure 2. Point B gives the highest cost with 36.62$/GJ as well as the lowest efficiency with 32.23%.

3) Calculate $d^- / (d^+ + d^-)$. d^- stands for the distance between any point in the Pareto front and the ideal point A , and d^+ is the distance between any point in the Pareto front and the non-ideal point B. The best optimum solution is chosen as the point which has the longest distance to the non-ideal solution (point B) and the shortest distance to the ideal solution (point A). Thus, the minimum value of $d^- / (d^+ + d^-)$ is the best optimum solution, which is labeled point C in figure 2. The cost is 34.74$/GJ, and the efficiency is 40.03%.

4.2. Case 2: Air leakage fault

To test the performance of the proposed optimization strategy in the case of air leakage fault, the hole diameter D_H is increased from 0 to 2mm to simulate the fault.

When the fault diagnosis module detects that the air leakage occurs, the switching signal S is changed from 1 to 2. The obtained optimization law in the air leakage fault is carried on the SOFC system. In comparison with the normal operation, the optimization results are presented in figure 3.

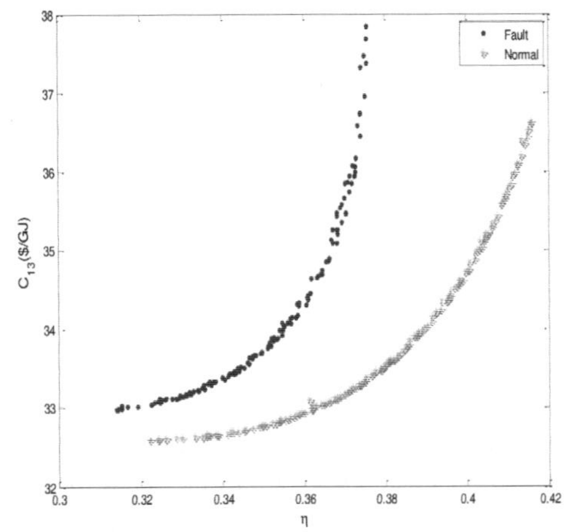

(a)The Pareto solution under air leakage fault

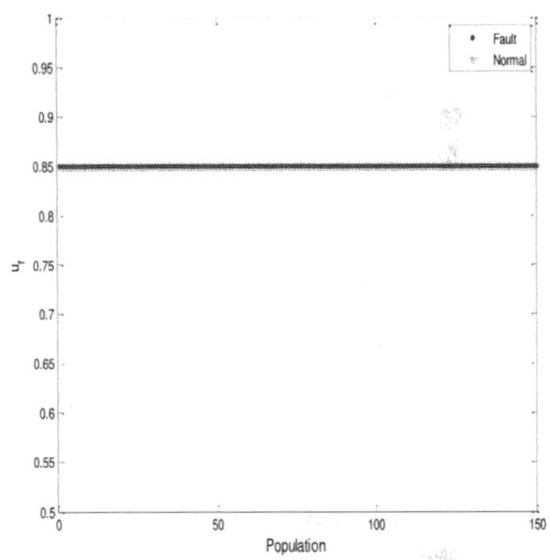

(b) The optimal fuel utilization

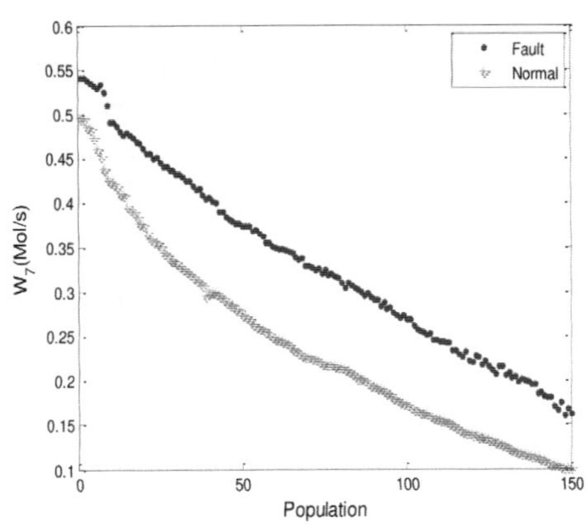

(c) The optimal air inlet flow rate

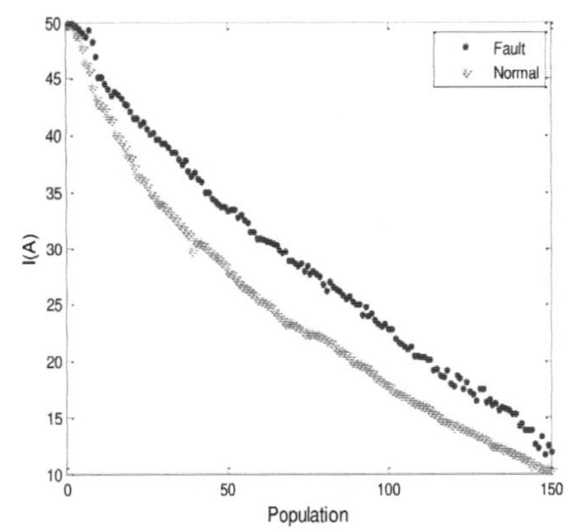

(d) The optimal current

Figure 3. Optimization results under air leakage fault

In figure 3(a), the blue line reflects the Pareto solution under the air leakage fault, while the red line describes the solution under the normal operation, which is the same with the blue line in figure 1(a). Because of the air leakage, the same efficiency will mean the greater costs compared to the normal operation. The decision variables variations under the air leakage fault are given in figure3(b,c,d). The fuel utilization still keeps 0.85, for an increase in the fuel utilization brings both the SOFC efficiency and the cost to the better optimization results. The current and the air inlet flow rate are still changed over their acceptable domains, however, are higher than the values under the SOFC normal operation. Based on the TOPSIS method previously introduced, the best optimum solution under the air leakage fault condition is marked point C in the figure 4.

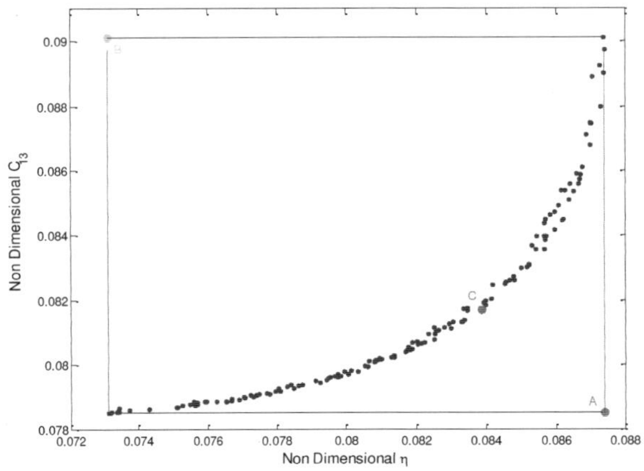

Figure 4. The optimal point under air leakage fault

The cost is 34.31$/GJ, and the efficiency is36.04%. The ideal and the non-ideal solutions are respectively labeled Point A and B. Point A represents the lowest cost with 32.97$/GJ as well as the highest efficiency with 37.54%, while Point B gives the highest cost with 37.84$/GJ as well as the lowest efficiency with 31.42%.

5. Conclusions

In this study, the improved multi-objective optimization strategy for the SOFC system has been proposed. The objective is to achieve the maximum efficiency and minimum cost in the air leakage fault state. The model-based fault diagnosis module is firstly employed to classify the current fault type, and then the switching module is used to adjust the backup optimizer. The NSPSO algorithm is presented to carry on the efficiency and cost optimization in the case of SOFC normal and air leakage fault. Compared with the NSGA algorithm, the trade-off curve between the efficiency and cost is better with the NSPSO. Based on the obtained Pareto solution, the TOPSIS method is applied to obtain the best optimum solution under the normal and air leakage fault. The optimization results can offer a help for designing the SOFC system with a high efficiency and a low cost.

6. **Reference**

[1] Ramin Roshandel, Atefeh Behzadi Forough, Two strategies for multi-objective optimisation of solid oxide fuel cell stacks, International Journal of Sustainable Energy,2014,33(4): 854–868

[2] Leyla Khani, Ali Saberi Mehr, Mortaz a Yari, S.M.S. Mahmoudi, Multi-objective optimization of an indirectly integrated solid oxide fuel cell-gas turbine cogeneration system,International Journal of Hydrogen Energy, 2016

[3] Mehdi Aminyavari, Alireza Haghighat Mamaghani, Ali Shirazi, Behzad Najafi , Fabio RinaldiExergetic, economic, and environmental evaluations and multi-objective optimization of

an internal-reforming SOFC-gas turbine cycle coupled with a Rankine cycle,Appl. Therm. Eng, 2016,108: 833–846

[4] S. Sanaye, A. Katebi, 4E analysis and multi objective optimization of a micro gas turbine and solid oxide fuel cell hybrid combined heat and power system,J. Power Sources 247 (2014) 294–306 .

[5] L. Barelli, E. Barluzzi, G. Bidini, Diagnosis methodology and technique for solid oxide fuel cells: A review[J], International Journal of hydrogen energy, 2013,38(12): 5060-5074.

[6] A. Greco, A. Sorce R. Littwin P. Costamagnac, L. Magistri,Reformer faults in SOFC systems: Experimental and modeling analysis, and simulated fault maps, International Journal of Hydrogen Energy, 2014,39(36):21700-21713.

[7] Q.A Huang, B.Wang, Wei Qu, Rob Hui, Impedance diagnosis of metal-supported SOFCs with SDC as electrolyte, Journal of Power Sources,2009, 191:297-303.

[8] N.Yousfi Steiner, D.Hissel, P.Moçotéguy, et al., Application of Fault Tree Analysis to Fuel Cell diagnosis, Fuel cells, 2012,12(2):302-309.

[9] XiaoJuan Wu, Danhui Gao, Fault tolerance control of SOFC systems based on nonlinear model predictive control, International Journal of Hydrogen Energy, 2017, 42:2288-2308.

[10] Leyla Khani, S.Mohammad S. Mahmoudi, Ata Chitsaz, Marc A. Rosen,Energy and exergoeconomic evaluation of a new power/cooling cogeneration system based on a solid oxide fuel cell, Energy ,2016,94(1):64-77

[11] Leyla Khani, Ali Saberi Mehr, Mortaza Yari, S.M.S. Mahmoudi,Multi-objective optimization of an indirectly integrated solid oxide fuel cell-gas turbine cogeneration system, International Journal of Hydrogen Energy,2016,1-19

[12] Hongliang Cao,Xi Li, Zhonghua Deng, Jian Li, Yi Qin, Thermal management oriented steady state analysis and optimization of a kW scale solid oxide fuel cell stand-alone system for maximum system efficiency, International Journal of Hydrogen Energy,2013, 38:12404-12417

[13] Meng Ni, Michael K.H. Leung, , Dennis Y.C. Leung, Parametric study of solid oxide fuel cell performance, Energy Conversion and Management,2007,48(5):1525–1535

[14] S. Campanari, P. Iora,Definition and sensitivity analysis of a finite volume SOFC model for a tubular cell geometry, Journal of Power Sources, 2004,132,113-126

[15] Lin Zhang,Xi Li,Jianhua Jiang,Shuanghong Li,Jie Yang,Jian Li,Dynamic modeling and analysis of a 5-kW solid oxide fuel cell system from the perspectives of cooperative control of thermal safety and high efficiency, International Journal of Hydrogen Energy, 2015,40,456-476

[16] Dong Yuhua,Gao Huilin,Zhou Jing,Feng Yaorong ,Evaluation of gas release rate through holes in pipelines, Journal of Loss Prevention in the Process Industries,2002,15,423-428

[17] M Salazar-Lechuga,JE Rowe, Particle swarm optimization and fitness sharing to solve multi-objective optimization problems, IEEE Congress on Evolutionary Computation, Piscataway, NJ, USA, 2005, 2:1204-1211

[18] Zhongliang Yue, a method for decision-making based on determining weights of decision makers using TOPSIS. Applied mathematical modelling, 2011,35:1926-1936

Acknowledgments

This work is supported by the National Natural Science Foundation of China (Grant No. 61304114), and the National Natural Science Foundation of China (Grant No. 61573162).

Experimental analysis on lithium iron phosphate battery over-discharged to failure

Dongxu Ouyang[1] and Jian Wang[*2]

State Key Laboratory of Fire Science, University of Science and Technology of China, Hefei, 230022, China
E-mail: [1] ouyang11@mail.ustc.edu.cn; [2] wangj@ustc.edu.cn

Abstract. In this paper, a series of experiments were performed to investigate the thermal and electrical characteristics of a commercial lithium ion battery (LIB) over-discharged to failure. Specific information including voltage, current, capacity and battery surface temperature were measured and analyzed. According to the results, it is demonstrated that batteries behave obvious temperature rise during the over-discharge process and the temperature rise increase with the increasing charge rate. Besides, the LFP (lithium iron phosphate) exhibits gentler temperature rise than the NMC (nickel manganese cobalt oxide) battery in the over-discharge process. And the discharge rate is found there no huge effect on the lost capacity of battery when it is over-discharged to failure.

1. Introduction

As a result of their excellent performances such as high energy density, less pollution, stable performance and long-life cycle, etc, LIBs have become one of the most promising power sources and widely used for both portable electronics and transportation [1-2]. Meanwhile, one of the major problems remaining to be solved is the safety issue, namely, the failure of LIB under extreme conditions including high temperature, low temperature, overcharge and over-discharge, etc [3-5]. Very often, when a LIB is used in a battery pack, it will experience overcharge or over-discharge to an extent, because there always exist capacity variations among different batteries. It is extremely difficult to control the state of charge (SOC) of each single LIB in a battery pack exactly the same, consequently, a single LIB in a battery pack inevitably experiences overcharge or over-discharge to a certain degree.

Within a variety of LIBs, the cathode materials based on LFP have been proven superior than the traditional materials such as lithium cobaltate (LCO) and lithium manganate (LMO) because of their stable performance and high security [6-8]. Even though, LFP battery is also sensitive to abusive conditions, where, over-discharge is a common abusive condition and has a great influence on the electrochemical performance of LIB including capacity fading, temperature rise and even thermal runaway. Liu et al. [9] found that the temperature variation of over-discharged cell changed significantly with cycle number, and the temperature rise during over-discharge was much faster than that before over-discharge, suggesting that much higher heat generation occurred during the over-discharge stage. It is possible that micro-shorting resulted from the copper dendrite formation and SEI decomposition. Zhang et al. [10] discovered that over-discharge caused the capacity fading of battery capacity and the fading degree would be worse with the increase of over-discharge depth. Besides, to have a better understand on the mechanism of over-discharge destroying battery, Zheng et al. [11] conducted a set of experiments and their results revealed that batteries over-discharged to

Content from this work may be used under the terms of the Creative Commons Attribution 3.0 licence. Any further distribution of this work must maintain attribution to the author(s) and the title of the work, journal citation and DOI.
Published under licence by IOP Publishing Ltd

0.5-0.0 V experienced serious irreversible capacity losses of 12.56-24.88%, namely, a serious loss of active lithium and anode materials occurred during over-discharge process. Guo et al. [12] investigated the entire over-discharge process and over-discharge-induced internal short circuit of large-format LIBs based on NMC cathode and graphite anode. According to the SEM and XRD results, they found that if the terminal SOC was lower than −12%, then the battery suffered from internal short circuit which was the result of Cu foil dissolution and deposition on electrodes. He et al. [13] systematically investigated the failure of commercial LIBs under different degrees of over-discharge (100-120% DOD). They proposed a possible failure mechanism of LIB during over-discharge process that the Cu was oxidized into Cu cations on the anode side, then, these Cu cations diffused through the separator to reach the surface of the cathode and, finally, these Cu cations were reduced to the Cu metallic form at the cathode. Eventually, the copper bridge formed which caused micro-shorting.
However, previous works mainly focus on the slight over-discharge and little research has been done to investigate the characteristics of LIB over-discharged to failure, especially, the effect of discharge rate on the process has not been concerned. In order to fill in this gap, this paper conducted an experimental analysis to investigate the thermal characteristics of a commercial LIB (LFP) over-discharged to failure. Specific information including voltage, current, capacity and battery surface temperature were measured and analyzed to provide necessary basic data for this issue.

2. Experiments

2.1. Batteries

The batteries used in current study were cylindrical SONY 18650 with a diameter of 18 mm, a height of 65 mm and a nominal capacity of 1300 mAh. Its cathode materials were based on LFP and that for anode were graphite. Besides, its cut-off voltages for charge and discharge were 4.2 V and 2.5 V, respectively. Three major components including safety valve, jellyroll and shell casing constituted the batteries. The electrochemical reactions of LFP battery during charge and discharge can be presented as:

$$LiFePO_4 + 6C \leftrightarrow Li_{1-x}FePO_4 + Li_xC_6$$

2.2 Apparatuses and experimental design

As shown in Fig. 1, experiments were carried out in a well-ventilated cone chamber with a dimension of 1.2 m × 1.2 m × 1.2 m. Batteries were placed upon a supporting mesh made of iron wire in tests. A K-type thermocouple with a diameter of 1 mm was attached to the battery to measure its surface temperature and the temperature was recorded at the data acquisition equipment (NI cDAQ-9174) by a computer. Charging/discharging battery was achieved by a cycle equipment (NEWARE CT-3008) with a diameter of 5 V/6 A. The electrical performances including voltage, current and capacity of batteries were obtained through the cycle equipment.
Experiments consisted of 4 discharge rates were carried out to explore the variations including voltage, current, capacity and surface temperature of LIBs during over-discharge to failure. Before tests, each battery was discharged at a 2C rate to 2.5 V, namely, 0% SOC. And then, it was placed still for 24 hours to ensure it remained stable. The experimental configurations are further listed in Table 1, where each configuration was repeated three times under the same condition to guarantee the repetition of results.

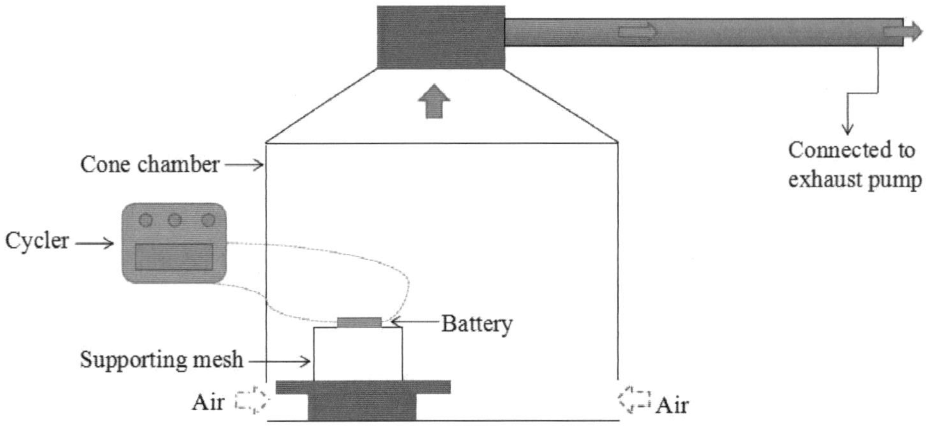

Fig. 1. Schematic of the experimental setup

Table 1. The experimental configurations

Test No.	Treatment	Rate (C)
1		1
2	over-discharge	2
3		3
4		4

3. Results and discussion

The typical curves of surface temperature, voltage and current versus capacity during the over-discharge process for the battery discharged at 2C rate are depicted in Fig. 2. Before experiment, the battery was discharged by a CC (constant current) of 2C rate to 0% SOC. As seen, the over-discharge process can be divided into three stages briefly. At stage (a), the battery was in the state of slight over-discharge. Its surface temperature increased steeply and the voltage decreased rapidly too. At the end of stage (a), the battery achieved the peak temperature of the whole over-discharge process, 30.9 °C. Then, it entered stage (b) where the surface temperature and voltage declined gradually, while the current kept stable. With the processing of over-discharge, the discharge current began to descend at a rapid speed, while the drop rate of voltage became slower at stage (c). It was the result of the serious over-discharge of battery so that the CC discharge could not be retained. When the over-discharge process approached the end, the voltage and current declined to 0.22 V and 0 mA respectively. After the over-discharge treatment, the battery could not be charged over 80% SOC again, the failure threshold for LIB [14], which implies that the battery failed. It is induced by the excessive loss of lithium ions from the anode which resulted in the damage of graphite structure and then lithium ions could not deposit into the anode efficiently any more [15].

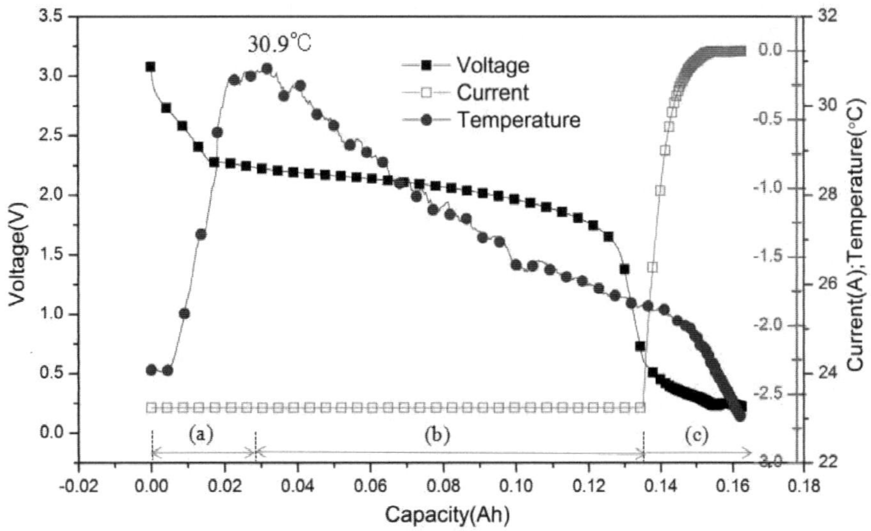

Fig. 2. Typical curves of surface temperature, voltage and current *vs* capacity during over-discharge

Further, Fig. 3 presents the battery surface temperature histories during over-discharge process where the over-discharge treatment was carried out at a CC of 1C, 2C, 3C and 4C, respectively. It can be found that with the increasing discharge rate, the peak temperature (T_p) of battery during over-discharge increased and possessed a peak temperature of 29.1, 30.9, 32.0 and 32.4 °C, respectively. Besides, the time to peak temperature (t_p) for the four batteries were 452, 412 378 and 346 s separately, namely, it was earlier to get to the peak temperature for the battery with a larger discharge rate. It implies that the battery over-discharged by a larger rate exhibited severer temperature rise and it was prone to heat up.

Combined by the the results of previous experiments, Fig. 4 displays the peak temperatures of two commercial LIBs (LFP and NMC) during over-discharge process under different rates. As shown, the peak temperatures of LFP batteries during the over-discharge process were lower than that of NMC batteries especially for the high rate condition. Namely, the NMC exists severer temperature rise than the LFP in the over-discharge process and this finding corresponds to the result of our paper published before [16], which also confirmed that the LFP has higher security compared to the NMC in some extent.

When battery was over-discharged to failure, its capacity would decrease in comparison with the original state. Fig. 5 shows the lost capacities of LFP and NMC batteries when they were over-discharged to failure under different rates. It is interesting to find that their lost capacities floated around 0.16 Ah, in other words, the -12.3% SOC. This finding is identical to the result of Guo et al. [8] and it indicates that when battery is over-discharged over a critic, it will suffer irretrievable damage even failure. The discharge rate does not have huge effect on the critic, and their critics are similar for LFP and NMC.

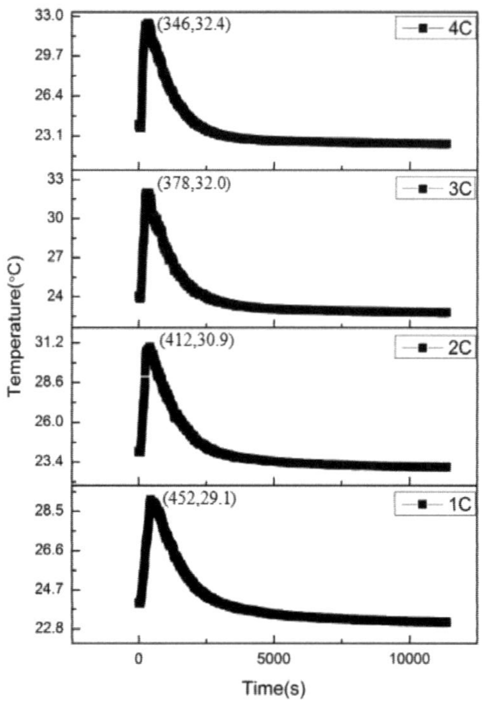

Fig. 3. Typical curves of surface temperatures of batteries and air temperature *vs* time during over-discharge

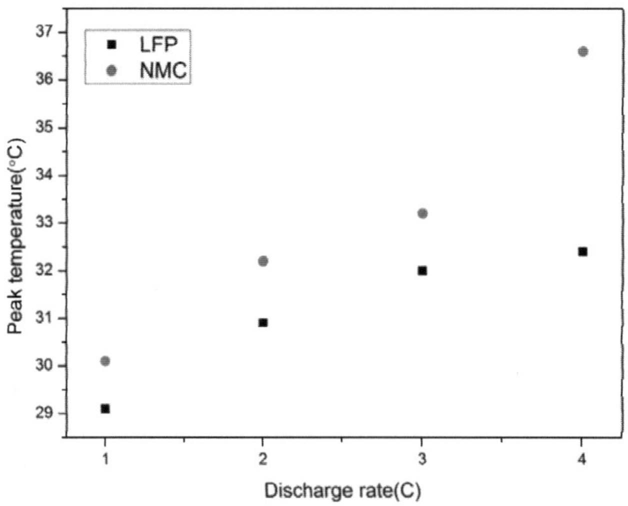

Fig. 4. The peak temperatures of LFP and NMC batteries in over-discharge

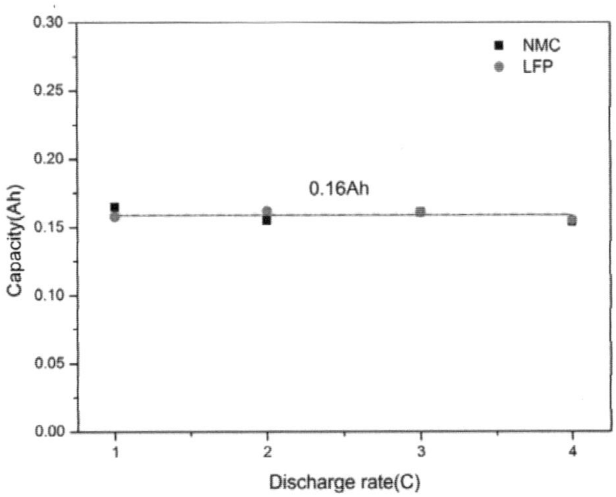

Fig. 5. Relationship between lost capacities and discharge rates

4. Conclusion

An experimental analysis was conducted to investigate the thermal and electrical characteristics of the LFP batteries over-discharged to failure. Results show that batteries have obvious temperature rise during the over-discharge process and the temperature rise will be worse and quicker with the increase of discharge rate. After comparing, it is demonstrated that the NMC battery presents severer temperature rise than that of the LFP in over-discharge process. In addition, discharge rate is found there no huge effect on the lost capacity of battery when it is over-discharged to failure.

Reference

[1] Bandhauer T M, Garimella S and Fuller T F 2011 *Journal of the Electrochemical Society* **158(3)** R1-25
[2] Noh H J, Youn S and Yoon C S 2012 *Journal of Power Sources* **233** 121-30
[3] Doughty D H and Roth E P 2012 *The Electrochemical Society Interface* **21(2)** 37-44
[4] Belov D and Yang M H 2008 *Journal of Solid State Electrochemistry* **12(7-8)** 885-94
[5] Maleki H and Howard J N 2006 *Journal of power sources* **160(2)** 1395-402
[6] Zhou X, Wang F and Zhu Y 2011 *Journal of Materials Chemistry* **21(10)** 3353-58
[7] Lu L, Han X and Li J 2013 *Journal of power sources* **226** 272-88
[8] Guo C X, Wang M and Chen T 2011 *Advanced Energy Materials* **1(5)** 736-41
[9] Liu Y, Liu Q and Li Z 2014 *Journal of The Electrochemical Society* **161(4)** A620-32
[10] Zhang L, Ma Y and Cheng X 2015 *Journal of power sources* **293** 1006-15
[11] Zheng Y, Qian K and Luo D 2016 *RSC advances* **6(36)** 30474-83
[12] Guo R, Lu L and Ouyang M 2016 *Scientific reports* **6** 30248
[13] He H, Liu Y and Liu Q 2013 *Journal of The Electrochemical Society* **160(6)** A793-804
[14] Petzl M, Kasper M and Danzer M A 2015 *Journal of Power Sources* **275** 799-807
[15] Liu Y and Xie J 2015 *Journal of The Electrochemical Society* **162(10)** A2208-17
[16] Ouyang D, Liu J and Chen M 2017 *Applied Sciences* **7(12)** 1314

Acknowledgments
This work was supported by the National Key R&D Program of China (No. 2018YFC0809500) and the National Nature Science Foundation of China (No. 51376172).

IOP Conf. Series: Earth and Environmental Science **257** (2019) 012044 doi:10.1088/1755-1315/257/1/012044

Scheme and research of realizing regional substation by using private cloud technology

Sun Shucai[1,2], Zhu Chenpeng[1,2] and Xie Yehua[1,2]

1 NARI Group Corporation/ State Grid Electric Power Research Institute, Nanjing, CHN
2 NARI Technology Development Co.,Ltd, Nanjing, CHN
E-mail: sunshucai@sgepri.sgcc.com.cn

Abstract. In recent years, the technology of Internet of things and information communication technology have developed rapidly. With the construction of intelligent substation, the intelligent level of main equipment is constantly improving, and the power data network is constantly improving, which provides the basic conditions for the realization of regional private cloud. In this paper, the concept, advantages, implementation scheme, construction process and advanced application of regional substation are proposed. And the application mode and scenario are discussed.

1. Private cloud technology

1.1. Cloud technology

Cloud technology is a kind of hosting technology that unifies hardware, software, network and other resources in WAN or LAN to realize data computing, storage, processing and sharing. Cloud services can be divided into infrastructure as a service (IaaS), platform as a service (PaaS) and software as a service (SaaS) [1] according to the service level of the cloud.

IaaS: Infrastructure-as-a-service is an infrastructure that consumers can access through the Internet from a complete computer infrastructure.

PaaS: Platform-as-a-service, PaaS actually refers to the software development platform as a service, submitted to the user in SaaS mode. Therefore, PaaS is also an application of SaaS mode[2]. However, the emergence of PaaS can speed up the development of SaaS, especially speed up the development of SaaS applications.

SaaS: Software-as-a-service is a mode of providing software through the Internet. Instead of purchasing software, users rent Web-based software from the provider to manage business activities, written in the style of a submission to *IOP Conference Series: Earth and Environmental Science*, show the best layout for your paper using Microsoft Word. If you don't wish to use the Word template provided, please use the following page setup measurements.

1.2. Virtualization

Virtualization is the foundation of cloud services and an important feature of cloud services[3]. Through virtualization technology, cloud service platform can integrate multiple servers into a linked cluster, forming a pool of resources to achieve common computing, sharing resources, and maximizing the use of idle computing resources within the system showing as figure 1.

Content from this work may be used under the terms of the Creative Commons Attribution 3.0 licence. Any further distribution of this work must maintain attribution to the author(s) and the title of the work, journal citation and DOI.
Published under licence by IOP Publishing Ltd

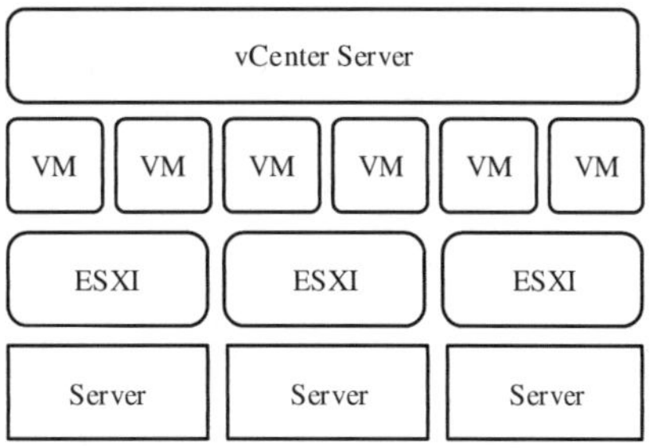

Figure 1. Exemplary example of virtualization with VMware products as an example.

1.3. Mass data storage

The traditional storage method uses a dedicated storage server (or directly use the application server) to store all data. The storage server becomes the bottleneck of the system performance, and can not meet the needs of large-scale storage applications[4]. Data is usually stored in place, with low security, limited scalability and weak sharing capability.

Figure 2 shows that mass data storage adopts scalable system structure, uses multiple storage servers to share the storage load, and uses addressing servers to locate and store information. It has high reliability, availability, access efficiency and is easy to expand.

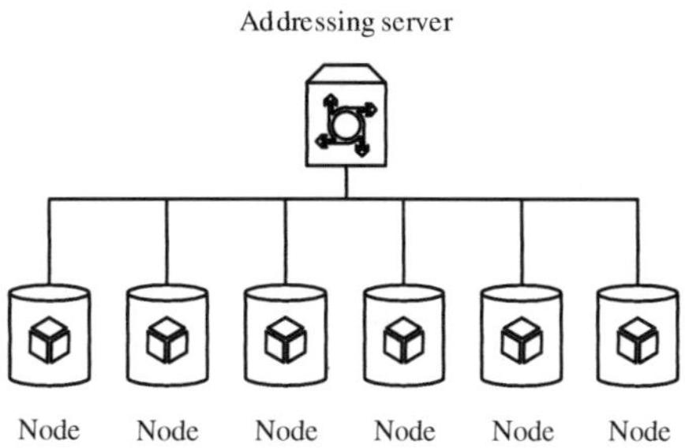

Figure 2. Schematic diagram of mass storage structure.

1.4. Mass data storage

Big data refers to the data set that can not be captured, managed and processed by conventional software tools within a certain period of time. Big data technology refers to the ability to get valuable information quickly from all kinds of data. The basic characteristics of large data are huge volume, diverse data types, fast processing speed and low value density.

2. Regional Substation

2.1. Regional substation

The regional substation described in this paper refers to the monitoring center in a specific power supply area. Regional substation realizes the unified management and control of multi-entity substation and independent equipment in the area through high-speed communication network[5].

It needs to be clarified that the concept of regional substation is limited to the regionalization of control information system. The actual power supply system is still implemented in the traditional way.

2.2. Schematic diagram of structure(figure 3)

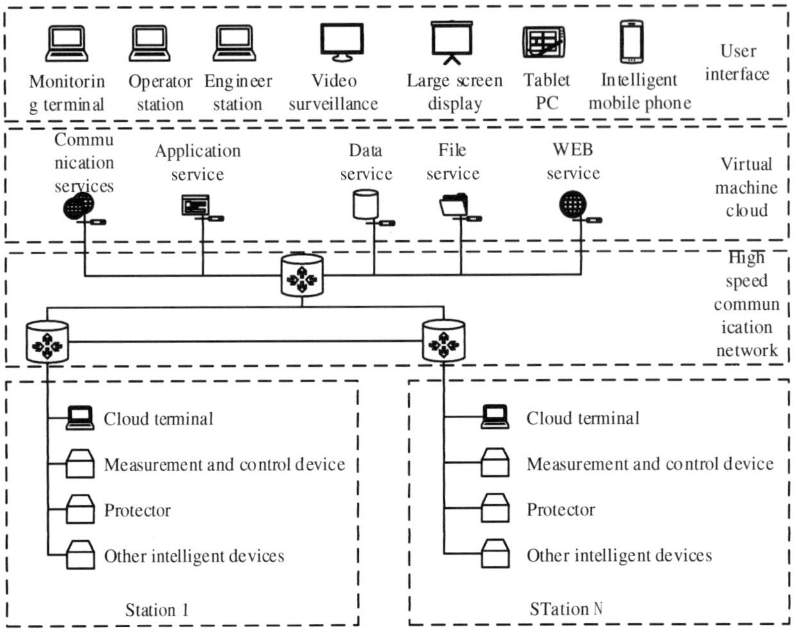

Figure 3. Schematic diagram of structure.

2.3. Advantage

2.3.1.Centralized equipment management, optimize resource allocation. The regional substation adopts centralized station building mode. All the servers and storage devices are centralized in the regional central computer room.

In this mode, virtualization technology is applied to virtualize all servers, building server resource pool to implement IaaS.

Traditional substation servers have low utilization rate of general resources. Individual high-utilization sites are prone to hardware failures, and uneven distribution of resources, failing to achieve resource sharing.

Local storage, which is stored on the hard disk of the application server or in a separate storage server, has a large amount of storage resources in a long-term idle state and low utilization rate[6].

The table 1 compares the two modes for different server numbers. Suppose the single station resource average demand is 40, the single server resource is 100, and the resource warning line is 90%.

Table 1. Resourcecomparison between two modes of different server numbers.

The server	Resource demand	Traditional substation		Regional substation		Saving rate
		Configure	Utilize	Configure	Utilize	
1	40	100	40%	100	40%	0%
2	80	200	40%	100	80%	50%
3	120	300	40%	200	60%	33%
4	160	400	40%	200	80%	50%
5	200	500	40%	300	67%	40%
6	240	600	40%	300	80%	50%
7	280	700	40%	400	70%	43%
8	320	800	40%	400	80%	50%
9	360	900	40%	400	90%	56%
10	400	1000	40%	500	80%	50%

Comparing with the traditional substation construction mode, the regional substation mode can save about 50% of the server resources. In the case of unchanged total investment, this 50% is the so-called system margin, which can be used as a hot standby resource to improve the reliability of the system operation.

In addition, from the budget point of view, the server budget is limited in the case of a single station, configuring the general server, with poor performance and reliability. In the regional mode, multi-server budget overlay considering the performance and reliability requirements, select better servers to improve the system performance and security quality.

2.3.2. Optimizing data aggregation scheme to reduce repeated investment in infrastructure projects. As mentioned above, the traditional storage server has become the bottleneck of system performance, and can not meet the needs of large-scale storage applications.

In traditional substation, the hot standby mode of dual (multi) machines is generally adopted. The application can only use the performance of a single device[7]. The equipment in hot standby state can only realize the function of backup, failing to achieve performance overlap or load balance.

Real-time generated data is sampled and stored in a fixed time (1 or 5 minutes), and the remaining data is disposed by discarding. The system does not have the ability of mass storage and analysis. When the main station side carries on the data analysis, data is send by convention. The timing collection point pattern obtains the original data. The network structure under this mode is as figure 4:

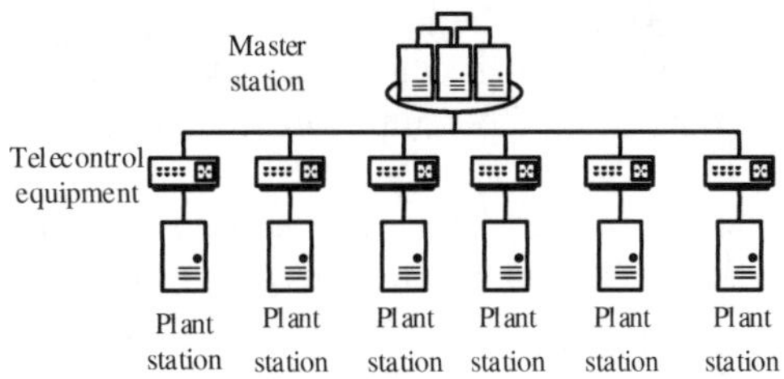

Figure 4. Network structure diagram of traditional mode data acquisition.

In the traditional mode, structured data upload can only be carried out by remote control. Unstructured data need to be encoded and processed before uploading, and the size of data is limited.

The regional substation mode can effectively solve this problem. Because of the high performance of the regional substation, it can completely meet the needs of distributed data processing[8]. In this mode, the distributed data storage mode is adopted to avoid a large number of data migration.

The network structure at this time is as figure 5:

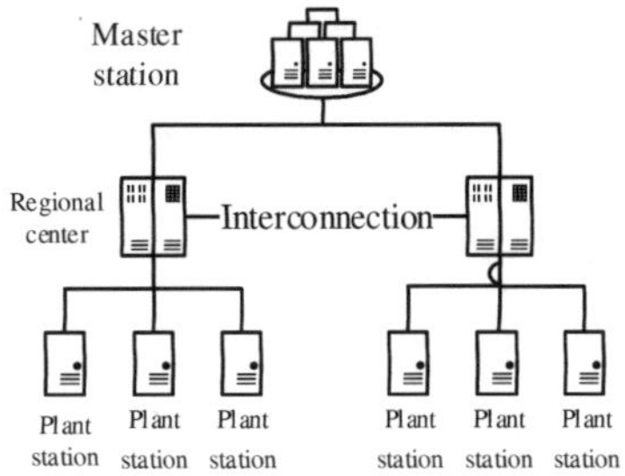

Figure 5. Network structure of data acquisition for regional patterns.

In regional mode, the master station only needs to obtain the original data from the center of the interval, or adopts the distributed computing model[9]. The calculation task is delegated to the calculation node in the center of the interval, and the results are fed back to the main station.

The network communication between the plant and the regional center is direct, and the data is directly connected. As long as the network environment is stable and high-speed, there is no obstacle to data acquisition.

2.3.3.Improving data sharing and interoperability and realizing regional big data analysis. The Using the regional substation mode, the regional center is equivalent to the distributed data center for the main station. By establishing a high-speed Internet, the data interconnection between the regional centers and the regional center can be easily realized, which provides the basic conditions for distributed computing and large-scale data analysis.On this basis, the protection system at the regional level can be realized, even cross-domain, multi-domain linkage protection system.
To achieve rapid recovery and system expansion, provide a unified disaster recovery plan[3]. By establishing a regional center, we can make use of the characteristics of virtual machine to provide perfect host hot standby, mirror, snapshot, support real-time migration, fast recovery, dynamic expansion and other functions.
Through regional interconnection, data can be backed up in different places, system redundancy disaster preparedness and even system-level substitution in extreme cases can be realized. Compared with the traditional substation, the system-level redundant backup is realized, and the data RAID, multi-copy and remote backup are realized, which greatly improves the stability and availability of the system.

2.3.4.Centralized operation and maintenance, reduce maintenance difficulty and costs. Under the traditional mode, the operation and maintenance of substations are generally divided into one shift (operators), two shifts (telecontrol, relay personnel).
In the regional mode, because the system is built on the cloud, the secondary operators can easily access and maintain the monitoring system through a dedicated channel or VPN using the cloud terminal system. In addition to routine on-site equipment maintenance, the rest of the cases can be adopted.
In the future, the TT&C and protection equipment can also be moved to the regional center. All the data can be uploaded by intelligent terminals and merging units[10]. At this time, the routine maintenance of the protection equipment can also be completed in the regional center.
In the regional center, operators can collect real-time data and video from cloud terminals and mobile devices for on-duty monitoring. Combined with intelligent push mechanism, when key information is generated, it can be pushed to personal mobile devices on time to reduce the link of information transmission, shorten the time of information transmission and improve the processing efficiency.

3. Construction Process

In this paper, according to the network construction and informatization level, the construction of regional substations is divided into four stages. The details are as follows:

3.1. Centralized station, hardware virtualization

In the first stage of the construction of regional substation, the centralization of monitoring system, server and storage equipment is mainly completed. The virtualization method is adopted to realize the private cloud, the hardware level without plant-station boundary, and the virtual machine technology is used to construct the monitoring system and storage system.
At this stage, no modification is made to the measuring and protecting devices at the scene. Only on-site servers and remote devices are cancelled, and all two devices are directly connected to the Internet.

3.2. Two equipment centralization

This stage mainly solves the problem that the secondary equipment in the first stage is still running in place. The realization of this stage depends on the construction of the network. It is necessary to build a communication network with high bandwidth and low delay.
The intelligent terminals and merging units are directly used in the scene. The protection and control devices are all transferred to the regional center, and data acquisition is achieved through high-speed communication network.

3.3. Platform virtualization

With the advancement of information technology, the main monitoring system should change the traditional single machine installation and operation mode, and enter the cloud platform operation mode.

Operating system level is not divided into plant and station, plant and station logical isolation through the cloud platform software architecture to achieve PaaS or SaaS.

3.4. Platform virtualization

By further promoting the interconnection and interoperability of communication networks and combining with the implementation of PaaS/SaaS, data access and migration between regional centers become easier. At this time, the demarcation between regions is no longer obvious, and platform-level interconnection, interoperability, cooperation and disaster preparedness are realized.

4. Advanced Application

After the establishment of an interoperable regional center, you can use the resources of the cloud platform to implement some advanced applications, as follows:

4.1. Fault prediction

Through real-time acquisition and analysis of regional data changes, large data analysis technology has been adopted to establish a fault prediction model to achieve fault early warning.

4.2. Behavior analysis

Access personnel single soldier system, analysis personnel operation habits, data sampling induction. The warning is issued to discover violations, and to prevent improper behavior through linkage operation, so as to achieve intelligent five prevention.

4.3. Real time failover

By establishing snapshots, redundant backups, hot standby and other ways, the important systems are provided with real-time recovery, substitution and other support to achieve real-time failover.

5. Conclusion

The scheme has been successfully applied in many engineering projects. The practical operation has proved the feasibility of the scheme. With the development of Internet of things and network infrastructure, regional substation technology no longer has technical barriers. This paper discusses the concept, advantages, implementation scheme, construction process and advanced application of regional substation. Some viewpoints may not be mature, and are welcome to be corrected.

6. References

[1] Zhao Junhua and Wen Fushuan 2010 *Cloud computing: Constructing the core computing platform of future power system. Power system automation* pp 1-8

[2] Gao Haipeng 2016 *Application of big data in communication industry. Telecom World* pp 92-93

[3] Zhu Zheng and Gu Zhongjian *Application of cloud computing in power system data backup business. Power grid technology* pp 43-50

[4] Zhu Liyin and Gao *Liang 2015 Analysis of information sharing characteristics of process bus in Smart Substation Based on station area protection principle. Electrical technology* pp 87-92

[5] Song Xuankun 2013 *A new generation of intelligent substation level protection system. Power construction* pp 24-29

[6] Ma Jing and Wang Zengping 2015 *Backup protection in station area based on current protection fusion factor. Power system protection and control* pp 1-9

[7] Chen Lei amd Zhang Kanjun 2013 *Review of research on substation protection in Smart Substation. East China electric power* pp 947-953

[8] Jiang Hongliang and Lu Feipeng 2017 *Relay protection strategy for substation backup protection in Smart Substation Based on graph theory. Electrical technology* pp 24-29

IOP Conf. Series: Earth and Environmental Science **257** (2019) 012044 doi:10.1088/1755-1315/257/1/012044

[9] Wang Chengshan and Wu Zheng 2014 *Research on Key Technologies of microgrid. Transactions of China Electrotechnical Society* pp 1-12
[10] Li Zhigang and Shuangping 2015 *Cloud comprehensive evaluation method for distribution network reliability. Transactions of China Electrotechnical Society* pp 418-421

Discussion on module-based hot-spot suppression in a PV generation system

S Yang[1,2], K Itako[1], T Kudoh[1], K Koh[1], and Q Ge[2]

1 Department of Electrical and Electronic Engineering, Kanagawa Institute of Technology, Kanagawa, Japan.
2 School of Hydraulic, Energy and Engineering, Yangzhou University, Yangzhou, China.
E-mail: itako@ele.kanagawa-it.ac.jp

Abstract. This paper presents a novel control system for the hot-spot detection and suppression in a photovoltaic (PV) generation system. In this novel system, a distributed structure of a PV string is employed by installing a buck-boost converter to each module. Following this structure, a systematic control strategy is proposed for the module-based hot-spot detection. Moreover, the safe operation is adopted to suppress the prolonged high temperature when the hot-spot is detected. On the other hand, the normal modules can work at their maximum power point tracking (MPPT), respectively. On this occasion, the PV string could be smart because every module could function in their optimal state which is determined to their respective conditions. In order to validate this novel control system, the experiments have been carried out in a laboratory prototype. The verified results show that the PV generation system can operate safely at module level by using this novel control system.

1. Introduction

The classical hot-spot phenomena usually exist in the c-Silicon PV generation system when under partial shading conditions (PSCs) [1]-[4]. The partially shaded PV modules are usually heated up homogenously due to the minor reverse current when those modules function in the reverse bias. More importantly, it has been reported recently that one kind of typical hot-spot phenomena which usually resulting from the low-resistance defects (LRDs) may lead to more severe destruction to the malfunction modules within a short period of time [5]-[7]. For monitoring and suppression of this typical phenomenon, the mechanism of which has been analysed in conjunction with the working modes of the PV modules range from the open circuit to the short circuit. According to the feature of LRDs induced hot-spots, we proposed a hybrid control method to detect and suppress this hot-spot phenomenon. The hybrid method involves the transient scan process to allow the online detection during power generation. Besides, if these hot-spots are detected, the heat-avoided control strategy is adopted to ensure the safe operation by shifting the operation from the maximum power point to the safe area, thus the high temperature can be controlled to the normal state.

In prior work [8], we have investigated this issue on the basis of a PV string. Unfortunately, during the experimental trials, we found that it is very difficult to figure out the malfunction modules by means of our control method. In addition, the safe operation will lead to comparatively large amount of power loss, which is related to the length of a PV string. Therefore, the newly developed technique is not suitable to be employed in the large-scale installed system at its current form. As a result, this article presents a novel research in order to deal with these critical problems. In particular, we propose a novel control system in conjunction with the distributed structure of a PV string.

Content from this work may be used under the terms of the Creative Commons Attribution 3.0 licence. Any further distribution of this work must maintain attribution to the author(s) and the title of the work, journal citation and DOI.
Published under licence by IOP Publishing Ltd

IOP Conf. Series: Earth and Environmental Science **257** (2019) 012045 doi:10.1088/1755-1315/257/1/012045

2. Theoretical analysis

In this section, the review of our prior work and new control system are mainly introduced.

2.1. Review of prior work

In order to easier comprehend the formation of LRDs induced hot-spot phenomenon, the simplified model of defective cell is shown in figure 1. The letter X represents the shadow ratio in theory, such that the maximum current that can be generated by this cell turns to be (1-X) I_{SC}, here, I_{SC} is the short circuit current. Moreover, it can be seen that compared with the normal cell, the defective cell will produce a new current path with low resistance, the fundamental mechanism of which is recognized as multilevel trap-assisted tunnelling effect (TAT) [9]-[10]. Thus, so long as the shaded defective cell is under reverse bias, and the main circuit current I_{PV} is greater than (1-X) I_{SC}, meaning that the reverse current I_{LR} will flow through the PN junction easily due to the presence of LRDs.

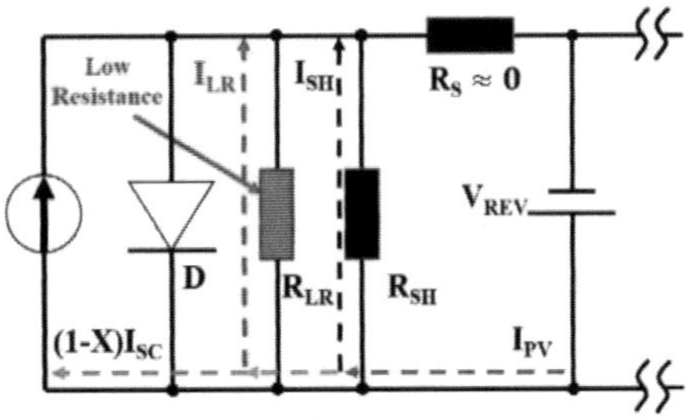

Figure 1. The simplified model of the defective cell

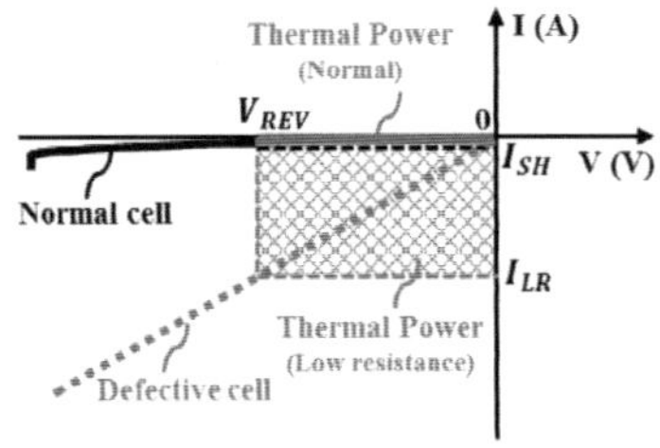

Figure 2. Thermal power produced at reverse bias

The thermal power produced at reverse bias is shown in figure 2. It can be seen that when under the same reverse voltage V_{REV}, the normal cell induced thermal power is limited to a low level. Whereas, on the other hand, the thermal power produced by defective cell is positively proportional to the square of the reverse current I_{LR} (or voltage V_{REV}). Therefore, it indicates that LRDs induced hot-spot phenomenon is much more severe that the classical one. It is also proved that if the system is in absence of any alarm or protection, the defective module will endure the prolonged high temperature.

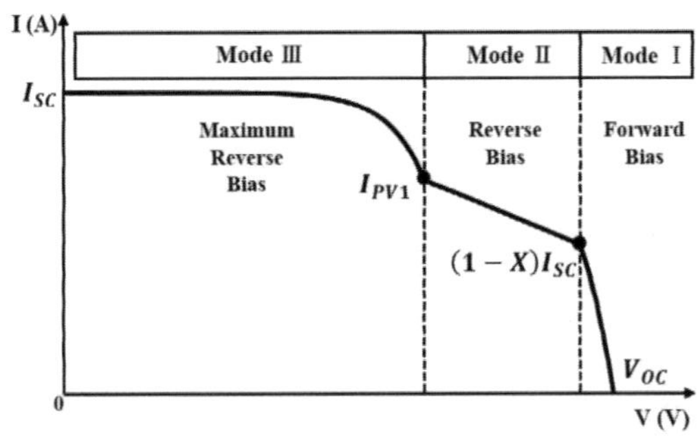

Figure 3. Schematic working modes of defective module

Figure 3 presents the detailed working modes of the system according to the magnitude of the current ranging from the open circuit to short circuit. In this case, we analysed that the module works at reverse bias in the Mode II and III for maximum power point tracking (MPPT) will produce nonzero thermal power corresponding to the simplified model. In contrast, in the Mode I, the module works at forward bias where the reverse current is prevented to zero, that is, the system is permitted to operate safely. Therefore, the conjunction between Mode I and II is treated as a safe point to suppress the undesirable thermal power.

2.2. New control system

2.2.1. Configuration of the control system. Based on the prior work, we propose a new control system to realize the aim at module-level hot-spot detection and suppression.

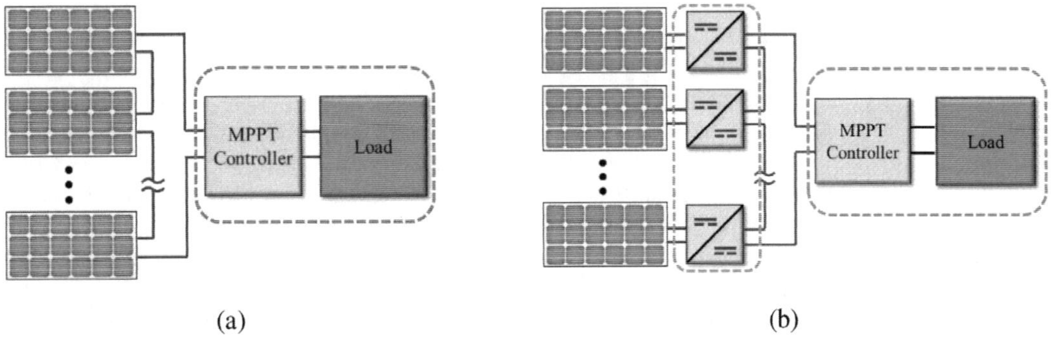

Figure 4. Schematic diagram of a standalone PV system,
(a) the conventional system, (b) the proposed system.

The conventional structure of a PV string is shown in figure 4(a). Especially, it can be seen in figure 4(b), the distributed structure of a PV string is adopted. In addition, the Two-Switch Buck-Boost (TSBB) converter in figure 5 is employed after each module.

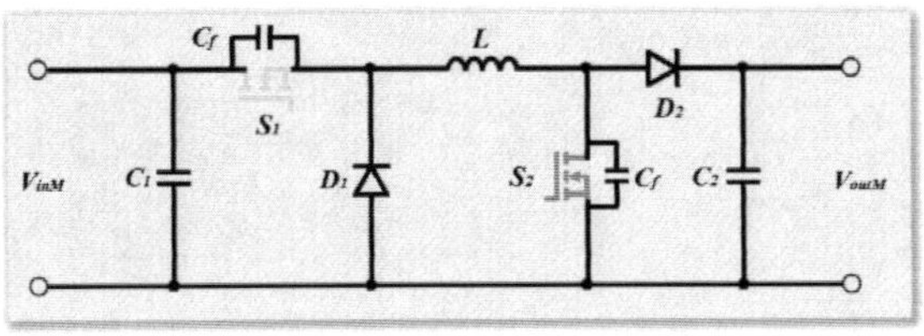

Figure 5. Schematic diagram of a TSBB converter

Table 1. Operation modes of TSBB

Mode	S1	S2
Buck	Switching	Off
Boost	On	Switching
Buck-Boost	Switching	Switching

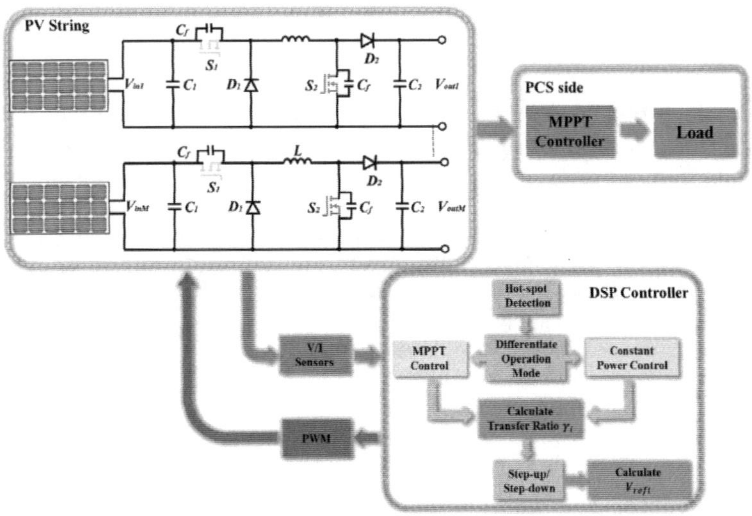

Figure 6. Schematic block diagram of control system

The normalized operation modes of TSBB is shown in Table 1, it is well recognized that due to feature of this converter, the wide operational range can be provided. As a result, we selected this converter for this research. Moreover, the block diagram of the control system is given in figure 6. It mainly consists of a PV string, a DSP controller, and the Power conditioning system (PCS). The DSP controller plays a role in data fetch, data analysis, and PWM generation. Especially, the data analysis includes hot-spot detection, operation modes determination, and parameter calculation.

2.2.2. Control scheme. In order to comprehensively explain the control process in detail, we mainly discussed two main situations involving hot-spot suppression and normal operation. Supposing that a PV string consists of (m+n) modules, especially the hot-spot phenomenon occurs in m modules. The rest of n modules are normal modules. In the case of the hot-spot suppression, once the hot-spot is detected, the constant power control is recommended, and the safe operation power can be saved simultaneously. Thus, the malfunctioning modules can operate at their constant power control by shifting the operational point from the MPP to a safe point, respectively. Similarly, on the other hand, in the case of normal modules, the optimal control is employed in this research. This control permits the normal modules work at MPPT to maximize the output energy by scan process, irrespective of other modules, or conditions. In summary, this systematic control scheme can provide the adaptive operation for each module.

3. Experimental results

The practical feasibility of this novel control system validated in a laboratory prototype shown in figure 7 by experiments that two-series connected PV modules are presented in this section.

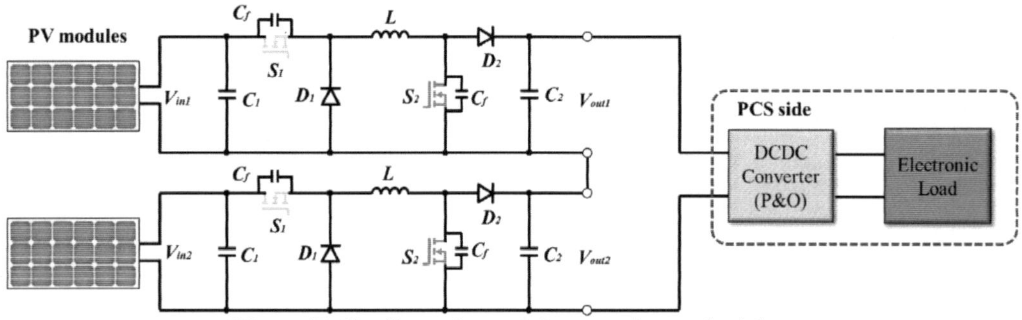

Figure 7. Configurations of the experimental trial.

4. Conditions

Before carrying out the experiments, it is necessary to determine the parameters of this control system. The specification of PV modules is shown in table 2, whereas the parameters of TSBB and experimental conditions are given in table 3 and 4, respectively.

Table 2. Specification of PV modules

Parameter	Value
Maximum output	50[W]
Open-circuit voltage	20.5[V]
Short-circuit current	3.35[A]
Optimal operating voltage	16.4[V]
Optimal operating current	3.05[A]

Table 3. Specification of TSBB

Parameter	Value
Filter C_1	4.7[μF]
Filter C_2	1000[μF]
Filter C_f	0.1[μF]
Inductance	7.5[mH]

Table 4. Specification of experimental conditions

Parameter	Value
Irradiance	650[W/m^2]
Temperature of Module	40.3[°C]

IOP Conf. Series: Earth and Environmental Science **257** (2019) 012045 doi:10.1088/1755-1315/257/1/012045

5. Results

The steady waveforms acquired from the experiments are shown in figure 8. In this experimental trial, the scan period is set to 200 ms, whereas the duration between detection is set to 30 s.

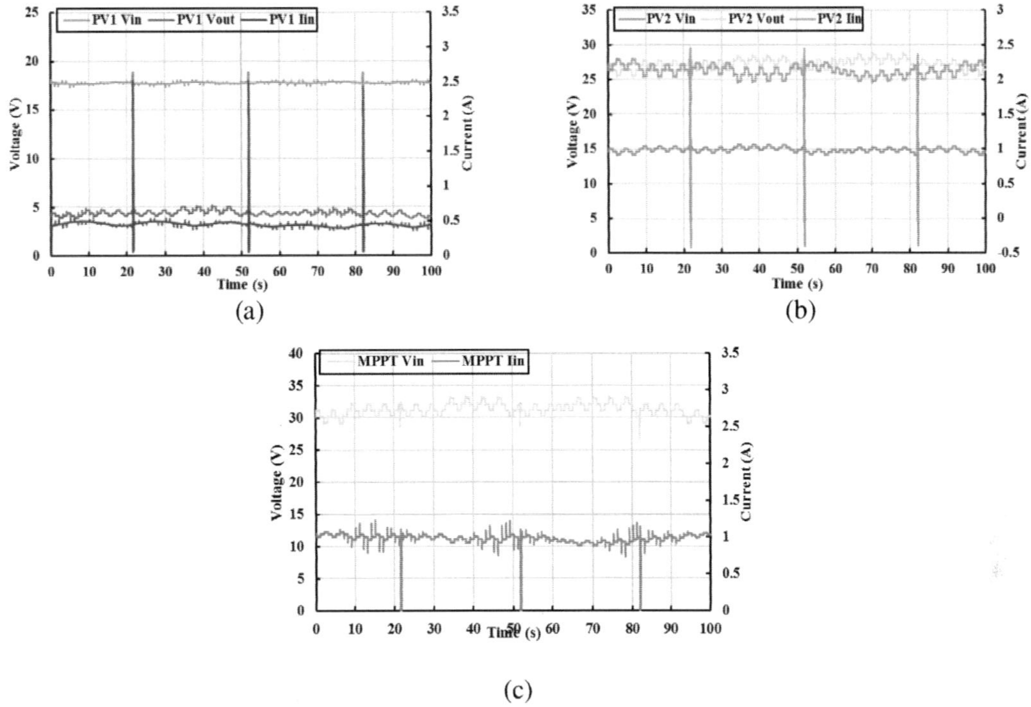

Figure 8. Waveforms acquired from experimental trial, (a) constant power control (hot-spot suppression), (b) optimal control, (c) PCS side (MPPT controller)

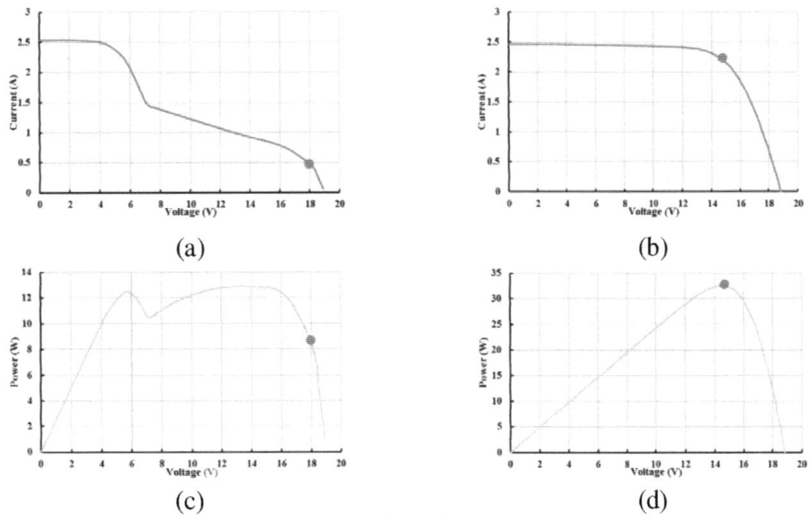

Figure 9. I(P)-V curves acquired from real-time scan process,
(a) and (c) are PV1, (b) and (d) are PV2.

It can be seen in figure 8 (a), the PV module 1 operates at constant power control when the defective cell is shadowed (e.g. a 3/4 shadow for one cell). The input voltage is approximately limited to the safe operation voltage which can be referred in figure 9 (a) and (c), in this case, the malfunctioning module is working at mode I (forward bias). In addition, the output voltage is working at MPPT with respect to the PCS operation. In figure 8 (b), the PV module 2 operates at its MPPT, which can be seen in figure 9 (b) and (d). Besides, the PCS side is always working at its MPPT for the maximum output. According to this experimental trial, we found that the distributed structure of a PV string has a potential to provide the hot-spot detection at the module level, more significantly, different operation for each module to make sure that the string is working at the optimal state.

6. Conclusion

We proposed a novel control system for hot-spot detection and suppression by employing the distributed TSBB converters. The main feature of this system is summarized as follows.

- Online hot-spot detection at module level,
- Independent safe operation of the malfunction module in a PV string is permitted.

From the experimental trial, in the case of hot-spot suppression, the PV 1 operates at the Buck mode, which may lead to low conversion efficiency. It is a trade-off problem to be considered that how to achieve a reasonable balance between the safe operation and high efficiency. As a result, to improve the conversion efficiency of the converter will be our future work.

7. References

[1] T. Ghanbari, "Permanent partial shading detection for protection of photovoltaic panels against hot spotting," *IET Renew. Power Gener.*, vol.11, no.1, pp. 123–131, Nov. 2017.

[2] J. Poon, P. Jain, C. Spanos, S. K. Panda, and S. R. Sanders, "Photovoltaic condition monitoring using real-time adaptive parameter identification," in *Proc. IEEE Energy Convs. Congr. and Expo.*, 2017, pp. 1119-1124.

[3] D. Rossi, M. Omana, D. Giaffreda, and C. Metra, "Modeling and detection of hotspot in shaded photovoltaic cells," IEEE Trans. VLSI Syst., vol. 23, no. 6, pp. 1031-1039, Jun. 2015.

[4] M. Dhimish, V. Holmes, P. Mather, and M. Sibley, "Novel hot spot mitigation technique to enhance photovoltaic solar panels output power performance," *Solar Energy*, vol. 179, pp. 72-79, Jun. 2018. Available: https:// doi.org/10.1016/j.solmat.2018.02.019.

[5] M. Simon, and E. L. Meyer, "Detection and analysis of hot-spot forma-tion in solar cells," *Solar Energy*, vol. 94, pp. 106-113, Feb. 2010. Available: https:// doi.org/10.1016/j.solmat.2009.09.016.

[6] A. Bouraiou, et.al, "Experimental investigation of observed defects in crystalline silicon PV modules under outdoor hot dry climatic conditions in Algeria," *Solar Energy*, vol. 159, pp. 475-487, Jan. 2018. Available: https://doi.org/10.1016/j.solener.2017.11.018

[7] P. Guerriero, G. Cuozzo, and S. Daliento, "Health diagnostics of PV panels by means of single cell analysis of thermographic images," in *Proc. 16th IEEE Int. Conf. Environ. and Electri. Engi.*, 2016, pp. 1-6.

[8] S. Yang, K. Itako, T. Kudoh, K. S. Koh, and Q. Ge, "Monitoring and Suppression of the Typical Hot-Spot Phenomenon Resulting from Low-Resistance Defects in a PV String," *IEEE J. Photovolt.*, vol. 8, no. 6, pp. 1809-1817, Nov. 2018. DOI: 10.1109/JPHOTOV.2018.2861734

[9] W. K. Loke, S. F. Yoon, S. Wicaksono, K. H. Tan, and K. L. Lew, "Defect-induced trap-assisted tunneling current in GaInNAs grown on GaAs substrate," *J. Appl. Phys.*, Jul. 2007. Available: https://doi.org/ 10.1063/1.2775908.

[10] S. Steingrube, O. Breitenstein, K. Ramspeck, S. Glunz, A. Schenk, and P. P. Altermatt, "Expla nation of commonly observed shunt currents in c-Si solar cells by means of recombination statistics beyond the Shockley-Read-Hall approximation," *J. Appl. Phys.*, May. 2011. Available: https:// doi.org/ 10.1063/1.3607310.

Cooling enhancement of photovoltaic cell via the use of phase change materials in a different designed container shape

Tachakun Sarikarin[1], Tanakorn Wongwuttanasatian[2] and Amnart Suksri[2]

1 Insulation Technology and High Voltage Engineering Laboratory, Faculty of Engineering, Khon Kaen University, Khon Kaen, 40002, Thailand
2 Center for Alternative Energy Research and Development, Khon Kaen University, Khon Kaen, 40002, Thailand
E-mail: tachakun.pri@gmail.com, tanwon@kku.ac.th. and samnar@.kku.ac.th.

Abstract. During the electricity generation of photovoltaic (PV) cells, large fraction of solar radiation gets converted into heat which raises the temperature and decreases the electrical efficiency. In order to reduce the heat accumulation, there are several methods that can be applied to cooling the solar panel. One of the simple and efficient approaches is to use the phase change materials (PCM) as a heat absorber. This research is the designed and constructed a housing container for filling up palm wax that is used as a PCM. The designed PCM containers are groove type, tube type, and fin type. Experiment was to compare the PV cell that installed the designed with PCM (test module) and with non-installed PCM containers (reference module). Exposure time for solar radiation is carried out from 9:00 AM to 5:00 PM at an atmospheric condition in Khon Kaen, Province of Thailand (latitude, 16° 25' 50" N and longitude, 102° 37' 0" E). Results showed that when the time is at 11:00 AM to 1:00 PM, solar radiation yield the most constant value. It was found that the designed fin type achieved the most outstanding heat reduction when compare with other designed prototype. The designed fin type has also reduced an average temperature by 6.167 °C and the average electrical efficiency is increased by 4.858 per cent.

1. Introduction

PV cell is one of the economically feasible renewable technologies for electricity generation. Heated PV cell causes drop of electrical voltage and decreases the electrical power at 0.65% per 1 °K [1]. It was also reported that, the heat in the PV cell causes the electrical power to decrease by 0.8 % per 1 °K [2] [3]. S. Sargunanathan [4] studied about the cooling method. There are 6 cooling methods used, they are Heat Pipe, Passive Cooling, Active Cooling, Liquid Immersion, Bionic and Additional PCM layer. The PCM method is another approach that many researchers have carried out an investigation. Literature review as shown in Table 1.

It can be seen from literatures that using the PCM method can lowered the temperature that accumulated on solar PV. This research is the designed and constructed the PCM container. The designed PCM containers are groove type, tube type, and fin type. The experiment was to compare between PV cell that is installed the PCM (test module) with non-installed PCM containers (reference module) and carried out during 9:00 AM to 5:00 PM at an atmospheric conditions in Khon Kaen Province of Thailand (latitude, 16° 25' 50" N and longitude, 102° 37' 0" E). The temperature measurements are measured at the front surface of the PV panel and electrical parameters as well as solar radiation were recorded every 3 minutes.

Content from this work may be used under the terms of the Creative Commons Attribution 3.0 licence. Any further distribution of this work must maintain attribution to the author(s) and the title of the work, journal citation and DOI.
Published under licence by IOP Publishing Ltd

Table 1. Literature review.

Authors	Method	Results
A. Sweidan [5]	Water pipe were installed at the backside of solar panel and connects a water pipe to the PCM tank	The electrical power is increased by 4 kW
S. harma [6]	Rubitherm 42 (RT42) was used as a PCM and installed them in a box at the back side of solar panel	The temperature was decreased by 3.8 °C and the electrical efficiency was increased by 7.7 per cent
T. Cui [7]	PCM box and a Fresnel lens for combining the solar light.	The electrical efficiency increases by 12.74 per cent
F. Hachem [8]	They have made the combined PCM container, the combined PCM container attribute to 70 per cent of PCM total mass. The material of container was inserted with copper 20 per cent and graphite 10 per cent of space used	The temperature was found to decrease by 6.3 °C and the electrical power increases by 5.8 per cent.
C. J. Ho [9]	Two layers of PCM with different melting points were employed	The electrical efficiency increases by 2.1 per cent.
R. Stropnik [10]	Rubitherm (RT28) PCM box were installed at the back side	The temperature was decrease by 35.6 °C and the electrical efficiency increases by 7.3 per cent (average of 1 year)
S. Khanna [11]	Simulated the temperature inside of the PCM box by ANSYS Fluent	The temperature is decreased by 6.9 °C and the electrical efficiency increases by 0.9 per cent.
A. Hasan [12]	Installation of the paraffin box at the back side of the solar panel	The electrical power increases up to 5.9 per cent

2. Methodology

2.1. PCM container

The PCM container is designed with the goal of increasing the surface area and to maximize the heat exchange media. The PCM containers were made by stainless steel for filling up the PCM (palm wax) of 3 kilograms. The surface areas of container are, groove type (4,158.8 cm^2), tube type (4,346.8 cm^2) and fin type (5,402 cm^2). The groove type dimension is 370 x 310 x 42 mm. The width of the fin is at 30 mm with 8 containers in rows as shown in Figure 1. The tube type dimension is 370 x 310 x 47 mm. It is buried with stainless steel pipes inside the container. The diameter is 20 mm with 12 slots of PCM container as shown in Figure 2. The fin type dimension is 370 x 310 x 47 mm. The width of the fin is 13 mm with 15 slots of PCM container as shown in Figure 3.

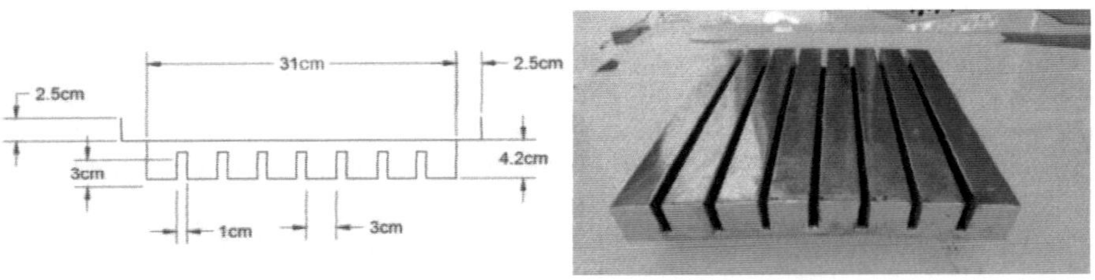

Figure 1. The groove type PCM container.

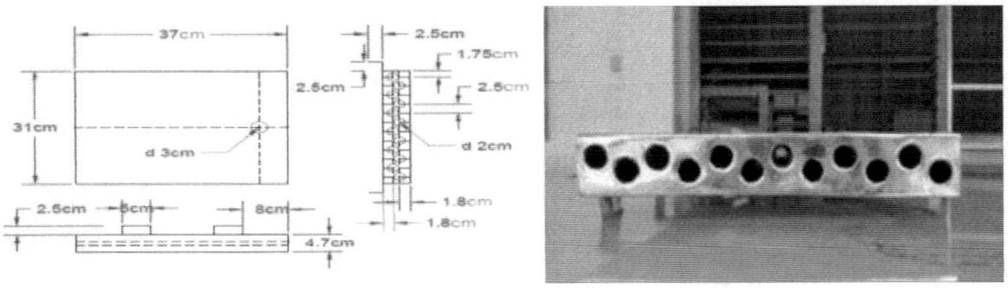

Figure 2. The tube type PCM container.

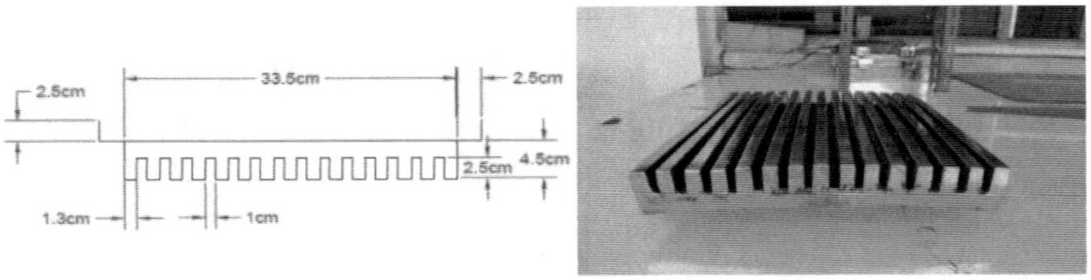

Figure 3. The fin type PCM container.

2.2. Palm wax as a PCM

Palm wax is a fat obtained from the palm oil production process. Its characteristic is in solid state phase at room temperature with white scaly flakes as shown in Figure 4. When it is heated at high temperature, it changes the phase into liquid state. Table 2 shows the characteristic of typical palm wax.

Table 2. The property of palm wax.

Physical state	Flakes
Melting point [°C]	48 - 59
Density [kg/m3]	850 - 900
Materials to avoid	Acids, Strong oxidizing
Acute toxicity (Oral)	LD50 (rat) : >2000 mg/g

Figure 4. Palm wax (solid state).

2.3. Experimental setup

The photovoltaic cell used in this research is polycrystalline type. Its characteristic is shown in Table 3. The experiment was to compare the test module with the reference module and conducted during 9:00 AM to 5:00 PM at an atmospheric condition. Both of the PV modules were installed facing south with an inclined angle of 15 degrees with respect to horizontal plane. Solar radiation is chosen during 11:00 AM to 1:00 PM due to its stability period as presented in Figure 5. The temperatures were measured at 3 spots on front of the PV surface panel using type K (WR 22) thermocouple. The electrical voltage was measured by a multimeter (UNI-T® ut 106). Solar radiation was measured by a pyranometer (CMP3, KIPP & ZONEN) as shown in Figure 6. All parameters were recorded every 3 minutes. The electrical power and the electrical efficiency are calculated using equations below.

$$P = V^2 / R \quad (1)$$

$$Eff = P_{out} / P_{in} \quad (2)$$

Where: P - is Power (Watts), V- is Voltage (Volts), R- is Resistance (ohms), Eff – is efficiency P_{out} - is Output Power (Watts) and P_{in} - is Intput Power (Watts)

IOP Conf. Series: Earth and Environmental Science **257** (2019) 012046 doi:10.1088/1755-1315/257/1/012046

Table 3. PV cell polycrystalline type.

Maximum power	20 W
Maximum power voltage	17.6 V
Maximum power current	1.14 A
Open circuit voltage	21.4 V
Short circuit current	1.57 A
Dimension (L x W x D)	485 x 350 x 20 mm
Weight	2.0 kg

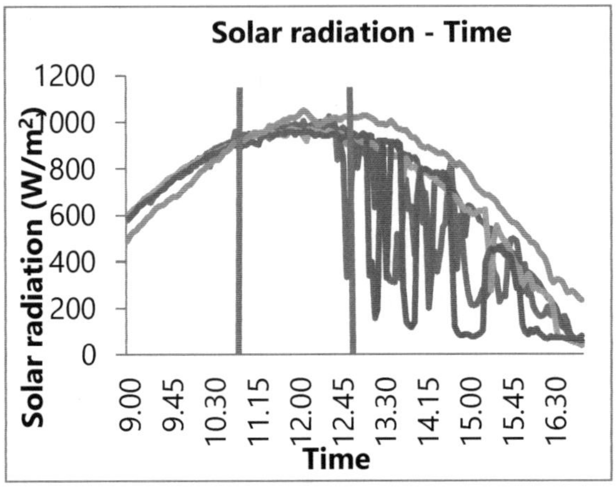

Figure 5. Solar radiation.

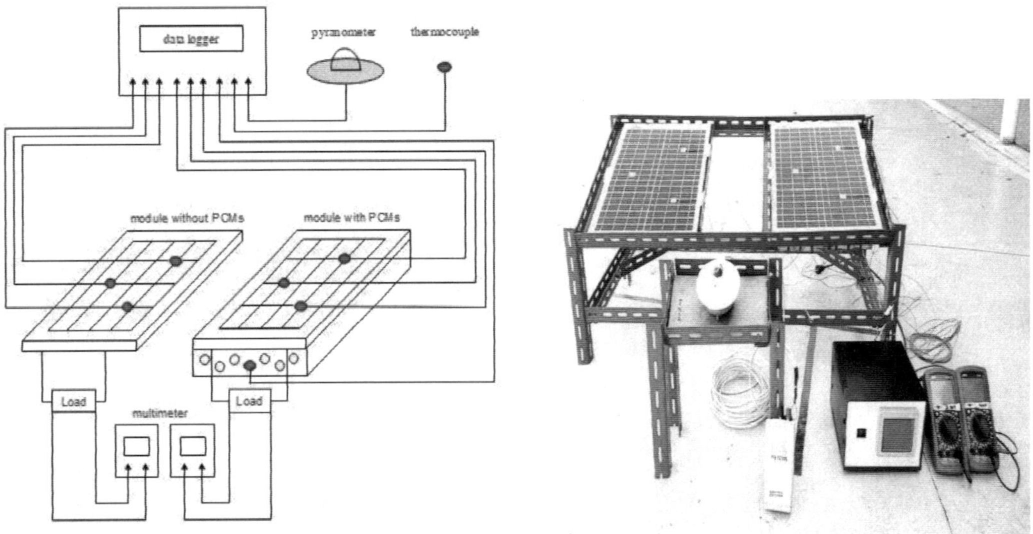

Figure 6. Experimental setup.

3. Results and Discussions

3.1. Groove type

Figure 7 shows the temperature profile of the groove type test module (blue line). It is found that the tested module has lowered the temperature than the reference module (yellow line). Also, Figure 8 shows the electrical power of the tested module (red line) and the reference module (blue line). It can be seen that there are slightly improvement on power that can be obtained. Table 4 revealed the temperature that is decreased by 4.764 °C and the electrical power increased by 0.466 Watts, resulting to an increased of electrical efficiency by 4.042 per cent.

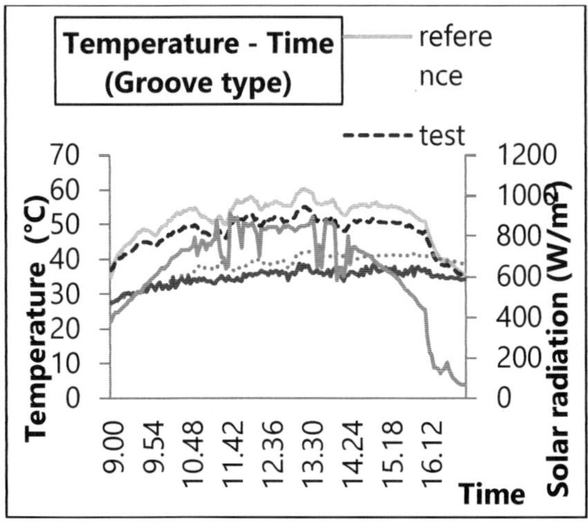

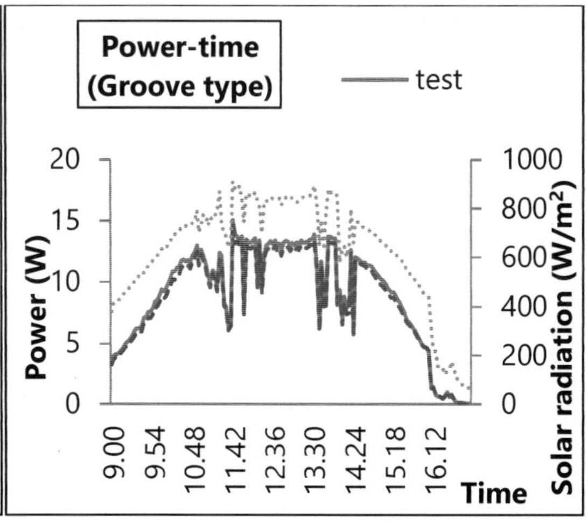

Figure 7. Temperature profile of groove type. **Figure 8.** Electrical power output of groove type.

Table 4. Results of a groove type.

No. of test	Temp (ref)	Temp (test)	ΔTemp	Power (test)	Power (ref)	ΔPower	eff (test)	eff (ref)	Δeff	%Δeff
	°C	°C	°C	W	W	W				
1	58.673	53.875	4.798	10.854	10.440	0.415	0.084	0.081	0.003	4.279
2	47.755	43.602	4.154	14.417	13.892	0.525	0.114	0.110	0.004	3.803
3	57.221	51.982	5.239	11.739	11.277	0.462	0.093	0.089	0.004	4.311
4	54.263	49.225	5.038	12.554	12.084	0.470	0.104	0.100	0.004	3.854
5	54.507	49.998	4.508	11.918	11.485	0.434	0.109	0.105	0.004	3.807
6	56.482	51.628	4.854	12.206	11.7148	0.4914	0.110	0.106	0.004	4.196
	Average		**4.765**			**0.466**				**4.042**

3.2. Tube type

Figure 9 shows the temperature profile of the tube type test module (blue line). It is found that the tested module has lowered the temperature than the reference module (yellow line). Also, Figure 10 shows the electrical power of the tested module (red line) and the reference module (blue line). It can be seen that little amount of power were achieved. Table 5 revealed the temperature that is decreased by 5.265 °C and the electrical power increased by 0.505 Watts, resulting to an increase in electrical efficiency by 4.307 per cent.

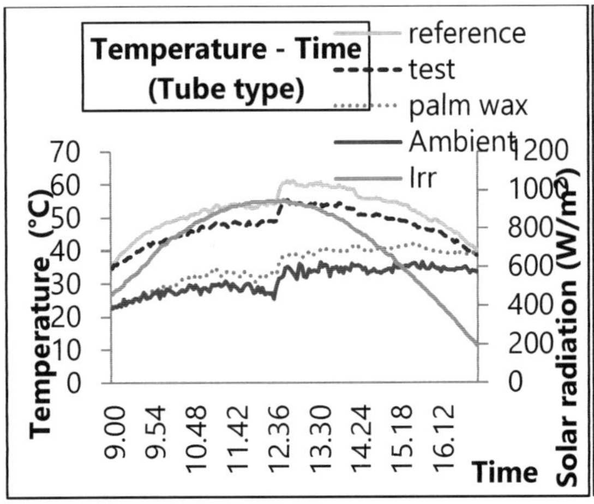

Figure 9. Temperature profile of a tube type.

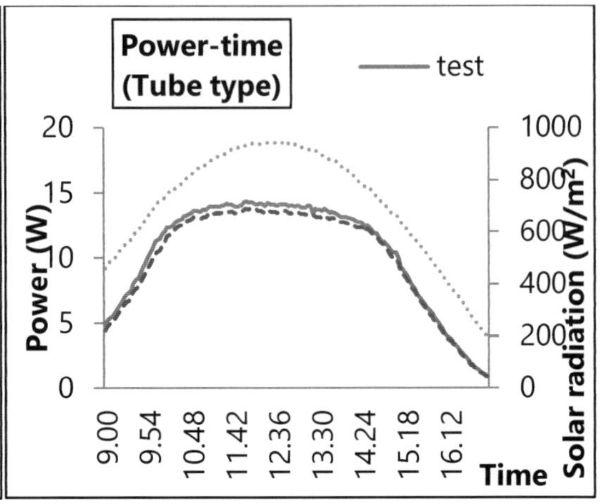

Figure 10. Electrical power output of a tube type.

Table 5. Results of a tube type.

No. of test	Temp (ref)	Temp (test)	ΔTemp	Power (test)	Power (ref)	ΔPower	eff (test)	eff (ref)	Δeff	%Δeff
	°C	°C	°C	W	W	W				
1	52.885	48.301	4.585	10.661	10.232	0.430	0.090	0.086	0.004	4.782
2	57.004	51.536	5.468	11.141	10.776	0.366	0.084	0.081	0.003	4.175
3	50.400	45.038	5.362	13.025	12.475	0.549	0.103	0.099	0.004	4.459
4	53.358	48.646	4.712	14.496	13.931	0.565	0.114	0.109	0.004	4.057
5	54.502	49.399	5.103	13.760	13.199	0.562	0.115	0.111	0.005	4.254
6	54.781	49.327	5.454	14.065	13.509	0.556	0.114	0.109	0.004	4.117
Average			**5.265**			**0.505**				**4.307**

3.3. Fin type

Figure 11 shows the temperature profile of the fin type test module (blue line). It is found that the tested module has lowered the temperature than the reference module (yellow line). Also, Figure 12 shows the electrical power of the tested module (red line) and the reference module (blue line). As with some earlier cases, it can be seen that output power has improved slightly. Table 6 revealed the temperature that is decreased by 6.167 °C and the electrical power increased by 0.596 Watts, resulting to an increase in electrical efficiency by 4.858 per cent.

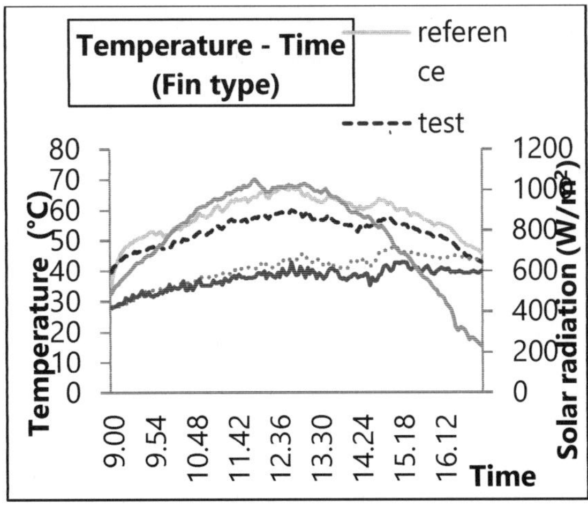

Figure 11. Temperature profile of a fin type.

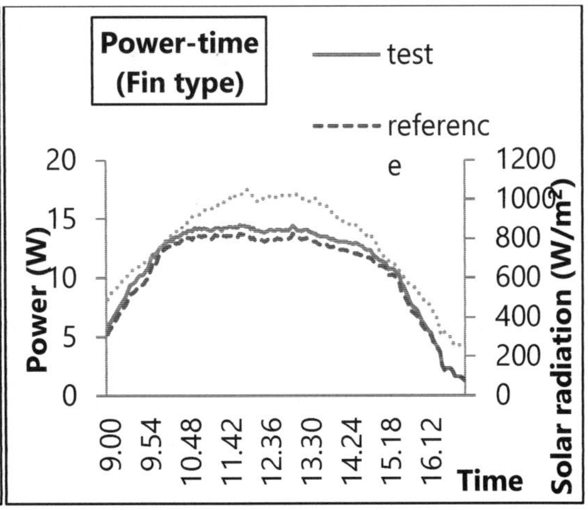

Figure 12. Electrical power output of a fin type.

Table 6. Results of a fin type.

No. of test	Temp (ref) °C	Temp (test) °C	ΔTemp °C	Power (test) W	Power (ref) W	ΔPower W	eff (test)	eff (ref)	Δeff	%Δeff
1	61.062	54.685	6.377	13.208	12.656	0.552	0.102	0.098	0.004	4.423
2	59.108	53.474	5.634	10.676	10.254	0.423	0.083	0.079	0.004	5.194
3	57.265	51.213	6.052	13.608	13.037	0.570	0.104	0.100	0.004	4.375
4	54.682	49.263	6.033	12.775	12.188	0.587	0.107	0.102	0.005	4.942
5	63.638	56.837	6.801	14.144	13.409	0.736	0.105	0.100	0.005	5.491
6	52.221	46.115	6.107	15.780	15.072	0.708	0.139	0.133	0.006	4.723
Average			**6.167**			**0.596**				**4.858**

This research uses palm wax as PCM for heat absorption. The PCM container is designed to increase the heat exchange capability. Utilizing both methods can enhance the heat transfer to the environment. In all cases, the temperature of the test module has been found to be lowered than the reference module. When the temperature of the PV module is lowered, electrical efficiency is increased. The results showed that the designed fin type has achieved the best performance on PCM container. This is because of larger surface area (5,402 cm^2). Currently, Thailand produces electricity from PV cell at 2,651 MW per year. Considering that if this cooling method is applied to it, electrical efficiency can be increased by 4.858 per cent. So that, there will be more electricity production by 128.79 MW per year.

4. Conclusion

The PCM container is designed and constructed to enhance heat exchange capability. When the temperature of a PV cell is lowered, it affects the electrical generating efficiency. The results showed that, during stable solar radiation period, the reference module heats up to a higher temperature than the tested module. The designed fin type has been found to be the best performance on PCM container. It was found that the temperature reduction can be reached by 6.167 °C and the electrical efficiency is increased by 4.858 per cent. This is due to the incremental of surface area which is 29.89 per cent more when compare with the groove type. Also, when compared with the tube type, fin type has more surface area than the tube type of 24.28 per cent. Hence, PCM container designed is the key parameter that attributes most on cooling enhancement. The PCM using palm wax has found to achieve the break even because it is inexpensive and can be easily found elsewhere in Thailand.

5. References

[1] E. Radziemska 2003 The effect of temperature on the power drop in crystalline silicon solar cells *Renewable Energy* **28** 1–12

[2] E. Radziemska and E. Klugmann 2002 Thermally affected parameters of the current-voltage characteristics of silicon photocell *Energy Conversion, and Management* **43** 1889-1900

[3] O. Dupré, R. Vaillon, and M. A. Green 2015 Physics of the temperature coefficients of solar cells *Solar Energy Materials & Solar Cells* **140** 92–100

[4] S. Sargunanathan, A. Elango, and S. T. Mohideen 2016 Performance enhancement of solar photovoltaic cells using effective cooling methods: A review *Renewable and Sustainable Energy Reviews* **64** 382–393

[5] A. Sweidan, N. Ghaddar, and K. Ghali 2016 Optimized design and operation of heat-pipe photovoltaic thermal system with phase change material for thermal storage Journal of Renewable and Sustainable Energy **8** 23501

[6] S. Sharma, A. Tahir, K. S. Reddy, and T. K. Mallick 2016 Performance enhancement of a Building-Integrated Concentrating Photovoltaic system using phase change material Solar Energy Materials & Solar Cells **149** 29-39

[7] T. Cui, Y. Xuan, E. Yin, Q. Li, and D. Li 2017 Experimental investigation on potential of a concentrated photovoltaic-thermoelectric system with phase change materials Energy **122** 94e102

[8] F. Hachem, B. Abdulhay, M. Ramadan, H. El, M. Gad, E. Rab, and M. Khaled 2017 Improving the performance of photovoltaic cells using pure and combined phase change materials Experiments and transient energy balance Renewable Energy **107** 567e575

[9] C. J. Ho, B. Jou, and C. Lai 2017 Thermal and electrical performance of a PV module integrated with double layers of water-saturated MEPCM Applied Thermal Engineering **123** 1120–1133

[10] R. Stropnik 2016 Increasing the efficiency of PV panel with the use of PCM Renewable Energy **97** 671e679

[11] S. Khanna, K. S. Reddy, and T. K. Mallick 2017 Performance analysis of tilted photovoltaic system integrated with phase change material under varying operating conditions Energy **133** 887-899

[12] A. Hasan, J. Sarwar, H. Alnoman, and S. Abdelbaqi 2017 Yearly energy performance of a photovoltaic-phase change material (PV-PCM) system in hot climate Solar Energy **146** 417-429

Acknowledgment

Authors would like to thank Insulation Technology and High Voltage Engineering Laboratory, Faculty of Engineering, Khon Kaen University for supporting this research. Laboratory in Chemical Engineering, Faculty of Engineering, Khon Kaen University for testing all of the specimens in this research and **Center for Alternative Energy Research and Development**, **Khon Kaen University**. Lastly, **RUN: Research University Network Project**, Funding Granted by **RUN** – Energy cluster, **Thammasat University**.

IOP Conf. Series: Earth and Environmental Science **257** (2019) 012047 doi:10.1088/1755-1315/257/1/012047

Implementation and research of a substation command interaction method using third party validation

Sun Shucai[1,2], Zhu Chenpeng[1,2] and Xie Yehua[1,2]

1 NARI Group Corporation/ State Grid Electric Power Research Institute, Nanjing, CHN
2 NARI Technology Development Co.,Ltd, Nanjing, CHN
E-mail: sunshucai@sgepri.sgcc.com.cn

Abstract. This paper describes a method to improve the security of control signal transmission in substation by introducing third party authentication and applying encryption technology. We can use asymmetric encryption technology to ensure the safety of data transmission process. A third party authentication device is introduced to verify the rights and identities of both sides of the command and prevent the interception and tampering of command information. Use dynamic key and authorization expiration system to further protect control security. Finally, the character code is checked by the command and the error data is filtered effectively. This method can effectively prevent signals from being truncated or tampered and ensure the security of data. Restricting the operation crowd, isolating illegal users, protecting information security, greatly improve the information security of substation control process.

1. Background technology

1.1. SSL Encryption technology

Secure Socket Layer enables communication between Client and server not to be eavesdropped by an attacker, and always authenticates the server and the user optionally. Secure Socket Layer has completed encryption algorithm, communication key negotiation and server authentication before application layer communication. The data transmitted by the application layer is encrypted to ensure the privacy of communication. The data transmitted by the application layer is encrypted to ensure the privacy of communication[1].

1.2. Asymmetric encryption algorithm

Asymmetric encryption algorithm, also known as "public key encryption algorithm", requires two keys: public key and private key. The public key and the private key are a pair. The data is encrypted with the public key and can be decrypted only with the corresponding private key. Encrypting data with a private key can only be decrypted with the corresponding public key. Encryption and decryption use two different keys. This algorithm is called asymmetric encryption algorithm.

The strength and security of the algorithm depend on the algorithm and key. The speed of encryption and decryption is not as fast as that of symmetric algorithm. But there is only one key in the symmetric algorithm. Decryption needs to know the key first, to guarantee the security. To ensure the security of the key, asymmetric encryption algorithm has two kinds of keys, one of which is public [2].

1.3. Hash algorithm

Content from this work may be used under the terms of the Creative Commons Attribution 3.0 licence. Any further distribution of this work must maintain attribution to the author(s) and the title of the work, journal citation and DOI.
Published under licence by IOP Publishing Ltd

Hashing algorithm can generate fixed length output for different length of input messages. Hash algorithm is a one-way encryption algorithm, which can not be decrypted by encrypted text to get plaintext. Even for minor changes in plaintext, the same hash operation will get completely different results. Based on this characteristic, hash algorithm is usually used to verify the integrity of information[3].

2. System Deployment and Object Definition

The method described in this paper is to introduce third party authentication and apply encryption technology. This method can improve the security of signal transmission in substation, prevent signal interception or tampering, ensure the security of data, limit the operation crowd, isolate illegal users, protect information security and other functions[4].

The simple system deployment structure is as figure 1:

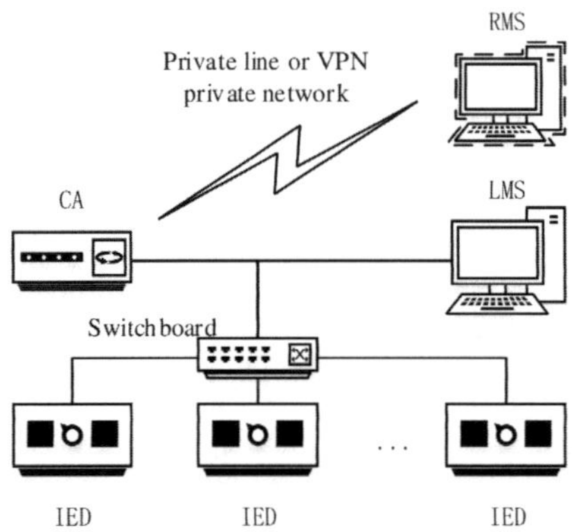

Figure 1. Simple schematic diagram of system deployment structure.

2.1. Authorized authentication device (hereinafter referred to as CA)

The device is a trusted third-party verification device used in this method. As the guarantee of data security and the safe transfer of control commands, it has the following main functions:

- The function of acquiring authentication authority through smart card or digital certificate.
- The function of issuing, managing and authenticating data certificates.
- The function of SSL encryption communication.
- The function of checking information integrity.
- The function of recording and auditing the authorized information.

2.2. Local monitoring system (hereinafter referred to as LMS)

LMS is the local management system for monitoring and operating IED(generally referred to as SCADA systems, this article only emphasizes its local control role),as the control command initiator in this method[5]. Compared with the traditional system, it needs to be extended to improve its security, and has the function of processing encrypted information, as follows:

- Upgrade the authentication mechanism to bind it to the authorized authentication device, and use the digital security certificate to improve the security of the account.
- Upgrade the data communication program, so that it can use SSL encryption mechanism to interact with the authorized authentication device.
- Calculate the hash code of information and verify the integrity of information.

2.3. Remote monitoring system (hereinafter referred to as RMS)

The system is a remote management system for monitoring and operating IEDs (generally referred to as the dispatching master station system), which interacts with LMS through dedicated lines or private networks. In this method, the command of RMS is directly interacted with CA, and is the initiator of the remote control command[6].

2.4. Intelligent equipment (hereinafter referred to as IED)
The device is the smart unit that handles the execution of the control command in this method. It is the receiver of the control command. As the receiver and processor of the final information, it is a very important part of the whole control process[7]. The device has the following functions:

- Receive, store and forward hash feature codes sent by LMS.
- Upgrade the data communication program, so that it can use SSL encryption mechanism to interact with the authorized authentication device.
- Calculate the hash code of information and verify the integrity of information.

3. Identity Authentication Mechanism

3.1. Authentication method based on CA certificate
In order to ensure its security from intrusion, CA, as the default trusted device of this method, is developed based on embedded system[8] and only runs the necessary functions. The digital certificate is solidified on the hardware, which is used for external authentication and is unique.
For local control, a list of trusted smart cards is set up to obtain the authentication key of the device authorization. Local operation needs to provide a smart card before CA can grant remote control operation privileges, which can effectively control the scope of local operation terminals and personnel.
For remote control, a list of trusted devices is established, corresponding digital certificates are stored to verify the identity, and reliable data channels are used to transmit authentication information first[9]. After CA authenticates the identity reliably, it opens remote control privileges and effectively limits the scope of remote control terminal.

3.2. Authentication method based on LMS certificate
LMS itself has a user authentication mechanism, and its control operation authority is granted by CA. LMS submits authentication information to CA before issuing an operation instruction. CA verifies the identity of LMS by verifying the digital signature according to the recorded information. Identity trust is temporary. After more than one operation cycle, CA will reclaim control authority and issue new digital certificates.

3.3. Authentication method based on RMS certificate
RMS itself is secure and has user authentication mechanism. Control rights within the substation are granted by CA. RMS actively exchanges digital certificates with CA before issuing an operation instruction. After cross-validation, CA opens control rights to RMS. Control rights are temporary, and beyond the agreed time, CA will reclaim control rights and require RMS to update digital certificates.

3.4. Authentication method based on IED certificate
IED can update its own digital certificates according to needs. The identity of IED is managed by CA, which establishes a library of intelligent devices, records equipment information, and is responsible for requesting and verifying digital certificates.

4. Interaction Process
This article takes the local control process as an example, as follows:
The use of sections to divide the text of the paper is optional and left as a decision for the author. Where the author wishes to divide the paper into sections the formatting shown in figure 2 should be used.

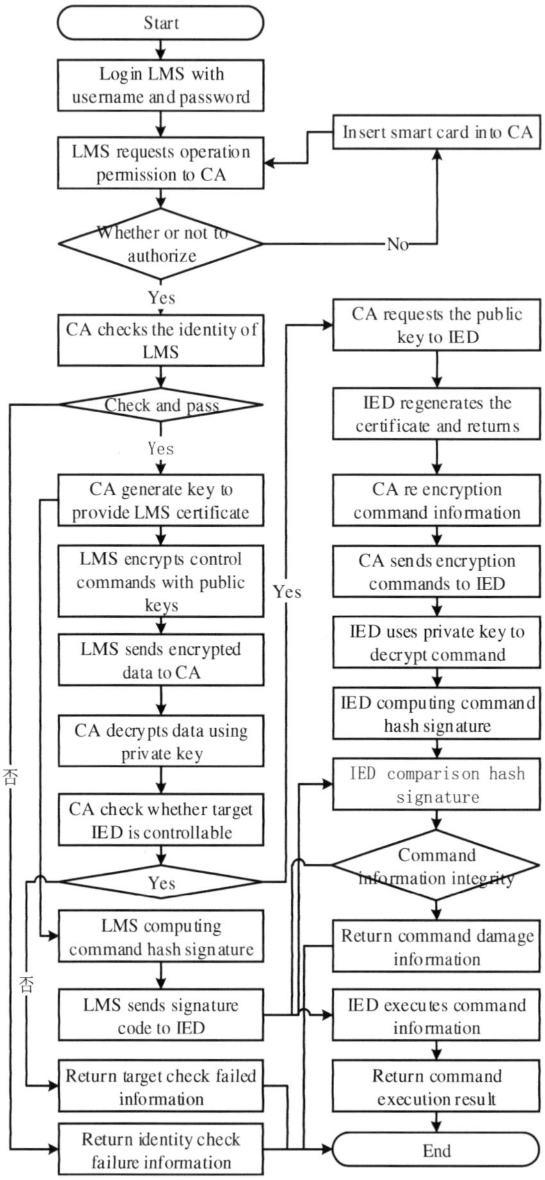

Figure 2. This method controls the flow chart of interaction process.

4.1. Initiation of control instructions

4.1.1. Get CA permission. The operator first needs to hold smart card and use smart card to get the access right of CA.

4.1.2. Sign in LMS. Use user name and password to log on to LMS. The account used needs to have control authority. operator first needs to hold smart card and use smart card to get the access right of CA.

4.1.3. Application control operation authority. The operator operates on the LMS, and the program automatically sends the request to CA. CA verifies the identity information of the monitoring device, returns a one-time public key when the verification passes, and terminates the control process if the verification fails[10].

4.1.4. Encryption control command. The LMS control program encrypts the control command using the received public key, and generates hash signatures for the control command for integrity checking. LMS submits the encrypted control command to CA and sends the hash feature code to the IED that needs to be operated.f

4.2. Transfer of control instructions

4.2.1. Declassified control command. After receiving the encrypted control command, CA uses its own private key to decrypt.

4.2.2. Check device control authority. CA checks whether the target IED is in the control range according to the control command.

4.2.3. Get the target IED public key. According to the parsed control command, the CA sends a command to update the certificate to the target IED. The IED regenerates the certificate and passes the public key to the CA.

4.2.4. Re-encrypt control command. CA encrypts the control command with the public key obtained and sends the encrypted data to the target IED.

4.3. Control command execution

4.3.1. Declassified control command. IED uses private key to decrypt control commands transmitted by CA.

4.3.2. Check control command. IED calculates the hash signature of the control command and compares it with the hash signature sent by LMS to verify the integrity of the command.

4.3.3. Execution control command. IED performs operations according to the contents of the command.

4.3.4. Re-encrypt control command. CA encrypts the control command with the public key obtained and sends the encrypted data to the target IED.

5. Examples

The following example is illustrated by a simple remote control closing process.

- Controller logs in LMS system with username and password, chooses the equipment object to be controlled, and issues remote control closing command.
- The LMS system detects remote operation and starts operation permission detection. The CA verifies the validity of the smart card or digital certificate. CA reads the certificate data, obtains the detailed information through the system solidified decryption key, carries on the authorization comparison user name, the password, the smart card number and so on information. These information agree that the identity of the operator is legal and verified[11]. LMS requests a communication certificate from CA, which detects a request initiated by LMS and randomly generates a set of asymmetric keys such as figure 3 showing the following key information:

```
m1 =
130ebebd67b16a9ab2c53a437badbf8f01a80c750095a7fcfe95742c3d5
ed1abb318babc5cb5d9350fee4da65ee074f65e1758117e6945f0fcfc81
37528053ce9d1da8618890dee24e5e0bf8c87795bb1d09eddd544640824
ee0dd0ea9fd908d27b0f8a1ae5c37f3647fbf2f5795500ad76c195b3387
d0458a8f51b701472301
e1 = 10001
d1 =
12e8da920d4599458e84ec5ef1656161807f427d05eb79182b7418259d6
f6c14364d1f5caf9130c8d9d9d6ea71d1bdbc87781a46a16bcb9e672814
fed3b9c96ddffe0a1b0955ae68055c8f92fef518a04fc32a2ea8390e617
cc5556a251f9ae9eee70a32e579cb3e9f298848a9b3aaf634f5930ffbf7
4473f7cb6c0cefee1751
```

Figure 3. Schematic key information.

- After LMS receives the key, it begins to transmit data. LMS generates command plaintext based on operation, assuming that the plaintext P1 is as table 1:

Table 1. Motioned remote command.

01	01	01
Destination IED address	Target remote control outlet	The remote control type is closing

Encrypted ciphertext *s1*as figure 4

```
011c74f981f6c6697f22b9d6fabad9cea0d7e4f6fdd05630e1beff1da07
0bcd85cd2872ed30c414ffc0c7f7e3587c3e2afc53b0a278fdb9cb1f6e1
f64a0e052695d1252bb67cfb3ced31e13650cd96cad54b182a48a623af1
2117cf603cf274554b207b2c03048ff5681003e9136c25b74729377a9f2
ee877cc45cd0c4ebd9e9
```

Figure 4. Implied ciphertext s1.

Hashing feature codes *l1* are obtained through hash feature algorithm (such as figure 5).

```
21ef05aed5af92469a50b35623d52101
```

Figure 5. Schematic hash feature code.

Where S1 is sent to CA, and L1 is sent to target IED.

- CA receives the ciphertext S1 sent by LMS, decrypts the ciphertext using D1 and e1, and queries the valid IED table of the local record after obtaining the target IED address[12]. If it is valid, it is considered controllable. If it is invalid, it returns the failure information of the target check.
- After the CA determines the controllability, it requests the encryption key to the target IED, the IED generates a temporary certificate, and feeds back the new *m2* and *e2* to the CA. The CA re-encrypts the plaintext and passes the ciphertext *s2* to the IED.
- As the figure 6 shows, IED decrypts the ciphertext *s2* using the reserved *d2* and *e2*, calculates the hash signature *l2* of the decrypted plaintext *p2*, and compares it with the previously received signature *l1*. The same result means that the command information is correct and can be executed. If the result is different, the command is considered damaged and the order is re-issued.

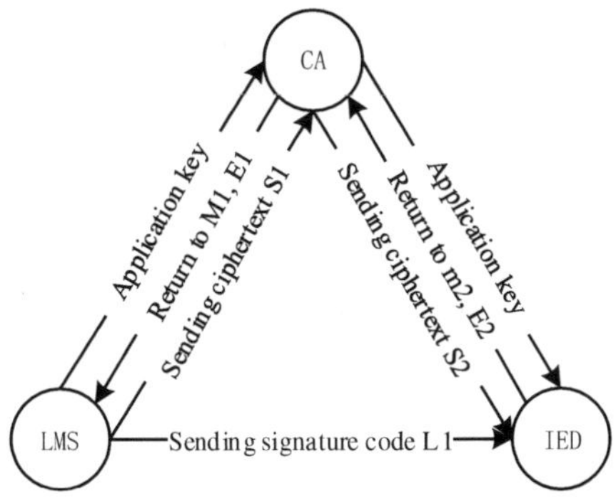

Figure 6. The data exchange process of this example.

6. Conclusion

The method described in this paper can effectively control the control rights and scope of IED equipment, prevent command information from being intercepted and tampered with in the interactive process, verify the integrity of command information, prevent illegal users from gaining operational authority, prevent the execution of wrong commands, and greatly improve the information security of substation control process, especially in unreliable network environment. This method is not restricted to substations, and other field contract samples that involve control information interaction.

7. References

[1] Guo Zhengrong and Zhou Cheng 2004 *Working process and application of SSL protocol[J]. Network security technology and Application* pp 55-57

[2] Chen Hongxing and Zhou Yuanlan 2012 *Research on network information security countermeasures based on asymmetric encryption. Network security technology and Application* pp 20-22

[3] Shen Changxiang and Zhang Huanguo 2010 *Research and development of Trusted Computing [J]. Chinese Science：Information Science* pp 139-166

[4] Sun Yong and Chen Wei 2006 *Trusted computing in embedded systems[J]. Information security and communication secrecy* pp 50-52

[5] He Changduan and Chen Shuguo 2016 *Analysis of abnormal network communication in Smart Substation [J]. Electrical technology* pp 135-138

[6] Song Lei and Luo Qiliang 2004 *Encryption scheme for real-time data communication in power system [J]. Power system automation* pp 76-81

[7] Jia Jing 1999 *Security and secrecy of information system* p 150

[8] Li Xichun and Yao Weigang 2008 *Application and security of power data network. Jiangsu Electrical Engineering* pp 8-10

[9] Liu Shimin and Fan Rui 2015 *Research and application of intelligent substation online monitoring system. Electrical technology* pp 132-134

[10] Xu Weifeng and Zhang Hong 2007 *Construction of network information security for electric power enterprises.Shaanxi Electric Power* pp 75-77

[11] Zhang Liang and Xu Dianguo 2014 *An improved ant colony routing algorithm for low voltage power line communication. Transactions of China Electrotechnical Society* pp 318-324

[12] Gu Zhiru and Tan Zhouwen 2014 *Noise suppression of narrowband OFDM communication system in Smart Grid. Transactions of China Electrotechnical Society* pp 269-276

Power acquisition effect for appropriate placement of MPPT units in a PV array

Yusuke Takeda[1], and Kazutaka Itako[1]

1 Kanagawa Institute of Technology, 243-0292 Japan
E-mail: itako@ele.kanagawa-it.ac.jp

Abstract. The PV generation system operates in maximum power point of PV array by MPPT control of power conditioning system (PCS). However, there is a problem that maximum power of the PV array configuration greatly decreases by partial shadow. For this problem, the authors previously proposed the method which connects MPPT unit to every PV panel between PCS and PV array. Moreover, the effect of this method was demonstrated by simulations and experiments. In the present day, this method is in practical use. However, the appropriate placement of MPPT units was still not clear. The authors previously investigated the appropriate placement of MPPT units from the viewpoint of acquisition of electric power and system stability, simulating P-V and I-V characteristics of some representative patterns connected MPPT unit to each PV panel in the PV array configuration. In this paper, power increase rate of the proposed PV array system is evaluated in experiment, compared to the conventional PV array system. It is clarified that maximum power of the proposed PV array system is 1.26 times, compared with maximum power of the conventional PV array system.

1. Introduction

The PV generation system operates in maximum power point of PV array by MPPT control of power conditioning system (PCS). However, there is a problem that maximum power of the PV array configuration greatly decreases by partial shadow.

For this problem, various MPPT methods have been investigated[1]-[3]. The authors previously proposed the method which connects MPPT unit to every PV panel between PCS and PV array. Moreover, the effect of this method was demonstrated by simulations and experiments[4]-[5]. In the present day, this method is in practical use. However, the appropriate placement of MPPT units was still not clear. The authors previously investigated the appropriate placement of MPPT units from the viewpoint of acquisition of electric power and system stability, simulating P-V and I-V characteristics of some representative patterns connected MPPT unit to each PV panel in the PV array configuration[6]. In this paper, power increase rate of the proposed PV array system is evaluated in experiment, compared to the conventional PV array system.

2. Circuit analysis of PV array with MPPT units

2.1. System configuration

Figure 1 shows the PV array configuration. The PV array is configured by some strings. The strings are composed of some PV panels and MPPT units connected to PV panel.

Content from this work may be used under the terms of the Creative Commons Attribution 3.0 licence. Any further distribution of this work must maintain attribution to the author(s) and the title of the work, journal citation and DOI.
Published under licence by IOP Publishing Ltd

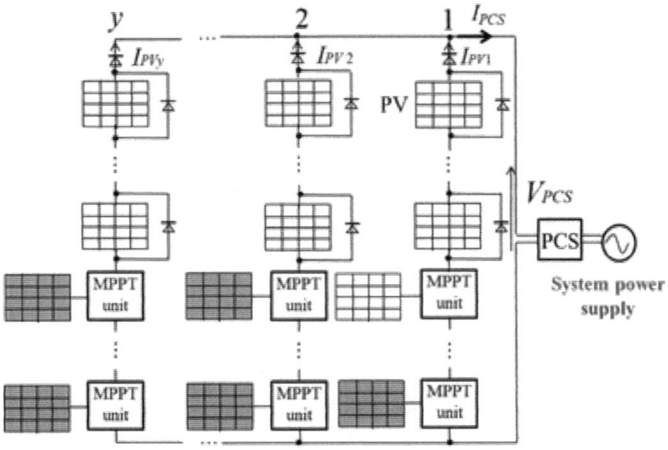

Figure 1. PV array configuration

2.2. Control method of MPPT unit

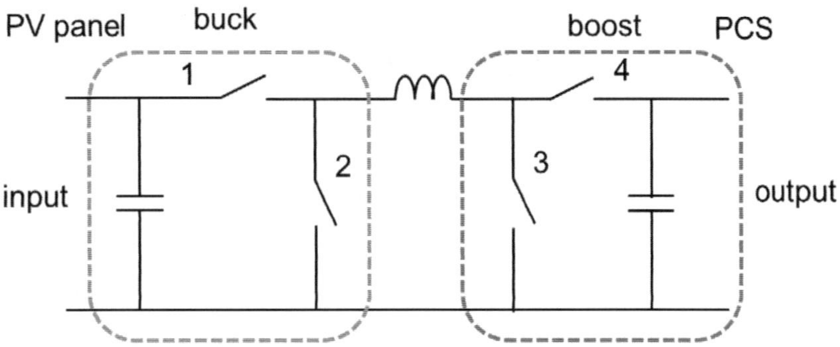

Figure 2. Back-boost DC-DC converter

Figure 2 shows back-boost DC-DC converter in MPPT unit. In MPPT unit, the P&O (Perturb and Observe) method is used as MPPT control(Figure 3). Figure 3-(a) shows conceptual diagram of P&O method. It is executed by periodically perturbing (incrementing or decrementing) the array terminal voltage and comparing the PV output power with that of the previous perturbation cycle. If the power is increasing the perturbation will continue in the same direction in the next cycle, otherwise the perturbation direction will be reversed. The flowchart of this method is represented by Figure 3-(b).
Figure 4 shows equivalent circuit of the PV array configuration (Fig. 1). Where, M is the number of MPPT unit (MPPT unit group), N is the number of PV panel (PV panel group), y is the number of string, I_{PV} is string current, VPV is output voltage of each PV panel, V_{MPPTk} is output voltage of MPPT, V_{PCS} is input voltage of PCS and X_k(0~1.0) is degree of a shadow on the PV panel.

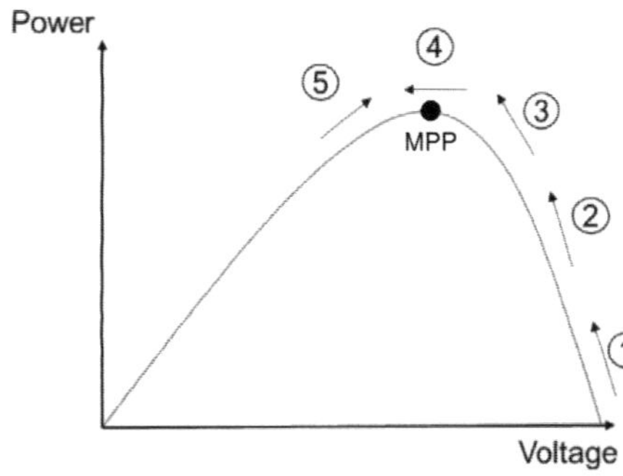

(a) Conceptual diagram

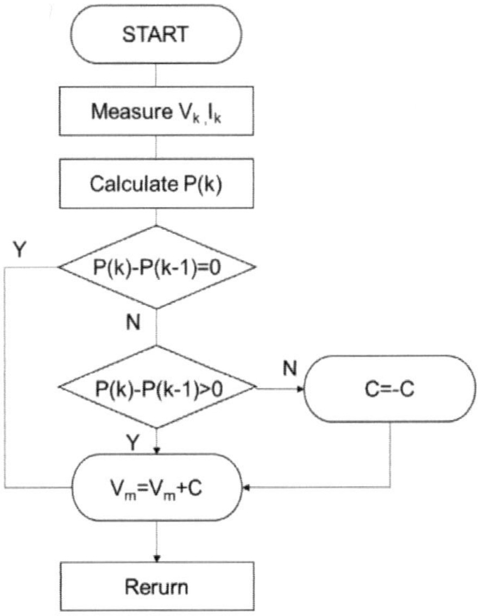

(b) Flowchart

Figure 3. P&O method

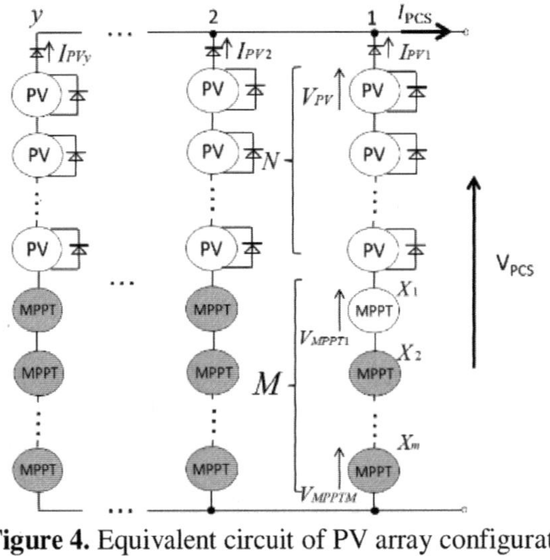

Figure 4. Equivalent circuit of PV array configuration

2.3. Operation of each mode

MPPT unit is able to extract the maximum power from the PV panel, so the MPPT unit is regarded as a constant power source. Each string has three circuit modes (Mode1~3), considering the bypass diode and the blocking diode.

In these strings, there are three circuit constitutions, as shown in the following.

(a) string(a)

The string is composed of only PV panel with MPPT unit.

(b) string(b)

The string is composed of only PV panel. In this string, a shadow does not exist for all PV panels.

(c) string(c)

The string is composed of both the PV panel with MPPT unit and the PV panel.

IOP Conf. Series: Earth and Environmental Science **257** (2019) 012048 doi:10.1088/1755-1315/257/1/012048

(1) Mode1(Normal Mode)(ISC≧IPV>0)
This Mode exists in all string constitution (string(a)~(c)). In this Mode, bypass diode and blocking diode do not operate. Where, IPV is short current.

(2) Mode 2(Bypass Mode)(IPV>ISC)
This Mode exists in string(c). In this Mode, bypass diode operates. Therefore, output voltage of PV is zero. In string(b), because the shadow does not exist for PV panels, this mode dose not exists.

(3) Mode 3(Blocking Mode)(NVOC<VPCS, IPV=0)
This Mode exists in string(b). Where, NVOC is total open voltage of PV panel. VOC is the open voltage of PV panel.

3. Conception of appropriate placement

The following shows a conceptual diagram of P-V characteristics in the conventional PV array system and of P-V characteristics in the appropriate placement of MPPT units in the PV array system using MPPT unit. Where, the PV array is configured by 6 strings, assuming the household PV array system. Where, one string has 6 series PV panels. The string has the constitution of string(a), (b) and (c). Grey part in Fig.5 shows the shadow.

3.1. Conventional PV array system

Figure 5-(a) shows the conventional PV array system. Figure 5-(b) shows the P-V characteristics of conventional PV array system. In this figure, the sum P_0 of maximum power of each PV panel is shown in dotted red line. P_0 is the potential power of the PV array in Figure 5-(a). P_{max} is maximum power of conventional PV array system. In this case, this PV array system is not able to extract the power of PV panel with shadow. It is difficult to track maximum power point, because multi-peeks in the P-V characteristics exist for V_{PCS}. As a result, this PV array system is not able to operate at the maximum power. In this PV array system, output power is decreased by mismatch of the string maximum power points. When a shadow exists in PV panel, PCS is not able to extract the potential power P_0 from this PV array system.

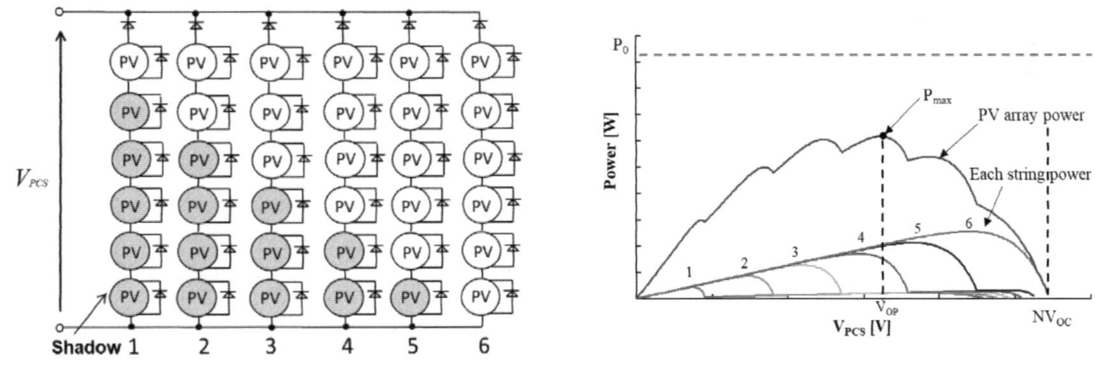

(a) Conventional PV array system (b) P-V characteristics

Figure 5. Conventional PV array system

3.2. Best placement of MPPT units

Figure 6-(a) shows the best placement of MPPT units found by simulation of some representative patterns to solve the above mentioned problem. This figure shows the PV array configured in one string(b) and the other strings(a). Figure 6-(b) shows P-V characteristics. MPPT control of PCS is able to definitely find one power peak point of string(b), because the power of string(a) is constant for V_{PCS}.

Therefore, the system is stable and PCS is able to operate in maximum power point. P_{max} is the same as P_0. PV array system is able to extract the potential power P_0 in principle.
From the viewpoint of acquisition of electric power and system stability, the best placement of MPPT units is able to be simply described as shown in Figure 7. This equivalent circuit is composed of one string(b) to determine the MPPT operation point of PCS and the string(a) group to eliminate the mismatch of the string maximum power points. In this PV array system, PCS is able to extract the potential power P_0.

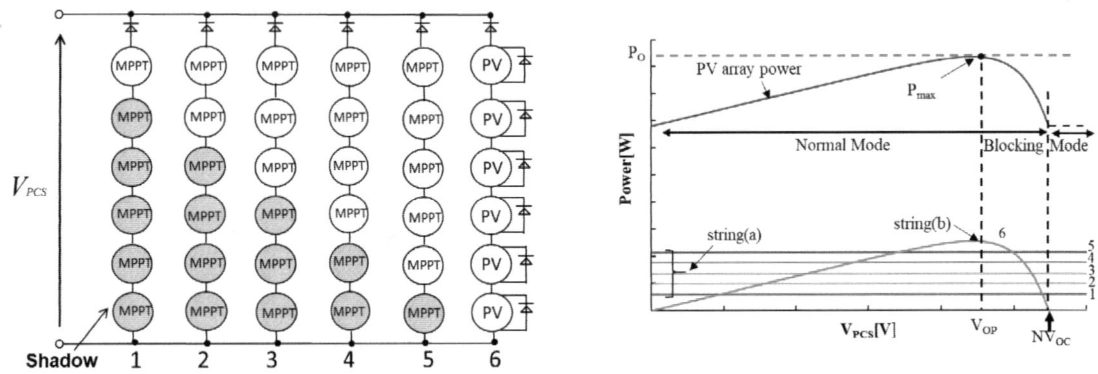

(a) Best placement of MPPT units (b) P-V characteristics

Figure 6. Best placement of MPPT units

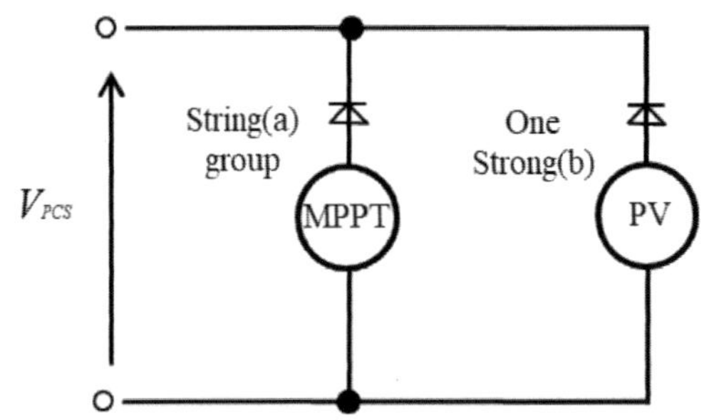

Figure 7. Equivalent circuit of best array configuration

4. Experiment

In order to confirm the validity of equivalent circuit(Figure 7.), the following experiment is carried out. The PV array is configured by 2 strings, assuming the fundamental PV array system(Figures 8 and 9). Where, one string has 2 series PV panels. The string has the constitution of string(a) and (b). Gray part in PV2 of Figures 8 and 9 shows the shadow, and efficiency of the MPPT unit is 86.4[%]. In this experiment, PV simulator is used instead of the PV panel. Each parameter of the PV simulator is shown in Table 1. In PV simulator of PV2, Short-circuit current is set low in order to emulate a shadow condition.

Table 1. Each parameters of the PV simulator

	Maximum output operating voltage [V]	Maximum output operating current [A]	Maximum output power [W]	Open-circuit voltage [V]	Short-circuit current [A]
PV1	14.8	3.72	55.1	21.8	4.0
PV2	14.1	0.79	11.1	18.1	1.08
PV3	16.0	3.11	49.8	20.4	3.46
PV4	16.0	3.11	49.8	20.4	3.46

IOP Conf. Series: Earth and Environmental Science **257** (2019) 012048 doi:10.1088/1755-1315/257/1/012048

4.1. Conventional PV array system

Figure 8-(a) shows the conventional PV array system. Figure 8-(b) shows the P-V characteristics of conventional PV array system. In this figure, the potential power P_0 is 165.8[W]. Maximum power P_{max} of conventional PV array system is 124.0[W]. In this case, this PV array system is not able to extract the power of PV panel with shadow. It is difficult to track maximum power point, because multi-peeks in the P-V characteristics exist for V_{PCS}. As a result, this PV array system is not able to operate at the maximum power. Where, useful utilization factor U is defined as the following equation (1).

$$U = \frac{P_{max}}{P_0} \times 100[\%] \qquad (1)$$

Therefore, U is calculated as 74.8[%]. In this PV array system, output power is decreased by mismatch of the string maximum power points. When a shadow exists in PV array, PCS is not able to extract the potential power P_0 from this PV array system.

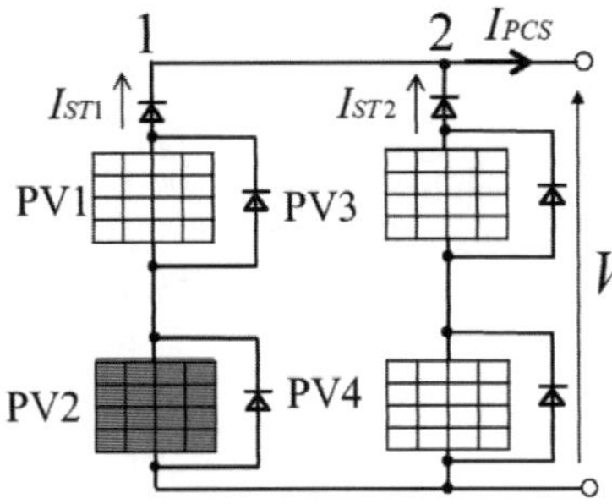

(a) Conventional PV array system

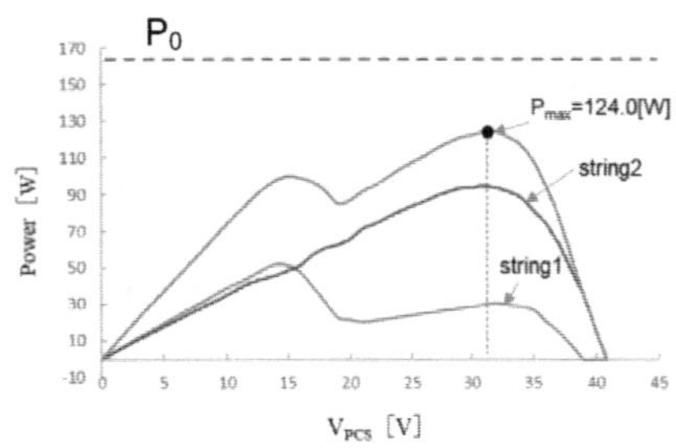

(b) P-V characteristics

Figure 8. Conventional PV array system

4.2. Proposed PV array system

Figure 9-(a) shows proposed PV array system (best placement of MPPT units). This figure shows the PV array configured in one string(b) and the other string(a). Figure 9-(b) shows P-V characteristics of proposed PV array system. MPPT control of PCS is able to definitely find one power peak point of string(b), because the power of string(a) is constant for V_{PCS}. Therefore, the system is stable and PCS is able to operate in maximum power point. In this experiment condition, P_{max} is 156.8[W]. U is calculated as 94.6[%]. When a shadow exists in PV array, PCS is able to extract the potential power P_0 from this PV array system.

In this experiment condition, P_{max} of the proposed PV array system is 1.26 times, compared with P_{max} of conventional PV array system. In proposed PV array system, MPPT control of PCS is able to definitely find one power peak point. Therefore, the system is stable and PCS is able to operate in maximum power point.

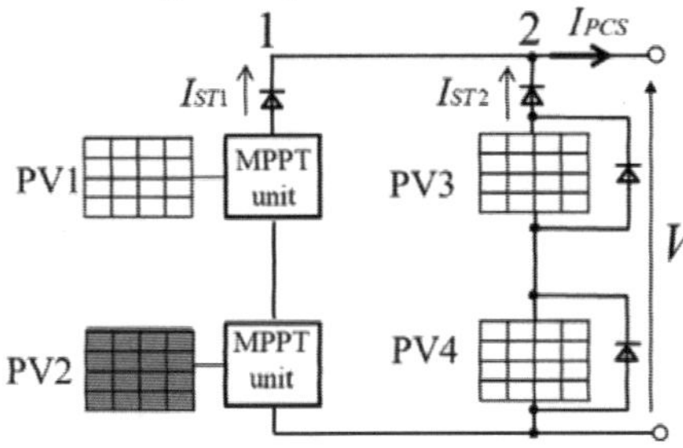

(a) Proposed PV array system

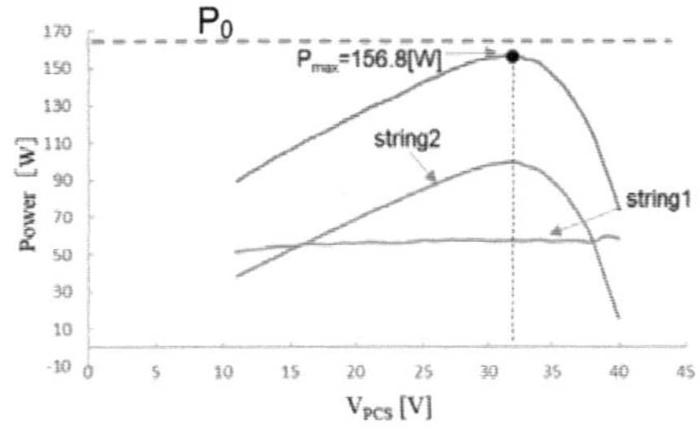

(b) P-V characteristics

Figure 9. Proposed PV array system

5. Conclusion

The authors previously proposed the appropriate placement of MPPT was proposed from the viewpoint of acquisition of electric power and system stability, simulating P-V and I-V characteristics of some representative patterns connected MPPT unit to each PV panel in the PV array configuration. power increase rate of the proposed PV array system was evaluated in experiment, compared to the conventional PV array system. In this experiment condition, it is clarified that maximum power of the proposed PV array system is 1.26 times, compared with maximum power of the conventional PV array system.

6. References

[1] G. Petrone, G. Spagnuolo, and M. Vitelli, "An Analog Technique for Distributed MPPT PV Applications", *IEEE Trans. Ind. Electron.* Vol. 59, no. 12, pp. 4713 – 4722. Dec. 2012

[2] M. A. Ghasemi, A. Ramyar, and H. I.-Eini, "MPPT Method for PV Systems Under Partially Shaded Conditions by Approximating I–V Curve," *IEEE Trans. Ind. Electron.* Vol. 65, no.5, pp.3966 – 3975, May. 2018

[3] Mo. Lasheen, A. K. A. Rahman, M. A.-Salam, S. Ookawara, "Adaptive reference voltage-based MPPT technique for PV applications," *IET Renewable Power Generation.* Vol.11, no.5, pp. 715 – 722, Apr. 201

[4] C.Huang, K.Itako, T.Mori and Q.Ge, "MPPT Control System for PV Generation System with Mismatch Modules", *Journal of Energy and Power Engineering(USA),*9, 2015, pp.83-90.

[5] C.Huang, K.Itako and T.Mori, Q.Ge,"MPPT Control Method Using Buck Type DC-DC Converter for PV Generation System with Mismatched Modules" *Proceedings of ICEE 2014*

[6] Y.Takeda, K.Yoshihara and K.Itako. " Best Solution for Placement of MPPT Units in a PV Array" *Proceedings of Grand Renewable Energy 2018*

Experimental study on luggage-laden pedestrian movement in narrow seat aisle

Shenshi Huang[1,2], Ruichao Wei[1,2], Siuming Lo[2], and Shouxiang Lu[1]

1 State Key Laboratory of Fire Science, University of Science and Technology of China, Hefei, China
2 Department of Civil and Architectural Engineering, City University of Hong Kong, Hong Kong
E-mail: kermit@mail.ustc.edu.cn

Abstract. The experiment of one-dimensional luggage-laden pedestrian movement in narrow seat aisle were performed in this study, to investigate the hindrance effect of seat configuration on pedestrian flow. In a seat arrangement similar to a vehicle carriage, experiments of individual walking and single-file pedestrian flow were carried out respectively. The result shows that carrying luggage has little effect on pedestrian walking ability. However, Carrying large luggage such as trolley case can significantly increase the distance headway of pedestrians. It is also found that, although the pedestrian dynamic pattern in narrow seat aisle was similar to that in open environment, the randomness of narrow seat aisle environment shows obviously greater.

1. Introduction

The topic on safe evacuation of high-speed train car has received widespread attention. As for the passenger evacuation progress inside the car, luggage-laden passenger and narrow seat aisle are the two important obstructive factors. Previous studies [1, 2] had pointed out that the pedestrians with luggage tend to walk slower and occupy a larger space when boarding and alighting, and this impact will be more serious when passengers passing through the narrow seat aisle.

Present studies on pedestrian evacuation of seat area mostly focused on the evacuation time, egress rate, and occupant density. The pedestrian dynamic is rarely of concern. In the past years, researches on pedestrian dynamics used to concern about the pedestrian walking ability [3, 4] and the interaction between individual pedestrian [5, 6]. It is pointed out that the pedestrian individual walking speed can be affect by many factors such as particularity of individuals, and environment. Also, when there are pedestrians walking, they will usually keep a certain distance between each other, meanwhile the movement speed will be affected by others. In order to better understand pedestrian evacuation of seat area, it is necessary to analyse the pedestrian dynamic parameters in such environment.

This study focuses on the movement of luggage-laden pedestrian. Experiments of individual walking and single-file pedestrian flow are carried out in the narrow seat aisle, and the pedestrian walking speed and distance headway are analysed. Based on the discussion of luggage impacts on pedestrian movement in the narrow seat aisle, this study will give the useful benchmark for evacuation simulation of similar scenarios as well as provide reference to safety design of new train cars.

2. Experimental conditions

The experimental set-up was similar to our previous studies [7, 8]. The impact of luggage on pedestrian movement was investigated by rearrange the seat and luggage configurations. Experiment

Content from this work may be used under the terms of the Creative Commons Attribution 3.0 licence. Any further distribution of this work must maintain attribution to the author(s) and the title of the work, journal citation and DOI.
Published under licence by IOP Publishing Ltd

was carried out in the multi-function room, and the aisle was set up using six seat rows with spacing of 1.0 m. The aisle width was fixed as 0.5 m, which is the most common width in vehicle cars. Pedestrians were asked to walk through this 5-meter-long seat aisle, carrying different kinds of luggage. All the trolley cases weight no more than 10 kg and all the bags weight no more than 5 kg. Normal walking and fast walking conditions were designed, to simulate
Cameras were set up to record the whole experiment process. By using the *PeTrack* [9, 10] software, the position of each pedestrian and the trajectories could be extracted, as illustrated in Figure 1. Based on the measured position, the pedestrian walking speed, distance headway, and other parameters can be obtained.

(a)

(b)

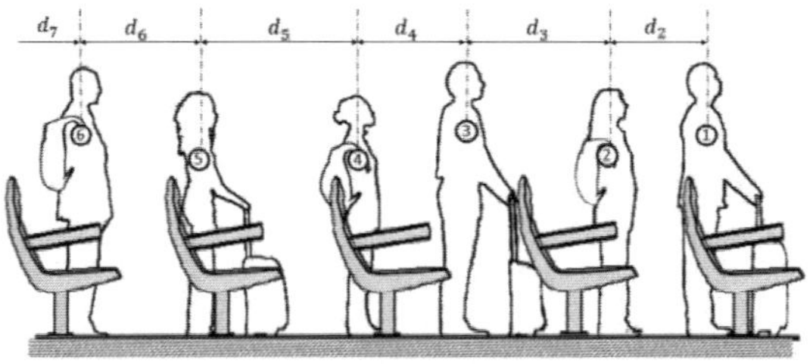

Figure 1. (a) The camera shots of luggage-laden pedestrian movement in this experiment; (b) Illustration of the pedestrian motion in seat aisle and distance headway.

This experiment includes two parts: individual pedestrian walking and single-file pedestrian movement. In the individual walking part, the pedestrian individual walking behaviours were measured. In the single-file pedestrian movement part, a "stop" instruction and a "go" instruction were designed, and the interaction between luggage-laden pedestrians were observed. The detail of experimental conditions have been list out in Table 1. Moreover, two repetitions of each test case were performed, in order to reduce the randomness of results.

Table 1. The detail of experimental conditions pedestrian movement in narrow seat aisle.

Case No.	seat aisle width	pedestrian type	luggage
D1	0.50 m	individual	none
D2	0.50 m	individual	bag
D3	0.50 m	individual	trolley case
D4	0.50 m	single-file	none
D5	0.50 m	single-file	bag
D6	0.50 m	single-file	trolley case
D7	0.50 m	single-file	trolley case and bag
D8	0.50 m	single-file	G1 none + G2 bag
D9	0.50 m	single-file	G1 trolley case + G2 bag
D10	0.50 m	single-file	G1 trolley case + G2 none

G1: 50% of all participants, with the Pedestrian No. of 1, 3, 5, 7, 9, 12, 14, 16, 18, 20, 22, 24, 26;
G2: the other 50% of all participants, with Pedestrian No. of 2, 4, 6, 8, 10, 11, 13, 15, 17, 19, 21, 23, 25.

Similar to our previous studies, different methods were applied to the analysis of different pedestrian types. The individual walking parameters were measured as spatial mean values, representing the basic attributes of the pedestrians. On the other hand, the single-file walking parameters were described by instantaneous values, in order to present the whole process of pedestrian movement under the action of instructions.

3. Results and discussions

3.1. Individual pedestrian movement

The average walking parameters of each individual in each test case is described in Figure 2, where the mean values are shown as scatter points and the distributions are present by box chart.

It is clearly observed that, the bag has little impact on walking behaviours, no matter on walking speed, step size, and step frequency. Under normal walking condition, the pedestrians were able to walk at the speed about 1.5 m/s in average, with about 0.7 m per step. Under fast walking condition, the pedestrian can walk at about 1.7 m/s, and the average step size increases to about 0.8 m. This result is not very different from that of open space [8].

However, in the case of pedestrian carrying large luggage, the walking parameters would be much different. The statistical mean value of waking speed decreases and the variance increases significantly. Due to the obstructive effect of narrow aisle, pedestrians cannot easily carry the trolley case. The average walking speed was observed to be 1.2 m/s for normal walking and 1.6 m/s for fast walking cases. Besides, with carrying trolley case, the mean step size reduces to about 80% while the step frequency slightly increases. It indicates that carry large piece of luggage has a significant impact on pedestrian's walking ability.

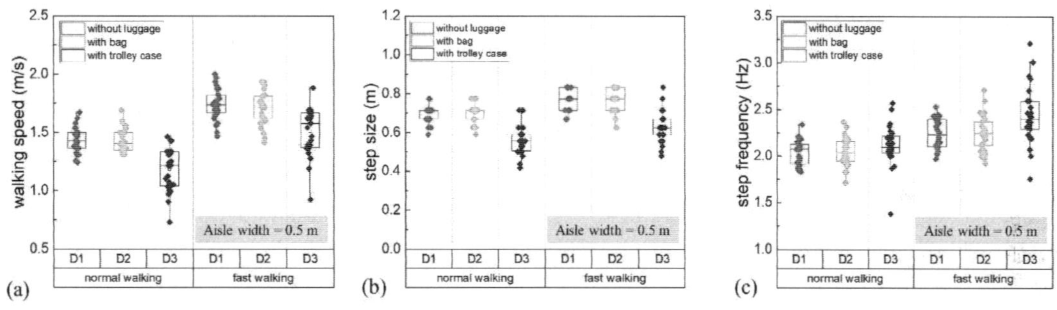

Figure 2. Distribution of (a) average walking speed, (b) mean step size, and (c) step frequency of individual luggage-laden pedestrian walking through the seat aisle.

3.2. Single-file pedestrian movement

The x-t curve of typical single-file pedestrian movement cases are shown in Figure 3, where the instantaneous moving speed is represent by the slope and the distance headway of each pedestrian can be indicated by the difference in height of curve clusters As seen from Figure 3, the x-t curves are quite similar, which includes three movement phase corresponding to the designed instructions. In the first and the third phase, when the pedestrian queue was moving smoothly, the walking parameters such as maximum speed, distance headway were observed. In the second phase, when the pedestrians were lining up and waiting, the safety margin can be measured. Moreover, the propagation velocity of the action speed can also be calculated based on the curve offsets.

The summary of walking speed and distance headway have been listed out in Table 2. The result shows that, when pedestrian queue moves through the narrow seat aisle (width = 0.5 m), they are able to walk at 1.33-1.51 m/s at normal speed condition and 1.63-1.77 m/s at fast speed condition respectively. It is found that the luggage has little effect on reducing the walking speed, no matter it is a bag or a trolley case. On the contrary, significant differences can be observed in distance headway.

IOP Conf. Series: Earth and Environmental Science **257** (2019) 012049 doi:10.1088/1755-1315/257/1/012049

When pedestrians are walking down the aisle one by one, they used to keep a distance of about 1.25 m to their predecessor. Due to the fact that the size of bag is small, the increase in pedestrian spacing is not obvious. However, in the tests with large luggage (trolley case), the distance headway of each pedestrian individual increases by 0.4-0.5 m. This phenomenon can be more clearly observed in the cases of fast walking. In addition, it is also obtained from the curve that, when the pedestrians stop or start to walk, the average acceleration is about 0.62 m/s^2 for normal walking condition and 0.85 m/s^2 for fast walking condition respectively. Carrying luggage is found to have little impact on acceleration.

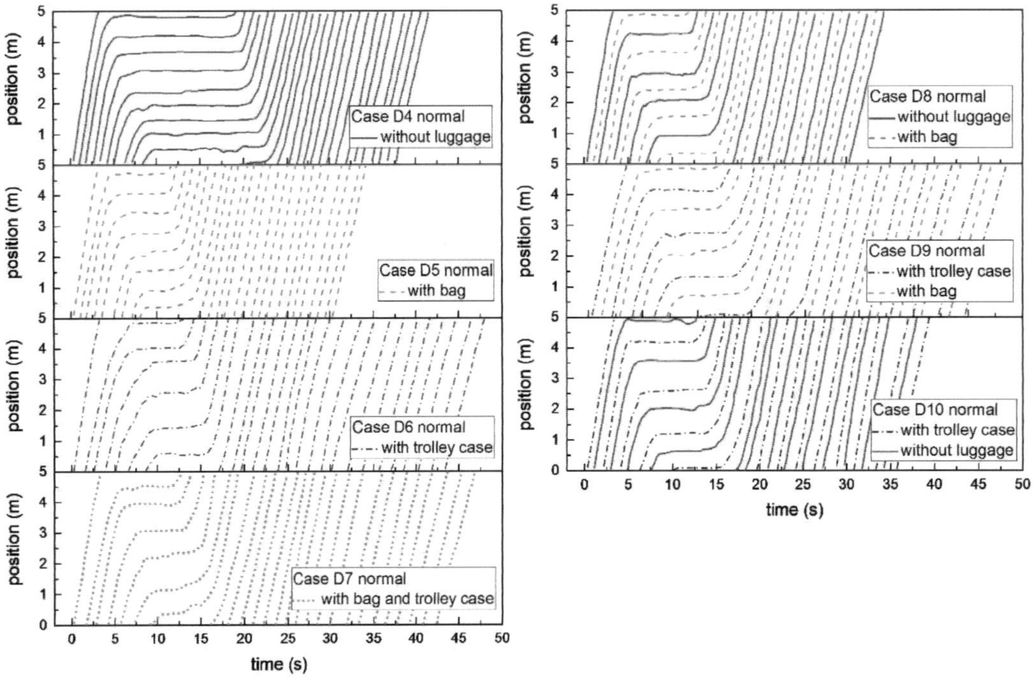

Figure 3. The x-t curves of the luggage-laden single-file pedestrian movement in each test cases.

3.3. Impact of narrow seat aisle

The experiment tests in this study can be well compared with the previous study [7, 8], as the main difference is the environment. This study focuses on the pedestrian movement in narrow seat aisle (width = 0.5 m), so that the movement is more hindered. Table 2 compares the results of pedestrian movement behaviour under the condition of narrow seat aisle and open environment. Not surprisingly, it is found that in the narrow seat aisle, the pedestrians tend to walk at a slower speed. For normal walking condition, there is a 4% drop on velocity; for fast walking condition, there is a 9% drop. By comparing the result of distance headway, it can be concluded that pedestrian in seat aisle prefer to keep slightly large distance to their predecessors in the seat aisle. And it can be observed by comparing each case pair that, trolley case has a much greater impact on distance headway. It means that when the pedestrians with large luggage are walking in the seat aisle, they might feel more crowded and uncomfortable, so that they prefer to keep a larger distance and safe margin headway.

By analysing the instantaneous data of each pedestrian, a fundamental diagram can be obtained using the mean value of each experiment test case. The fundamental diagram can graphically show the relationship between pedestrian walking speed and occupant density.

Table 2. Summary of maximum walking speed and distance headway of pedestrians in seat aisle (width = 0.5 m) and open environment.

Case No.	maximum speed (m/s)		distance headway (m)		Case No.	maximum speed* (m/s)		distance headway (m)	
	normal	fast	normal	fast		normal	fast	normal	fast
D4	1.42	1.77	1.24	1.25	E4	1.48	1.84	1.26	1.35
D5	1.41	1.73	1.29	1.29	E5	1.49	1.77	1.28	1.31
D6	1.51	1.65	1.68	1.68	E6	1.46	1.85	1.51	1.59
D7	1.33	1.67	1.74	1.79	E7	1.42	1.91	1.47	1.65
D8	1.35	1.72	1.21	1.32	E8	1.55	1.84	1.31	1.35
D9	1.38	1.71	1.46	1.59	E9	1.44	1.91	1.34	1.48
D10	1.47	1.63	1.56	1.56	E10	1.43	1.95	1.28	1.54
Average	1.41	1.70	1.45	1.50	Average	1.47	1.87	1.35	1.47

As illustrated in Figure 4(a), the average safety margin of the pedestrians in such seat aisle is about 0.7 m. It means that the pedestrians are willing to move forward only when there is 0.7 m distance free ahead. The slope of deceleration and acceleration phase curve is approximately 2.5 s^{-1}, which mean that the pedestrians may walk 2.5 m/s faster with every meter distance greater headway. It is sound that when the distance headway is larger enough, the speed will stabilize at the maximum. The critical distance can be 1.3 m for normal walking condition and 1.5 m for fast walking condition. Figure 4(b) shows the relationship of average walking speed and linear density, compared with previous study in open environment [8]. Based on the fact the curves are almost overlapping, it can be conclude that the pedestrian dynamic pattern of these cases are similar. The scatter point of seat aisle experiment are distributed more discrete, indicating that the randomness of pedestrian behaviour is greater.

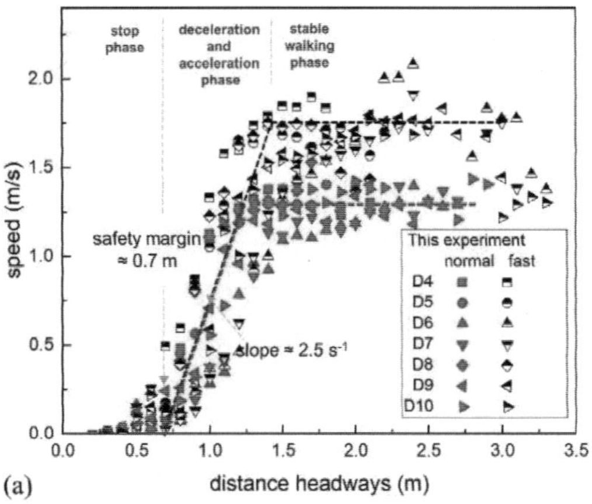

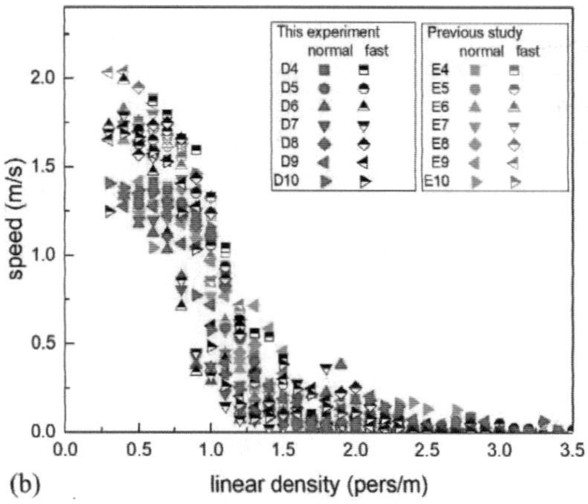

Figure 4. The (a) speed-distance plot and (b) speed-density plot of single-file pedestrian movement in seat aisle (width = 0.5 m) and previous study in open environment.

4. Conclusion

The luggage-laden pedestrian dynamic pattern in narrow seat aisle is investigated in this study. By conducting a series of pedestrian walking experiment. The pedestrian walking speed and distance headway are analysed. It is found that the individual pedestrian walks at about 1.5 m/s for normal walking condition and 1.7 m/s for fast walking condition. When there is a single-file pedestrian walking through the seat aisle, they used to keep a distance of about 1.25 m to their processor.

Carrying large luggage such as trolley case will significantly increase the distance headways, about 0.4-0.5 m. By drawing the fundamental diagram and making comparison with previous study, pedestrian dynamic pattern in narrow seat aisle is found to be similar to that in open environment. Nevertheless, the hindrance effect of seat configuration make the randomness of pedestrian behaviour obviously greater.

5. Reference

[1] Fang, Z.M., et al., Observation, simulation and optimization of the movement of passengers with baggage in railway station. International Journal of Modern Physics C, 2015. 26(11): p. 1550124.

[2] Young, S., Evaluation of pedestrian walking speeds in airport terminals. Transportation Research Record: Journal of the Transportation Research Board, 1999(1674): p. 20-26.

[3] Sun, J., et al., An experimental study on individual walking speed during ship evacuation with the combined effect of heeling and trim. Ocean Engineering, 2017.

[4] Jelic, A., et al., Properties of pedestrians walking in line. II. Stepping behavior. Phys Rev E Stat Nonlin Soft Matter Phys, 2012. 86(4 Pt 2): p. 046111.

[5] Cao, S., et al., Pedestrian dynamics in single-file movement of crowd with different age compositions. Phys Rev E Stat Nonlin Soft Matter Phys, 2016. 94(1-1): p. 012312.

[6] Jelic, A., et al., Properties of pedestrians walking in line: fundamental diagrams. Phys Rev E Stat Nonlin Soft Matter Phys, 2012. 85(3 Pt 2): p. 036111.

[7] Huang, S., et al., Experimental study of individual and single-file pedestrian movement in narrow seat aisle. Physica A: Statistical Mechanics and its Applications, 2018.

[8] Huang, S., et al., Experimental study on one-dimensional movement of luggage-laden pedestrian. Physica A: Statistical Mechanics and its Applications, 2018.

[9] Boltes, M., et al., Automatic extraction of pedestrian trajectories from video recordings, in Pedestrian and Evacuation Dynamics 2008. 2010, Springer. p. 43-54.

[10] Boltes, M. and A. Seyfried, Collecting pedestrian trajectories. Neurocomputing, 2013. 100: p. 127-133.

Acknowledgements
This research was supported by The National Key Research and Development Program of China (2016YFB1200403).

AUTHOR INDEX